Methods in Molecular Biology

Series Editor
John M. Walker
School of Life Sciences
University of Hertfordshire
Hatfield, Hertfordshire, AL10 9AB, UK

For further volumes:
http://www.springer.com/series/7651

Gene Correction

Methods and Protocols

Edited by

Francesca Storici

School of Biology, Georgia Institute of Technology, Atlanta, GA, USA

Humana Press

Editor
Francesca Storici
School of Biology
Georgia Institute of Technology
Atlanta, GA, USA

Videos to this book can be accessed at http://www.springerimages.com/videos/978-1-62703-760-0

ISSN 1064-3745 ISSN 1940-6029 (electronic)
ISBN 978-1-4939-6321-8 ISBN 978-1-62703-761-7 (eBook)
DOI 10.1007/978-1-62703-761-7
Springer New York Heidelberg Dordrecht London

Cover Art Caption: Molecular model of human genomic DNA with adenine mutation (in red) corrected to guanine (in green). A guanine to adenine mutation in the beta globin gene is a common cause of beta thalassemic disease. For more information on this gene correction, see Chapter 8 on Triplex-Mediated Genome Targeting by Faisal Reza and Peter M. Glazer.

Printed on acid-free paper

Humana Press is a brand of Springer
Springer is part of Springer Science+Business Media (www.springer.com)

Dedication

I dedicate this book to my research mentors.

To Carlo Bruschi, who initiated me to the field of genetic engineering using the Flp/FRT system of the yeast 2 μ plasmid.

To Dmitry Gordenin, who did his best to divert me from gene targeting work and focus my research more on mechanistic studies. I learned that gene targeting discloses new avenues for mechanistic research and vice versa.

To Jude Samulski, who helped me understand the relevance of translational research.

To Mike Resnick, who opened my mind to new ways of thinking about old hurdles, and who suggested me as the editor for this book.

Preface

The scope of "Gene Correction: Methods and Protocols" is to provide a user-friendly, well-detailed, and up-to-date collection of many strategies and methodologies utilized for generating specific sequence changes in the DNA of cells in the laboratory and for tackling the major problems that the field of gene correction is facing. Now that DNA sequencing technology has become sensitive and reliable enough to enter routine clinical practice, it is easy to identify genetic defects in genomic DNA. Considering that there are thousands of genetic diseases that are caused by a single sequence defect in a gene, it is obvious that the best way to prevent or cure a genetic disease is by correcting the defective gene that is causing it. Thus, it is becoming more and more important to have the knowledge and the tools to edit DNA at will. As our skills to manipulate the genetic material of cells progressed dramatically in the last decade, we acquired novel techniques and remarkably enhanced our capacity to genetically engineer genes for the purpose of better understanding the molecular mechanisms of life, and also for directly fixing mutations that cause innumerable devastating and incurable diseases in humans. Nevertheless, editing the genetic information of DNA is a challenging task. The goal of gene correction goes far beyond the process of making a desired change in a chosen target gene in the most efficient way. It is essential that the product of the modified gene should then be functional, the DNA correction stable and the engineering process accurate and restrained to the target to minimize unwanted DNA, cellular, and/or tissue damage.

The strategies for gene modification are currently numerous and diverse and are subjected to continuous evolution, improvement, and optimization. This book brings together many experts in the field of gene correction to disclose a wide and varied array of specific gene correction protocols for engineering mutations in DNA, for delivering correcting DNA to target cells and for improving the accuracy and safety of the gene correction process. This book is aimed at an audience of scientists of all backgrounds interested in the area of gene targeting/recombination/therapy. The methodologies presented in this volume are carefully explained and detailed so that they can be easily learned and applied by researchers who are not initially familiar with the procedures. The objective is from scratch to success: starting with a comprehensive listing of the Materials, every chapter contains a step-by-step guidance through the Methods and a series of useful tips provided in the form of Notes intercalated into the text.

The book is informally divided into four sections based on topic. Because each chapter could belong in more than one section, at the end of each section I have added a list of those chapters that provide additional protocols for gene correction specific to the topic of that section. Thus, each section goes beyond the subject matter presented in the selected chapters, and better helps the reader to find the material of interest. Gene correction can be accomplished in many different organisms and cell types. The first section (Part I) presents a sample of gene correction approaches in hosts as different as *Pseudomonas*, *Drosophila*, chicken cells, and human pluripotent stem cells. Approaches for gene correction in these and many other different host organisms and cell types are presented throughout the book in several other chapters; hence, these are reported at the end of the first section for useful reference. Similarly structured, the second section (Part II) centers on some of the most effective instruments for gene correction, comprising both nonviral and viral tools. The third section (Part III) contains protocols that emphasize the impact of inducing a break in the target DNA to stimulate gene correction, exploiting the positive features of break-induced gene targeting, and addressing its negative aspects. Finally, *ad hoc* gene correction protocols developed to correct mutations associated with specific genetic diseases are presented in the fourth section (Part IV).

I am passionate about gene correction because it gives us the tools for both repairing and mutating DNA, for discovering gene functions and for engineering new genetic variants. As Nobel laureate for gene targeting in mice, Mario Capecchi once said, "gene targeting gives us complete freedom in choosing which gene to alter and how to alter it." The preparation of this book has been an exciting experience. I learned a lot from reading and reviewing the chapters. I think all the methods and protocols collected in this volume are a precious resource for the current and future gene "targeters"....there is still a long way to go!

The participating authors deserve great appreciation for the valuable contribution, effort, and patience they offered for the preparation of this volume. I am extremely thankful to all contributors. I would like to thank very much John Walker for his constant assistance and advice for this book. I am also grateful to all the staff at Springer and Humana Press for their work in assembling the chapters and producing this book.

Atlanta, GA, USA ***Francesca Storici***

Contents

Contributors

HAYDER H. ABDUL-RAZAK • *School of Biological Sciences, Royal Holloway-University of London, Egham, Surrey, UK*

MD. ROWSHON ALAM • *Girindus America, Inc., Cincinnati, OH, USA*

CHARLOTTE ANDRIEU-SOLER • *INSERM, Centre de Recherche des Cordeliers, Université René Descartes Sorbonne Paris Cité, Paris, France*

CHRISTOPHER J. ANTICO • *Wallace H. Coulter Department of Biomedical Engineering, Georgia Institute of Technology, Atlanta, GA, USA*

GANG BAO • *Wallace H. Coulter Department of Biomedical Engineering, Georgia Institute of Technology, Atlanta, GA, USA*

MELISSA A. BARTEL • *Department of Chemical and Biomolecular Engineering, University of California, Berkeley, CA, USA*

FRANCINE BEHAR-COHEN • *INSERM, Centre de Recherche des Cordeliers, Université René Descartes Sorbonne Paris Cité, Paris, France; Assistance Publique Hôpitaux de Paris, Hôtel-Dieu de Paris, Paris, France*

DWIGHT A. BELLINGER • *Francis Owen Blood Research Laboratory, University of North Carolina, Chapel Hill, NC, USA*

CARMEN BERTONI • *Department of Neurology, David Geffen School of Medicine, University of California Los Angeles, Los Angeles, CA, USA*

LINGZHI CAI • *Department of Medicine/Endocrinology, University of Pittsburgh, Pittsburgh, PA, USA*

MICHAEL T. CERTO • *Center of Immunity and Immunotherapies and Ben Towne Center for Childhood Cancer Research, Seattle Children's Research Institute, Seattle, WA, USA*

GEORGE CHURCH • *Department of Genetics, Harvard Medical School, Boston, MA, USA*

PAULA R. CLEMENS • *Department of Neurology, University of Pittsburgh, Pittsburgh, PA, USA; Neurology Service, Department of Veterans Affairs Medical Center, Pittsburgh, PA, USA*

MARIO COOPER • *Department of Pediatrics, University of Florida, Gainesville, FL, USA*

THOMAS J. CRADICK • *Wallace H. Coulter Department of Biomedical Engineering, Georgia Institute of Technology, Atlanta, GA, USA*

MARILYN DERNIGOGHOSSIAN • *INSERM, Centre de Recherche des Cordeliers, Université René Descartes Sorbonne Paris Cité, Paris, France*

ELI J. FINE • *Wallace H. Coulter Department of Biomedical Engineering, Georgia Institute of Technology, Atlanta, GA, USA*

DANIEL L. GALVAN • *Department of Medicine, Baylor College of Medicine, Houston, TX, USA*

FREDERICK S. GIMBLE • *Department of Biochemistry, Purdue University, West Lafayette, IN, USA*

PETER M. GLAZER • *Department of Therapeutic Radiology, Yale University School of Medicine, New Haven, CT, USA; Department of Genetics, Yale University School of Medicine, New Haven, CT, USA*

MANUEL A.F.V. GONÇALVES • *Department of Molecular Cell Biology, Leiden University Medical Center, Leiden, The Netherlands*
PHILIP D. GREGORY • *Sangamo BioSciences, Inc., Pt. Richmond Tech Center, Richmond, CA, USA*
DIETER C. GRUENERT • *Department of Otolaryngology-Head and Neck Surgery, Eli and Edythe Broad Center for Regenerative Medicine and Stem Cell Research, Institute for Human Genetics, Helen Diller Family Comprehensive Cancer Center, Cardiovascular Research Institute, University of California, San Francisco, San Francisco, CA, USA; University of Vermont School of Medicine, Burlington, VT, USA*
ROLAND W. HERZOG • *Department of Pediatrics, University of Florida, Gainesville, FL, USA*
MATTHEW L. HIRSCH • *Gene Therapy Center, University of North Carolina, Chapel Hill, NC, USA; Department of Ophthalmology, University of North Carolina, Chapel Hill, NC, USA; Department of Cell and Developmental Biology, University of North Carolina, Chapel Hill, NC, USA*
MICHAEL C. HOLMES • *Sangamo BioSciences, Inc., Pt. Richmond Tech Center, Richmond, CA, USA*
SARA E. HOWDEN • *Department of Cell and Regenerative Biology, University of Wisconsin School of Medicine and Public Health, Madison, WI, USA*
PATRICK D. HSU • *Broad Institute of MIT and Harvard, Massachusetts Institute of Technology, Cambridge, USA*
RAKESH JOSHI • *Department of Biochemistry, Siebens-Drake Medical Research Institute, Schulich School of Medicine and Dentistry, University of Western Ontario, London, ON, Canada*
CLAUDIA S. KETTLUN • *Department of Medicine, Baylor College of Medicine, Houston, TX, USA*
CAROLINE KIM-KISELAK • *Biological and Biomedical Sciences Program, Harvard Medical School, Boston, MA, USA*
BHANU MUNIL KOPPANATI • *Department of Neurology, University of Pittsburgh, Pittsburgh, PA, USA*
ARTHUR KRIGEL • *INSERM, Centre de Recherche des Cordeliers, Université René Descartes Sorbonne Paris Cité, Paris, France*
HONG LI • *LaserGen, Inc, Houston, TX, USA*
YANNI LIN • *Wallace H. Coulter Department of Biomedical Engineering, Georgia Institute of Technology, Atlanta, GA, USA*
CORY R. LINDSAY • *Department of Pathology and Laboratory Medicine, Raymond and Ruth Perelman School of Medicine, University of Pennsylvania, Philadelphia, PA, USA; Abramson Family Cancer Research Institute, Raymond and Ruth Perelman School of Medicine, University of Pennsylvania, Philadelphia, PA, USA*
ANDREA LUCHETTI • *Medical Genetics Section, Department of Biomedicine and Prevention, School of Medicine, Tor Vergata University, Rome, Italy*
ARIANNA MALGIERI • *Medical Genetics Section, Department of Biomedicine and Prevention, School of Medicine, Tor Vergata University, Rome, Italy*
PRASHANT MALI • *Department of Genetics, Harvard Medical School, Boston, MA, USA*
ELIZABETH P. MERRICKS • *Francis Owen Blood Research Laboratory, University of North Carolina, Chapel Hill, NC, USA*

MICHAEL J. METZGER • *Basic Sciences, Fred Hutchinson Cancer Research Center, Seattle, WA, USA; Department of Biochemistry and Molecular Biophysics, Columbia University, New York, NY, USA*

KOHNOSUKE MITANI • *Gene Therapy Division, Research Center for Genomic Medicine, Saitama Medical University, Hidaka, Saitama, Japan*

GUSTAVO MOSTOSLAVSKY • *Department of Medicine, Center for Regenerative Medicine (CReM), Boston University School of Medicine, Boston, MA, USA*

AKIRA MOTEGI • *Department of Radiation Genetics, Kyoto University Graduate School of Medicine, Sakyo-ku, Kyoto, Japan*

SAYANDIP MUKHERJEE • *Molecular Immunology Unit, Centre for Immunodeficiency, UCL Institute of Child Health, London, UK*

TIMOTHY C. NICHOLS • *Francis Owen Blood Research Laboratory, University of North Carolina, Chapel Hill, NC, USA*

LAETITIA P.L. PELASCINI • *Department of Molecular Cell Biology, Leiden University Medical Center, Leiden, The Netherlands*

NEENA K. PYZOCHA • *Broad Institute of MIT and Harvard, Massachusetts Institute of Technology, Cambridge, USA; Department of Biology, Massachusetts Institute of Technology, Cambridge, USA*

F. ANN RAN • *Broad Institute of MIT and Harvard, Massachusetts Institute of Technology, Cambridge, MA, USA*

ROBIN A. RAYMER • *Francis Owen Blood Research Laboratory, University of North Carolina, Chapel Hill, NC, USA*

FAISAL REZA • *Departments of Therapeutic Radiology, Yale University School of Medicine, New Haven, CT, USA*

CÉLINE J. ROCCA • *School of Biological Sciences, Royal Holloway-University of London, Egham, Surrey, UK*

YIKANG S. RONG • *Laboratory of Biochemistry and Molecular Biology, National Cancer Institute (NCI), NIH, Bethesda, MD, USA*

DAVID B. ROTH • *Department of Pathology and Laboratory Medicine, Raymond and Ruth Perelman School of Medicine, University of Pennsylvania, Philadelphia, PA, USA; Abramson Family Cancer Research Institute, Raymond and Ruth Perelman School of Medicine, University of Pennsylvania, Philadelphia, PA, USA*

R. JUDE SAMULSKI • *Gene Therapy Center, University of North Carolina at Chapel Hill, Chapel Hill, NC, USA; Department of Pharmacology, University of North Carolina at Chapel Hill, Chapel Hill, NC, USA*

FEDERICA SANGIUOLO • *Medical Genetics Section, Department of Biomedicine and Prevention, School of Medicine, Tor Vergata University, Rome, Italy*

R. GEOFFREY SARGENT • *Department of Otolaryngology-Head and Neck Surgery, University of California, San Francisco, San Francisco, CA, USA*

DAVID V. SCHAFFER • *Department of Chemical and Biomolecular Engineering, University of California, Berkeley, CA, USA; Department of Bioengineering, University of California, Berkeley, CA, USA; The Helen Wills Neuroscience Institute, University of California, Berkeley, CA, USA*

ALEXANDER SCHLACHTERMAN • *Department of Medicine, Drexel University, Philadelphia, PA, USA*

WILLIAM SCHREINER • *Laboratory of Biochemistry and Molecular Biology, National Cancer Institute (NCI), NIH, Bethesda, MD, USA*

MICHAEL M. SEIDMAN • *National Institute on Aging (NIA), NIH, Baltimore, MD, USA*
ALEXANDRA SHERMAN • *Department of Pediatrics, University of Florida, Gainesville, FL, USA*
FRANCESCA STORICI • *School of Biology, Georgia Institute of Technology, Atlanta, GA, USA*
SAMANTHA STUCKEY • *School of Biology, Georgia Institute of Technology, Atlanta, GA, USA*
SHINGO SUZUKI • *Department of Otolaryngology-Head and Neck Surgery, University of California, San Francisco, San Francisco, CA, USA; Department of Molecular Medicine, Graduate School of Pharmaceutical Sciences, Kumamoto University, Kumamoto, Japan*
BRYAN SWINGLE • *Plant–Microbe Interactions Research Unit, Department of Agriculture, Agricultural Research Service, Ithaca, NY, USA*
MINORU TAKATA • *Department of Late Effects, Kyoto University Radiation Biology Center, Sakyo-ku, Kyoto, Japan*
ARUN KALLIAT THAZHATHVEETIL • *Department of Chemistry, Northwestern University, Evanston, IL, USA*
DOLLY THOMAS • *Section of Gastroenterology, Department of Medicine, Center for Regenerative Medicine (CReM), Boston University School of Medicine, Boston, MA, USA*
JAMES A. THOMSON • *Department of Cell and Regenerative Biology, University of Wisconsin School of Medicine and Public Health, Madison, WI, USA*
ADRIAN J. THRASHER • *Molecular Immunology Unit, Centre for Immunodeficiency, UCL Institute of Child Health, London, UK*
MATTHEW H. WILSON • *Department of Medicine and Center for Cell and Gene Therapy, Baylor College of Medicine and Michael E. DeBakey VA Medical Center, Houston, TX, USA*
RAFAEL J. YÁÑEZ-MUÑOZ • *School of Biological Sciences, Royal Holloway-University of London, Egham, Surrey, UK*
LUHAN YANG • *Department of Genetics, Harvard Medical School, Boston, MA, USA; Biological and Biomedical Sciences Program, Harvard Medical School, Boston, MA, USA*
FENG ZHANG • *Broad Institute of MIT and Harvard, Cambridge, MA, USA; McGovern Institute for Brain Research, MIT Cambridge, MA, USA; Department of Brain and Cognitive Sciences, MIT, Cambridge, MA, USA; Department of Biological Engineering, MIT, Cambridge, MA, USA*
YI ZHANG • *Laboratory of Biochemistry and Molecular Biology, National Cancer Institute (NCI), NIH, Bethesda, MD, USA*

Part I

Approaches for Gene Correction from Bacteria to Human Cells

Additional protocols described in this book for gene correction using in vivo, ex-vivo, in vitro or in silico systems

- bacteria: Chapters 2, 4, 14
- yeast: Chapter 20
- hamster cells: Chapters 7, 8
- rat neural stem cells: Chapter 11
- mouse cells: Chapters 5, 6, 21
- mouse hematopoietic stem cells: Chapter 29
- mouse induced pluripotent stem cells: Chapter 28
- mouse fetal skeletal muscle: Chapter 26
- mouse retina: Chapter 25
- mouse liver: Chapter 27
- dog liver: Chapter 27
- human cells: Chapters 7, 8, 9, 12, 15, 16, 17, 19, 22, 23
- primary human CD34+ hematopoietic progenitor cells: Chapter 8
- human embryonic pluripotent stem cells: Chapters 10, 11
- human induced pluripotent stem cells: Chapters 10, 16, 18
- in silico: Chapters 13, 24

Chapter 1

RecTE$_{Psy}$-Mediated Recombineering in *Pseudomonas syringae*

Bryan Swingle

Abstract

A recently developed *Pseudomonas syringae* recombineering system simplifies the procedure for installing specific mutations at a chosen genomic locus. The procedure involves transforming *P. syringae* cells expressing recombineering functions with a PCR product that contains desired changes flanked by sequences homologous to a target location. Cells transformed with the substrate undergo homologous recombination between the genomic DNA and the recombineering substrate. The recombinants are found by selection for traits carried by the recombineering substrate, usually antibiotic resistance.

Key words *Pseudomonas syringae*, Recombineering, RecTE, Protocol, Homologous recombination, Electroporation, Genetic engineering, Site-specific mutagenesis

1 Introduction

The ability to change DNA sequences is fundamental to molecular genetics and the branches of engineering that have arisen from this discipline. Most methods for site-directed mutagenesis of bacterial genomes use homologous recombination because it achieves a high degree of location specificity through base pairing and can incorporate of a wide range of changes. The traditional marker exchange method for site-directed genome mutagenesis involves constructing a plasmid containing the desired change flanked by sequences homologous to the target location. This strategy is effective but slow because plasmid construction requires molecular cloning and several validation steps for plasmid and mutant constructs. A revolutionary approach was introduced in 1998 [1, 2] that markedly improved the speed and types of mutations that could be introduced into bacterial genomes. This new method, dubbed recombineering, is based on phage-encoded functions that orchestrate recombination of linear DNA molecules directly with the target genome. Until this point, transformation and recombination of linear DNA with bacterial genomes was considered

Francesca Storici (ed.), *Gene Correction: Methods and Protocols*, Methods in Molecular Biology, vol. 1114,
DOI 10.1007/978-1-62703-761-7_1,

impractical because in the absence of phage-encoded recombination functions, these events are exceedingly rare, primarily because cellular nucleases degrade the incoming DNA [3, 4] and because of the need for extensive homology between the recombining molecules [5]. However, Kenan Murphy [1] and Youming Zhang and colleagues [2] independently demonstrated that phage-encoded recombination functions could be used to facilitate recombination of linear DNA molecules introduced directly into the cell by electroporation. This discovery eliminated the need for in vitro molecular cloning steps and made it possible to install mutations directly in the genome of living cells.

The functions that make recombineering possible were first characterized in *E. coli* using lambda phage's Red genes (*exo*, *bet,* and *gam*) and Rac prophage (*recTE*) genes. These phage-encoded functions catalyze recombination independently of the cell endogenous RecA-dependent pathways by processing the linear substrate DNA and facilitating base paring with the target molecule. The initial processing steps involve converting the transformed linear DNA into a single-stranded intermediate by the lambda Exo or RecE 5′→3′ exonucleases [6, 7]. The newly formed ssDNA molecule is coated with the associated ssDNA annealing protein (lambda Beta or RecT), which protects the ssDNA molecule from degradation and promotes/stabilizes annealing at the target location [8, 9]. Gam is a third function encoded by lambda Red, which provides additional protection to the transformed recombineering substrate by inhibiting the exonuclease activity of the RecBCD complex. However, Gam is not necessary, but provides a modest increase recombination frequency [10].

Lambda Red and Rac RecET are very efficient for recombineering in *E. coli* and other enterics, but their ability to function in more distantly related bacteria is unpredictable. For example, we have been unable to observe Red-dependent recombination in *P. syringae*, but Red recombineering has been reported for *Pseudomonas aeruginosa* [11]. The nature of this host specificity has not been identified but presumably involves an interaction between the phage proteins and host-encoded functions, possibly the cell's DNA replication machinery [12]. It was these observations of host specificity that led to the hypothesis that recombineering functions identified in phage associated with a particular species would be more likely to function in those organisms or closely related species. Guided by this hypothesis we identified *recTE*$_{Psy}$ genes in a putative prophage or remnant in the *P. syringae* B728a genome [13] and demonstrated that they encode functional orthologs of the well-characterized lambda Red Beta/Exo and Rac RecET enzymes [13]. The *recTE*$_{Psy}$ genes facilitate recombination of linear DNA introduced directly into *P. syringae* cells by electroporation and have allowed development of a new recombineering system for use in this species. Notably, similar approaches have

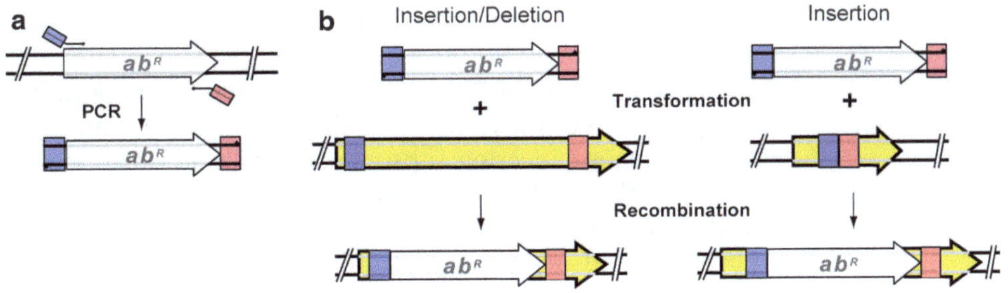

Fig. 1 Recombineering substrates and targets. (**a**) PCR is used to amplify an antibiotic resistance gene (ab^R). PCR primers include 80 nt of homology to a genomic target at the 5′ ends (*blue* and *red rectangles*). (**b**) *P. syringae* cells expressing $RecTE_{Psy}$ are transformed with PCR product using electroporation and recombine with target location (*yellow arrow*) via the pair of 80 bp genomic homologies located at the ends of the PCR product. Depending on how the homologies are located in the genome, insertions can delete a genomic DNA region

been used to identify and develop recombineering systems in *Mycobacteria* [14–16] and lactic acid bacilli [17, 18].

Single-stranded or double-stranded DNA substrates can be used for $RecTE_{Psy}$ recombineering in *P. syringae*. Deciding which form of substrate to use depends on the experimental objectives. Single-stranded substrates typically require fewer laboratory steps prior to transformation, because made-to-order oligonucleotides only need to be dissolved to prepare for recombineering. Additionally, oligos usually generate more recombinant clones than double-stranded substrates, possibly because preprocessing by the exonuclease is not necessary to begin the recombination reaction [10]. However, the length of commercially available oligos limits the types of changes that can be included on these substrates, making it impractical to introduce most genes that provide a selectable phenotype. The ability to use dsDNA substrates obviates some of the limitations and makes a range of practical mutations possible, including insertion/deletions that incorporate antibiotic resistance (or other) cassettes used to select for mutant strains. dsDNA substrates can be made using PCR to amplify an antibiotic resistance gene with primers that have 80 bases of sequence homologous to the target at the 5′ ends (Fig 1a). Depending on the relative position of the homologous target sequences in the genome, recombination results in deletions and/or insertions (Fig 1b).

The principal advantage of recombineering is that the procedure is quick and straightforward. In a typical experiment, *P. syringae* cells expressing $RecTE_{Psy}$ are transformed with substrate DNA using electroporation, and recombinants are selected for growth on media containing appropriate antibiotics. $RecTE_{Psy}$ are supplied from a plasmid (pUCP24/recTE) that provides constitutive expression of these genes from the *nptII* promoter and also carries the *sacB* gene, which can be used for efficient counterselection of the plasmid. Current *P. syringae* recombineering frequencies are too low to find recombinant clones without introducing a phenotypic change. So, in most cases a gene encoding antibiotic resistance is introduced along with the desired mutation, but prototrophic selections are also possible [19]. After transformation with the recombineering substrate, the cells are incubated in rich media broth to allow the cells to recover and express the genes needed to survive the selection (i.e., antibiotic resistance). The transformation outgrowth culture is then plated on solid growth medium and incubated for several days. After colonies appear the presence of the mutation is confirmed using PCR and sequencing.

2 Materials

1. *Pseudomonas syringae* pv. *tomato* DC3000 transformed with pUCP24/recTE.
2. Plasmid DNA or other source of genes encoding antibiotic resistance.
3. 10 mg/ml gentamicin.
4. Oligonucleotides to amplify the antibiotic resistance gene and incorporate sequences homologous to target location. Dissolve lyophilized oligo in sterile water to a final concentration of 100 μM.
5. Taq polymerase.
6. PCR spin columns to clean and concentrate PCR product.
7. Phosphate stock (100×): 0.86 M K_2HPO_4 (filter sterilize).
8. KB broth [20]: 2 % proteose peptone #3, 1.6 mM $MgSO_4 \cdot 7H_2O$, and 1 % glycerol. Autoclave. After media has cooled, adjust to 1× phosphate stock (final concentration).
9. KB agar [20]: 2 % proteose peptone #3, 1.6 mM $MgSO_4 \cdot 7H_2O$, 1 % glycerol, 8.6 mM K_2HPO_4, and 1.8 % agar.
10. Sucrose: 300 mM in H_2O (filter sterilize).
11. Sterile dH_2O.
12. Gene pulser (Bio-Rad Laboratories).
13. 0.2 cm electroporation cuvette.
14. Glucose: 20 % solution in H_2O (filter sterilize).
15. Mg^{+2} stock: 1 M $MgCl_2 \cdot 6H_2O$ and 1 M $MgSO_4 \cdot 7H_2O$ (filter sterilize).
16. SOC broth [21]: 2 % Bacto tryptone, 0.5 % yeast extract, and 9.92 mM NaCl. Autoclave. After the media has cooled adjust to 0.2 % glucose and 1× Mg^{+2} stock (final concentration).

3 Methods

3.1 Experimental Design

The flexibility of recombineering stems largely from the ability to produce substrates that can anneal to any genomic target location (*see* **Note 1**) and introduce different types of mutations. Currently, the most practical *P. syringae* recombineering involves introducing an antibiotic resistance cassette. The substrates used in this type of reaction can be designed to generate an insertion with or without deletion of genomic sequences depending on how the homologies are positioned in the genome sequence (Fig. 1). Once the structure of the desired product has been established, PCR primers are designed and obtained that incorporate the homologies to guide the correct recombinant product. The sequence of the PCR primers should consist of approximately 80 nt of homology to the genomic target at the 5′ ends followed by and ~25 nt of homology to the antibiotic resistance gene (Fig 1a).

3.2 Substrate Generation

Conventional PCR is used to amplify the PCR substrate.

1. Combine 25 μl of ExTaq PCR mix, template DNA containing antibiotic resistance gene (e.g. plasmid), 1 μl of each primer (100 μM stock concentration) and adjust final volume to 50 μl of sterile water.
2. Incubate reaction for 25 cycles (95 °C, 1′; 55 °C, 30″, 72 °C, 1′).
3. Confirm that the substrate has been amplified and a significant amount of product of the correct size is present by gel electrophoresis.
4. Use a spin column or ethanol precipitation to clean and concentrate PCR product. A concentration of 20 μg/ml or greater is adequate for recombineering.

3.3 Transform Cells Expressing the $RecTE_{Psy}$ with the Recombineering Substrate

1. Grow *P. syringae* pv. *tomato* DC3000 pUCP24/recTE in KB medium [20] supplemented with 10 μg/ml gentamicin overnight at 28–30 °C (*see* **Note 2**).
2. Dilute overnight culture 1:10 in 125 ml of fresh KB broth with 10 μg/ml gentamicin and grow to an OD600 of 0.6–0.8 (*see* **Note 3**).
3. When the culture has grown to an OD_{600} of 0.8–1.0, harvest cells by centrifugation (5,000 × *g*) at 20 °C; wash twice with equal volume of room temperature 300 mM sucrose (*see* **Note 4**).
4. Pellet cells by centrifugation at 5,000 × *g* and resuspend in 1/60th the original culture volume 300 mM sucrose and dispense 100 μl into 1.5 ml test tubes.
5. Add 5 μl of PCR product (100–500 ng) to electro-competent cell suspension. Mix by pipetting several times.
6. Transfer mixture to a 0.2 cm electroporation cuvette.

7. Electroporate cells with substrate at 2.5 kV, 25 μF, and 200 Ω.
8. Immediately add electroporated cells to 5 ml of SOC broth and incubate at 28–30 °C overnight with vigorous shaking.

3.4 Select for Recombinants

1. Plate 200 μl aliquots of electroporation outgrowth culture on KB agar supplemented with the appropriate antibiotic.
2. Incubate at 28–30 °C for 3–4 days (*see* **Note 5**).

3.5 Confirm Presence of Mutation

1. Use colony PCR to test for presence of the mutation. If insertion of the desired mutation alters the length of the region, use a pair of primers that anneal to sequences flanking the recombinant allele.
2. This PCR product is sequenced to confirm that recombinant allele conforms to the desired change.

3.6 Eliminate the Recombineering Plasmid

1. Test recombinant clones for presence of pUCP24/recTE, by determining whether clones are resistant to gentamicin. In most cases cells will lose pUCP24/recTE during the post-electroporation culturing steps, which are done without selection for the plasmid expressing the recombineering functions (*see* **Note 3**).
2. If the recombineering plasmid is still present after identifying recombinant clones, a simple counterselection step can be used to identify plasmid free cells. The pUCP24/recTE plasmid encodes the *sacB* gene, which causes toxic accumulation of levan in cells grown on sucrose [22]. To select for cells that have lost the plasmid, resuspend cells confirmed to have acquired the mutant allele in 1 ml Kb broth, grow for 4–6 h, and then spread on Kb agar containing 10 % sucrose.
3. Incubate at 28 °C for 2–3 days.
4. Confirm loss of pUCP24/recTE by testing for gentamicin resistance.

4 Notes

1. The *P. syringae* pv. *tomato* DC3000 sequence is available for download at www.pseudomonas-syringae.org.
2. The sequence for pUCP24/recTE is available at Genbank accession: HM368666. The physical plasmid can be obtained by contacting the author. *P. syringae* strains transformed with pUCP24/recTE are grown in 10 μg/ml gentamicin to maintain selection for the recombineering plasmid.
3. An overnight culture of *P. syringae pv. tomato* DC3000 pUCP24/recTE usually grows to an OD_{600} of 4.0, a 1:10 dilution

in fresh medium achieves desired OD_{600} of 0.4. This 1:10 dilution is much more dense than a typical subinoculation (which is usually in the range of 1:40). However, we have found that it is necessary to increase the proportion of inoculum in order to attain the desired growth (OD_{600} of 0.8–1.0) in the course of the workday. The pUCP24/recTE reduces the growth rate of *P. syringae* pv. *tomato* DC3000. We are not sure why the growth rate is affected, but suspect that it might be due to a degree of toxicity related to constitutive expression of the $recTE_{Psy}$ genes. Also consistent with this observation, we have found that pUCP24/recTE is rapidly lost from *P. syringae* cells in the absence of selection.

4. For routine work, satisfactory results can be obtained using frozen competent cells. To prepare frozen competent cells, wash cells two additional times in 10 % glycerol prior to the final resuspension in 1/60th the original culture volume of 10 % glycerol and freeze in 100 μl aliquots at −80 °C. Thaw aliquots on ice for 10 min prior to electroporation.
5. Colonies formed from recombinant clones usually take longer to grow than those composed of wild-type *P. syringae*. Colonies should be visible at 3–4 days.

Acknowledgments

Mention of trade names or commercial products in this publication is solely for the purpose of providing specific information and does not imply recommendation or endorsement by the U.S. Department of Agriculture. The USDA is an equal opportunity provider and employer.

References

1. Murphy KC (1998) Use of bacteriophage lambda recombination functions to promote gene replacement in escherichia coli. J Bacteriol 180(8):2063–2071
2. Zhang Y, Buchholz F, Muyrers JP, Stewart AF (1998) A new logic for DNA engineering using recombination in escherichia coli. Nat Genet 20(2):123–128
3. Dutra BE, Sutera VA Jr, Lovett ST (2007) RecA-independent recombination is efficient but limited by exonucleases. Proc Natl Acad Sci U S A 104(1):216–221
4. Winans SC, Elledge SJ, Krueger JH, Walker GC (1985) Site-directed insertion and deletion mutagenesis with cloned fragments in Escherichia coli. J Bacteriol 161(3):1219–1221
5. Lovett ST, Hurley RL, Sutera VA Jr, Aubuchon RH, Lebedeva MA (2002) Crossing over between regions of limited homology in Escherichia coli. RecA-dependent and RecA-independent pathways. Genetics 160(3): 851–859
6. Cassuto E, Radding CM (1971) Mechanism for the action of lambda exonuclease in genetic recombination. Nat New Biol 229(1):13–16
7. Little JW (1967) An exonuclease induced by bacteriophage lambda. II. Nature of the enzymatic reaction. J Biol Chem 242(4):679–686
8. Kmiec E, Holloman WK (1981) Beta protein of bacteriophage lambda promotes renaturation of DNA. J Biol Chem 256(24): 12636–12639

9. Karakousis G, Ye N, Li Z, Chiu SK, Reddy G, Radding CM (1998) The beta protein of phage lambda binds preferentially to an intermediate in DNA renaturation. J Mol Biol 276(4):721–731
10. Ellis HM, Yu D, DiTizio T, Court DL (2001) High efficiency mutagenesis, repair, and engineering of chromosomal DNA using single-stranded oligonucleotides. Proc Natl Acad Sci U S A 98(12):6742–6746
11. Lesic B, Rahme LG (2008) Use of the lambda Red recombinase system to rapidly generate mutants in Pseudomonas aeruginosa. BMC Mol Biol 9:20
12. Datta S, Costantino N, Zhou X, Court DL (2008) Identification and analysis of recombineering functions from Gram-negative and Gram-positive bacteria and their phages. Proc Natl Acad Sci U S A 105(5): 1626–1631
13. Swingle B, Bao Z, Markel E, Chambers A, Cartinhour S (2010) Recombineering using RecTE from pseudomonas syringae. Appl Environ Microbiol 76(15):4960–4968
14. van Kessel JC, Hatfull GF (2007) Recombineering in mycobacterium tuberculosis. Nat Methods 4(2):147–152
15. van Kessel JC, Hatfull GF (2008) Efficient point mutagenesis in mycobacteria using single-stranded DNA recombineering: characterization of antimycobacterial drug targets. Mol Microbiol 67(5):1094–1107
16. van Kessel JC, Marinelli LJ, Hatfull GF (2008) Recombineering mycobacteria and their phages. Nat Rev Microbiol 6(11):851–857
17. van Pijkeren JP, Britton RA (2012) High efficiency recombineering in lactic acid bacteria. Nucleic Acids Res 40(10):e76. doi:gks147 [pii] 10.1093/nar/gks147
18. van Pijkeren JP, Neoh KM, Sirias D, Findley AS, Britton RA (2012) Exploring optimization parameters to increase ssDNA recombineering in Lactococcus lactis and Lactobacillus reuteri. Bioengineered 3(4):209–217. doi:21049 [pii]
19. Sharan SK, Thomason LC, Kuznetsov SG, Court DL (2009) Recombineering: a homologous recombination-based method of genetic engineering. Nat Protoc 4(2):206–223
20. King EO, Ward MK, Raney DE (1954) Two simple media for the demonstration of pyocyanin and fluorescein. J Lab Clin Med 44(2):301–307
21. Hanahan D (1983) Studies on transformation of Escherichia coli with plasmids. J Mol Biol 166(4):557–580
22. Ried JL, Collmer A (1987) An nptI-sacB-sacR cartridge for constructing directed, unmarked mutations in gram-negative bacteria by marker exchange-eviction mutagenesis. Gene 57(2–3): 239–246

Chapter 2

Genome Manipulations with Bacterial Recombineering and Site-Specific Integration in Drosophila

Yi Zhang, William Schreiner, and Yikang S. Rong

Abstract

Gene targeting is a vital tool for modern biology. The ability to efficiently and repeatedly target the same locus is made more efficient by the site-specific integrase mediated repeated targeting (SIRT) method, which combines homologous recombination, site-specific integration, and bacterial recombineering to conduct targeted modifications of individual loci. Here we describe the recombineering designs and procedures for the introduction of epitope tags, in-frame deletion mutations, and point mutations into plasmids that can later be used for SIRT.

Key words Bacterial recombineering, Gene targeting, Drosophila, Site-specific recombination, Genome engineering

1 Introduction

Gene targeting by homologous recombination revolutionized the study of gene function by enabling genetic manipulations of endogenous loci in vivo. By combining homologous gene targeting and site-specific recombination, we recently developed the "Site-specific Integrase mediated Repeated Targeting" (SIRT) method for *Drosophila melanogaster*, which facilitates repeated rounds of targeted manipulation of a single locus [1]. In SIRT, homologous recombination is used to place an *attP* attachment site of the phage phiC31 integrase in the vicinity of the target locus. All subsequent modifications to the same gene are introduced as plasmids carrying the modifications and the *attB* attachment site. These plasmids are directly injected into *attP*-containing embryos expressing the phiC31 integrase. The integrase mediates an exchange between the two *att* sites, which results in plasmid integration precisely at the chromosomal *attP* site. Figure 1 provides

Yi Zhang and William Schreiner contributed equally to this manuscript.

Francesca Storici (ed.), *Gene Correction: Methods and Protocols*, Methods in Molecular Biology, vol. 1114,
DOI 10.1007/978-1-62703-761-7_2, © Springer Science+Business Media, LLC 2014

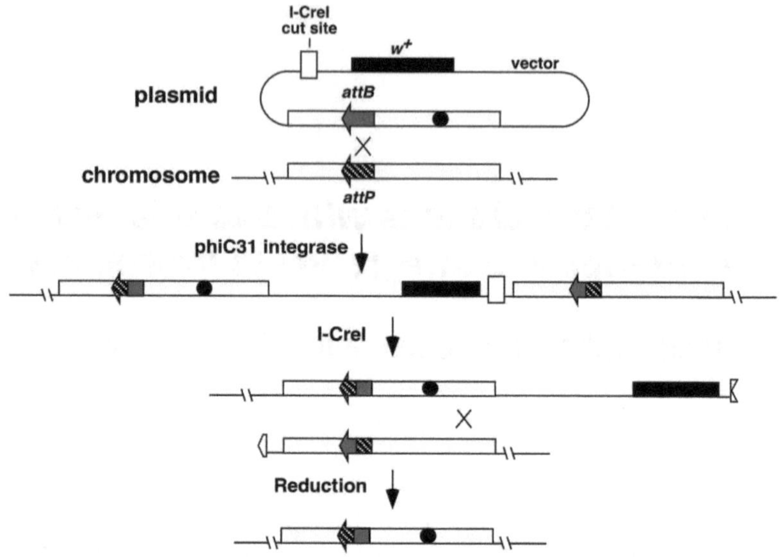

Fig. 1 SIRT. The *top diagram* depicts phiC31-mediated site-specific integration. An *attP* attachment site is first targeted to a locus of interest on the chromosome by traditional ends in gene targeting [9]. The *open rectangular box* represents the locus of interest. A plasmid carrying the desired modification (*filled circle*), an *attB* attachment site, a cut site of the I-CreI endonuclease, and a *white*(w^+) marker gene (*filled rectangular box*), is introduced into *attP*-containing flies that express phiC31 integrase. Plasmid integration into the chromosome, which is mediated by an exchange ("X") between the *att* sites, results in the duplication of the target locus and the integration of the w^+ marker that gives rise to eye pigmentation. The *bottom diagram* depicts I-CreI-mediated reduction of the target duplication. I-CreI expression is induced to generate a DNA break at its cut site. This break induces recombination between the two target copies. If recombination occurs at the position denoted by the "X", the reduction product will harbor the desired modification. Flies with successful reduction events will be white-eyed due to the loss of the w^+ marker

a detail description of the steps in SIRT. We have successfully used SIRT to modify multiple genes important for telomere maintenance in Drosophila (e.g., 1, 2). These modifications include deletion of a locus, small in-frame deletions, point mutations, and insertions of epitope tags.

To construct plasmids for SIRT, we extensively utilize the bacterial recombineering technology, which is based on recombination induced by the expression of the RED system from phage lambda [3]. In this chapter, using the *verrocchio* (*ver*, 4), *hiphop* [5], and *caravaggio* (*cav*, 6) loci as specific examples, we describe recombineering designs and procedures for introducing epitope tags, in-frame deletion mutations, and point mutations. Once the final plasmid is generated, it can be introduced into flies by SIRT protocols previously described [1, 7].

Although we mainly use SIRT for gene disruption, the same practice can be used to achieve gene correction. Even though our SIRT method was specifically developed for Drosophila, all of its components are functional in other eukaryotes making it easily adaptable in other model organisms.

2 Materials

2.1 PCR Primers

Primers used for recombineering are generally longer than 70 bp as each primer would minimally include 50 bp homologous sequence to the target gene and 20 bp homologous sequence to an antibiotic resistant gene. In addition, restriction enzyme cut sites are often included in the primers to facilitate excision of the antibiotic marker. For specific design of primers, *see* discussion in the Subheading 3. Purification of the primers is not necessary.

2.2 Bacterial Strains and Culturing Materials

1. A strain that is competent for recombineering. We used the SW102 strain [8]. For other available strains, *see* http://redrecombineering.ncifcrf.gov/.
2. Standard Bacterial cloning strains such as DH5α and DH10β.
3. LB liquid medium.
4. SOC medium.
5. Antibiotics.
6. Bacterial Culture Tubes.
7. Bacterial Culture flasks.
8. Bacterial electroporator.
9. Electroporation cuvettes.
10. Shaking water bath.

2.3 Enzymes and Buffers

1. DNA polymerase for PCR with proofreading activities.
2. Restriction enzymes and buffers.
3. DNA ligase and buffer.

2.4 Molecular Biology Kits

1. MiniPrep.
2. PCR purification.
3. TOPO TA cloning.

3 Methods

3.1 A Two-Step Scheme to Modify Multi-copy Plasmids Using Recombineering

For the purpose of vector construction, recombineering enables integration of exogenously provided DNA into a homologous region on the plasmid. This integration is facilitated by the expression of the lambda RED system. In typical recombineering experiments (some

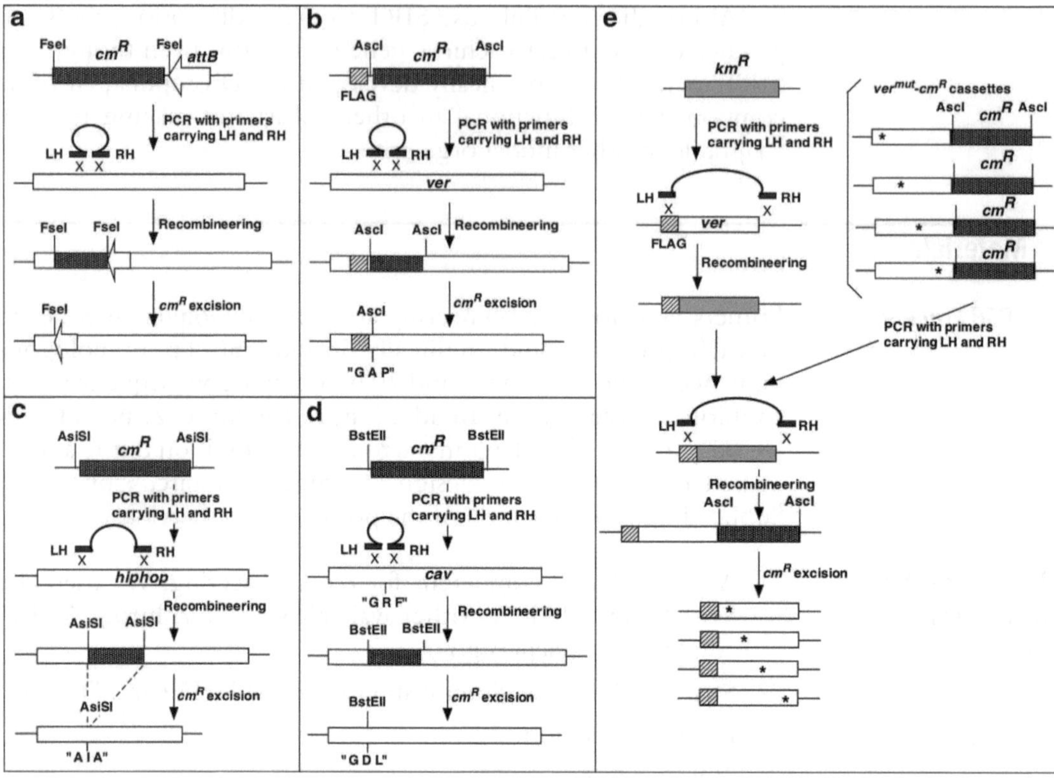

Fig. 2 Vector construction by recombineering. (**a**) depicts a scheme for inserting *attB*. At the top is the *cm*^{R}*-attB cassette* as a template for generating the PCR products for recombineering. The PCR product is represented as an Ω-shaped object with the two homology arms (LH for left homology and RH for right homology). The two "X" symbols represent points of exchange between the PCR product and the *ver* region (*open rectangular box*) on the plasmid (not shown). After recombineering, *cm*^{R} along with *attB* are inserted into *ver*. The *cm*^{R} gene is excised with the FseI restriction enzyme. (**b**) depicts a scheme for introducing an epitope tag. At the top is the FLAG-*cm*^{R} template for PCR. At the last step, *cm*^{R} is later excised with AscI, which cut site encodes the peptide of "G A P" that serves as a spacer between FLAG (*stripped box*) and Ver. (**c**) depicts a scheme for creating an internal deletion mutation. At the top is the *cm*^{R} template for PCR. The LH and RH regions are separated by the *hiphop*-coding region to be deleted. After recombineering, the deleted *hiphop* region is replaced with the *cm*^{R} gene, which is excised by *AsiSI*. The *AsiSI* site encodes the peptide of "A I A". (**d**) depicts a scheme for mutating specific residues. At the top is the *cm*^{R} template for PCR. The LH and RH regions flank the region that encodes "G R F" in *cav*. After excision of cm^R with BstEII, "G R F" is mutated to "G D L". (**e**) depicts a scheme for introducing multiple mutations into a single locus. In the first step (*top left*), the entire *ver* coding region is replaced with a *km*^{R} gene. In the second step (*top right*), a *cm*^{R} gene is placed next to every *ver* mutations (four shown). These *cm*^{R}-*ver*^{mut} cassettes serve as PCR templates for the final recombineering step (*bottom*), in which *km*^{R} is replaced with *cm*^{R}-*ver*^{mut} followed by *cm*^{R} excision. Each mutant allele is also tagged with FLAG

of which are depicted in Fig. 2), the exogenous DNA is a PCR product containing the desired DNA fragment to be inserted flanked by short (50 bp) homology on both sides of the fragment (LH for left homology and RH for right homology in Fig. 2). The sequences of these flanking homologous pieces are identical to those on the plasmid that flank the future insertional site. When dealing with multicopy plasmids, it is highly unlikely that every plasmid molecule will acquire an insertion considering that recombination between a PCR product and a circular plasmid is not very efficient. Thus it is essential to use antibiotic selection to recover the desired clone.

We designed a two-step scheme to accomplish efficient modification of multicopy plasmids. In the first step, the desired DNA fragment is physically linked with a selectable marker [e.g., chloramphenicol resistant gene (cm^R)]. This can be done by PCR with overlapping primers or PCR followed by DNA ligation. Our preferred method is to use recombineering to insert cm^R next to the desired DNA fragment already cloned into a common cloning vector, such as pCR2.1 from the TOPO TA cloning kit. Unique restriction cut sites are included in the PCR primers to flank the cm^R marker for its excision in the final cloning step. In the second step, a PCR reaction amplifies the desired DNA fragment along with cm^R. It integrates, via recombineering, at the desired position on the final plasmid vector, which is then subjected to restriction digestion to excise the cm^R marker. This is followed by intramolecular ligation giving rise to the final product.

3.2 A General Recombineering Protocol

Here we give a detailed protocol for recombineering, which is based on protocols at the Web site:

http://web.ncifcrf.gov/research/brb/recombineeringInformation.aspx. The readers are encouraged to derive their own modifications of the standard protocol described on the Web site.

Step 1. Prepare PCR insert for recombineering

Amplify the cm^R cassette with the desired DNA element using a DNA polymerase with proofreading activities. Use agarose gel electrophoresis to check the specificity and efficiency of the PCR. Purify the PCR product using a commercially available PCR purification kit. Digest the PCR product with DpnI enzyme to destroy template DNA from the plasmid. DpnI only digests methylated DNA so that PCR products are protected. Purify the PCR product after digestion and elute in distilled water.

Step 2. Prepare bacterial cells competent for recombineering

2.1. Inoculate SW102 cells in 5 ml LB growth media with 12.5 μg/ml of tetracycline (tet) at 30 °C overnight (*see* **Note 1**).

2.2. Transfer 500 μl of overnight culture to 25 ml fresh, pre-warmed LB + tet media in a flask larger than 100 ml, and inoculate at 30 °C until OD_{600} reaches 0.4–0.6 (about 3 h) (*see* **Note 2**).

2.3. Heat shock the SW102 culture in a 42 °C shaking water bath for exactly 15 min. This is to induce expression of lambda RED.

2.4. Chill the culture by placing the flask in ice slurry with frequent mixing. Once the culture is sufficiently chilled (after a few minutes in ice slurry), transfer 10 ml of the cells to prechilled 15 ml culture tubes. Pellet the cells by centrifugation at 2,000 × *g* for 5 min at 0 °C (*see* **Note 3**).

2.5. Decant the supernatant. Resuspend the cell pellet in 1 ml of ice-cold, sterile water. Swirl the tube in icy water to achieve fast chilling of the cells. Fill the tube with water and mix by inverting the tube several times. Pellet cells by centrifugation. Repeat this wash three more times.

2.6. Carefully pour off the supernatant as the pellet at this stage is very loose, and invert the tubes on a paper towel for a few seconds before returning them to ice. Resuspend the cells in the residual liquid, and keep the tubes on ice until transformation by electroporation.

Step 3. Bacterial transformation
Add ~50 ng of the target plasmid and ~100 ng of the purified PCR product from **step 1** to a cuvette prechilled on ice. The total volume should not exceed 10 % of the cell volume. Add 25–50 μl of the competent SW102 cells from **step 2** into the same cuvette, and perform bacterial electroporation according to manufacturer's instructions. Add 500 μl of SOC medium, incubate for 1 h at 30 °C, and plate the entire culture on LB + cm (17 μg/ml) plates. Incubate overnight at 30 °C.

Step 4. Post-transformation cleanup and validation

4.1. Isolate DNA from small cultures inoculated from several cm^R colonies grown overnight at 30 °C. Perform restriction digests to confirm the overall structure of the recovered plasmids by comparing it to a similar digestion of the starting plasmid. Most colonies would yield a DNA mixture of both the desired plasmid with a cm^R insertion and the original unmodified plasmid (*see* **Note 4**).

4.2. To clean up the mixture, dilute the miniprep DNA 1:200–1,000 in water. Transform 1 μl of the diluted DNA into cells of a standard bacterial cloning strain and plate on LB + cm plates. Purify DNA from several colonies per plate, and perform restriction digest to identify colonies that no longer contain the original non-cm^R plasmid.

Step 5. cm^R marker excision
Digest several nanograms of the "cleaned-up" plasmid with the restriction enzyme, which cut sites have been previously engineered to flank the cm^R marker. Perform ligation after heat inactivation of the restriction enzyme. Transform the ligation reactions into cells of a standard bacterial cloning strain, and plate the cells on plates with the appropriate antibiotics for the target plasmid [e.g., ampicillin (amp)]. Perform restriction digest validation of several clones and DNA sequencing to validate intact DNA elements if necessary.

3.3 An Alternative Recombineering Protocol

The protocol described in Subheading 3.2 has been referred to as the "co-transformation" protocol in which the target plasmid and the donor PCR products are transformed simultaneously into bacterial cells made competent for recombineering. However, due to the large amount of DNA required and the relatively small volume of total DNA allowed, this might not be practical for all situations. Alternatively, one can introduce the target plasmid and the donor PCR products in two transformation steps.

Step 1. Preparing competent SW102 cells
When used for transformation of the target plasmid, SW102 cells should not be heat shocked to induce lambda RED. Cells can be made competent according to **steps 2.1–2.6** in Subheading 3.2 with **step 2.3** omitted. In **step 4** instead of using only 10 ml of the cells, the entire culture can be used for preparing competent cells as unused cells will be stored for future uses (*see* **Note 5**).

Step 2. Target plasmid transformation
Transform a few nanograms of the target plasmid into competent SW102 cells by electroporation as described in **step 3** of Subheading 3.2, and plate the cells on plate with the appropriate antibiotics for the target plasmid (e.g., amp).

Step 3. Recombineering with PCR products
Perform recombineering as described in **steps 2** and **3** in Subheading 3.2 with the following changes:

1. For **steps 2.1** and **2.2**, when culturing SW102 cells with the target plasmid, use the appropriate antibiotics for the target plasmid.
2. For **step 3**, omit DNA from the target plasmid and reduce the amount of the PCR product to about 50 ng.

3.4 Generating the Master Construct with an attB Attachment Site

In performing SIRT-mediated gene manipulation in Drosophila, all genetic modifications of the target gene are to be introduced as plasmids carrying the *attB* attachment site. We routinely construct a master clone with *attB* inserted at the desired position in the plasmid (i.e., identical to the position where *attP* has been introduced onto the chromosome, Fig. 1). All subsequent modifications are made to this master clone.

The construction of this *attB*-containing master clone is accomplished with the two-step scheme described in Subheading 3.1. Below we described this cloning step in detail using an example in which we constructed a master clone for modifying the Drosophila *ver* gene. Figure 2a is a schematic representation of this experiment.

For PCR amplification of *attB*, we use an existing cm^R cassette in which an *attB* site was cloned adjacent to a cm^R gene that is flanked by FseI restriction cut sites [7]. The primers used are:

1. ver4401Cm:
   aataagtaaaaattagcaggggcgtagtcaaaacaactgaaaatttgtaaGGCCG-GCCctgtggaacaccc

   The 50 bp sequence in lower case is homologous to the left side of the position where we plan to insert *attB* (LH in Fig. 2a). The *Fse*I site is in upper case. Sequence in italics is homologous to cm^R (*see* **Note 6**).
2. ver4500attB:
   cagggtcacattaatttgcagaaccgcgcaatattttcttttaaccccCGACAT-GCCCGCCGTGACCG

   The 50 bp sequence in lower case is homologous to the right side of the *attB* insertion site at *ver* (RH in Fig. 2a). The 20 bp sequence in upper case is homologous to *attB*.

After PCR amplification of the cm^R-*attB* cassette with the above primers, the PCR products are transformed into SW102 cells that carry the target plasmid pTV[ver], which contains a genomic fragment of *ver* sub-cloned into a generic gene targeting vector. Using recombineering protocols described in Subheading 3.2 or 3.3, recover clones with cm^R-*attB* inserted into pTV[ver]. After cleaning up (**step 4** in Subheading 3.2), cm^R was excised by an FseI digestion followed by plasmid re-ligation. This generates the plasmid pTV[ver-attB] (*see* **Note** 7).

3.5 Inserting an Epitope Tag

SIRT can be used to introduce epitope tags to an endogenous locus. Again using the *ver* locus as an example, we describe a protocol to insert a FLAG tag to the N-terminus of Ver. Figure 2b is a schematic representation of this experiment.

Step 1. Generating a plasmid containing a Flag-cm cassette

1.1. Use the following pair of primers to amplify a DNA fragment that contains a FLAG-encoding fragment followed by a cm^R gene that is flanked two *Asc*I cut sites.

1. Flag-AscI-Cm:
   gactacaaagacgatgacgacaagGGCGCGCCagccagtatacactc-cgcta Sequence in lower case encodes FLAG. The AscI site is in upper case. The italicized sequence in lower case is homologous to cm^R.
2. AscI-Cm:
   GGCGCGCC ctgtggaacacctacatctg

   The AscI site is in upper case. The italicized sequence in lower case is homologous to cm^R.

1.2. Clone the above PCR product using the TOPO cloning kit from Invitrogen according to manufacturer's instruction. Use LB + cm plates to select for the correct clones, which are subjected to sequencing to confirm the integrity of the FLAG tag and AscI cut sites.

Step 2: Inserting a FLAG tag N-terminal to Ver

2.1. Use the following primers to amplify a *Flag-cm*^{R} fragment flanked by small regions of homology. The template was the *Flag-cm*^{R} cassette constructed at **step 1** of Subheading 3.5.

1. ver4640-Flag-L:
gcactgcaataaagaaatcccctttgaaatcgcagactaagcaaatagaatg-GACTACAAAGACGATGACGAC

Sequence in lower case is homologous to *ver*(LH in Fig. 2b). Sequence in upper case is homologous to FLAG.

2. ver4691-Cm-R:
acgaagttatccagctggctttctatgtcctcgaaactctgattaaaatc**T**GGC-GCG CCctgtggaacac.

Sequence in lower case is homologous to *ver* (RH in Fig. 2b). In this primer, the AscI cut site (GGCGCGCC) is preceded with a "T" in bold. Since the AscI cut site is 8 bp in length, an extra "T" has been added to ensure that the FLAG tag is in frame with the rest of the Ver protein. The sequence **T**GGCGCGCC when translated in the reverse direction encodes a peptide of Gly Ala Pro ("G A P" in Fig. 2b), which also serves as a spacer between FLAG and Ver (*see* **Note 8**).

2.2. Using the recombineering protocols described in Subheading 3.2 or 3.3, insert this *Flag-cm*^{R} fragment into the master clone of pTV[ver-attB] generated from Subheading 3.4. Use AscI to excise the *cm*^{R} gene. This gives rise to the plasmid pTV[FLAG-ver-attB].

3.6 Creating In-Frame Deletion Mutations

An efficient way to identify critical domains for protein function is to create in-frame deletion mutations that encode truncated proteins missing different domains. Using the *hiphop* locus as an example, we describe a protocol to create an internal deletion eliminating about one third of the protein (Fig. 2c). The scheme involves replacing the DNA fragment to-be-deleted with a *cm*^{R} gene flanked by restriction sites using recombineering. The *cm*^{R} gene is then excised resulting in replacing the deleted region with the restriction site (*see* **Note 9**).

Step 1. Use the following primers to amplify a *cm*^{R} gene flanked by 50 bp fragments homologous to the *hiphop* locus. The 50 bp homologous pieces (LH and RH in Fig. 2c) are taken from positions in the HipHop-coding region immediately adjacent to either side of the future deletion as shown in Fig. 2c.

1. HipHopA2As/SICm-Forward
gccagggagactgccgcgagcattacggacgtcagcggcagtcagtcatcgGC-GATCGCaggagggacagctgatagaa

Sequence in lower case is homologous to *hiphop* (LH in Fig. 2c). The *AsiSI* cut site is in upper case. The lower case and italicized sequence is homologous to *cm*^{R}.

2. HipHopA2AsiSICm-Forward
gtgccattcacggcgttcaaactgagattcgagttggggtcgtagtcgtc**A**GC-GATCGCcctgtggaacacctacatct

Sequence in lower case is homologous to *hiphop* (RH in Fig. 2c). The *AsiSI* cut site is in upper case. The "A" in bold preceding the *AsiSI* cut site is included to preserve reading frame. The sequence **A**GCGATCGC when translated in the reverse orientation encodes the peptide Ala Ile Ala ("A I A" in Fig. 2c). The lower case and italicized sequence is homologous to cm^R.

Step 2. Using the recombineering protocols described in Subheading 3.2 or 3.3, insert this cm^R fragment into the master clone of pTV[hiphop-attB]. Use *AsiSI* to excise cm^R.

3.7 Site Directed Mutagenesis with Recombineering

Besides making large deletion of the coding region, mutating conserved residues by site-directed mutagenesis is another common way to dissect protein function. Using the *cav* locus as an example, we describe a protocol to mutate individual residues (Fig. 2d). The scheme involves replacing individual residue(s) of interest with a cm^R gene flanked by restriction sites. The cm^R gene is excised following recombineering essentially replacing the target residue(s) with the restriction site (*see* **Note 9**).

Step 1. Use the following primers to amplify a cm^R gene flanked by 50 bp fragments homologous to the *cav* locus.

1. HOAP89R:DBstEICm-Forward
atgaccgcttggaattgtctgtgggaagccaaaaagaggtttgaagcaaaaGGT-GACCaggagggacagctgatagaa

Sequence in lower case is homologous to *cav*(LH in Fig. 2d). The *BstEII* cut site is in upper case. The lower case and italicized sequence is homologous to cm^R. The *BstEII* enzyme is a 7 bp cutter with the sequence GGTNACC where N represents any base. Due to this ambiguous nucleotide, various amino acid combinations can be created.

2. HOAP89R:DBstEIICm-Reverse
gcgcaccgctttcatatacattcggttgatgaatctctcagacttgttcac**AA**GGTCACCcctgtggaacacctacatct

Sequence in lower case is homologous to *cav*(RH in Fig. 2d). The *BstEII* cut site is in upper case. The "AA" in bold is included to preserve reading frame. The lower case and italicized sequence is homologous to cm^R.

We wish to change the amino acid Gly Arg Phe ("G R F" in Fig. 2d) of the Cav protein to Gly Asp Leu ("G D L" in Fig. 2d) where Gly and Arg are conserved residues. We choose to add two adenosines immediately upstream of the *BstEII* cut site in the reverse primer. This puts the coding sequence in frame and creates our desired mutation. The sequence AAGGTCACC when translated in the reverse orientation encodes Gly Asp Leu.

Step 2. Using the recombineering protocols described in Subheading 3.2 or 3.3, insert this cm^R PCR product into the master clone. Use BstEII enzyme to excise the cm^R gene.

3.8 One-Step Introduction of Multiple Mutations

The site-directed mutagenesis protocol described in Subheading 3.7 is limited by the number of suitable restriction enzymes. This limitation is placed not only on the residues that can be mutated, also on the exact amino acid to which a particular residue can be mutated. Situations exist in which a series of random mutations have been selected based on functional assays ex vivo and that the introduction of these mutations into the intact organism would be rather illuminating on gene function. Using the *ver* locus as an example, we describe a protocol for introducing a series of point mutations that we recovered from a yeast two-hybrid assay into the master clone of pTV[Flag-Ver-attB] generated in Subheading 3.5, and doing so with a single set of primers.

Figure 2e is a schematic representation of this experiment. The scheme involves first replacing the entire *ver* coding region in the pTV[Flag-ver-attB] master clone with a kanamycin-resistant (km^R gene. Secondly, a series of cassettes are constructed in which a cm^R gene, excisable by restriction digest, is placed next to the *ver* coding region for each *ver* mutation (ver^{mut}) cloned into the pBTM vector. Thirdly, the ver^{mut}-cm^R cassettes are introduced as PCR products by recombineering, replacing the km^R gene in the master clone. After cm^R excision, a series of plasmids are generated each containing a different mutation.

It is necessary to replace the coding region of *ver* in the master clone of pTV[Flag-ver-attB] in the first step. Otherwise the ver^{mut}-cm^R PCR products would share extensive homology (the entire *ver* coding region) with pTV[Flag-ver-attB], which would make it difficult to predict the exact point of exchange between the plasmid and the PCR product ("X" in Fig. 2e). This would necessitate a cumbersome screening step by DNA sequencing to identify clones with the desired mutations. By limiting the exchange points to a 50 bp region to either side of the *ver* coding region, our scheme ensures the recovery of mutations in all clones after recombineering.

Step 1. Replacing ver with km^R in the plasmid pTV[Flag-ver-attB]

1.1. Use the following primers to amplify a km^R gene using the pCR2.1 vector as the PCR template.

1. ver4654L1-Flag-Km:
gaaatcgcagactaagcaaatagaatggactacaaagacgatgacgacaagT-GCTAAAGGAAGCGGAACAC

   Sequence in lower case is homologous to the 5′ region of *ver* including the FLAG tag in pTV[Flag-ver-attB] (LH in Fig. 2e). Sequence in upper case is homologous to km^R.

2. ver5320-R1-Km:

tttgaatttttattaccagtaaaatttcaatacaaaaaaccaacgatactaG-GTGAGCAAAAACAGGAAGG

Sequence in lower case is homologous to the 3′ region of *ver* (RH in Fig. 2e). Sequence in upper case is homologous to *km*^{R}.

1.2. Using the recombineering protocols described in Subheading 3.2 or 3.3, replace the *ver* coding region in pTV[Flag-ver-attB] with this *km*^{R} fragment. Use km as the selectable marker for recombineering (*see* **Note 10**).

Step 2. Generating a cm^R *cassette for each ver mutation*

2.1. Use the following primers to amplify a *cm*^{R}.

1. ver-end-L2-Cm-F:

gacccagctttcttgtacaaagtggttgatggggatccgtcgacctg-cagGGCGCGCC*agccagtatacac*

2. ver-end-R2-Cm-R:

tttaataataaaaatcataaatcataagaaattcgcccggaattagcttgg GGCGCGCC*ctgtggaacacc*

In these primers, sequence in lower case is homologous to vector sequences right after the stop codon of *ver*. The AscI sites are in upper case. The italicized sequence in lower case is homologous to *cm*^{R} (*see* **Note 11**).

2.2. Using the recombineering protocols described in Subheading 3.2 or 3.3, insert this *cm*^{R} fragment into pBTM clones with the *ver* mutations.

Step 3: Generating the final plasmid of pTV[FLAG-ver^{mut,}*-attB]*

3.1. Use the following primers to amplify the *ver*^{mut}*-cm*^{R} cassette for each *ver*^{mut}, using template plasmids generated in **step 2** of Subheading 3.8.

1. 4654L1-Flag-ver-F:

gaaatcgcagactaagcaaatagaatggactacaaagacgatgacga-caagGATTTTAATCAGAGTTTCGAG

Sequence in lower case is homologous to the 5′ region of *ver* including the FLAG tag in pTV[FLAG-Ver-attB] (LH in Fig. 2e). Sequence in upper case is homologous to the start of *ver* coding region downstream of the start codon.

2. ver5320-R1-cm:

tttgaatttttattaccagtaaaatttcaatacaaaaaaccaacgatactaG-GCGCGCC*ctgtggaacacc*

Sequence in lower case is homologous to the 3′ region of *ver* (RH in Fig. 2e). The AscI cut site is in upper case. Italicized sequence in lower case is homologous to *cm*^{R}.

3.2. Using the recombineering protocols described in Subheading 3.2 or 3.3, replace the km^R gene in the construct generated in **step 1** with these ver^{mut}-cm^R cassettes. Use AscI to excise cm^R. Sequence the clones to ensure the integrity of the *ver* coding region and the presence of the desired point mutations.

4 Notes

1. It is important to grow SW102 cells under 32 °C. Higher temperature will result in premature activation of the lambda RED system.
2. We have performed successful recombineering experiments with an OD value as low as 0.3 or as high as 0.7.
3. From this step on, keep the cells on ice at all time and use prechilled solutions.
4. We usually use the *EcoRI* enzyme since the cm^R marker introduces an additional *EcoRI* site.
5. SW102 cells made this way can be stored at −80 °C for future uses. Substitute water with 10 % sterile glycerol at the final washing step, and aliquot unused cells into tubes for storage at −80 °C.
6. Both FseI site and the sequence in italics will anneal to the cm^R-*attB* cassette.
7. *Fse*I is not stable at −20 °C, and should be stored at −80 °C.
8. Because the *Flag*-cm^R cassette does not carry an ATG codon, this fragment has to be inserted downstream of the endogenous ATG codon.
9. Care needs to be taken to preserve the correct reading frame when using a restriction enzyme that does not have a 6 bp cut site.
10. This recombineering reaction involves the replacement of a DNA fragment with another. It is very important to sequence several clones to ensure the integrity of the recombineering junctions.
11. This cm^R-homologous sequence is shorter than one would normally use for PCR amplification since we use a cm^R gene already flanked by AscI as the PCR template so that the *Asc*I site in the oligos also serves as a part of the primer.

Acknowledgment

Research in our laboratory is supported by the intramural research program of the National Cancer Institute.

References

1. Gao G, McMahon C, Chen J, Rong YS (2008) A powerful method combining homologous recombination and site-specific recombination for targeted mutagenesis in Drosophila. Proc Natl Acad Sci U S A 105:13999–14004
2. Gao G, Cheng Y, Wesolowska N, Rong YS (2011) Paternal imprint essential for the inheritance of telomere identity in Drosophila. Proc Natl Acad Sci U S A 108:4932–4937
3. Sharan SK, Thomason LC, Kuznetsov SG, Court DL (2009) Recombineering: a homologous recombination-based method of genetic engineering. Nat Protoc 4:206–223
4. Raffa GD, Raimondo D, Sorino C, Cugusi S, Cenci G, Cacchione S, Gatti M (2010) Verrocchio, a Drosophila OB fold-containing protein, is a component of the terminin telomere-capping complex. Genes Dev 24:1596–1601
5. Gao G, Walser JC, Beaucher ML, Morciano P, Wesolowska N, Chen J, Rong YS (2010) HipHop interacts with HOAP and HP1 to protect Drosophila telomeres in a sequence-independent manner. EMBO J 29:819–829
6. Cenci G, Siriaco G, Raffa GD, Kellum R, Gatti M (2003) The Drosophila HOAP protein is required for telomere capping the Drosophila HOAP protein is required for telomere capping. Nat Cell Biol 5:82–84
7. Gao G, Wesolowska N, Rong YS (2009) SIRT combines homologous recombination, site-specific integration, and bacterial recombineering for targeted mutagenesis in Drosophila. Cold Spring Harb Protoc. 2009(6):pdb.prot5236.
8. Warming S, Costantino N, Court DL, Jenkins NA, Copeland NG (2005) Simple and highly efficient BAC recombineering using galK selection. Nucleic Acids Res 33(4):e36
9. Rong YS, Titen SW, Xie HB, Golic MM, Bastiani M, Bandyopadhyay P, Olivera BM, Brodsky M, Rubin GM, Golic KG (2002) Targeted mutagenesis by homologous recombination in D. melanogaster. Genes Dev 16:1568–1581

Chapter 3

Multiple Genetic Manipulations of DT40 Cell Line

Akira Motegi and Minoru Takata

Abstract

Reverse genetics is gaining importance in the field of modern biological sciences. Gene disruption and the use of siRNAs are the favored techniques for current research. Many researchers, however, are aware that the data from siRNA experiments are frequently inconsistent and that epistatic analysis of multiple genes using siRNAs is barely feasible. In recognition of the drawbacks of the siRNA technique, many researchers, especially in the field of DNA repair, are now introducing multiple genetic disruption techniques using the chicken DT40 cell line into their research. Thus, recent publications increasingly include data utilizing DT40 cells. In this chapter, we describe the current standard methods of multiple genetic manipulation in DT40 cells. We place a particular emphasis on describing the basic concepts and theoretical background of the genetic manipulation of DT40 cells for researchers who are new to such techniques.

Key words Chicken DT40, Gene targeting, Marker recycling, Transgene, Epistatic analysis

1 Introduction

Reverse genetics is gaining importance in modern biological science, as evidenced by the prevailing usage of genetic disruption and siRNA techniques. Despite recent improvements in the use of siRNA, many researchers recognize the intrinsic drawbacks of the technology, such as off-target effects and the imperfect penetration of target effects. In fact, in a genome-wide analysis, Adamson et al. recently demonstrated that Rad51, the central mediator of recombination, is actually one of the most frequent targets of siRNA off-target effects [1]. This observation should serve as a warning against the use of siRNA in the study of the DNA-damage response. Quantitative comparison of data obtained using siRNAs for different target genes is virtually impossible due to a number of reasons, including the differing knockdown efficiencies, differing siRNA kinetics, and the limited capacity of the siRNA-processing machinery [2, 3].

By contrast, DT40 is the only vertebrate cell line where the genome can be accurately manipulated by highly efficient

Francesca Storici (ed.), *Gene Correction: Methods and Protocols*, Methods in Molecular Biology, vol. 1114,
DOI 10.1007/978-1-62703-761-7_3, © Springer Science+Business Media, LLC 2014

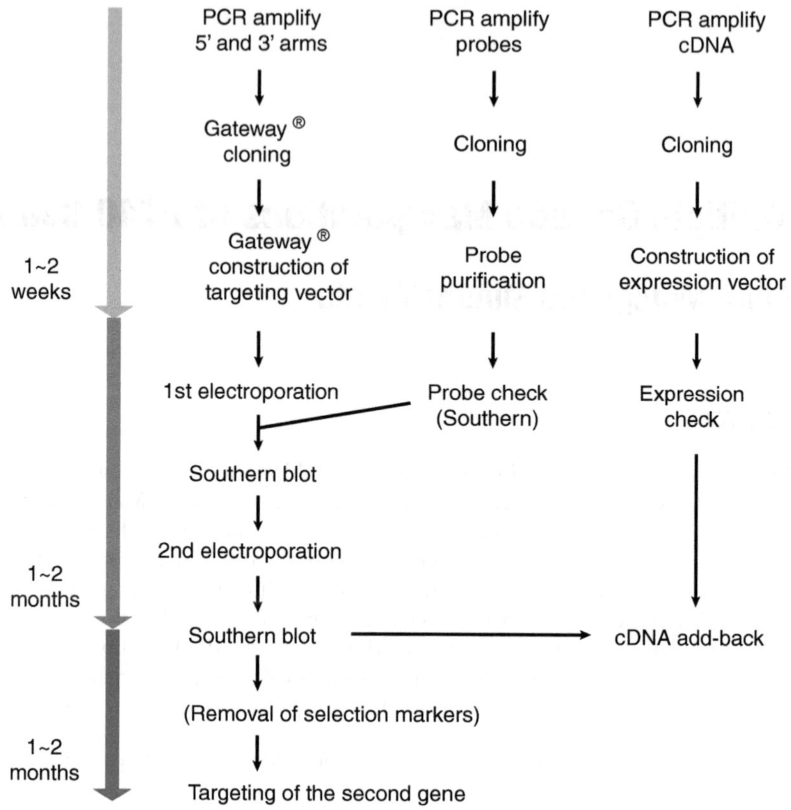

Fig. 1 Outline of gene targeting in DT40 cells

targeted-integration methods. This unique property of DT40 cells, along with a relatively stable karyotype, affords a unique opportunity to manipulate multiple genes of interest and to analyze the epistatic relationships in vertebrate cells. For these reasons, many researchers are now introducing the DT40 system in their research and a growing number of research articles include data obtained from genetically manipulated DT40 cells.

The DT40 cell line was originally established from an avian leukovirus (ALV)-transformed B cell lymphoma developed in the chicken organ bursa of Fabricious, where avian B lymphocytes differentiate [4]. Subsequent demonstration of the high ratio of targeted vs. non-targeted integration events among transformed populations [5] prompted the use of this cell line in reverse genetics. So far, hundreds of mutant cells have been generated worldwide, creating a valuable resource for the detailed genetic analysis in vertebrates.

In this chapter, we describe the current standard methods for multiple genetic manipulations in DT40 cells, including targeted-gene disruption, the recycling of selection markers, and the rescue of mutant cells by adding back cDNAs as transgenes (Fig. 1).

For further information on the genetic manipulation, phenotypic analyses, and historical aspects of DT40 cells, please refer to the cited book and review articles [6–10].

2 Materials

2.1 Construction of Gene-Targeting Plasmids

1. High-fidelity DNA polymerase.
2. MultiSite Gateway® Three Fragment Vector Construction Kit (Invitrogen), including BP Clonase® II and LR Clonase® II Plus.

2.2 DT40 Cell Culture

1. CO_2 incubator set at 39.5 °C with 5 % CO_2 supply (*see* **Note 1**).
2. DT40 culture medium: RPMI1640 supplemented with heat-inactivated 10 % fetal bovine serum (FBS), 1 % chicken serum (CS), 1 % L-glutamine, 50 μM β-mercaptoethanol (2ME), and 1 % Penicillin G-Streptomycin (optional).

2.3 Electroporation

1. Electroporator (Gene Pulser XCell System, Bio-Rad).
2. Electroporation cuvettes (4 mm gap).
3. 0. 5 mg/mL Puromycin (×200 stock, Sigma).
4. 10 mg/mL Blasticidin S (×200 stock, Funakoshi).
5. 50 mg/mL Histidinol (×200 stock, Sigma).
6. 50 mg/mL G418 (×25 stock, Nacalai Tesque).
7. 100 mg/mL PBS Hygromycin B (×50 stock, Nacalai Tesque).
8. 5 mg/mL Mycophenolic acid (×250, Wako).

2.4 Extraction, Digestion and Electrophoresis of Genomic DNA

1. Lysis Buffer: 200 mM NaCl, 20 mM EDTA, 40 mM Tris–HCl (pH 8.0), and 0.5 % SDS. Add 5 μL 2ME and 10 μL Proteinase K (Sigma) per mL Lysis Buffer before use.
2. Saturated NaCl solution.
3. Biodyne® B Nylon membrane, pore size 0.45 μm (PALL).
4. 0.4 N NaOH.
5. 2× SSC: 300 mM NaCl and 30 mM Na citrate (pH 7.0).

2.5 Southern Blotting

1. α-^{32}P-dCTP, 370 MBq/mL (Perkin Elmer).
2. Megaprime DNA labeling kit (GE).
3. Sephadex G-50 column DNA Grade (GE).
4. Hybridization Buffer: 0.5 M Na_2HPO_4 (pH 7.4), 1 mM EDTA, 1 % BSA, and 7 % SDS.
5. Washing Buffer: 40 mM Na_2HPO_4 (pH 7.4), 1 mM EDTA, and 1 % SDS.
6. Imaging Plates (BAS-MS, Fujifilm).
7. Imaging Plate Reader (Fujifilm).

3 Methods

3.1 Design of Gene-Targeting Constructs

1. Retrieve chicken cDNA and genomic DNA sequences from public databases such as NCBI and UCSC genome browser Web sites (ICGSC Gallus_gallus-4.0, Nov. 2011) (*see* **Notes 2** and **3**).
2. Find a coding sequence encompassing a region essential for specific enzymatic activities or protein–protein interactions (*see* **Note 4**).
3. Set left and right arms of gene-targeting constructs on both sides of above selected targeting region (Fig. 2a). We usually design them with minimal length of 1.5 kb and 3.0 kb for the shorter and longer arms, respectively (*see* **Note 5**).

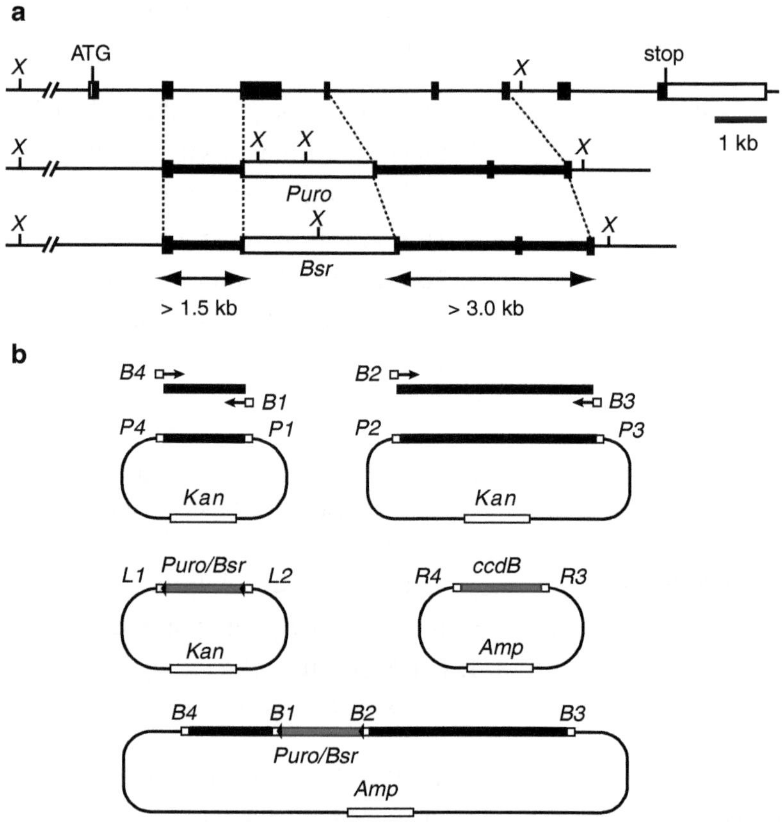

Fig. 2 Construction of gene-targeting constructs. (**a**) Schematic representation of typical wild-type allele (*top*) and targeted alleles (*middle* and *bottom*). Xs represent restriction sites. (**b**) Construction of gene-targeting plasmids by Gateway®. 5′ and 3′ arms (*black thick bars*) are amplified by PCR with primers attached with *att* sites (*open squares*) and cloned into pDONR plasmids by BP reactions. Arms' plasmids are recombined with pENTR-Puro (or Bsr) and pDEST DTA-MLS (*middle four circles*) to generate the complete targeting plasmid (*bottom circle*) by LR reaction. Filled *arrowheads* represent mutated loxP sites

4. Select 0.4–1.0 kb genomic fragments outside of arms as probes for Southern blotting (*see* **Notes 6** and 7).
5. In parallel to **steps 3** and **4**, pick up several restriction enzymes that are predicted to separate wild-type and targeted alleles in Southern blotting. Restriction sites for linearizing the targeting plasmids also need to be selected at this step.

3.2 Construction of Gene-Targeting Plasmids

We use MultiSite Gateway® system (Invitrogen) to clone and assemble genomic fragments into the complete gene-targeting plasmids (Fig. 2b). This method was originally developed by Iiizumi et al. and will take 1–2 weeks to complete constructions [11] (*see* **Note 8**). For further detail of the Gateway® system, please refer to the manufacturer's manuals.

1. PCR-amplify arms and probes with DT40 genomic DNA as a template. Primers are 25–30 nucleotides in length without any additional sequences.
2. Gel-separate and column-purify the PCR products.
3. Re-amplify the arm fragments with primers attached with specific recombination signals.
4. Clone 5′ and 3′ arms into pDONR® P4-P1R and pDONR® P2R-P3, respectively, by using BP Clonase® II.
5. Determine restriction patterns of obtained genomic fragments with enzymes selected in Subheading 3.1, **step 5** (*see* **Note 9**).
6. Assemble 5′ and 3′ arms, a selection marker cassette (Puro, Bsr, or His is the routine choice), and the backbone plasmid pDEST DTA-MLS into a complete gene-targeting construct by LR reaction (*see* **Note 10**).
7. Clone probe fragments into the cloning vector and verify the sequence by base sequencing (*see* **Note 11**).

3.3 Electroporation

1. Linearize 30 μg targeting plasmids with a restriction enzyme selected in Subheading 3.1, **step 5**.
2. Ethanol-precipitate and resuspend DNA in 50 μL sterile PBS.
3. Warm up 20 mL complete medium in the CO_2 incubator.
4. Spin down 5×10^6 cells by centrifuging at $270 \times g$ for 5 min.
5. Wash the cells with 5 mL sterile PBS once.
6. Resuspend the cells in 450 μL PBS and transfer to a cuvette.
7. Add DNA and incubate on ice for 10 min.
8. Electroporate at 550 V, 25 μF.
9. Incubate the cuvette on ice for 10 min.
10. Transfer the cells to a dish with pre-warmed medium.
11. Culture the cells for 16 h.

12. Add a selection drug of choice and dilute the cells to a total volume of 80 mL.
13. Plate the cells in four 96-well plates (200 μL/well).

3.4 Extraction of Genomic DNA

1. 1 week later, pick up colonies and culture them in 6-well plates (5 mL/well) (*see* **Note 12**).
2. When the cells are grown to subconfluency, take 0.9 mL, mix with 0.1 mL DMSO (final concentration of 10 %) and freeze at −80 °C.
3. Pellet the cells in 4 mL culture by centrifuging at 270 × *g* for 3 min.
4. Remove supernatants completely and break pellets by finger tapping.
5. Add 500 μL Lysis Buffer (supplemented with 2ME and Proteinase K) and vortex for 10 s.
6. Incubate at 55 °C for overnight.
7. Add 250 μL saturated NaCl solution, vortex for 10 s, and leave on ice for 15 min.
8. Centrifuge at 200 × *g* for 10 min at 4 °C.
9. Carefully transfer supernatant to a new tube.
10. Add 750 μL 100 % ethanol and invert ~50 times. You will see thick strings with successful genomic DNA extraction.
11. Centrifuge at 30 × *g* for 1 min.
12. Remove supernatant and rinse with 800 μL of 70 % ethanol.
13. Spin again, aspirate supernatant, and air-dry.
14. Dissolve in 50 μL TE (*see* **Note 13**).

3.5 Digestion and Electrophoresis of Genomic DNA

1. Digest 15 μL of dissolved genomic DNA with the choice of restriction enzyme.
2. Make 0.7 % agarose gel in 1× TAE (without ethidium bromide for better resolution).
3. Load about half of digested DNA premixed with 5× loading dye (*see* **Note 14**).
4. Electrophorate at 50 V for 2–3 h (depending on the sizes of separating fragments).
5. Stain a gel with ethidium bromide and take a picture with a scale.
6. Set up a capillary transfer platform with a piece of 17 Chr filter paper with its sides dropped into 0.4 N NaOH.
7. Put a gel upside down atop of the platform (so as DNA near the bottom of the gel locates near the membrane), place a 3MM filter paper, a stuck of paper towel, flat plate and weight in this order.

8. Take out a membrane, label the sides and rinse with 2× SSC buffer twice.
9. Sandwich a membrane with clean filter papers and bake at 80 °C for 2 h.

3.6 Southern Blotting

3.6.1 Labeling Probe DNA

1. Prepare probe DNA either by PCR or excision from the plasmids and column purification.
2. Label 25 ng of probe DNA according to the manufacturer's protocol (*see* **Notes 15** and **16**).
3. Gel filtrate labeled DNA by using Sephadex G-50 column.
4. Denature purified DNA by boiling for 5 min and then chill on ice for 5 min.

3.6.2 Hybridization

1. Prehybridize membranes with 25 mL Hybridization Buffer at 62 °C for 30 min.
2. Replace with 25 mL Hybridization Buffer.
3. Add labeled probe DNA.
4. Hybridize at 62 °C for O/N.
5. Wash membranes with 100 mL Washing Buffer at 62 °C for 5 min three times.
6. Take out membranes, wrap with plastic wrap and expose to the Imaging Plate.
7. Read by BAS.

3.7 Targeting of the Second Allele

1. Expand targeted clones from frozen stocks.
2. Make some cell stocks and repeat the targeting procedure (*see* **Note 17**).

3.8 Checking Expression by RT-PCR

1. Design sets of primers that locate on different exons (*see* **Note 18**).
2. Prepare cDNA from wild-type and mutant cells and perform PCR.

3.9 Removal of Selection Markers

Selection markers flanked by mutated loxP sites can be removed by transient expression of the Cre recombinase [12]. This allows us to use efficient selection markers such as Puro, Bsr, and His in disrupting the second gene (*see* **Note 19**).

1. Spin down 5×10^6 cells.
2. Wash the cells with 5 mL PBS.
3. Electroporate the circular MreCreMer Recombinase expression plasmid at 250 V, 950 μF (*see* **Note 20**).
4. Transfer the cells to 10 mL pre-warmed medium with 200 nM Tamoxifen.
5. Incubate the cells for 2 days.

6. Plate into 96-well plates (~0.3cell/well).
7. 1 week later, pick up ~100 colonies.
8. Divide each clone into medium with or without selection drugs.
9. Choose clones that become sensitive to selection drugs.

3.10 Generation of cDNA Add-Back Clones

Even if karyotype of DT40 cells is more stable than other types of cancer cells, a modest level of karyotypic variations had been reported [13]. Thus, confirming the reversion of phenotypes by the cDNA add-back construct is essential (*see* **Note 21**).

1. PCR-amplify cDNA by RT-PCR.
2. Clone the amplified fragment and confirm the sequence by base sequencing (*see* **Note 22**).
3. Subclone the fragments into the expression plasmid of choice (*see* **Note 23**).
4. Electroporate the expression plasmid into mutant cells.
5. Select several clones with transgene by the selection drug on the plasmid (*see* **Note 24**).

4 Notes

1. DT40 cells can be grown at 34–43 °C with reduced proliferation rates [14]. However, 39.5 °C is recommended since this is the best optimal for growth and all genetic manipulations and phenotypic analyses have been established at this temperature.
2. Some chicken counterparts of human/mouse genes cannot be found in databases. This could be because either they really do not exist or they are not covered even in the latest version of chicken genome assembly. Indeed, not a few gene products without any trace of sequences in databases have been identified by mass spectrometry.
3. Typical karyotype of DT40 cells is 11 autosomal macrochromosomes (disomic chromosomes 1, 3, 4, and 5 and trisomic chromosome 2), 67 autosomal minichromosomes (32 disomies and 1 trisomy), and the ZW sex chromosomes. Note that numbers of alleles are different according to where they are located.
4. Ideally, removal of a whole gene would exclude a possibility of partial disruption of the gene. However, deletion of longer genomic fragments is less efficient and therefore we usually aim to disrupt up to several kb (the maximum size with reasonable efficiency is ~10 kb).
5. We usually place all primers for gene-targeting constructs on exons for ensuring the disruption of particular exons. The other reason for this is that sequence complexity of exons is generally

higher than those of introns and thus you have better specificities in PCR. Note that primers need to be located within introns when you are designing knock-in constructs with minimum modifications other than intended changes in coding sequence. For example, *see* [15]. We also avoid amplifying genomic fragments spanning over long gaps in the genome database because PCR over gaps is sometimes less efficient probably due to the intrinsic difficulties in replication and thus in PCR amplification.

6. Probes inside of arms detect randomly integrated events, which could not be differentiated from targeted events by size.
7. Longer probes have more chance to overlap with repeat sequences, which could cause smearing of signals. Running BLAST or RepeatMasker with selected sequences could help avoiding such repeats. Shorter probes have lower signals and less specificity.
8. Targeting constructs can be also generated by using conventional molecular biology techniques [7].
9. Chicken genome project has been done with the genome derived from a single inbred female, but not from the DT40 cell line. Therefore, actual restriction patterns need to be determined with obtained genomic DNA fragments. Also, multiple bands could be observed with wild-type DT40 cells due to the allelic variations.
10. Use absolutely LR Clonase® II Plus for overcoming the low efficiency of 4-fragments assembly.
11. Size-excised and column-purified PCR products from cloned probes generally give cleaner signals than the direct PCR products from genomic DNA.
12. Usually, dozens of colonies grow. Note that "high targeting efficiency" does not mean high electroporation efficiency.
13. Based on a calculation from the number of cells in 4 mL culture (~4×10^6 cells) and chicken genome size (2×10^9), approximately 10–15 μg of genomic DNA can be obtained.
14. This amount of genomic DNA corresponds to ~2 μg, which contains ~10 pg of a 10 kb genomic fragment. This should be well above the detection limit of Southern blotting (several pg~sub-pg at best).
15. Non-RI detection is an option, but the sensitivity is generally several folds less than RI method.
16. Including sub-nmol of size marker DNA in the labeling reaction visualizes them in the final images.
17. Unlike mouse ES cells, simultaneous disruption of two alleles is not possible in DT40 cells.

18. Mutated mRNAs usually undergo nonsense-mediated mRNA decay and thus phenotypes of mutant cells can be equal to those of null mutants. Sometimes, however, upstream or downstream of the targeted regions could be detected after targeting and may be partially functional.
19. Targeting of the second gene can be done with other selection markers such as Hyg, Ecogpt, or G418.
20. MerCreMer is the Cre recombinase fused with the ligand-binding domain of the estrogen receptor Mer in both N- and C-termini. This fusion protein efficiently translocates into the nucleus in the presence of the estrogen ligand Tamoxifen [16].
21. cDNA transgenes could not fully rescue the endogenous genes, most likely due to the differences in the strength or expression timing of promoters.
22. Chicken ORFs frequently do not align well with mammalian counterparts in both N- and C-termini. This could be the real diversity between species or the misalignments of exons in the genomic databases.
23. Attaching epitope tags to cDNA may facilitate later analyses, since anti-human or mouse antibodies do not cross-react chicken proteins often.
24. We use pCMV-IRES-GFP, by which the level of cDNA expression can be conveniently monitored by bicistronically expressed GFP [17].

Acknowledgments

We thank Dr. Masamichi Ishiai (Kyoto Univ.) for critical reading of the manuscript. This work was supported in part by Grants-in-aid from the Ministry of Education, Science, Sports, and Culture of Japan (to AM and MT) and by Mochida Memorial Foundation for Medical and Pharmaceutical Research (to AM).

References

1. Adamson B, Smogorzewska A, Sigoillot FD et al (2012) A genome-wide homologous recombination screen identifies the RNA-binding protein RBMX as a component of the DNA-damage response. Nat Cell Biol 14(3):318–328
2. Jackson AL, Linsley PS (2010) Recognizing and avoiding siRNA off-target effects for target identification and therapeutic application. Nat Rev Drug Discov 9(1):57–67
3. Singh S, Narang AS, Mahato RI (2011) Subcellular fate and off-target effects of siRNA, shRNA, and miRNA. Pharmacuet Res 28(12):2996–3015
4. Baba TW, Giroir BP, Humphries EH (1985) Cell lines derived from avian lymphomas exhibit two distinct phenotypes. Virology 144(1):139–151
5. Buerstedde JM, Takeda S (1991) Increased ratio of targeted to random integration after

transfection of chicken B cell lines. Cell 67(1): 179–188

6. Buerstedde J-M, Takeda S (2006) Reviews and protocols in DT40 research, vol 40, Subcellular biochemistry. Springer, New York, NY
7. Ishiai M, Uchida E, Takata M (2012) Establishment of the DNA repair-defective mutants in DT40 cells. Methods Mol Biol 920:39–49
8. Kitao H, Hirano S, Takata M (2011) Evaluation of homologous recombinational repair in chicken B lymphoma cell line, DT40. Methods Mol Biol 745:293–309
9. Yamazoe M, Sonoda E, Hochegger H et al (2004) Reverse genetic studies of the DNA damage response in the chicken B lymphocyte line DT40. DNA Repair 3(8–9):1175–1185
10. Sonoda E, Morrison C, Yamashita YM et al (2001) Reverse genetic studies of homologous DNA recombination using the chicken B-lymphocyte line, DT40. Phil Trans Roy Soc Lond B Biol Sci 356(1405):111–117
11. Iiizumi S, Nomura Y, So S et al (2006) Simple one-week method to construct gene-targeting vectors: application to production of human knockout cell lines. Biotechniques 41(3): 311–316
12. Arakawa H, Lodygin D, Buerstedde JM (2001) Mutant loxP vectors for selectable marker recycle and conditional knock-outs. BMC Biotech 1:7
13. Chang H, Delany ME (2004) Karyotype stability of the DT40 chicken B cell line: macrochromosome variation and cytogenetic mosaicism. Chromosome Res 12(3): 299–307
14. Nakai A, Ishikawa T (2001) Cell cycle transition under stress conditions controlled by vertebrate heat shock factors. EMBO J 20(11): 2885–2895
15. Arakawa H, Moldovan GL, Saribasak H et al (2006) A role for PCNA ubiquitination in immunoglobulin hypermutation. PLoS Biol 4(11):e366
16. Zhang Y, Wienands J, Zurn C et al (1998) Induction of the antigen receptor expression on B lymphocytes results in rapid competence for signaling of SLP-65 and Syk. EMBO J 17(24):7304–7310
17. Fujimori A, Tachiiri S, Sonoda E et al (2001) Rad52 partially substitutes for the Rad51 paralog XRCC3 in maintaining chromosomal integrity in vertebrate cells. EMBO J 20(19): 5513–5520

Chapter 4

Gene Targeting of Human Pluripotent Stem Cells by Homologous Recombination

Sara E. Howden and James A. Thomson

Abstract

The ability of human embryonic stem cells and induced pluripotent stem cells to differentiate into all adult cell types greatly facilitates the study of human development, disease pathogenesis, and the generation of screening systems to identify novel therapeutic agents. Autologous cell therapies based on patient-derived induced pluripotent stem cells also hold great promise for the treatment and correction of many inherited and acquired diseases. The full potential of human pluripotent stem cells can be unleashed by genetically modifying a chosen locus with minimal impact on the remaining genome, which can be achieved by targeting genes by homologous recombination. This chapter will describe a protocol for gene modification of pluripotent stem cells by homologous recombination and several methods for the screening and identification of successfully modified clones.

Key words Gene targeting, Homologous recombination, Transfection, Induced pluripotent stem cells, Embryonic stem cells, Bacterial artificial chromosome

1 Introduction

Pluripotent stem cells offer enormous potential for modeling disease, drug discovery, and transplantation medicine due to their ability to differentiate into any given cell type. Induced pluripotent stem (iPS) cells, generated by introducing defined factors to reprogram terminally differentiated somatic cells [1,2], are particularly advantageous for the development of autologous or customized cellular therapies to treat or correct many inherited and acquired diseases. Complications associated with immunorejection can be avoided through the generation and subsequent disease correction of patient-specific iPS cells, which can be differentiated into relevant cell types for the repopulation and regeneration of a defective tissue or organ. Thus, the ability to genetically modify pluripotent stem cells represents a powerful tool, as it can enable the genetic correction of a known disease-causing mutation prior to the downstream therapeutic application of any given patient-derived iPS cell

Francesca Storici (ed.), *Gene Correction: Methods and Protocols*, Methods in Molecular Biology, vol. 1114,
DOI 10.1007/978-1-62703-761-7_4,

line. Gene targeting by homologous recombination is the ideal approach for the correction of genetic defects as it enables replacement of the defective allele with a normal functional one without disturbing the remaining genome.

In addition to gene correction of patient-specific iPS cells, homologous recombination can also be used to introduce specific mutations or reporter genes into just about any locus of interest, which can aid in the study of the underlying mechanisms of a particular disease or cellular processes.

Here we describe detailed protocols for maintaining and expanding pluripotent stem cells in feeder-free culture conditions, introducing gene-targeting constructs for homologous recombination, and several methods for screening and identifying successfully targeted clones. We will provide examples that demonstrate the correction of a single base pair mutation in the *OAT* gene in an iPS line derived from a patient with gyrate atrophy [3]. However, these methods can be applied to the correction of any gene of interest or the "knock-in" or "knockout" of specific sequences in human pluripotent stem cells.

2 Materials

2.1 Tissue Culture

2.1.1 Equipment

1. Humidified incubator at 37 °C with 5 % CO_2.
2. Laminar flow tissue culture hood.
3. Centrifuge (for 15 and 50 mL tubes).
4. Inverted microscope.
5. Filtered glass or plastic 5 and 10 mL pipets.
6. Filtered pipet tips.
7. Glass Pasteur pipets and media waste trap for aspirating media.
8. 15 and 50 mL conical centrifuge tubes.
9. Nunc 10 cm petri dishes and 6-well and 24-well plates (Fisher Scientific).
10. Hood equipped with inverted microscope for picking clones.

2.1.2 Maintenance and Expansion of Human iPS Cells

1. mTeSR1 medium (STEMCELL technologies).
2. Matrigel (BD Biosciences).
3. PBS (Life Technologies).
4. 0.5 mM EDTA solution (made up in PBS).

2.1.3 Gene Targeting of iPS Cells

1. Linear gene-targeting construct (20–50 μg DNA per transfection).
2. Gene Pulser II electroporator or equivalent (Bio-Rad).

3. 0.4 cm cuvettes (Bio-Rad).
4. TrypLE Express (Life Technologies).
5. DMEM-F12 (Life Technologies).
6. Y-27632 dihydrochloride (Tocris Bioscience).
7. Puromycin (Sigma) or Geneticin (Life Technologies).

2.2 Screening of Drug-Resistant Colonies

2.2.1 Polymerase Chain Reaction and Reverse Transcription

1. Thermal cycler.
2. Agarose and gel running apparatus.
3. DNeasy Blood and Tissue Kit (QIAGEN).
4. Taq polymerase (QIAGEN).
5. 10 mM dNTP mix (NEB).
6. 3′ and 5′ primers.
7. RNeasy Mini Kit (QIAGEN).
8. SuperScript III First-Strand Synthesis System for RT-PCR Kit (Life Technologies).
9. QIAQuick PCR Purification Kit (QIAGEN)

2.2.2 TaqMan Copy Number Assay

1. Real-time PCR machine.
2. TaqMan Copy Number Assay Mix, specific to target gene (Life Technologies).
3. TaqMan Copy Number Reference Assay (Life Technologies).
4. TaqMan Genotyping Master Mix (Life Technologies).
5. CopyCaller software.

2.2.3 Fluorescent In Situ Hybridization

1. Nick Translation Kit (Abbott Molecular).
2. Vysis Spectrum-Green or Spectrum-Red labeled dUTP (Abbott Molecular).
3. Human COT-1 DNA (Abbott Molecular).
4. Colcemid (Life Technologies).
5. 0.56 % KCl solution, made up in water.
6. Carnoy's fixative solution: 3:1 (v/v) methanol/glacial acetic acid, freshly prepared.
7. Microscope slides.
8. Coplin jars.
9. Water bath.
10. 20× SSC: 3.0 M NaCl, 0.3 M sodium citrate.
11. Deionized formamide (Ambion), make up to 70 % with 2× SSC.
12. 70, 85, and 100 % ethanol solutions.
13. LSI/WCP Hybridization Buffer (Abbott Molecular).

14. Coverslips.
15. Rubber cement.
16. Air tight container humidified with damp paper towel.
17. Vectashield with DAPI Mounting Medium (Vector Labs).
18. Upright epifluorescence microscope.

2.2.4 Southern Blot

1. PCR DIG Probe Synthesis Kit (Roche Applied Science).
2. 5′ and 3′ PCR primers for amplifying probe.
3. Nucleon BACC2 Genomic DNA Extraction Kit (GE Healthcare).
4. Depurination buffer: 0.25 M HCl.
5. Denaturation buffer: 0.5 M NaOH and 1.5 M NaCl.
6. Neutralization buffer: 0.5 M Tris–HCl (pH 7.5) and 1.5 M NaCl.
7. Positively charged nylon membrane (Roche Applied Science).
8. Whatman 3MM paper (Fisher Scientific).
9. Hybridization buffer (Roche Applied Science).
10. Hybridization bottles and oven (fitted with rotisserie).
11. Low-stringency buffer: 2× SSC containing 0.1 % SDS.
12. High-stringency buffer: 0.5× SSC containing 0.1 % SDS.
13. Maleic acid buffer: 0.1 M Maleic acid and 0.15 M NaCl (adjust with NaOH to pH 7.5) (available from Roche Applied Science in the DIG Wash and Block Buffer Set).
14. Washing Buffer: 0.1 M Maleic acid and 0.15 M NaCl (pH 7.5) with 0.3 % (v/v) Tween 20 (available from Roche Applied Science in the DIG Wash and Block Buffer Set).
15. Blocking Solution (Roche Applied Science, available in DIG Wash and Block Buffer Set).
16. Anti-digoxigenin-alkaline phosphatase antibody (Roche Applied Science).
17. Ready-to-use CDP-Star (Roche Applied Science).
18. Phosphorimager or chemiluminescent detection film.

3 Methods

3.1 Maintenance and Expansion of Pluripotent Stem Cells

Human iPS and ES cells can be maintained indefinitely on Matrigel-coated plates in mTeSR1 medium and should be passaged every 3–4 days. Cells should not be allowed to reach >90 % confluency, and culture media should be changed daily to prevent excessive cell death and/or unwanted differentiation.

In the following protocol for passaging cells, use the smaller volume for one well of a 6-well plate and the larger volume for a

10 cm dish (for 24-well plate, halve the volumes indicated for 6-well plate format):

1. Prepare Matrigel-coated plates at least 30 min in advance. Matrigel should be thawed on ice and divided into small aliquots (typically 200–500 μL/tube). Add 0.2 mg Matrigel (*see* **Note 1**) per mL of cold DMEM-F12, mix well, and add 1 mL per well of a 6-well plate or 4 mL for each 10 cm dish. Plates can be kept in 37 °C incubator for up to 2 weeks.
2. To passage cells, remove spent medium and wash cells with 1–5 mL room temperature 0.5 mM EDTA (in PBS) solution and aspirate with glass Pasteur pipet.
3. Add 1–5 mL 0.5 mM EDTA solution and incubate cells for 2–5 min in 37 °C incubator. Following EDTA incubation the cells should still be loosely attached to the plate in distinct colonies, but cells should appear somewhat dissociated from one another when observed under the microscope (*see* **Note 2**).
4. During the incubation step, remove residual media from a Matrigel-coated plate or dish and add 2–8 mL of fresh mTesR medium.
5. Carefully remove EDTA solution and add 2–10 mL mTesR medium to remove cells from the plate. Pipet up and down once or twice to break up cells into smaller clumps, and add 2 mL to new 10 cm dish or divide evenly over 6-well plate.

3.2 Gene Targeting

This chapter will not provide an in-depth methods section for the generation of gene-targeting constructs. However we recommend the use of BAC-based vectors for gene-targeting experiments since BAC clones are readily accessible from publicly available libraries (distributed by Life Technologics), and the UCSC genome browser (http://genome.ucsc.edu) can be used to identify clones that carry any particular target gene of interest. A suitable gene-targeting construct can be generated by inserting a selection cassette using Red/ET recombination. For selection in pluripotent stem cells, we recommend the use of a puromycin or neomycin resistance genes driven by a PGK or Ef1a promoter (*see* **Note 3**), and for selection in bacteria, we recommend either kanamycin or ampicillin resistance genes. The selection cassette may also be flanked by loxP sites to permit its subsequent removal following a successful gene-targeting event. Red/ET recombination kits containing all the necessary reagents required for BAC-recombination, including suitable selection cassettes and a detailed protocol, can be acquired from Gene Bridges (www.genebridges.com).

To generate the gene-targeting construct used to correct the human *OAT* gene, we obtained a BAC clone (RP11-113M14) that contains the entire *OAT* locus. A loxP-flanked cassette comprising of a puromycin resistance gene driven by the murine PGK promoter

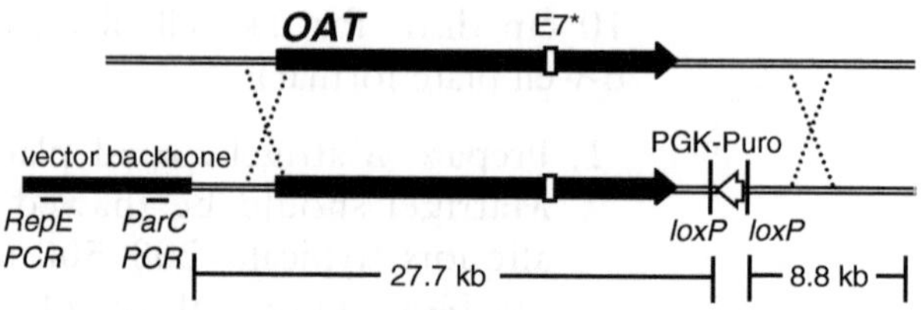

Fig. 1 Schematic diagram of the BAC-based gene-targeting vector used to correct a single base pair mutation in the *OAT* gene. A loxP-flanked selection cassette (PGK-Puro) was inserted approximately 2 kb downstream of the *OAT* coding region by recombineering. The sizes of the homologous arms are shown. The point mutation in exon 7 (E7*), the regions amplified by PCR to identify random integrants are also indicated

and a kanamycin resistance gene was inserted just downstream of the *OAT* coding sequence. To reduce size, the BAC was then digested with *Xho*I, and a 38 kb fragment containing the entire *OAT* gene and newly inserted selection cassette was subcloned into the *Sal*I site of a second BAC vector (Fig. 1). Reducing construct size is not completely necessary; however one should keep in mind that although a positive correlation exists between the length of homologous sequences and gene-targeting efficiency, a negative correlation exists between size of the vector and its transfection efficiency in mammalian cells. We therefore suggest aiming for a gene-targeting vector with 10–30 kb homology arms flanking the selection cassette.

Prior to transfection, BAC DNA can be prepared using a standard alkaline lysis procedure for plasmid purification followed by cesium chloride density gradient extraction. Alternatively, kits designed specifically for the purification of BACs and other large plasmids are also available from numerous suppliers.

At least 1 day prior to transfection, 50–100 μg of the gene-targeting vector (per transfection) should be linearized with a restriction endonuclease that cuts exclusively within the vector backbone (*see* **Note 4**). Following digestion, the DNA should be purified by phenol/chloroform extraction followed by ethanol precipitation and then resuspended in TE buffer at a concentration of 0.25–1.5 μg/μL.

Cells should be passaged 2 days prior to transfection and in 10 cm dish format.

A single 10 cm dish is usually enough for 1–2 transfections:

1. Remove culture media from 10 cm dish containing cells and add 5 mL of pre-warmed TrypLE Express and place cells in 37 °C incubator for 2–3 min.
2. Remove TrypLE Express before cells begin lifting from the plate, and wash cells from the plate with 10 mL pre-warmed DMEM-F12.

3. Transfer cells into 15 mL tube and take a small aliquot for counting.
4. Pellet cells by centrifugation at 300 × *g* for 5 min.
5. Resuspend cells in DMEM-F12 to a final concentration of approximately 10^7 cells/mL.
6. Transfer 0.5 mL of the cell suspension to a 0.4 cm cuvette.
7. Make up 50–100 μg of the linearized gene-targeting vector to a total volume of 300 μL with PBS and add to cuvette.
8. Mix by pipetting up and down 2–3 times and electroporate with the following conditions: 320 V, 200 μF, and infinite resistance (*see* **Note 5**).
9. Transfer the cells to a new Matrigel-coated 10 cm dish containing 10 mL mTeSR1 supplemented with 10 μM of Rock-inhibitor compound, Y-27632.
10. Media should be changed the following day and daily thereafter (Rock inhibitor isn't necessary after the first day). Drug selection should commence 3–4 days post-transfection depending on cell density. Puromycin and Geneticin are typically used at a concentration of 0.5–2 μg/mL and 50–200 μg/mL, respectively. However, we recommend performing kill curve experiments with untransfected cells and increasing doses of selection to obtain optimal dosage for a given line or drug batch.

3.3 Colony Picking

Drug-resistant colonies are typically ready for picking between 10 and 15 days post-transfection. It is not unusual for a large portion of the colony to fail to attach to the new plate following picking. Thus, colonies should be broken up into several smaller clumps and, as a general rule of thumb, be picked when similar in size to colonies observed just prior to regular passaging of untransfected cells.

1. Remove residual media from a 24-well Matrigel-coated plate and add 1 mL mTeSR1 medium (without selection) per well.
2. Mark colonies ready for picking with a colony marker or felt tip pen.
3. In a sterile hood equipped with a microscope, use a 200 μL pipet with barrier tip to gently score, scrape, and aspirate a marked colony and transfer to a single well of a 24-well plate.
4. The cells from each picked colony should be ready for passaging after 4–7 days and expanded for further analysis.

3.4 Genomic Polymerase Chain Reaction

BAC-based gene-targeting constructs usually comprise long homology arms, and so identifying correctly targeted colonies by performing PCR across vector/host genome junctions is not typically possible. However, genomic PCR can be utilized as an initial screen for excluding a good proportion of, but not all of, the clones that have undergone random integration of the gene-targeting

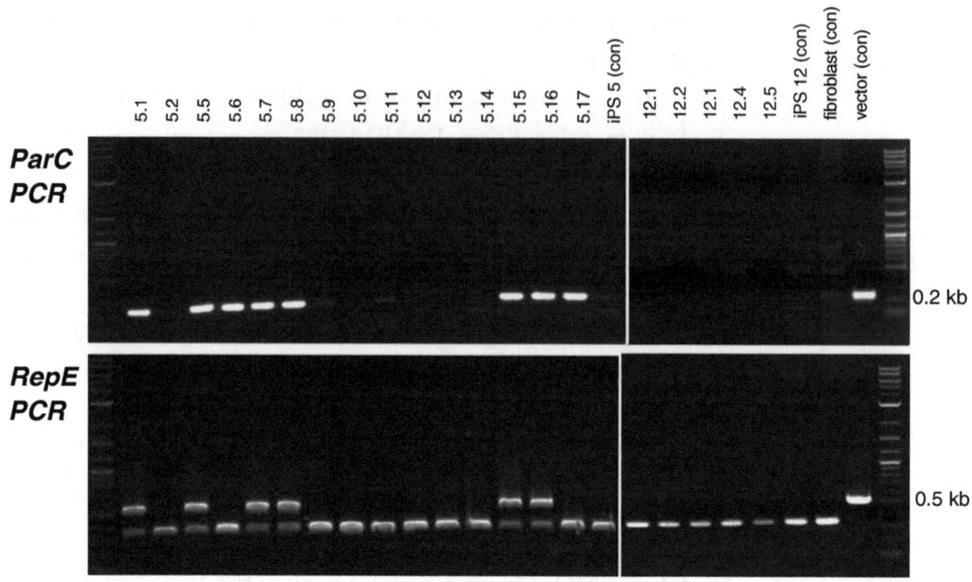

Fig. 2 PCR analysis of human iPS cell lines to detect gene-targeting vector backbone sequences. PCR was performed on the genomic DNA extracted from 20 drug-resistant human iPS cell lines following transfection of the gene-targeting vector in two patient iPS cell lines (5 and 12). For controls, PCR was also performed on the genomic DNA extracted from untransfected iPS cell lines (5 and 12), on the patient fibroblast line, as well as on the gene-targeting vector. The size of the bands is indicated

vector (Fig. 2). This is performed using primers specific to sequences on the vector backbone, since clones that have undergone a successful gene-targeting event should not contain such sequences. In the following protocol, we use primers specific to one of the partition genes (ParC) and the RepE operator (these sequences are present on all BAC vectors). Be sure to include appropriate positive and negative controls, such as the gene-targeting vector and genomic DNA from untransfected cells when setting up the reactions:

1. Extract genomic DNA using the DNeasy Blood and Tissue Kit (from QIAGEN) according to manufacturer's instructions.
2. Prepare the PCR reactions as follows (25 μL final volume): 100–200 ng genomic DNA, 2.5 μL 10× PCR buffer, 5 μL 5× Q-Solution, 3 μL 25 mM $MgSO_4$, 0.5 μL 10 mM dNTP mix, 0.5 μL 10 μM 5′ primer, 0.5 μL 10 μM 3′ primer, 0.25 μL Taq, and H_2O to 25 μL.
3. Run the samples in a thermal cycler as follows: hold at 94 °C for 4 min, then 35 cycles of 94 °C for 30 s, 57 °C for 30 s, and 72 °C for 60 s. Hold at 72 °C for 5 min and then maintain at 4 °C.
4. Analyze by electrophoresis by loading the entire reaction on a 1 % agarose gel, stained with ethidium bromide. Use appropriate molecular weight standards.

3.5 Reverse-Transcriptase Polymerase Chain Reaction and Sequencing Analysis

This analysis can be used to assess whether a modified gene is being expressed in a given clone and is most suitable for target genes known to be expressed in pluripotent stem cells. For target genes that are not expressed in pluripotent stem cells, an additional differentiation step must be performed before RNA can be extracted for analysis by reverse transcriptase (RT) PCR (which may not always be feasible). Depending on the design of the construct used for gene targeting, this analysis has the capacity to identify correctly targeted clones and distinguish them from clones arising from random integration. If the gene-targeting construct contains the entire coding region (and possibly some upstream regulatory sequences) of the target gene, expression from the gene-targeting construct can occur in clones arising from random integration, and it may be difficult to distinguish these from correctly targeted clones (Fig. 3a). On the other hand, by using a "gene-trap" approach, whereby upstream promoter sequences and/or part of the gene's 5′ coding region are lacking, it is unlikely that expression of the modified target gene will occur following its random integration into the host genome. In this case RT-PCR analysis can be used to identify correctly targeted clones. An example is shown in Fig. 3b, in which a "gene-trap" approach was used to introduce a single base pair mutation into the PRPF8 gene in a wild-type ES cell line (H9):

1. Extract RNA from drug-resistant clones using the RNeasy Mini Kit (by QIAGEN) according to the manufacturer's instructions.
2. Perform reverse transcription using SuperScript III First-Strand Synthesis System for RT-PCR Kit (from Life Technologies) according to the manufacturer's oligo(dT) protocol.
3. Perform PCR with 0.1–1 μL of cDNA generated from the reverse transcription reactions, using primers that flank the modified region of the target gene.
4. Column purify PCR products using the PCR Purification Kit (from QIAGEN).
5. Submit sample to sequencing facility for Sanger sequencing (can use one of the primers used for PCR reaction as a sequencing primer).
6. Analyze chromatograms using appropriate software (e.g. Chromas, 4 Peaks).

3.6 Copy Number Assay

Targeted integration relies on the replacement of a host allele with a modified one (contained on the gene-targeting vector), so that the copy number for the target gene will remain unchanged. Random integration of a gene-targeting vector, on the other hand, would result in an increase in copy number, due to the presence of the endogenous allele(s) and one or more exogenous alleles from

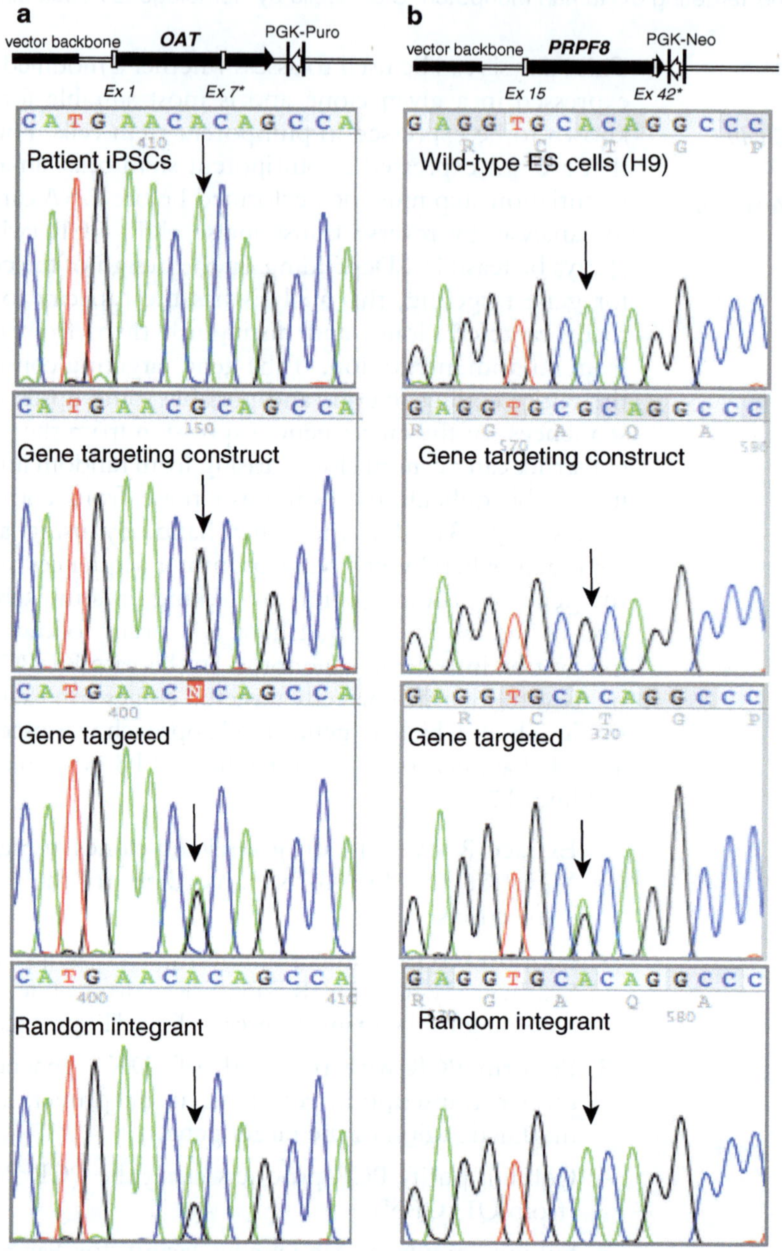

Fig. 3 Reverse-transcriptase polymerase chain reaction and sequencing analysis of gene-corrected pluripotent stem cell clones. (**a**) Chromatograms resulting from Sanger sequencing of *OAT* exon 7 PCR products amplified from the gene-targeting vector or cDNA from the non-gene-corrected patient iPS cells, a gene-targeted iPS cell line, and a line arising from random integration of the gene-targeting vector. *Arrows* indicate the location of the point mutation in the patient line. The *OAT* gene-targeting construct contains the entire *OAT* coding region and some upstream regulatory regions, which can permit expression of the wild-type transcript in nontargeted clones. (**b**) Chromatograms resulting from Sanger sequencing of *PRPF8* exon 42 PCR products amplified from the gene-targeting vector or cDNA from a wild-type ES cell line (H9), a gene-targeted ES cell line, and a line arising from random integration of the gene-targeting vector. *Arrows* indicate the location of the point mutation introduced by homologous recombination. The *PRPF8* gene-targeting construct contains only part of the *PRPF8* coding region (beginning at exon 15), and so only correctly targeted ES cell clones should express the mutant transcript

Table 1
Copy number assay reaction

Component	384 wells (μL)	96 wells (μL)
2× TaqMan Genotyping Master Mix	5	10
TaqMan Copy Number Assay, 20× working stock	0.5	1
TaqMan Copy Number Reference Assay, 20×	0.5	1
Nuclease-free water	2	4
Total volume	8	16

the gene-targeting vector. Thus, a real-time PCR-based copy number assay can be performed to distinguish targeted clones versus those arising from random integration. We recommend the use of the TaqMan Copy Number Assay System (from Life Technologies), in which a comprehensive collection of predesigned TaqMan probes that span the entire genome are available. Choose a probe positioned as closely as possible to the selection cassette that has been incorporated into the gene-targeting vector:

1. Extract genomic DNA from drug-resistant clones using DNeasy Blood and Tissue Kit (from QIAGEN).
2. Dilute each sample to 5 ng/μL using nuclease-free water or TE buffer.
3. Determine the number of reactions (use at least three replicates for each sample), and combine the required amounts of each reaction component in a microfuge tube according to Table 1.
4. Pipet 8 μL or 16 μL of the reaction mixture into the wells of a 384- or 96-well reaction plate, respectively.
5. Add 2 or 4 μL of the diluted genomic DNA to wells containing the reaction mixture.
6. Seal the reaction plate with optical adhesive film, vortex, and then centrifuge the reaction plate briefly.
7. Run the plate in a suitable real-time PCR instrument using the following parameters: 95 °C for 10 min, then 40 cycles of 95 °C for 15 s followed by 60 °C for 60 s.
8. Import results data into CopyCaller software to determine the approximate copy number for the target gene in each clone.

3.7 Fluorescent In Situ Hybridization

Because FISH can be a somewhat technically demanding and lengthy procedure, it is best reserved for confirming a suspected gene-targeting event in those clones that passed earlier screening methods (e.g., genomic PCR, RT-PCR, or a real-time PCR copy number assay). However, it is a good idea to include at least one drug-resistant clone thought to have arisen from random integration of the gene-targeting construct as a control.

3.7.1 Harvesting of Cells for Metaphase Spreads

Cells should be passaged 2 days before harvesting to ensure cells are in exponential growth. For the following protocol, harvest cells from two wells of a 6-well plate (at approximately 50 % confluency):

1. Remove spent medium and add new medium containing 100 ng/mL Colcemid to cells 2–3 h prior to harvest (*see* **Note 6**).
2. Remove medium and add 1 mL pre-warmed TrypLE Express to each well. Incubate at 37 °C for 4–5 min.
3. Add 3 mL DMEM-F12 to each well and pipet up and down a few times to achieve a single cell suspension. Pool cells and transfer to a 15 mL tube. Collect by centrifugation at 300 × *g* for 5 min.
4. Pour off supernatant, leaving approximately 0.5 mL. Flick tube or vortex to resuspend cells. Add 8 mL 0.56 % KCl and mix.
5. Incubate in 37 °C water bath for 8 min.
6. Collect cells by centrifugation at 300 × *g* for 10 min. Remove supernatant to 0.5 mL.
7. Resuspend cells using vortex to achieve a single cell suspension. With tube still on vortex, add dropwise 20 drops of fixative at ~1 drop/s (*see* **Note** 7). Slowly add fixative to 6 mL while mixing on vortex.
8. Centrifuge at 300 × *g* for 10 min.
9. Remove supernatant, vortex, and add 6 mL fixative (does not need to be dropwise).
10. Repeat **steps 8–9**.
11. Store fixed cell suspensions at −20 °C. If possible, store cells at −20 °C overnight before dropping onto slides.

3.7.2 Preparing the Slides

1. Remove fixed cells from freezer. Centrifuge for 10 min at 300 × *g* and remove supernatant.
2. Add fixative to ~0.5 mL.
3. Drop 15–20 μL of cell suspension at a height of approximately 10–15 cm onto microscope slide laid horizontally.
4. Check cell density, mitotic index, and quality of spreading and make adjustments as necessary (*see* **Note 8**).
5. Age slides on a 37 °C hot plate or at room temperature for 24 h before using for FISH.

3.7.3 Probe Synthesis

1. The gene-targeting vector (circular or linearized) can be directly labeled with a fluorescent fluorophore such as Vysis Spectrum-Red or Spectrum-Green dUTP using the Nick

Translation Kit (from Abbott Molecular) according to the manufacturers' instructions.

2. Allow the reaction to go for 4–6 h at 15 °C (longer than generally recommended). Check the length of the labeled DNA by running 2–3 μL of labeled DNA on a 1 % agarose gel. Fragment size should be mainly below 200 bp (large fragments do not hybridize effectively). Stop the reaction by heating or EDTA as recommended.
3. Add 20 μL of human COT-1 DNA and alcohol precipitate labeled DNA.
4. Dissolve product in 10–15 μL of TE buffer.

3.7.4 Denaturation and Hybridization

1. Place Coplin jar containing 70 % formamide/2× SSC in room temperature water bath and then set to 70 °C. Formamide will take 30–40 min to reach temperature.
2. Denature aged slides containing metaphase spreads by immersing in the 70 % formamide/2× SSC at 70 °C for 2 min.
3. Dehydrate slides by putting through ethanol series (incubate for 10–20 s each in 70 %, 85 %, and 95 % in separate Coplin jars) at room temperature and leave to air-dry.
4. Add 1–3 μL (*see* **Note 9**) to 7–9 μL LSI/WCP Hybridization Buffer, so that total volume is 10 μL. Denature by incubating in 70 °C water for 8 min.
5. Place probe on ice (may need to touch spin in centrifuge to collect contents).
6. Add 4 μL of probe to slide over spread cells. Cover with 13 mm coverslip and seal with rubber cement (may be applied using a bulb and glass Pasteur pipet).
7. Incubate overnight at 37 °C in a lunchbox style container lined with a damp (but not wet) paper towel.

3.7.5 Stringency Wash and Analysis

1. Bring two Coplin jars containing 2× SSC to temperature in a 70 °C water bath.
2. Remove rubber cement and carefully slide off coverslip.
3. Briefly immerse slide in 2× SSC at room temperature.
4. Immerse slide in 2× SSC at 70 °C for 10 min (5 min in each jar).
5. Dehydrate through ethanol series (as above) and allow slide to air-dry.
6. Apply a drop or two of Vectashield Mounting Medium (with DAPI) and blot off any excess with tissue or paper towel.
7. Analyze numerous metaphase and interphase nuclei for each clone by fluorescent microscopy equipped with 63× and/or 100× oil immersion objective and appropriate camera/software. Two signals should be consistently observed in a successfully

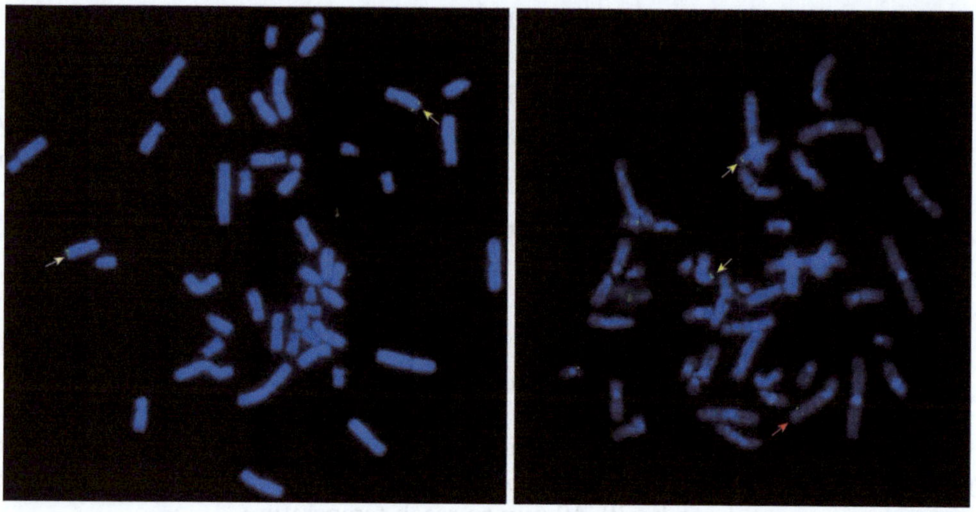

Fig. 4 Fluorescent in situ hybridization analysis of a gene-targeted iPS cell line (*left*) and a random integrant (*right*) using a fluorescently labeled probe specific to the *OAT* locus. *Yellow arrows* indicate endogenous signals on chromosome 10q; *red arrow* indicates a third signal caused by random integration of the gene-targeting construct

targeted clone (unless the target gene lies on a sex chromosome), whereas three or more signals should be detected in clones arising from random integration of the gene-targeting vector (Fig. 4).

3.8 Southern Blot

To use Southern blot as method for identifying successfully targeted clones, appropriately positioned restriction sites must be available either within and/or relatively close to the desired gene-targeting site. Depending on the gene-targeting strategy, this may not always be the case. It is also worth noting that the functionality of many restriction endonucleases is either blocked or hindered by CpG methylation. Nonetheless, Southern blot is a sure way of confirming a gene-targeting event in a given line and, unlike the other methods described above, can provide important information on the genomic integrity of the modified locus. Like FISH, this method is somewhat time-consuming and low throughput compared to other methods and is best reserved for clones where a potential gene-targeting event is suspected.

To confirm a gene-targeting event at the *OAT* locus in iPS cells derived from a patient with gyrate atrophy, genomic DNA from a potentially targeted clone was digested with *Eco*RI and *Spe*I. In the wild-type genome, this releases an approximately 12.5 kb band that harbors the last third of the *OAT* coding region along with some downstream noncoding sequences. A correctly targeted allele will release an approximately 11 kb band due to the presence of a *Spe*I site located within the selection cassette (that would have

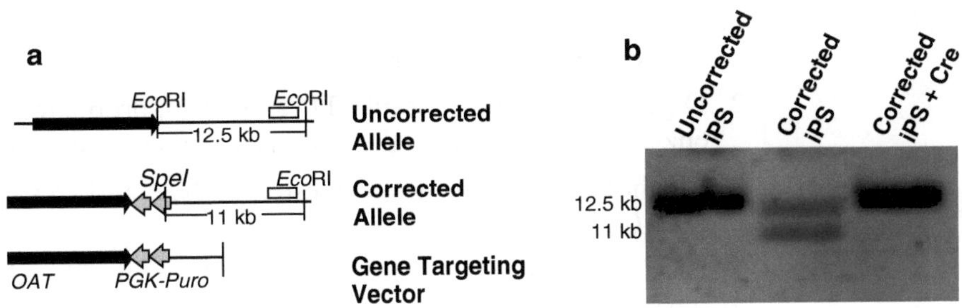

Fig. 5 Southern blot analysis to confirm correction of the *OAT* gene in a patient-specific iPS line. (**a**) Diagrammatic representation of the uncorrected (endogenous) and corrected alleles and expected band size following digestion with *Eco*RI and *Spe*I are shown. A probe specific to sequences downstream of the *OAT* coding region that lies outside of the gene-targeting vector (indicated by *white box*) was used to identify a correctly targeted clone. (**b**) Southern blot analysis of an uncorrected clone, a gene-corrected clone, and a gene-corrected clone following excision of the loxP-flanked puromycin cassette with Cre-recombinase

been introduced just downstream of the *OAT* coding region by a successful homologous recombination event). Using a probe specific to this fragment that binds outside of the gene-targeting vector allows one to distinguish a gene-targeted clone from a random integrant (Fig. 5).

3.8.1 Probe Synthesis

Identify a 400–800 bp sequence that lies next to but outside of the gene-targeting vector, with a balanced GC:AT ratio (try to avoid probes with >55 % GC content). For probe synthesis we recommend the use of the PCR DIG Probe Synthesis Kit (from Roche Applied Science). Genomic DNA or a second BAC vector that harbors the probe sequence can be used as a template:

1. In a PCR tube combine 10–100 ng genomic DNA or 0.1–1 ng BAC DNA, 5 μL 10× High-Fidelity PCR buffer (vial 3), 5 μL 10× PCR DIG mix (vial 2), 4 μL 10 μM 5′ primer, 4 μL 10 μM 3′ primer, 0.5 μL High-Fidelity Taq Polymerase (vial 1), and H_2O to 50 μL.
2. As a control, in a separate PCR tube, set up the above reaction but substitute 5 μL 10× dNTP mix (vial 4) for the 10× PCR DIG mix.
3. Run the samples in a thermal cycler as follows: hold at 94 °C for 4 min, then 10 cycles of 94 °C for 30 s, 60 °C for 30 s, and 72 °C for 60 s, followed by 20 cycles of 94 °C for 30 s, 60 °C for 30 s, 72 °C for 60 s, + additional 20 s for each successive cycle. Hold at 72 °C for 5 min, and then maintain at 4 °C.
4. Analyze 5 μL of the reaction by gel electrophoresis. If the probe reaction was successful, both the labeled experimental probe and the unlabeled control probe should be clearly visible on the gel, and the DIG-labeled probe should run slower than the control (which should be the expected size).

3.8.2 Electrophoresis and Blotting

1. Extract high-molecular-weight genomic DNA using the Nucleon BACC2 Kit (from GE Healthcare).
2. Digest 10–20 μg genomic DNA with appropriate restriction enzyme(s) according to the manufacturer's recommendations.
3. Alcohol precipitate the digested DNA and resuspend in 30–40 μL TE.
4. Load 5–10 μg on a 1 % agarose gel and run until bands are well separated. A DIG-labeled DNA Molecular Weight Marker may also be included as a size reference.
5. To assess the quality of the target DNA, stain the gel briefly in 0.25–0.50 μg/mL ethidium bromide and examine the gel under UV light (*see* **Note 10**). Excess agarose may be trimmed at this point.
6. Destain the gel in water for 15 min.
7. Incubate gel in Depurination buffer (0.25 M HCl) for 10 min with gentle agitation. Rinse briefly with water.
8. Incubate gel in Denaturation buffer (0.5 M NaOH, 1.5 M NaCl) for 2 × 15 min, with gentle shaking. Rinse briefly with water.
9. Incubate gel in Neutralization Solution (0.5 M Tris–HCl, pH 7.5; 1.5 M NaCl) for 2 × 15 min, with gentle shaking.
10. Equilibrate the gel for at least 10 min in 20× SSC.
11. Set up the transfer by placing a glass plate across a reservoir of 20× SSC. Prewet a long strip of Whatman paper (the wick) in 20× SSC and place atop of glass plate (the ends should be lying in the reservoir). Eliminate any air bubbles between the wick and plate.
12. Place the gel atop the soaked sheet of Whatman paper, and roll a sterile pipet over to remove air bubbles between the gel and paper.
13. Place thin strips of Parafilm on the edges of the gel to prevent lateral capillary action and improve the quality and resolution of the transferred bands.
14. Cut a piece of positively charged nylon membrane to the size of the gel, and place on the DNA-containing surface of the gel. Use a pipet to eliminate air bubbles as above.
15. On top of the membrane, place about 10 pieces of Whatman paper also cut to the size of the gel and prewet with 20× SSC.
16. Complete the assembly with a stack paper towels approximately 10–15 cm, a glass plate or tube rack, and a 200–500 g weight.
17. Allow the transfer to proceed overnight.
18. Fix the DNA to the blot using a UV-crosslinker or bake at 120 °C for 30 min.

19. Rinse the membrane in distilled water or 2× SSC and allow to air-dry.
20. Proceed to hybridization or store the dry blot between two sheets of Whatman 3MM paper in a sealed bag at 4 °C.

3.8.3 Hybridization and Stringency Washing

1. Transfer the membrane to a hybridization bottle and pre-hybridize with 20 mL pre-warmed hybridization buffer for 30 min at 37 °C in a suitable hybridization oven fitted with a rotisserie.
2. Add 40–60 μL of DIG-labeled probe to 50 μL TE buffer, and denature in boiling water bath for 5 min.
3. Place tube containing the probe on ice for at least 2 min.
4. Add probe to a tube containing 20 mL hybridization buffer pre-warmed to 37 °C.
5. Remove the pre-hybridization buffer, add the new buffer containing the probe to the blot, and hybridize overnight with constant agitation/rotation.
6. Remove hybridization solution. This can be saved in a tube at −20 °C for future use and can be reused 3–5 times.
7. Add 100 mL of low-stringency wash buffer (2× SSC containing 0.1 % SDS). Incubate at room temperature for 5 min with constant agitation.
8. Pour off the used buffer, and add 100 mL new low-stringency buffer, and incubate for another 5 min at room temperature.
9. Pour off the used low-stringency buffer and add 100 mL pre-heated high-stringency wash buffer (0.5× SSC containing 1 % SDS). Incubate at 65 °C for 15 min with constant agitation.
10. Pour off used high-stringency buffer and add another 100 mL preheated high-stringency wash buffer. Incubate at 65 °C for an additional 15 min.

3.8.4 Probe Detection

1. Transfer the membrane to a plastic container containing 100 mL Washing Buffer. Incubate at room temperature for 2 min with gentle agitation.
2. Discard Washing Buffer and add 100 mL Blocking Solution. Incubate for 30 min at room temperature with gentle agitation.
3. During this step make up the Antibody Solution by adding 5 μL of Anti-DIG-alkaline phosphatase antibody to 50 mL of fresh Blocking Solution.
4. Discard Blocking Solution and add Antibody Solution. Incubate at room temperature for 30 min with gentle agitation.
5. Discard Antibody Solution and wash membrane twice (2 × 15 min) with 100 mL portions of Washing Buffer.

6. Equilibrate membrane 3 min in 20 mL Detection Buffer.
7. Place the membrane in a development folder or plastic sheet protector, and add 10–20 drops of the chemiluminescent substrate, CSPD-star. Cover and ensure the substrate is spread evenly over the membrane. Incubate at room temperature for 5 min.
8. Squeeze out any excess liquid and expose the membrane to chemiluminescent detection film or analyze with a phosphorimager.

4 Notes

1. Matrigel concentration varies according to lot number. A lot-specific, product specification sheet with the exact protein concentration is provided by the supplier with each shipment.
2. Timing is critical when passaging cells with EDTA. Cells left too long will dissociate and come away from the plate before addition of the new culture medium, whereas cells that have not been incubated long enough will be very difficult to remove from the plate. If done correctly, the cells should easily come away from the plate upon addition of new culture medium and should not require pipetting up and down more than once or twice with the Pipet-Aid on the fastest setting. Cells that have been incubated too long and become free-floating can be collected by centrifugation before resuspension and plating.
3. Commonly used viral promoters, such as those derived from CMV and SV40, work very inefficiently in human ES and iPS cells and should be avoided.
4. A restriction site located toward one end of the genomic fragment may be used if a unique restriction site within the plasmid backbone cannot be identified.
5. Observe the time constant following the electroporation pulse, which should fall between 3 and 3.5 ms.
6. Cells overexposed to Colcemid will result in shorter and more compact chromosomes, which may be more difficult to analyze by FISH.
7. Adding fixative too quickly can cause cells to clump.
8. Improved chromosome spreading can be obtained by adding extra drops of fixative as the cells become visible on the slide.
9. A probe concentration that produces optimal signal to noise ratio may vary from probe to probe and may require some trial and error.
10. Do not add ethidium bromide to the gel before running as this can interfere with DNA migration.

References

1. Yu J, Vodyanik MA, Smuga-Otto K et al (2007) Induced pluripotent stem cell lines derived from human somatic cells. Science 318:1917–1920
2. Takahashi K, Tanabe K, Ohnuki M et al (2007) Induction of pluripotent stem cells from adult human fibroblasts by defined factors. Cell 131: 861–872
3. Howden SE, Gore A, Li Z et al (2011) Genetic correction and analysis of induced pluripotent stem cells from a patient with gyrate atrophy. Proc Natl Acad Sci U S A 108:6537–6542

References

Part II

Gene Correction Tools Using Non-Viral or Viral Systems

Additional protocols described in this book for gene correction using non-viral or viral tools

- plasmids: Chapters 2, 3, 4, 22
- PCR products: Chapter 1
- small fragment homologous recombination: Chapter 18
- oligonucleotides: Chapters 8, 20, 25, 26
- PNAs: Chapters 5, 8
- cellular NHEJ for gene knockout: Chapters 16, 17
- AAV vectors: Chapters 19, 27
- lentiviral vectors: Chapters 15, 21, 28, 29

Chapter 5

Methods for the Assessment of ssODN-Mediated Gene Correction Frequencies in Muscle Cells

Carmen Bertoni

Abstract

The past decade has seen the development of new technologies capable of editing the genome that have naturally led to exploring their therapeutic application for the treatment of many disorders. Among those, Duchenne muscular dystrophy (DMD) represents an ideal candidate for gene editing primarily due to the large size of dystrophin, the gene responsible for the disease, which limits the use of gene replacement approaches. Critical in the evaluation of the efficacy of the treatment is the development of a method that can accurately quantitate the frequencies of gene repair obtained in the dystrophin gene at both the genomic level as well as the mRNA level. The mdx^{5cv} mouse model of DMD offers an ideal system to precisely determine the frequencies of gene repair. Here we describe the methods used for determining those frequencies and the limitations associated with the use of gene correction for the treatment of DMD. Clinical approaches to muscle disorders using ssODNs will heavily rely on the optimization of the technology and will have to take into consideration the safety, efficacy and cost of the procedure in vision of systemic delivery of the therapeutic treatment.

Key words Gene editing, Gene repair, Single-stranded oligodeoxynucleotides, ssODN, Peptide nucleic acids, PNA, Transfection, Dystrophin, Duchenne muscular dystrophy, DMD, Muscle, Mdx^{5cv}

1 Introduction

Gene editing strategies have emerged as a promising approach to target and correct mutations at the genomic level. Since the initial discovery that small oligodeoxynucleotides (ODNs) containing a mismatch could pair to homologous regions of genomic DNA in mammalian cells and substitute single bases, there has been great interest in its potential clinical applications in genetic hereditary diseases that could benefit from this approach. Several methods have been developed in order to optimize and effectively implement the gene editing strategy in vitro as well as in vivo. These methods all employ different structures of targeting molecules, resulting in variable success rates. Initially, the approach focused on stimulating the endogenous gene repair mechanisms using chimeric (chODNs) or

Francesca Storici (ed.), *Gene Correction: Methods and Protocols*, Methods in Molecular Biology, vol. 1114, DOI 10.1007/978-1-62703-761-7_5, © Springer Science+Business Media, LLC 2014

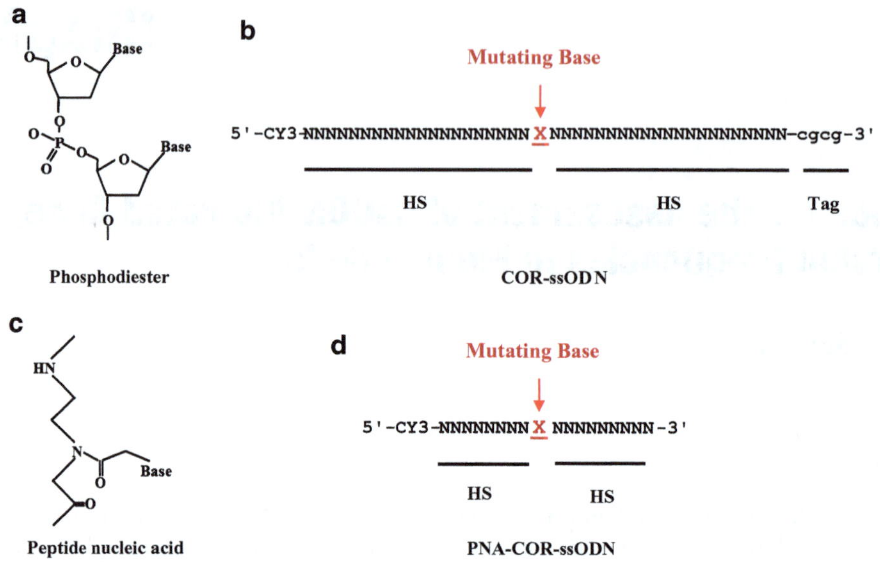

Fig. 1 Structure of the ssODNs tested in muscle cells. To date, the nucleotide modifications used to induce gene correction in muscle have consisted primarily of unmodified bases (**a**) or PNA (**c**). The linear sequence of targeting ssODN (**b**) and PNA-ssODN (**d**) vary primarily in their length. Both oligonucleotides share a sequence perfectly homologous to the region of DNA targeted for repair (HR). The single base pair mismatch on the targeting oligonucleotide, the mutating base, is *underlined* and is indicated with an *X* in *red*. The ssODN contains four phosphorothioate bases at its 3′ ends to increase its stability. The CY3 molecule at the 5′ end is used to follow the uptake of the oligonucleotides after transfection

single-stranded (ssODNs) oligonucleotides. The oligonucleotides are generally complementary to the region of the genomic DNA targeted for correction with the exception of a single base, the targeted base that contains the mismatch (Fig. 1). Upon introduction into the cells, the oligonucleotide anneals to the region of the DNA targeted for repair with the exception of the targeted base. The mismatch will activate the repair process and convert the base accordingly. Proof-of-concept studies have been performed in different cell types and have been limited primarily to the use of reporter genes. Correction in eukaryotic models of diseases has also been reported in vitro as well as in vivo.

Despite these successes, clinical trials for genetic diseases using correcting ODNs are still far from entering the clinical scenario. This is primarily due to the low frequencies of gene repair reported to date which limit their applicability in a clinical setting. Progress has been made in the development of new ssODNs that could be used to target certain diseases such as Duchenne muscular dystrophy (DMD). The disease is characterized by mutations in the dystrophin gene which lead to a complete absence of dystrophin expression in skeletal muscles of patients. It manifests in the first 5 years of the patient's life and leads to progressive muscle wasting and ultimately death by the age of thirty.

The dystrophin gene is one of largest genes identified to date. Mutations causing DMD have been mapped throughout the gene, rendering the optimization of chemical structures to be used for gene editing strategies a difficult task to accomplish. Oligonucleotide-mediated correction of the dystrophin gene has been successfully demonstrated in mouse models as well as a canine model for DMD [1–6]. The frequencies of gene repair vary widely depending on the base being targeted, the position of the mutation within the dystrophin gene and the structure of the oligonucleotide being used. Imperative in the optimization of an effective therapy for the disease is the use of a model that can efficiently quantitate the levels of gene correction and that can be used to study how different mechanisms of gene repair influence the correction process. The ideal genetic target to be tested with ssODNs is a single point mutation that generates a restriction site absent in the wild-type gene. Correction of the mutation mediated by ssODNs should disrupt the restriction site only in cells that have undergone repair. Restriction DNA analysis of genomic DNA (gDNA) or cDNA (obtained after reverse transcription of the dystrophin mRNA) will eliminate all non-corrected sequences leaving corrected sequences intact. Polymerase chain reaction (PCR) of digested products can then be used to quantitate the frequencies of gene repair and to compare those frequencies to those achieved using different sequences or different chemical structures (Fig. 2).

The *mdx*^{5cv} mouse model for DMD has been instrumental in determining the clinical applicability of ssODN-mediated gene repair for the treatment of DMD in muscle. This model contains a single base substitution in exon 10 of the dystrophin gene that creates a cryptic splice site. As a result, exon 10 is aberrantly spliced 53 base pairs (bp) upstream the normal dystrophin exon 10/intron 10 splice junction causing a complete absence of dystrophin expression. Correction of the *mdx*^{5cv} mutation can be assessed quantitatively at both the gDNA and the mRNA levels, rendering this model particularly appealing to test gene repair in muscle [4, 6].

The focus of this chapter is to provide the tools necessary to asses repair and to guide the execution of the experiments required to test ssODNs in muscle culture in vitro independently of the mutation targeted for repair. The same methods can be used to design and determine the efficacy of ssODNs for the treatment of other genetic disorders in which alteration of a single base can result in the correction of the genetic defect thus restoring normal function of the gene causing the disease. The protocols herein described can also be applied to test and compare other types of gene editing approaches including site specific genetic modifications induced by engineered nucleases such as zinc-finger nucleases (ZFNs), transcription activator-like effector nucleases (TALENs) and meganucleases which have recently emerged as powerful tools to induce repair of gene defects.

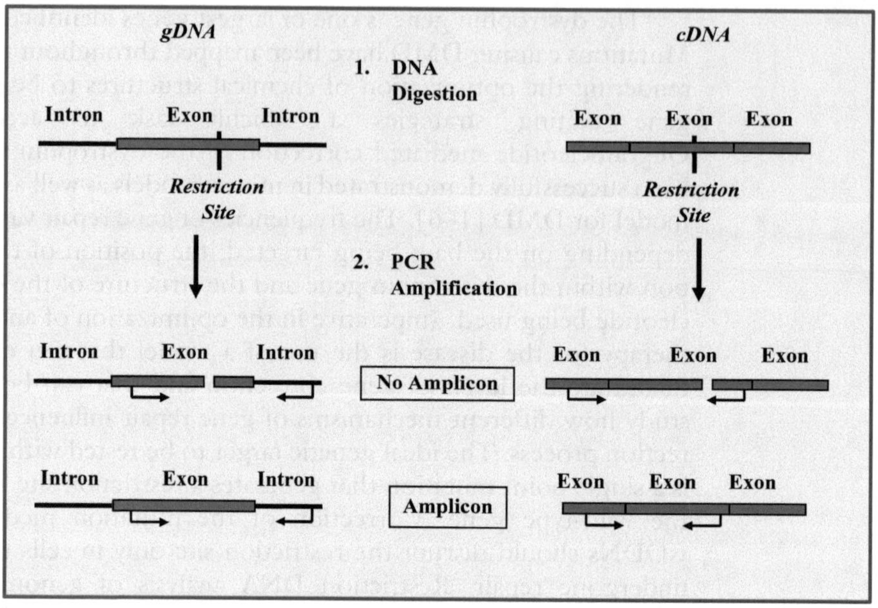

Fig. 2 Strategy used to quantitate the frequencies of gene repair in muscle cells. Correction at the genomic level can be assessed by digestion of gDNA isolated after ssODN treatment followed by a step of amplification performed using primers encompassing the intron/exon sequences surrounding the mutation targeted for repair. Analysis at the mRNA level first requires the mRNA to be reverse-transcribed to generate a cDNA complementary to the mature mRNA sequence. The cDNA is then subjected to restriction DNA to eliminate all non-corrected sequences. Detection of cDNA sequences that have undergone repair is carried out using primers encompassing one or more adjacent exons. Amplification should only be obtained from sequences refractory to endonuclease digestion and as the result of the correction process

2 Materials

2.1 ssODNs Design and Synthesis

ODNs can be purchased from a number of different vendors worldwide. The cost of synthesis varies depending on the structure of the oligonucleotide, its length and the modifications introduced on the ODN (*see* Subheading 3.1).

1. Oligonucleotides.
2. UltraPure™ DNase/RNase-Free Distilled Water Buffer (ultrapure water; Life Technologies).

2.2 Primary Culture of Skeletal Muscle

2.2.1 Isolation of Primary Cells

1. Sterile phosphate-buffered saline (PBS) containing 10 U/mL penicillin, 10 mg/mL streptomycin stored at room temperature.
2. Sterile 0.22 μm syringe filter (Nalgene).
3. Collagenase Type II stock solution: Dissolve 1 g of collagenase Type II (Life Technologies) in 200 mL of PBS, aliquot in 15-mL conical tubes and store at −20 °C. The solution is stable for up to 2 years.

4. Dispase stock solution: Dissolve 1 g powder dispase II (Life Technologies) in 100 mL of PBS to generate a stock solution of 11.7 U/mL. Filter the solution through a PES 500 mL filter, aliquot into 15-mL conical tubes (10 mL/tube) and store aliquots at −20 °C (up to 1 year).
5. Isolation medium: Ham's F-10 supplemented with 10 U/mL penicillin and 10 mg/mL streptomycin.
6. Sterile razor blades.
7. Sterile nylon cell strainers 40 μm pore sizes (BD Falcon).
8. Extra-Cellular Matrix (ECM, Sigma).
9. FGF basic human recombinant (bFGF) stock solution: Resuspend 50 μg in 2 mL PBS + 0.1 % BSA (sterile); aliquot in 50–100 μL tubes and store aliquots in −20 °C (*see* **Note 1**).
10. Microcentrifuge.
11. Sterile laminar flow hood.
12. Water bath set to 37 °C.

2.2.2 Propagation of Primary Cultures

1. Proliferation medium: Nutrient Mixture Ham's F-10 containing 20 % v/v fetal bovine serum, 10 U/mL penicillin, 10 mg/mL streptomycin, 5 ng/mL bFGF.
2. 0.05 % Trypsin-EDTA Dissociation Enzyme with Phenol Red (Life Technologies).
3. Trypsin 10× aliquots, diluted to 1× solution in PBS.
4. PBS tablets, 1× solution stored at 4 °C.
5. Differentiation medium: High Glucose Dulbecco's Modified Eagle's Medium (DMEM) containing 10 % horse serum (HS), 10 U/mL penicillin, 10 mg/mL streptomycin. Store at 4 °C and use within 1 month.
6. Trypan blue solution: 0.4 % Trypan blue in distilled water.
7. Sterile laminar flow hood.
8. Water bath set to 37 °C.
9. Microcentrifuge.
10. Hemocytometer.
11. Inverted microscope.

2.2.3 Preservation and Storage

1. 1-mL cryogenic vials (Nalgene).
2. Liquid nitrogen storage tank (−150 °C freezer).
3. Sterile freezing medium: 90 % calf serum and 10 % dimethyl sulfoxide (DMSO). Prepare freezing medium and immediately store on ice. Unused sterile freezing medium can be stored at 4 °C for up to 4 weeks.

2.3 Myoblasts Transfection and Propagation

2.3.1 Coating Plates for Transfection with ECM

1. Tissue-culture treated polystyrene 6-well plates (BD Falcon™) with flat bottom and low-evaporation lid.
2. ECM.
3. PBS.

2.3.2 Transfection with Oligonucleotides

1. Proliferation medium (see Subheading 2.2.2, item 1).
2. ssODNs stored at −80 °C.
3. Lipofectamine™ 2000 reagent (Life Technologies).
4. Opti-MEM serum-free medium.

2.4 Isolation of Genomic DNA and Quantification

1. Nuclei Lysis Solution (Promega).
2. Ultrapure water.
3. RNase solution: 4 mg/mL RNase A (Sigma) resuspended in ultrapure water.
4. Protein Precipitation Solution (Promega).
5. 70 % ethanol, room temperature (molecular biology grade).
6. Trypsin.
7. PBS.
8. Isopropyl alcohol, room temperature (molecular biology grade).
9. Liquid nitrogen.
10. 1.5-mL microcentrifuge tubes.
11. Water bath set at 95 °C.
12. Water bath set at 37 °C.
13. Vortex mixer.

2.5 RNA Isolation, Quantification, and First-Strand cDNA Synthesis

2.5.1 RNA Isolation

1. Ice-cold PBS.
2. TRIzol™ reagent (Life Technologies).
3. Chloroform (molecular biology grade).
4. DEPC-treated water (Ambion).
5. Isopropyl alcohol (molecular biology grade).
6. 75 % v/v ethanol (molecular biology grade) in DEPC-treated water.
7. 1.5-mL microfuge tubes (autoclaved).
8. Refrigerated centrifuge.
9. Aerosol-barrier tips.
10. Vortex mixer.
11. Microcentrifuge.

2.5.2 RNA Quantitation

1. NanoDrop Spectrophotometer (Thermo Scientific).
2. Agilent Bioanalyzer and RNA 6000 Nano chips for Agilent Bioanalyzer (optional).

2.5.3 Reverse Transcription and First-Strand Synthesis

1. Total RNA.
2. Ultrapure water.
3. 50 μM oligo(dT)20, 2 μM gene specific primer or random hexamers (50 ng/μL).
4. 10 mM dNTP mix.
5. 10× Reverse Transcriptase (RT) buffer.
6. 25 mM $MgCl_2$.
7. 0.1 M DTT.
8. RNAaseOUT or similar solution.
9. SSIII Single Step RT enzyme (Life Technologies).
10. RNase H (Life Technologies).
11. Water bath set at 65 °C.
12. Water bath set at 50 °C.
13. Ice.
14. Water bath set at 85 °C.
15. Water bath set at 37 °C.

2.6 Restriction Enzyme Digestion

1. Ultrapure water.
2. Purified genomic DNA.
3. Restriction enzyme.
4. 1× Enzyme buffer.
5. Vortex mixer.
6. Water bath set at 37 °C.
7. Water bath set at 65 °C for restriction enzyme inactivation (optional).

2.7 Quantitative PCR of Gene Correction Frequencies

1. Gene specific primers. All the primers are desalted and both UV absorbance and capillary electrophoresis are used to assess the quality of primer synthesis. Select the appropriate internal control gene for your experiment. Possible examples include 18S rRNA, 7S rRNA, U6 RNA, β-actin, or GAPDH.
2. Purified gDNA or cDNA (see Subheading 2.4 or 2.5 respectively).
3. SYBR Green PCR master mix, 200 reactions (Applied Biosystems) containing optimized amount of DNA polymerase, dNTP, reaction buffer, and dyes.
4. 96-well Optical Reaction Plate.

5. Optical Adhesive Covers.
6. DNA ladder (1 kb to 100 bp).
7. Real-Time PCR system.
8. Software for real-time PCR to analyze results.
9. Ice to maintain samples and reagents cold.

2.8 DNA Fractionation and Purification

2.8.1 Agarose Gel Fractionation of PCR Products

1. Agarose powder.
2. 50× TAE buffer: 24.2 % w/w Tris base, 0.0571 % v/v glacial acetic acid, 0.05 M EDTA (pH 8.0) in water.
3. 5× Loading buffer: 15 % w/w Ficoll (type 400), 0.1 % bromophenol blue, 0.15 % xylene blue in distilled water.
4. DNA size standards (100 bp, 1 kb).
5. High-voltage power supply.
6. Ethidium bromide.
7. UV transilluminator.
8. Gel Documentation Imager (gel-dock) or equivalent system suitable for the acquisition of gel images.

2.8.2 Purification of PCR Products from Agarose

1. Sterile scalpel to excise DNA bands.
2. QIAquick® Extraction Kit (Qiagen).
3. Water bath 55 °C.
4. Isopropyl alcohol.
5. Ethanol.
6. Vortex mixer.
7. Centrifuge.

3 Methods

3.1 ssODN Design and Synthesis

Correcting ssODNs can be homologous or complementary to the region of the gDNA targeted for repair. Whenever possible, analyses should be performed using oligonucleotides targeting both strands to account for possible strand bias between the regions of the gene targeted for repair. Alternatively, the best option is to test ssODNs targeting the noncoding strand of the gene since they have shown to be highly specific and do not interfere with mRNA transcription and processing in vivo [6]. Finally, ssODNs perfectly homologous to the targeting oligonucleotides but lacking the mismatch should also be included as controls.

Regular ssODNs are made of unmodified bases flanked by a CGCG repeat of phosphorothioate bases which are added to increase its intracellular stability toward endonucleases and exonucleases (Fig. 1). Consideration should be given to the stability of the oligonucleotides and their specific properties. ODNs made of unmodified bases require at least 40–45 bases in length to show effects (Fig. 1). Peptide nucleic acids (PNA) consist of nucleic acid bases attached to an achiral peptide backbone that is made up of *N*-(2-aminoethyl) glycine units. PNA have a higher binding affinity for DNA and show greater stringency in hybridization than DNA. As a result, PNA sequences of 18 bp or less are sufficient to form stable bonds with their target and to activate the repair process (Fig. 1) [7]. The mutating base is usually positioned within the 5 bp encompassing the middle sequence of the ssODN. A Cy3 can be incorporated at the end of the ssODN to follow its cellular uptake and distribution into cells upon delivery.

Critical in the design of the ssODN is to chose a sequence capable of hybridizing to the targeted sequence but not to other sequences in the gDNA, as this may result in nonspecific effects and a decrease in correction frequencies. The most widely used approach is to implement sequence alignment algorithms selection programs that use Basic Local Alignment Search Tool (BLAST). Each ssODN that meets the required level of sequence specificity is then evaluated further using MFOLD to assess its self-complementarity and the degree of secondary structure that it can form [8, 9].

Oligonucleotides produced by an automated DNA synthesizer need to be purified prior to use as they often contain improperly synthesized oligonucleotides and incomplete sequence products. The optimal purification method is ion exchange (IE) or reverse phase (RP) HPLC which can be performed by the vendor upon request. The purity and integrity of the ssODN can be confirmed by denaturing polyacrylamide-urea gel electrophoresis (PAGE) prior to commencing the tissue culture experiments.

ssODNs are usually shipped in lyophilized form and, in this form, are stable at room temperature for several days. Upon arrival, ssODNs can be resuspended in UltraPure™ DNase/RNase-Free Distilled Water Buffer (ultrapure water; Life Technologies) at a concentration of 150 pmol/μL, dispensed into aliquots, and lyophilized using a SpeedVac. Aliquots can be stored at −20 or −80 °C for several months.

3.2 Myogenic Cell Propagation and Storage

Myoblasts are prepared using the protocol first described by Rando and Blau with minor modifications [10]. All steps are performed in a sterile laminar flow hood using sterile tissue culture technique.

1. Coated plates: Dilute ECM (Sigma) 1:500 in PBS, aliquot 4 mL into sterile 100-mm tissue culture-treated plastic dishes and rock gently at room temperature for 8–14 h.
2. Dissect limb muscle from 1 to 4 days old mice and place into 100-mm dishes containing penicillin and streptomycin. Make sure to remove small pieces of skin and fat from muscles as those could potentially contaminate the culture and reduce its purity.

3. Using the razor blades, mince the muscle to very fine pieces and transfer to a 50-mL Falcon tube containing 15 mL isolation medium.
4. Add Collagenase (1:40) and dispase (1:20) to the 20 mL (final volume) solution and incubate at 37 °C for 30 min with moderate agitation.
5. Let the undigested tissue settle at the bottom of the tube for 10 min at room temperature.
6. Filter the supernatant through a cell strainer and spin the sample at 400×*g* for 5 min. Discard the supernatant and resuspend the pellet in 10 mL of proliferation medium.
7. Add 20 mL of isolation medium and repeat **steps 3** through **5**. Plate the cells on a new dish containing proliferation medium.
8. Change the medium after 24 h making sure to add bFGF to prevent cells from differentiating into myocites.
9. On day 3 (48 h after cell isolation), trypsinize the cells and preplate on a non-coated dish for 30 min. Follow the adherence of the cells under the microscope. The majority of the fibroblasts should adhere to the plastic, leaving the myoblasts in suspension. Collect the medium and transfer to a coated plate.
10. Allow the cells to grow for an additional 3 days without changing the medium. Cells will grow rapidly and should divide every 18–20 h. Make sure to supplement the medium with bFGF (2.5 ng/mL final concentration) every 24 h.
11. Monitor the cells every 24 h. Additional pre-plating may be necessary during the following days depending on the number of fibroblasts in the culture. Cells should be passaged every 5 days onto a new ECM-coated dish. Maintain the cells at a low confluency (50–60 %) to prevent differentiation and expand the culture by splitting the cells into different plates (*see* **Note 2**).
12. Freeze the cells at a concentration of 10^6 cells/mL in ice-cold freezing medium. Store cryovials at −80 °C for 2–3 days, then transfer them to −150 °C where they can be permanently stored until necessary.
13. When required, remove cryovials from liquid nitrogen storage and place on dry ice. Then, thaw the vials and place directly into 37 °C water to ensure rapid thawing of the cells and maximal viability. Once thawed, the cells should be transferred to a 15-mL tube with 10 mL of wash medium added. This should be done as quickly as possible to minimize any toxicity due to exposure to DMSO.
14. Pellet the cells by centrifugation at 400×*g* for 5 min, aspirate supernatant, and resuspend in 10 mL of pre-warmed proliferation medium (*see* **Note 3**).

15. Transfer 10 mL of cell solution to each 100-mm plate and incubate at 37 °C, 5 % CO_2. Check the cells the next day to ensure that they are still viable and are adherent to the base of the flask. Continue to incubate until 60–80 % confluence and passage as required.

3.3 Myoblast Transfection and Propagation

Passage number may affect transfection experiments and will limit the time cells can be maintained in culture prior to analysis of gene correction frequencies. Use cells that are at passage 8 or lower after isolation or restart the culture using a new vial of cells stored in liquid nitrogen.

1. Coat a 6-well plate with 2 mL of a solution containing ECM in PBS (1:500 v/v) as described in Subheading 3.2, **step 1**.
2. Trypsinize the cells and resuspend them in proliferation medium at a concentration of 1×10^4 cells/mL of medium.
3. Plate 2 mL per well in a 6-well plate, incubate cells at 37 °C, 5 % CO_2 and allow to attach for 10–12 h.
4. Aspirate the medium and replace with pre-warmed proliferation medium immediately before transfection. Return plates into the incubator at 37 °C, 5 % CO_2.
5. Add 5 μL of ssODN to 250 μL of Opti-MEM in a 1.5-mL microcentrifuge tube.
6. Prepare sufficient transfection reagent by combining 250 μL of Opti-MEM with 5 μL of Lipofectamine™ 2000 per each well to be transfected. Keep Lipofectamine™ 2000 on ice while preparing the transfection reagent to minimize loss of transfection efficiencies over time.
7. Add 250 μL of transfection reagent to the 1.5-mL microcentrifuge tube containing the ssODN and Opti-MEM. Mix gently and incubate at room temperature for 15–20 min (*see* **Note 4**).
8. Pipette the solution directly into the appropriate well containing cells and mix gently.
9. Incubate at 37 °C, 5 % CO_2 overnight.
10. To stop the transfection aspirate the medium and replace with warm proliferation medium.
11. Allow cells to grow and expand by changing medium every 20–24 h.
12. Split cells when 60–70 % confluent as described above (*see* Subheading 3.2). Cells can be maintained in culture for several weeks. Each transfection will give enough cells to perform analysis at both the DNA and mRNA levels in triplicate experiments. Analysis can be performed as early as 1–2 weeks after transfection (*see* **Note 5**).

13. If analyzing mRNA, seed 2.5×10^5 cells per well in a 6-well plate in Ham's F-10 containing penicillin and streptomycin but no bFGF and switch to differentiation medium 24 h later (cells at this stage should be 80–90 % confluent and should appear elongated, an indication that they have begun to fuse into myotubes).

3.4 Isolation of Genomic DNA

The Wizard® Genomic DNA Purification Kit is a rapid and convenient method to purify DNA from mammalian cells. The kit offers a variety of reagents that can be purchased individually depending on the need of the investigators. The purification follows four major steps: (a) lysis of cells and nuclei, (b) RNase digestion, (c) protein precipitation, and (d) precipitation and purification of the high molecular weight genomic DNA in solution. DNA purified using this system is suitable for a variety of applications including amplifications and restriction endonuclease analysis.

1. Trypsinize the cells and harvest them by centrifugation at $400 \times g$ for 5 min.
2. Resuspend in 500 μL of PBS and transfer them to a 1.5-mL microcentrifuge tube.
3. Centrifuge at $20{,}000 \times g$ for 10 s to pellet the cells.
4. Remove the supernatant, leaving behind the cell pellet plus 10–50 μL of residual liquid.
5. Add 200 μL PBS to wash the cells. Centrifuge as in **step 3**, and remove the PBS. Vortex vigorously to resuspend cells.
6. Add 600 μL of Nuclei Lysis Solution, and pipet to lyse the cells. Pipet until no visible cell clumps remain. Perform four consecutive freeze–thaw steps by submerging the tubes in liquid nitrogen for 5 min followed by heating at 95 °C for 5 min.
7. Add 3 μL of RNase H solution to the lysate and mix the sample by inverting the tube several times. Incubate the mixture for 30 min at 37 °C.
8. Allow sample to cool at room temperature for 5 min.
9. Add 200 μL of Protein Precipitation Solution and vortex vigorously at high speed for 30 s.
10. Chill sample on ice for 5 min.
11. Centrifuge at $12{,}000 \times g$ for 4 min to precipitate proteins.
12. Transfer the supernatant containing the DNA to a clean 1.5-mL microcentrifuge tube containing 600 μL of room-temperature isopropyl alcohol. Mix the samples immediately by inverting the tubes 30–50 times or until the DNA becomes visible.
13. Centrifuge for 15 min at $20{,}000 \times g$ at room temperature.
14. Remove the supernatant.

15. Add 600 μL of room-temperature 70 % ethanol to the pellet and invert the tube several times to wash the DNA.
16. Centrifuge at 20,000 × *g* for 15 min at room temperature.
17. Carefully aspirate the ethanol using a pipette. The DNA pellet may be loose and attention needs to be paid to avoid aspirating the pellet into the pipette.
18. Invert the tube on clean absorbent paper, and air-dry the pellet for 10–15 min.
19. Add 50 μL of ultrapure water and rehydrate the DNA by incubating at 4 °C overnight.
20. Store the DNA at −20 °C.

3.5 Isolation of RNA Quantification and Synthesis of First-Strand cDNA

3.5.1 Isolation of RNA

RNA is isolated in cells maintained in differentiation medium for 72 h using TRIzol™ reagent [4, 6]. Always wear gloves and eye protection and avoid contact with skin or clothing. Procedures should be performed in a chemical hood to avoid airway exposure to toxic fumes [11].

1. Trypsinize the cells and harvest them by centrifugation at 400 × *g* for 5 min.
2. Resuspend in 1 mL of PBS and transfer them to a 1.5-mL microcentrifuge tube.
3. Centrifuge at 20,000 × *g* for 10 s to pellet the cells.
4. Remove the supernatant and add 1 mL of TRIzol Reagent. Pass the cell lysate several times through a pipette.
5. Vortex for 30 s and incubate samples for 5 min at room temperature.
6. Add 200 μL of chloroform per 1 mL of TRIzol Reagent. Vortex samples vigorously for 30 s and incubate them at room temperature for 30 min.
7. Centrifuge the samples at 10,000 × *g* for 15 min at 4 °C.
8. Transfer the upper aqueous phase containing the RNA into a fresh tube leaving behind 40–50 μL of aqueous phase behind. Avoid aspirating the interphase or the phenol–chloroform phase as this results in contamination of the RNA preparation.
9. Add 500 μL of isopropanol and vortex samples for 5 s.
10. Incubate at room temperature for 10 min and centrifuge at 10,000 × *g* for 10 min at 4 °C. The RNA should be visible at the bottom of the tube.
11. Aspirate the supernatant and wash the RNA pellet with 1 mL of 75 % ethanol.
12. Vortex the samples for 10 s and centrifuge at 10,000 × *g* for 20 min at 4 °C.

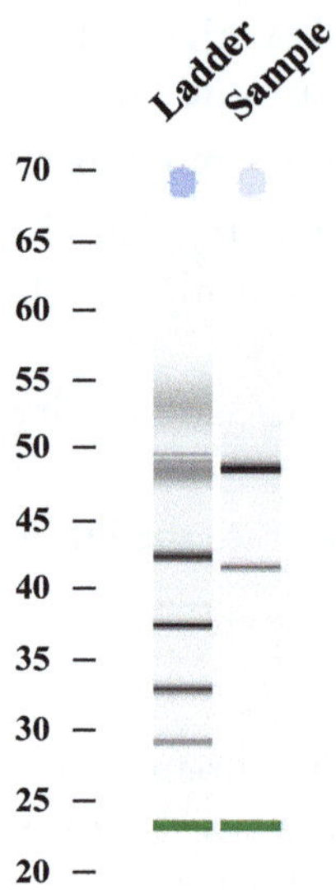

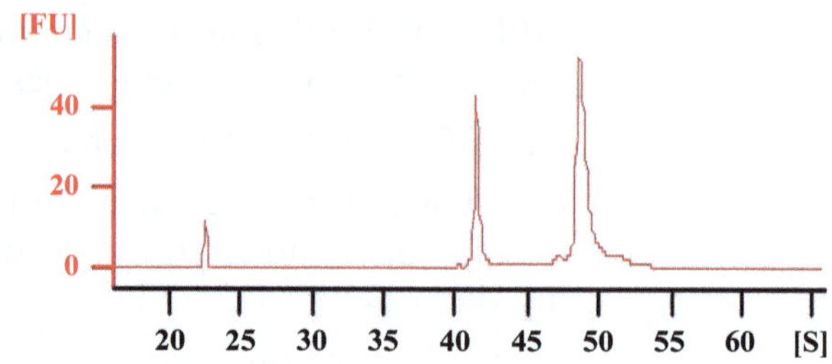

Overall Results for sample

RNA Area:	181.2
RNA Concentration:	179 ng/μl
rRNA Ratio [28s / 18s]:	2.2
RNA Integrity Number (RIN):	10 (B.02.07)
Result Flagging Label:	RIN:10

Fragment table for sample

Name	End Time [S]	Start Time [S]	Area	Total Area (%)
18S	40.58	42.65	38.0	21.0
28S	47.63	50.82	81.8	45.1

Fig. 3 Purity of RNA following isolation. Electropherogram for total RNA isolated from mouse myotubes maintained in differentiation medium for 72 h. A high-quality total RNA sample should contain two peaks corresponding to the 28S and 18S ribosomal RNAs. The third peak of smaller molecular weight shown in *green* corresponds to small RNAs. The ratio for the two peaks should be of approximately 2:1 for the 28S to 18S bands respectively. The version of the Agilent software used to check the sample is shown in parenthesis next to the RNA Integrity Number (RIN)

13. Aspirate the supernatant and invert the tube on clean absorbent paper.
14. Air-dry the pellet for 10–15 min. Do not let the RNA pellet dry completely as this will greatly decrease its solubility.
15. Resuspend samples in 50 μL of ultrapure water.
16. Store the RNA at −80 °C.

3.5.2 RNA Quantitation and Quality Check

1. Quantify 1 μL of each RNA sample using a NanoDrop spectrophotometer on the RNA-40 settings using water as a blank reference. The A260/A280 ratio should be above 1.8 (*see* **Note 6**).
2. The quality of RNA samples can be checked using the Agilent Bioanalyzer. When this is not available, a simple electrophoresis on a 2 % agarose gel is sufficient. RNA run on a gel should reveal the presence of two major bands corresponding to 18S and 28S ribosomal RNAs. The ratio of 28S/18S should be approximately 2 for intact RNA. Samples run on the Agilent Bioanalyzer should reveal the presence of a third peak of smaller molecular weight, corresponding to small RNAs (Fig. 3). RNA quality is based on the RNA integrity number (RIN) which is computed by the Agilent software. An RNA sample of good quality should have a RIN value of 7 or greater.

3.5.3 Synthesis of First-Strand cDNA

The amount of total RNA used as starting material can range from 2 μg up to 5 μg of total RNA per reaction. Synthesis can be scaled up as needed to up to five reactions in the same tube. The first-strand cDNA synthesis reaction can be primed using oligo(dT)$_{20}$, gene specific primers or random hexamers (*see* **Note 7**).

1. In microcentrifuge tubes, aliquot 2 μg of total RNA for each of the samples being analyzed.
2. Add 1 μL of a 50 μM solution of oligo(dT)$_{20}$ (or 1 μL of a 2 μM solution of reverse primer complementary to the 3′ region of the gene to be analyzed or 50 ng of random hexamers) and 1 μL of a 10 mM dNTP solution.
3. Bring the final volume in each tube to 10 μL.
4. Heat at 65 °C for 5 min and place on ice.
5. Prepare a cDNA synthesis master mix. For each of the reactions, add 2 μL 10× RT buffer, 4 μL of a 25 mM $MgCl_2$ solution, 2 μL of 0.1 M DTT solution, 1 μL RNase OUT (40 U/μL), and 1 μL SuperScript. III RT (200 U/μL).
6. Mix and centrifuge at 20,000 × *g* for 10 s to collect the solution at the bottom of the tube.
7. Add 10 μL of cDNA synthesis mix to the tube containing the RNA and the primer mixture.
8. Mix gently and centrifuge briefly to collect the solution as indicated above (**step 6**).
9. Incubate at 50 °C for 50 min.
10. Terminate the synthesis of cDNA by incubating at 85 °C for 5 min.
11. Chill on ice.
12. Centrifuge briefly and add 1 μL of RNase H to each tube.
13. Incubate at 37 °C for 20 min.
14. Store at −20 °C.

3.6 Restriction Enzyme Digestion

3.6.1 Digestion of Genomic DNA

A minimum of 250 ng DNA per reaction is recommended for any application looking at identifying rare correction events. DNA should be free of contaminants, such as phenol, chloroform, ethanol, detergents, or high salt concentrations, as these may interfere with restriction endonuclease activity. 1 U of restriction endonuclease completely digests 1 μg of substrate DNA in 1 h. However, genomic DNA generally requires more than 1 U/μg to be digested completely. It is recommended to add a tenfold excess of enzyme to the reactions in order to ensure complete digestion. Make sure that the restriction enzyme does not exceed more than 10 % of the total reaction volume, otherwise the glycerol, in which the enzyme is supplied with, may inhibit digestion.

Digestions are carried out in a volume comprised between 10 and 50 μL. Reaction volumes smaller than 10 μL are susceptible to pipetting errors and are not recommended.

1. Pipet reaction components into a microcentrifuge tube, vortex, and centrifuge at 20,000 × g for 10 s. The enzyme should be added last and kept on ice when not in the freezer. A reaction master mix consisting of water, buffer, and enzyme should be used when setting up a large number of digestions.
2. Centrifuge the tube briefly to collect the liquid at the bottom of the tube.
3. Incubate the reaction in a water bath, usually for 3–8 h at 37 °C.
4. Restriction enzymes can be inactivated by incubating the reactions at 65 °C for 20 min (optional).

3.6.2 Digestion of First-Strand cDNA

1. Combine 20 μL of cDNA with 2.5 μL of 10× buffer in a microcentrifuge tube.
2. Vortex briefly and spin down at top speed for 10 s.
3. Add 2.5 μL of restriction enzyme.
4. Incubate at 37 °C as indicated above (see Subheading 3.6.1, items 3 and 4).
5. Use 2.5 μL for each Q-PCR reaction.

3.7 Quantitative Analysis of Corrected Sequences

3.7.1 Amplification and Detection of PCR Products

Real-time quantitative polymerase chain reaction (qPCR) can precisely and accurately determine the frequencies of gene correction among samples treated with ssODNs. Care must be taken to avoid amplification of nonspecific products which could result in misinterpretation of the results. Typical amplification products should range between 100 and 300 bp in length. Amplification of gDNA should be carried out using primers encompassing intron/exon sequences to avoid amplification of pseudogenes and to ensure the specificity of the amplification product. Similarly, amplification of reverse transcribed mRNA genes should be carried out using primers that span at least two exons in order to avoid amplification of contaminant gDNA that could be present in the reactions (Fig. 2). Primers need to be optimized to ensure that they give low primer dimer formation (*see* **Note 8**).

SYBR Green I dye and TaqMan® hydrolysis probe are the most common fluorescent chemistries used to detect and quantitate amplification of specific genes in real time. The SYBR Green I method is cost effective because it does not require a fluorescently labeled probe/primer and is highly sensitive. The protocol described below is intended for use with SYBR Green I. A specialized thermo cycler with fluorescence detection modules is used to monitor and record fluorescence as amplification occurs (Fig. 4).

1. Calculate the number of reactions to be run for each of the specific genes (including the no-template controls) in the experimental conditions being tested. Each sample should be amplified in triplicate.
2. Prepare a PCR master mix in 1.5-mL microcentrifuge tubes for each of the genes to be amplified including the reference gene used as the internal standard. Combine 12.5 μL SYBR Green PCR Master, 5.5 μL of ultrapure water, 1 μL forward primer (50 μM solution), and 1 μL reverse primer (50 μM solution).
3. Mix the cocktail to obtain a homogenous solution.
4. Centrifuge at top speed for 5–10 s.
5. Aliquot 20 μL of the cocktail to each well.
6. Add 5 μL of DNA (or ultrapure water for the no-template controls) into the appropriate well.
7. Cover reaction plate with an adhesive cover.
8. Centrifuge the plate at 700 × *g* for 2 min to collect the liquid at the bottom.
9. Position the plate in the thermal cycler and begin the run according to the manufacturer's instructions (*see* **Note 9**).
10. When the run is completed, remove the plate from the instrument and store at −20 °C.

3.7.2 Data Analysis and Amplicon Quantification

After the qPCR amplifications are finished, software packages will provide the relative fluorescent units (RFU) and the threshold cycles (C_T) values (Fig. 4a). Quantification is based on the C_T which is defined as the number of cycles required for the fluorescence intensity to exceed a threshold set as the background level. More simply, this is the cycle where the PCR system starts to detect the exponential growth of the amplicons.

Gene repair can be assessed using the relative quantification method which determines the ratio between two test samples (samples A and B). This method can provide the relative efficiency of gene repair achieved using different ssODN chemistries and is particularly useful when comparing the gene correction frequencies achieved by the ssODN in dose response studies. The method can also estimate gene correction frequencies of the ssODN by comparing the C_T of the ssODN-treated samples to that of a sample containing known amounts of corrected template. Alternatively, gene correction frequencies in samples treated with targeting ssODNs can be determined using the absolute quantification method [12]. This method requires a standard curve generated using known concentrations of template with the sequence resulting after ssODN-mediated gene conversion. In the case of single base pair substitutions that are designed to result in a sequence identical to that of wild-type, the standard curve can be generated

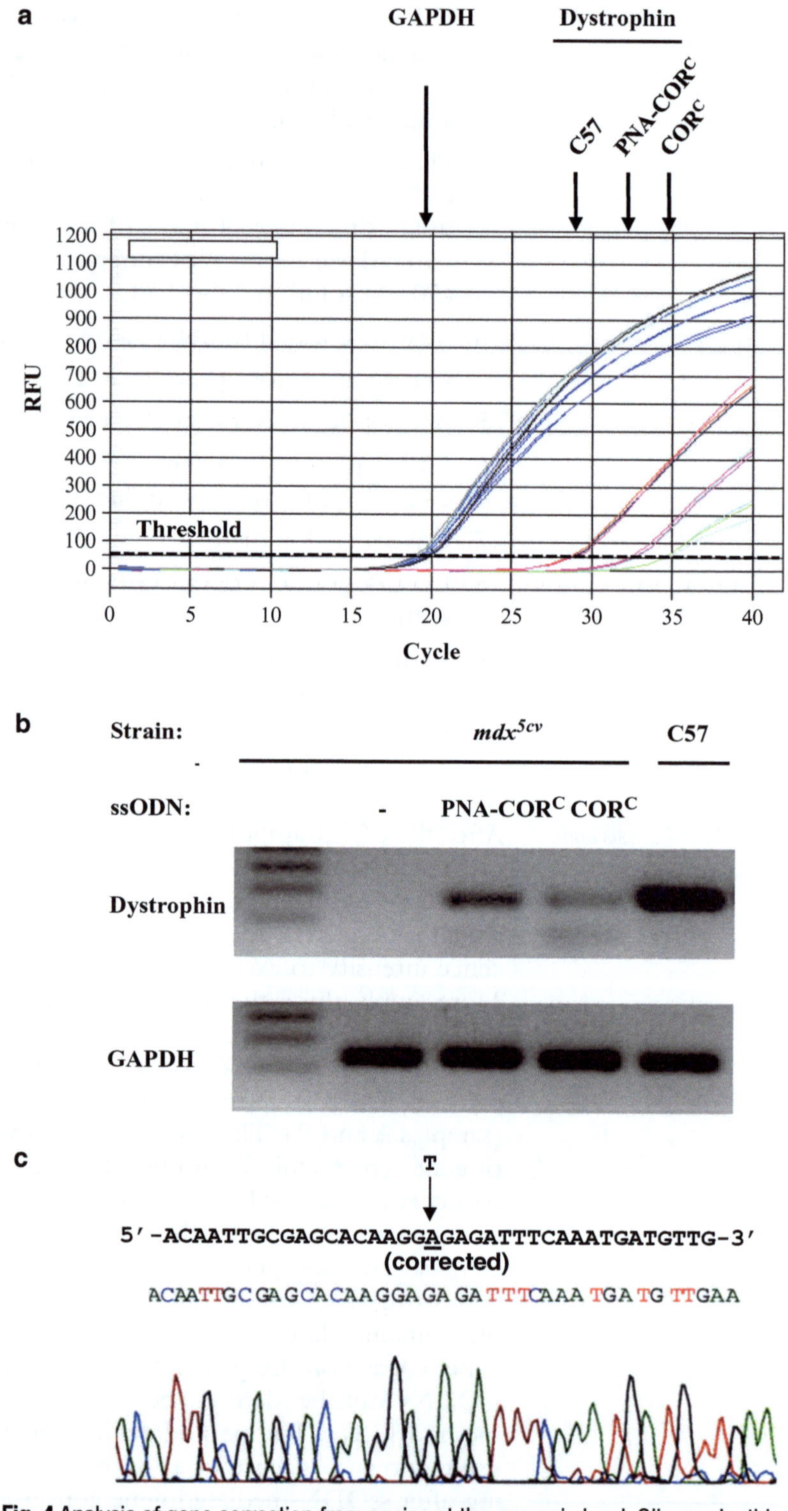

Fig. 4 Analysis of gene correction frequencies at the genomic level. Oligonucleotides designed to target and correct the *mdx*^{5cv} mutation (COR^C and PNA-COR^C) were transfected in *mdx*^{5cv} myoblasts and gDNA was isolated 2 weeks after transfection,

by using template isolated from C57BL/10 or control myotubes (*see* **Note 10**). Corrected and non-corrected template should be mixed in order to maintain the final concentrations of DNA in each test tube constant. If the analysis is performed at the mRNA level, the standard curve should be prepared prior to reverse transcription and first-strand cDNA synthesis (Fig. 5). Restriction enzyme digestion should then be used to cut the non-corrected sequence prior to amplification (*see* **Note 11**).

Different relative quantification methods are available [13]; each has advantages and disadvantages. The Livak method, also known as $2^{-\Delta\Delta C_T}$ method, is one of the easiest and perhaps the most used [14]. This method assumes that both target and reference genes are amplified at similar efficiencies. It is important to determine the amplification efficiencies of the target and the reference gene (*see* Subheading 3.7.2.1).

Quantities obtained from a qPCR experiment must first be normalized to an internal standard or reference gene that is used to control for the amount of sample present in the reaction.

1. Normalize the C_T of the gene of interest (GOI) to the C_T of the internal standard (std) using the formula:

$$\Delta C_T(\textit{sample } \mathrm{A}) = C_T(\textit{GOI},\mathrm{A}) - C_T(\textit{std},\mathrm{A})$$

$$\Delta C_T(\textit{sample } \mathrm{B}) = C_T(\textit{GOI},\mathrm{B}) - C_T(\textit{std},\mathrm{B})$$

where A is the sample containing known levels of corrected sequences (or calibrator) and B is the new sample to be tested for comparison (test sample).

2. Normalize the ΔC_T of the test to the ΔC_T of the calibrator as follow:

$$\Delta C_T = \Delta C_T(\textit{sample } \mathrm{B}) - \Delta C_T(\textit{sample } \mathrm{A})$$

Fig. 4 (continued) digested with the restriction enzyme *HphI* that is present in DNA of cells refractory to correction and then analyzed by qPCR. The region of the dystrophin gene targeted for repair was amplified using a forward primer homologous to a sequence of exon 10 upstream the mutation and a reverse primer complementary to a portion of the dystrophin intron 10. GAPDH was used as an internal standard and for normalization. (**a**) Results were plotted as the relative fluorescence units (RFU) versus the number of PCR cycles. SYBR green was used as the fluorophore for monitoring and quantifying the amplification reaction. The point of intersection between the amplification curve of each amplicon and the threshold line (*dotted line*) is indicated by an *arrow* and corresponds to the C_T. (**b**) A specific PCR product of identical molecular weight to that from wild-type cells was obtained only in cells treated with targeting oligonucleotides but not in untransfected cells. (**c**) Direct sequencing of the amplicons confirmed that the correction had occurred at the genomic level to reverse the A-to-T mutation of the mdx^{5cv} strain

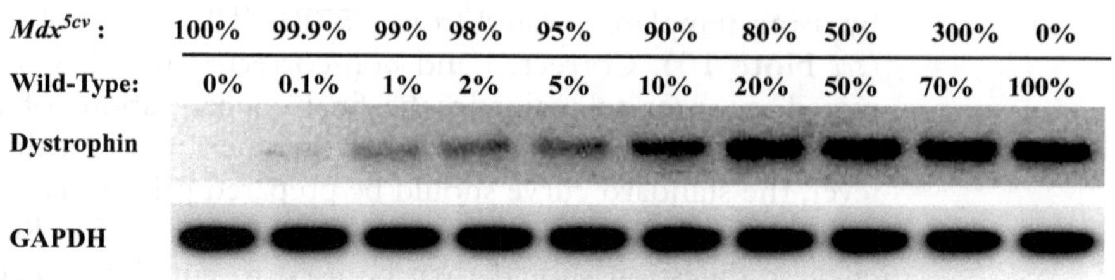

Fig. 5 Dystrophin expression in muscle cells. Example of a standard curve generated by mixing different amounts of total mRNA isolated from wild-type and *mdx*^{5cv} myotubes and PCR products were resolved on agarose gel. No amplification should be detected in untreated cells. All amplicons were normalized to GAPDH, which was used as an internal standard

3. Fold differences are calculated using the following formula:

$$Ratio\ (fold\ differences) = 2^{-(\Delta\Delta C_T)}$$

Calculating the Amplification Efficiency of the Target and the Reference Gene

1. Make tenfold serial dilutions of template and amplify the DNA using primers to both the gene of interest and internal control.
2. Plot the CT (y-axis) versus log cDNA dilution (x-axis) and determine the slope of the line [15].
3. Calculate the PCR efficiency using the equation:

$$E = 10^{-1/slope}$$

The PCR efficiency is normally expressed as a percentage:

$$\%Efficiency = (E - 1) \times 100\%$$

Efficiency close to 100 % is the best indicator of a robust and reproducible assay (*see* **Note 12**).

3.8 Fractionation, Purification and Analysis of PCR Products

3.8.1 Agarose Gel Electrophoresis

Agarose gel electrophoresis is the most common and easiest way to fractionate DNA fragments. The percentage of agarose to be used strictly depends on the size of the DNA fragment that needs to be purified and the application of the DNA following purification (e.g., sequencing and/or cloning). Usually, small molecular weight fragments (100–300 bp) are run and visualized on a 2 % w/v agarose gel. However, at this concentration, there is a risk of carryover of agarose and other impurities which may inhibit downstream reactions. Utilizing a 1.5 % agarose gel will greatly improve the purity and quality of the DNA purified after electrophoresis and to be used for the subsequent step of sequencing, without affecting the resolution of the PCR products analyzed.

Bands are visualized in an ethidium bromide stained gel on a UV light-box (a trans-illuminator), excised, purified and then sequenced to confirm the specificity of the amplification product and the accuracy of the quantification method (*see* **Note 13**).

1. Prepare the gel by weighing 6 g of agarose powder into 400 mL of 1× TAE buffer.
2. Heat the solution in a microwave for 2 min or until the powder is completely dissolved.
3. Let the solution stand for a few minutes to cool down and pour into a tray of appropriate size.
4. Allow the gel to polymerize for approximately 1 h.
5. Place the gel in a running tank and fill with 1× TAE buffer.
6. Add 7 μL of Loading buffer to the qPCR samples, mix and centrifuge at 20,000 × g for 5 s.
7. Load samples into the agarose gel.
8. Electrophorese at 100 V for 60 min.
9. After electrophoresis, stain the gel with ethidium bromide solution and photograph the gel using a gel-dock imaging system.

3.8.2 Purification and Analysis of DNA Fragments

Different methods are available for extracting DNA bands from an agarose gel. The QIAquick® purification system is a rapid and convenient system capable of removing most of the impurities while maintaining an optimal recovery of DNA. Alternatives are available and the purification system can be substituted with other methods of choice depending on the consumables available in the laboratory (*see* **Note 14**).

Correction is confirmed through sequencing of purified products and can be performed in house or samples can be shipped at room temperature to specialized facilities (recommended).

1. Excise the gel slice containing the DNA band with a clean, sharp scalpel. Remove excess agarose to ensure maximal DNA recovery of PCR products.
2. Weigh the gel slice in a 1.5-mL microcentrifuge tube and add 600 μL of Buffer QG.
3. Incubate at 50 °C for 20 min (or until the gel slice has completely dissolved), vortexing the tube every 3 min during the incubation to help dissolve the gel.
4. Add 300 μL of isopropanol to the sample and mix by vortexing for 15 s.
5. Follow the manufacturer's recommendations for the purification of DNA using the QIAquick® spin columns and the reagents provided in the kit (*see* **Note 15**).
6. To collect the DNA, place each QIAquick® spin column into a clean 1.5-mL microcentrifuge tube.

7. Add 30 μL of ultrapure water to the center of the column and centrifuge the column for 1 min (*see* **Note 16**).
8. Quantify 1 μL of each sample using a NanoDrop spectrophotometer on the DNA-50 settings using water as a blank reference.
9. Use 10–15 μL for sequencing and store the remaining solution at −20 °C.

4 Notes

1. Repeated freeze–thaw cycles avoided of bFGF should be as this results in a loss of activity. Once the aliquot has defrosted, the working tube can be stored at 4 °C for up to 1 month.
2. Lipofectamine™ 2000 can form stable complexes with DNA or RNA within 5 min of incubation at room temperature. Incubation for longer periods of time (up to 30 min) will not affect the transfection efficiency of myoblast culture. If handling a large number of samples, it is preferable to allow longer periods of incubation to minimize variability among independent experiments.
3. The quality of the preparation can be checked by plating some of the cells on 35-mm or 60-mm dishes. Allow the cells to reach 80–90 % confluence and switch the proliferating medium to fusion medium. Cells should start to elongate within the first 12 h and myotubes containing two or more nuclei should appear within the first 24 h. Myotubes should continue to grow over time. Cells can be maintained in differentiation medium for up to 5 days. However, by day 4 in differentiation medium, some myotubes will begin to die and lift off the plate.
4. If the vial contains a higher number of cells, they will need to be resuspended in a larger volume to give approximately 5×10^5 cells per 10 mL of medium.
5. Oligonucleotides can anneal to their complementary mRNA and interfere with transcription processes. Allowing the cells to replicate after transfection ensures that the results obtained are due exclusively to the correction process. The rate of cell division in myoblasts is between 18 and 20 h. At each cell division, ssODNs should distribute evenly among daughter cells and should gradually diminish over time. When combined with the degradation process of nucleic acids occurring naturally in cells, the number of ssODN molecules left into myoblasts should become negligible by 2 weeks after transfection.
6. The spectrophotometer can provide indications of the purity of the RNA. Proteins or phenol contaminations in the sample will result in a reduction of the A_{260}/A_{280} ratio. A pure sample should have a ratio of 2, but a ratio above 1.7 is generally acceptable.

7. Priming with oligo$(dT)_{20}$ ensures that only mature mRNA are reverse transcribed and therefore is the method of choice in most RT-PCR applications. However, most reverse transcriptases (including Superscript III) have limitations in the length of cDNA that they can produce. In cases where the coding sequence of the gene being analyzed is larger than 1.6 kb, as in the case of full-length dystrophin transcripts, the best option is to use a primer specific to the region of the gene being analyzed or to use random hexamers. If random hexamers are used, the ratio of random hexamers to RNA needs to be 5:1.
8. A number of different tools are available online that can facilitate the design and selection of primers. Primer Express v2.0 from ABI, Vector NTI, or free Web-based software like PrimerQuest on IDTdna.com and Primer3 (http://fokker.wi.mit.edu/primer3/input.htm) are few examples. It is preferable to design three forward and three reverse primers encompassing the region of the DNA to be analyzed and to test all nine possible combinations by qPCR. The set of forward and reverse primers that gives the lowest primer dimer formation, as determined by the C_T detected in the no-template control well, while maintaining optimal amplification of the target gene is selected for all subsequent experiments. Finally, ensure that the region of the reference gene to be amplified as the internal control does not contain the restriction site used to detect correction of the gene of interest, as this will prevent amplification and will produce no results.
9. A melt curve analysis should always be performed to verify the specificity of product formation, as it can help discriminate between primer dimers and multiple bands amplified in the samples. It is generated by increasing the temperature of each sample obtained at the end of the run by small increments and by monitoring the fluorescent signal at each step. As the temperature increases, the PCR products are denatured (melt), resulting in a decrease in fluorescence. The fluorescence is measured continuously and when the melting temperature (T_m) of a particular double-stranded DNA product is reached, is detected by the instrument and plotted as a negative first derivative of fluorescence intensity as a function of temperature. Products of different sizes will appear on the plot as distinct fluorescent peaks of different T_m. Consult the manufacturer's manual for details on how to perform a melting point analysis.
10. The generation of standard curves for gDNA and cDNA analyses is a laborious and time consuming process often subjected to error [16]. The use of the relative quantification method is highly recommended at least during the initial stage of testing and optimization of the transfection procedures.

11. If the analysis is performed using ssODNs targeting the *mdx*^{5cv} mutation, restriction endonuclease treatment is not necessary. Correction at the mRNA level can be assessed using a reverse primer complementary to a region of exon 10 spliced out as the result of the mutation [6].
12. An efficiency of 90–105 % is considered optimal. Low reaction efficiencies may be caused by poor primer design or by suboptimal reaction conditions. Reaction efficiencies greater than 100 % may indicate pipetting error in the serial dilutions or amplification of multiple products. If the reaction efficiency is below 90 % or greater than 105 %, the assay can be optimized by adjusting the concentration of $MgCl_2$ in the reaction or by adding DMSO (up to 10 %) to stabilize the primers. If the PCR efficiency fails to reach the desired percentage range, it is recommended to design new primers.
13. UV is dangerous; wear gloves, long-sleeves and face protection. Never look at UV with unprotected eyes. Set the transilluminator to long-wavelength UV (or low-power) and minimize the amount of time that the DNA is exposed.
14. The use of commercial kits that maximize the recovery yield is particularly important in circumstances where low amounts of PCR products are produced after amplification as the result of low frequencies of gene correction.
15. Ensure that the pH value of the ultrapure water is within pH 7.0 and 8.5, as pH outside this range may negatively affect recovery of the DNA.
16. Be sure to follow the manufacturer's recommendations for the suspension of buffers and solutions prior to commencing the purification steps.

Acknowledgments

The author would like to thank Farnoosh Nik-Ahd for technical assistance. This work was supported by funds from the Muscular Dystrophy Association (USA).

References

1. Rando TA, Disatnik MH, Zhou LZ (2000) Rescue of dystrophin expression in mdx mouse muscle by RNA/DNA oligonucleotides. Proc Natl Acad Sci U S A 97:5363–5368
2. Bartlett RJ, Stockinger S, Denis MM, Bartlett WT, Inverardi L, Le TT, thi Man N, Morris GE, Bogan DJ, Metcalf-Bogan J, Kornegay JN (2000) In vivo targeted repair of a point mutation in the canine dystrophin gene by a chimeric RNA/DNA oligonucleotide. Nat Biotechnol 18:615–622
3. Bertoni C, Rando TA (2002) Dystrophin gene repair in mdx muscle precursor cells in vitro and in vivo mediated by RNA-DNA chimeric oligonucleotides. Hum Gene Ther 13:707–718
4. Bertoni C, Morris GE, Rando TA (2005) Strand bias in oligonucleotide-mediated dystrophin gene editing. Hum Mol Genet 14:221–233

5. Bertoni C (2008) Clinical approaches in the treatment of Duchenne muscular dystrophy (DMD) using oligonucleotides. Front Biosci 13:517–527
6. Kayali R, Bury F, Ballard M, Bertoni C (2010) Site directed gene repair of the dystrophin gene mediated by PNA-ssODNs. Hum Mol Genet 19:3266–3281
7. Nielsen PE, Egholm M, Berg RH, Buchardt O (1991) Sequence-selective recognition of DNA by strand displacement with a thymine-substituted polyamide. Science 254:1497–1500
8. SantaLucia J Jr (1998) A unified view of polymer, dumbbell, and oligonucleotide DNA nearest-neighbor thermodynamics. Proc Natl Acad Sci U S A 95:1460–1465
9. Zuker M (2003) Mfold web server for nucleic acid folding and hybridization prediction. Nucleic Acids Res 31:3406–3415
10. Rando TA, Blau HM (1994) Primary mouse myoblast purification, characterization, and transplantation for cell-mediated gene therapy. J Cell Biol 125:1275–1287
11. Chomczynski P, Mackey K (1995) Short technical reports. Modification of the TRI reagent procedure for isolation of RNA from polysaccharide- and proteoglycan-rich sources. Biotechniques 19:942–945
12. Pfaffl MW (2001) A new mathematical model for relative quantification in real-time RT-PCR. Nucleic Acids Res 29:e45
13. Schmittgen TD, Livak KJ (2008) Analyzing qPCR data by the comparative C(T) method. Nat Protoc 3:1101–1108
14. Livak KJ, Schmittgen TD (2001) Analysis of relative gene expression data using real-time quantitative PCR and the 2(-Delta Delta C(T)) Method. Methods 25:402–408
15. Mygind T, Birkelund S, Birkebaek NH, Ostergaard L, Jensen JS, Christiansen G (2002) Determination of PCR efficiency in chelex-100 purified clinical samples and comparison of real-time quantitative PCR and conventional PCR for detection of Chlamydia pneumoniae. BMC Microbiol 2:17
16. Freeman WM, Walker SJ, Vrana KE (1999) Quantitative RT-PCR: pitfalls and potential. Biotechniques 26:112–115

[illegible]

Chapter 6

Small Fragment Homologous Replacement (SFHR): Sequence-Specific Modification of Genomic DNA in Eukaryotic Cells by Small DNA Fragments

Andrea Luchetti, Arianna Malgieri, and Federica Sangiuolo

Abstract

The sequence-specific correction of a mutated gene (e.g., point mutation) by the Small Fragment Homologous Replacement (SFHR) method is a highly attractive approach for gene therapy. Small DNA fragments (SDFs) were used in SFHR to modify endogenous genomic DNA in both human and murine cells. The advantage of this gene targeting approach is to maintain the physiologic expression pattern of targeted genes without altering the regulatory sequences (e.g., promoter, enhancer), but the application of this technique requires the knowledge of the sequence to be targeted.

In our recent study, an optimized SFHR protocol was used to replace the eGFP mutant sequence in SV-40-transformed mouse embryonic fibroblast (MEF-SV40), with the wild-type eGFP sequence. Nevertheless in the past, SFHR has been used to correct several mutant genes, each related to a specific genetic disease (e.g., spinal muscular atrophy, cystic fibrosis, severe combined immune deficiency). Several parameters can be modified to optimize the gene modification efficiency, as described in our recent study.

In this chapter we describe the main guidelines that should be followed in SFHR application, in order to increase technique efficiency.

Key words Small fragment homologous replacement (SFHR), Small DNA fragments (SDF), Gene targeting, Homologous replacement, Cell cycle

1 Introduction

The purpose of gene therapy approaches is to permanently replace or correct a defective gene, generally associated with an inherited disorder.

In situ stable correction of the defective endogenous gene allows the recovery of a normal gene function [1], offering significant advantages with respect to gene augmentation approaches.

Although being the most common gene therapy approach, gene augmentation has significant drawbacks since prolonged expression of the transgene requires integration into the genome of the host cell. On the contrary, a site-specific chromosomal

Francesca Storici (ed.), *Gene Correction: Methods and Protocols*, Methods in Molecular Biology, vol. 1114,
DOI 10.1007/978-1-62703-761-7_6,

modification, as in gene targeting, may lead to a long-term and genetically inheritable expression of the corrected gene, regardless to its size, without altering the sequences regulating it.

The availability of new and more efficient gene delivery methods (e.g., nucleofection and microinjection) has made the nonviral gene transfer a more attractive approach for gene therapy. If further developed, gene targeting strategies could gain a higher capacity of correction leading to fewer mutagenic side effects than those methods that are based on random integration of the normal genes into the genome [2].

Among different nonviral gene targeting strategies (Sleeping Beauty, chimeric RNA/DNA oligonucleotides) currently employed in laboratory, SFHR uses small DNA fragments (SDFs) to obtain homologous replacement in recipient cells [3–9].

In mitotic cells, homologous recombination (HR) is a basic mechanism to repair DNA damage and in particular DNA double-strand breaks (DSBs). Different observations suggest that gene targeting process could involve two different DNA-repair steps: in a first stage a homologous recombination event occurs, and then a mismatch correction of the newly formed DNA heteroduplexes takes place.

SFHR is a gene repair strategy relying on the introduction into cells of up to 1 kb-long small DNA fragments. These SDFs are synthesized in vitro and must have complete homology with the endogenous target sequence. After entering the cells, the double-stranded or the single-stranded fragment (dsSDF, ssSDF) pairs with its homologue and replaces the endogenous sequence (genomic or episomal) with the exogenous fragment through an, as yet, undefined mechanism [10–12] that probably involves homologous recombination [13].

SFHR was successfully used to target genomic mutations working in vitro or in vivo, in both human and mouse cells, demonstrating its ability to correct several disease-associated genes [14], such as CFTR (cystic fibrosis, OMIM #219700) [15–19], dystrophin gene [20, 21], SMN (spinal muscular atrophy, OMIM #253300) [22, 23], DNA-PKs (SCID) [24], β-globin (β-thalassemia, OMIM #613985) [25], and HPRT [26].

Among factors influencing targeting mechanism, changes in the chromatin structure during cell cycle, as well as cell mechanisms involved in genome structure maintenance, are key factors in SFHR efficiency [27–29].

Despite this, SFHR shows low correction efficiency, ranging from 0.01 % to 5 % in vitro and about 0.1 % in vivo [30]. In addition the absence of a selectable marker makes difficult to quantify and optimize the efficiency of SFHR-mediated modifications.

Recently, we were able to increase correction efficiency of SFHR using an in vitro reporter system in which several parameters were tested, such as amount of SDF, cell cycle stage, and methylation status. Each of these parameters should be considered every time gene targeting by SFHR is taken under consideration [31]. By exploiting our assay, we were able to quantify gene correction frequency of SFHR-modified cells, and we confirmed the inheritance of the modified allele in a cell clone via molecular analyses (e.g., allelic discrimination, Southern blot). Moreover, the involvement of PARP-1-mediated repair mechanism was revealed in the gene correction process [31].

In this chapter, we describe an optimized protocol of SFHR that provides high correction efficiency. It should be noted that the technical steps described here can be used as reference or starting point to develop *ad hoc* applications of SFHR to perform the desired gene modifications in the cell types of choice [31].

2 Materials

2.1 General Supplies, Solutions, and Culture Medium

All procedures require standard culture room facilities with a laminar flow cabin, a water bath set at 37 °C, and a 37 °C incubator with a water-saturated atmosphere containing 5 % CO_2. An inverted microscope with phase contrast and a centrifuge are also required.

1. Sterile plastic Pasteur pipettes.
2. Syringes and 0.22 μm syringe filters (*see* **Note 1**).
3. Adjustable volume pipettes with sterile-filtered tips.
4. 15 ml conical centrifuge tubes.
5. Bürker hemocytometer chamber.
6. 60 cm^2 culture dish.
7. 4 mg/ml Trypan blue (Sigma Aldrich).

2.2 SDF Preparation

1. DNA sequence for the wild type.
2. 5× PCR buffer (10 mM Tris–HCl, pH 8.4, 500 mM KCl, 20 mM Mg^{2+}, 0.01 % gelatin).
3. 10 mM dNTPs.
4. Oligonucleotide primers (30 μM).
5. Phusion® Hot Start High-Fidelity DNA Polymerase, 5 U/μL (Thermo Fisher Scientific).
6. 1 % high-resolution agarose gel (Euroclone).
7. Tris–borate buffer (TBE): 0.9 M Tris–borate, 2 mM EDTA pH 8.3.
8. Orange G loading buffer (Sigma Aldrich).
9. 1 kb DNA ladder (New England Biolabs).
10. Gel electrophoresis equipment and UV Transilluminator.

11. TA cloning kit (Invitrogen, San Diego, CA).
12. Stable-3 competent cells (Invitrogen).
13. Luria Broth (LB) Medium: 10 g tryptone, 5 g yeast extract, 10 g NaCl, ddH_2O to 1 L.
14. QIAGEN Plasmid Maxi Prep (Qiagen).
15. LB ampicillin plates: LB Medium, 14 g agar, 1 ml ampicillin (50 mg/ml).
16. 3 M Na-acetate, pH 5.2.
17. 100 % and 70 % ethanol.
18. QIAquick Gel extraction kit (Qiagen).
19. H_2O DNase-/RNase-free.

2.3 Cell Cycle Synchronization and Nucleofection

1. 100 mM Vinblastine (Sigma Aldrich).
2. 1,5-Isoquinolinediol (Sigma Aldrich).
3. H_2O DNase-/RNase-free.
4. Short DNA fragment (SDF).
5. Nucleofector device (Lonza).
6. Nucleofector® Kit containing specific Nucleofector solutions, supplements, certified cuvettes, and certified plastic pipettes (Lonza).
7. 75 cm^2 flask (Corning).
8. Trypsin (0.05 %)—with EDTA-4 Na.
9. 4 mg/ml Trypan blue (Sigma Aldrich).

2.4 RNA and DNA Extraction

2.4.1 RNA Extraction

1. 2×10^6 cell pellet.
2. Phosphate-buffered saline (PBS, Euroclone).
3. Trizol reagent (Invitrogen).
4. Chloroform (Sigma Aldrich).
5. H_2O DNase-/RNase-free DEPC, sterile and filtered.
6. Ambion® Turbo™ DNase (Ambion).
7. Tris-saturated phenol, chloroform:isoamyl alcohol (24:1).
8. Refrigerated Bench Centrifuge.
9. 100 % isopropanol.
10. 75 % ethanol in H_2O DEPC.

2.4.2 DNA Extraction

1. 2×10^6 cell pellet.
2. Flexigene kit (Qiagen).
3. H_2O DNase-/RNase-free.
4. Bench Centrifuge.
5. Thermoblock set to 65 °C.

2.5 FACS Analysis

1. Nucleic acid dye Topro-3 0.1 μM (Molecular Probes).
2. 1× PBS.
3. FACS Tube (Falcon).
4. Trypsin (0.05 %)—with EDTA-4 Na.
5. 1× PBS with 10 % FBS.

2.6 Molecular Analyses

Prepare solution using ultrapure water and analytical grade reagent.

2.6.1 Southern Blotting

1. Genomic DNA.
2. Restriction Endonuclease (specific for each kind of analysis).
3. Water bath.
4. UV cross-linker.
5. 0.8 % agarose gel.
6. Nylon Hybond N^+ membrane.
7. Citosine 5′-triphosphate (α-^{32}P), 4,000 Ci/mmol (Perkin-Elmer).
8. Amersham Nick Translation Kit (Amersham).
9. NucTrap® Probe Purification Columns (Stratagene).
10. ^{32}P-labeled probe.
11. 2× and 1× SSC (20× SSC is 175.3 g NaCl and 88.2 g sodium citrate in 1 L, pH 7.0) and 0.1 % SDS.
12. 0.75 M NaCl, 5 % dextran sulfate, 1 % SDS, Heparin Sodium Salt 7.80 U/mg per ml and 50 μg/ml herring sperm DNA.
13. X-ray film.

2.6.2 Restriction Fragment Length Polymorphism and Sequencing

1. Genomic DNA.
2. Primer pair outside the SDF homology region.
3. Thermocycler.
4. AmpliTaq Gold™ DNA Polymerase (Applied Biosystems).
5. Restriction Endonuclease (specific for each kind of analysis).
6. Water bath set to 37 °C.
7. 0.8 % agarose gel.
8. QIAquick gel extraction kit (Qiagen).
9. BigDye Terminator v3.1 Cycle Sequencing Kit (Applied Biosystems).
10. Capillary electrophoresis device.

2.6.3 Allelic Discrimination

1. Gel-purified RFLP Amplicon.
2. TaqMan Universal Master Mix (Applied Biosystems).
3. Target Gene Genotyping Probe (Applied Biosystems).
4. Real-Time 7500 FAST System.

3 Methods

3.1 Small DNA Fragments (SDFs)

3.1.1 SDF Design Strategy

1. SDF is generated by PCR amplification (*see* **Note 2**) using specific oligonucleotides for the target gene. Successively, its size and quality must be checked by electrophoresis loading 10 μL of the PCR reaction on a 1.8 % agarose gel (*see* **Note 3**).
2. The fragment is cloned into the TA cloning vector, according to manufacturer's instructions. Stable-competent bacterial cells are transformed with the resulting vector and plated onto LB agar plates containing the appropriate antibiotic for selection. Colonies are selected and screened by restriction enzyme digestion and sequence analysis.
3. Cells containing the vector with the correct sequence (pCR2.1-SDF) are grown and harvested by standard Maxi prep protocol and used as a template for subsequent PCR reactions to prepare the desired fragment (Fig. 1).
4. Preparative quantities of SDF are produced with 96-plate 100 μL/well PCR amplifications. For a 100 μL reaction: 1 μL of plasmid vector (1 ng/μL), 20 μL 5× buffer-GC, 2 μL dNTPs, 1 μL of each oligonucleotide primer (30 μM), 1.5 μL of DMSO, 0.5 μL Taq, 73 μL ddH_2O. The PCR amplification conditions are as follows for an Applied Biosystems 9700 machine: initial denaturation, 98 °C, 30 s; amplification, 98 °C, 10 s (denaturation)/55 °C, 20 s (annealing)/68 °C, 30 s (extension) for 30 cycles, final extension 72 °C.

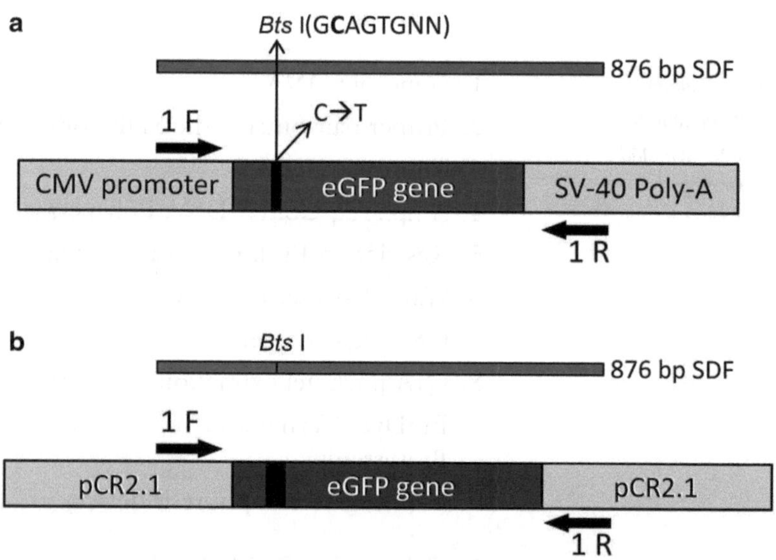

Fig. 1 Example of SDF design. (**a**) SDF sequence is homologous to the entire wild-type eGFP coding sequence amplified with primer pair 1F/1R and cloned in pCR2.1. (**b**) SDF-PCR-WT, 876 bp long was generated by PCR amplification using the cloning vector pCR2.1 as template [31]

3.1.2 Purification

1. Individual PCR amplifications are pooled and the DNA fragment precipitated by the addition of one-tenth vol 3 M Na-acetate and 2.5× the volume of 100 % ethanol. The mixture is placed at −20 °C overnight. DNA is pelleted by centrifugation in Eppendorf microcentrifuge tubes at high speed (16,000 × *g*) for 30 min. The resulting pellet is washed once with 70 % ethanol and then air-dried for 10 min. The DNA pellet is suspended in pyrogen-free water.
2. Load concentrated PCR product on a 23 cm long 1 % high-resolution agarose gel and run gel at 100 V for 5 h.
3. Place gel on UV transilluminator and with a clean scalpel cut the band (*see* **Note 4**).
4. Purify product with QIAquick gel extraction kit.
5. Elute fragment in pre-warmed (50 °C) ddH_2O and concentrate PCR product repeating step 1.
6. Resuspend pellet with ddH_2O to a final concentration of 2 μg/μL (*see* **Notes 5** and **6**). Check DNA quality and molecular weight by electrophoresis.
7. Store at −20 °C until usage.

3.2 Cell Culture

1. Before opening the cryovials, wipe them with 70 % ethanol to avoid contamination of the cells. Quickly thaw the cryovial in a 37 °C water bath being careful not to submerge the entire vial. Before ice melts, remove the vial from the water bath. Add 10 ml of fresh medium to dilute DMSO and centrifuge for 5 min at 200 × *g*. Resuspend cells, seed in a 75 cm^2 flask, and place in a 37 °C, 5 % CO_2, humidified incubator.
2. Replaced medium according to cell type used.
3. Remove medium from the cultured cells and wash cells twice with PBS using at least the same volume of PBS of culture medium. Incubate cells about 1–3 min at 37 °C with 4 ml of trypsin. If necessary, prolong the incubation time for 2 more minutes at 37 °C.
4. Neutralize trypsinization reaction with fresh medium once the cells have been detached and singularized. Centrifuge 5 min at 200 × *g* count and seed 3×10^5 cells in a 75 cm^2 flask.

3.3 Nucleofection

1. To set up cultures, calculate and cultivate the final required number of cells (1.7×10^6 cells per sample).
2. Two days before nucleofection, harvest the cells by trypsinization (*see* Subheading 3.2) and centrifuge cells at 200 × *g* for 5 min at room temperature. Remove and discard supernatant and resuspend cells in 1× PBS.
3. Count the trypsinized cells, determine cell density, and seed 1.5 × 106 cell in a 150 cm^2 plate.

4. Add Vinblastine at 25 nM 18 h before transfection to synchronize cell in G_2/M phase (*see* **Note 7**).
5. Harvest the cells by trypsinization (*see* Subheading 3.2) (*see* **Note 8**).
6. Count viable cell with Trypan blue and centrifuge the required number of cells (1.7×10^6 cells per sample) at $200 \times g$ for 10 min at 4 °C. Carefully remove all and discard supernatants using a pipette. At room temperature, add the entire Nucleofector supplement to the Nucleofector solution and pre-warm to room temperature (*see* **Note 9**).
7. Resuspend the cell pellet carefully in 100 μL of Nucleofector solution maintained at room temperature per sample by pipetting up and down twice. Combine 100 μL of cell suspension with 20 μg of SDF corresponding to 12×10^6 SDF-PCR-WT/cell. Transfer cell/DNA suspension into nucleofection cuvette; sample must cover the bottom of the cuvette without air bubbles (*see* **Note 10**).
8. Select Nucleofector program T-20 for MEF from Lonza (*see* **Note 11**). Insert the cuvette with cell/DNA suspension into the Nucleofector cuvette holder and apply the selected program, according to instrument and nucleofection-buffer specifications. Take the cuvette out of the holder once the program is finished and immediately add 500 μL of the 37 °C warmed culture medium to the cuvette, and using the supplied pipettes, gently transfer the sample into the 75 cm^2 flask (final volume 10 ml media per flask).
9. Incubate the cells in a humidified, 37 °C/5 % CO_2 incubator until analysis.
10. To increase correction efficiency 1,5-Isoquinolinediol (Sigma Aldrich, Milan, Italy), an inhibitor of Poly-(ADP-ribose) synthetase-1 is added soon after transfection at 0.622 mM for 24 h (*see* **Note 12**).

3.4 Analysis

For the following analysis, used to confirm genomic-modification stability, all steps require DNA extraction with Flexigene kit or RNA extraction with Trizol.

3.4.1 Flow Cytometry

1. FACS analysis was performed 3–5 days after transfection.
2. Detach cells via trypsinization (*see* Subheading 3.2).
3. Rinse cells from culture plate and transfer into a 1.5-ml reaction tube.
4. Centrifuge at $200 \times g$ for 5 min at 4 °C.
5. Remove supernatant and resuspend pellet in 200 μL PBS containing 0.1 μM of nucleic acid dye Topro-3, to stain dead cells, and mix well.

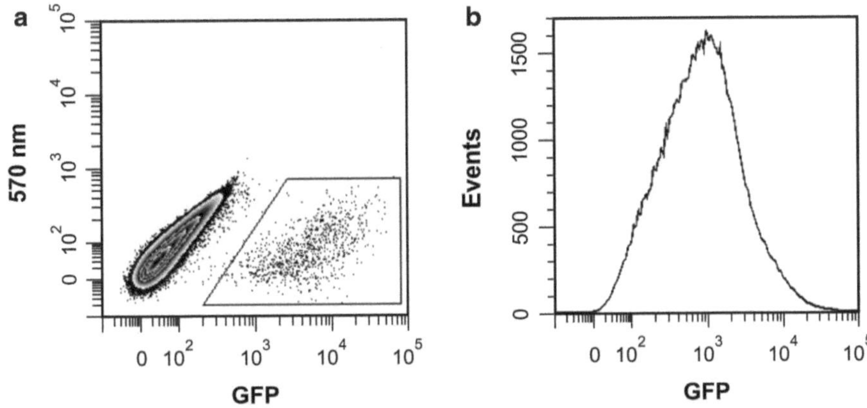

Fig. 2 Modification efficiency in cells transfected with 12×10^6 SDF-PCR-WT/cell. Positive cells (0.5 %) were sorted and soon after reanalyzed (*right panel*) to asses population purity (>99 %) [**31**]

6. Analyze samples by flow cytometry.
7. 5×10^5 positive cells are sorted (Fig. 2) and plated on 60 cm^2 plate to obtain a homogeneous fluorescent cell population (*see* **Note 13**).

3.4.2 Genotyping Analysis

Successful replacement of endogenous mutated sequences with wild-type sequences is first analyzed by PCR with TaqMan MGB Custom Probe able to discriminate between wild-type and mutant alleles:

1. 1×10^6 cell is used to extract genomic DNA following QIAGEN extraction kit procedure.
2. Genomic DNA extracted from sorted transfected cells is amplified with oligonucleotide primers designed upstream and downstream the region homologous to SDF, generating an analytical amplicon (*see* **Note 14**). Reaction mixture for analytical amplicon generation is as follows: 100 ng genomic DNA, 10 μL 10× buffer, 2 μL dNTPs, 1 μL of each oligonucleotide primer (30 μM), 0.5 μL Taq, and ddH_2O to a final volume of 100 μl. The PCR amplification conditions are as follows for an Applied Biosystems 9700 machine: initial Taq activation 95 °C for 10 min and 95 °C 30 s (denaturation), 55 °C 30 s (annealing), and 72 °C 30 s (extension) for 35 cycles, final extension 72 °C 10 min.
3. PCR amplification products are separated by electrophoresis on a 1 % agarose gel and gel purified.
4. Reaction mixture for the assay is as follows: 1 μl of gel-purified PCR product, 10 μl 2× TaqMan® Universal Master Mix, 1 μl 20× TaqMan Genotyping Assay, and water to a final volume of 20 μl. Cycle conditions were 50 °C for 2 min, 95 °C for 10 min, 40 cycles of 95 °C for 30 s, and 62 °C for 1 min,

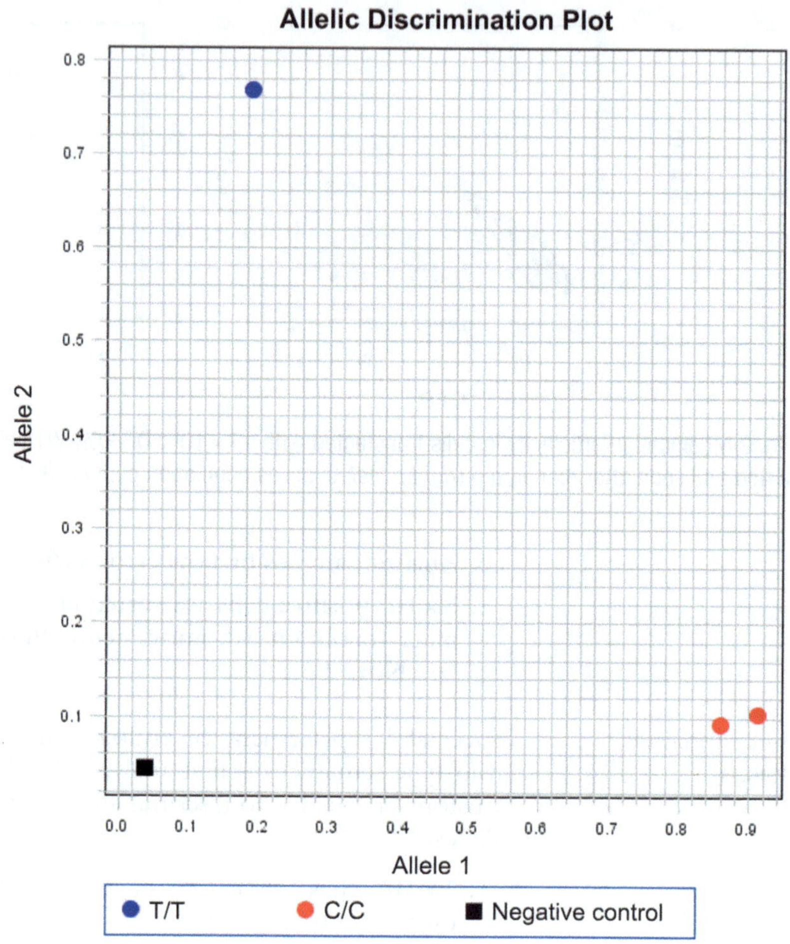

Fig. 3 Allelic discrimination plot. *Red* and *blue dots* represent wild-type (sorted positive cell and a positive control) and mutated genotype, respectively [31]

performed in a 96-well optical plate. Each plate contained positive control and a negative control.

5. Manually score genotypes using Sequence Detection Software 2.0.5 (Applied Biosystems) (Fig. 3).

3.4.3 Southern Blot Analysis

1. 10 μg digested genomic DNA
2. Electrophoresis on 0.8 % agarose gel.
3. Transfer DNA to a nylon Hybond N+ membrane (Amersham-Pharmacia Biotech, Piscataway, NJ, USA)
4. Place filter 3 min under a UV cross-linker.
5. Prepare α-^{32}P-CTP-labeled probe following Amersham Nick Translation Kit.
6. Prehybridize membrane for 2 h at 65 °C with a solution of 0.75 M NaCl, 5 % dextran sulfate, 1 % SDS, salt 7.80 U/mg per ml of Heparin Sodium, and 50 μg/ml of herring sperm DNA.

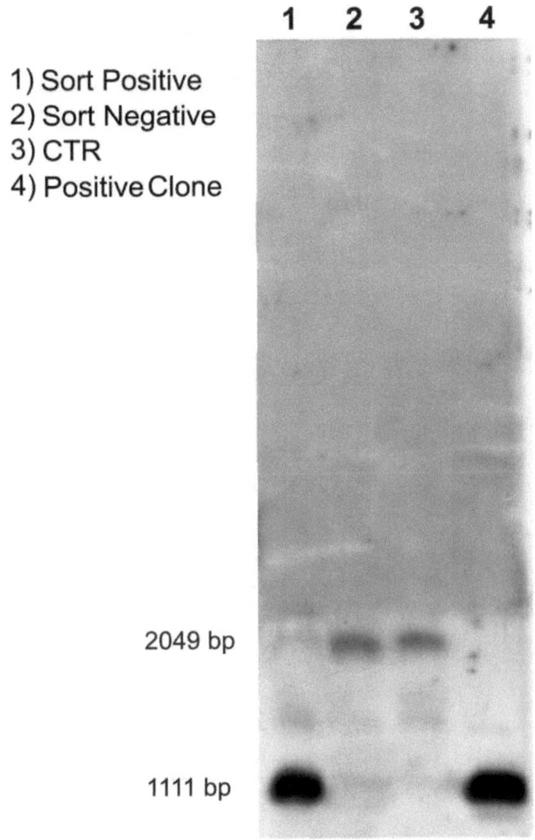

Fig. 4 Southern blot analysis. In our study a 566 bp probe was used, recognizing a region of eGFP gene. *BtsI* site recovery highlights the correction of the eGFP gene. After *SalI/BtsI* genomic DNA digestion, two different restriction patterns can be obtained, according to the presence/absence of *BtsI* restriction site. The 1,111 bp band is obtained only in cells in which *BtsI* site is present (sorted positive cell and positive clone) [31]. The 2,049 bp band is obtained when *BtsI* site is not recovered such as in sorted negative (lane 2) and in control cells (lane 3)

7. Add α-^{32}P-CTP-labeled oligonucleotide to obtain 750,000 CPM/ml and hybridize overnight at 65 °C.
8. Membranes are then washed twice at 65 °C in 2× SSC/0.1 % SDS for 30 min and once at room temperature in 1× SSC/0.1 % SDS for 30 min.
9. The membranes are then exposed to X-ray film, and positively hybridized bands can be visualized autoradiographically (Fig. 4).

3.4.4 Sequencing Analysis

1. Gel-purified Analytical Amplicon.
2. Direct sequencing with BigDye Terminator v3.1 Cycle Sequencing Kit (Applied Biosystems, Foster City, CA, USA) using the same primer pair as for analysis amplification.

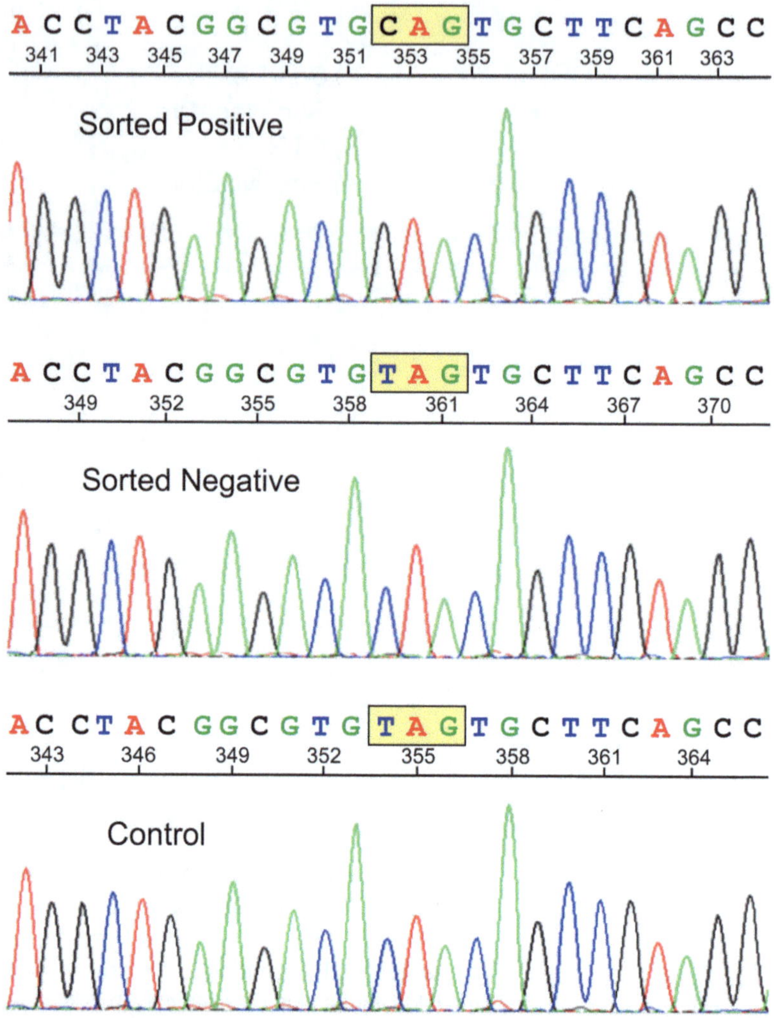

Fig. 5 Electropherogram analysis. Sequence analysis of transfected cells (sorted positive, sorted negative, and CTR). The site-specific T-to-C conversion was present only in sorted positive cells (*see* **Note 3** for SDF design) [31]

3. Purification of sequence-PCR with BigDye XTerminator™ Purification Kit.
4. Capillary electrophoresis separation (Fig. 5).

3.4.5 Restriction Enzyme Analysis

1. Gel-purified Analytical Amplicon.
2. Site-specific Endonuclease.
3. 16 h digestion.
4. Load reaction on 1.8 % agarose gel.
5. Visualize gel on a transilluminator (Fig. 6).

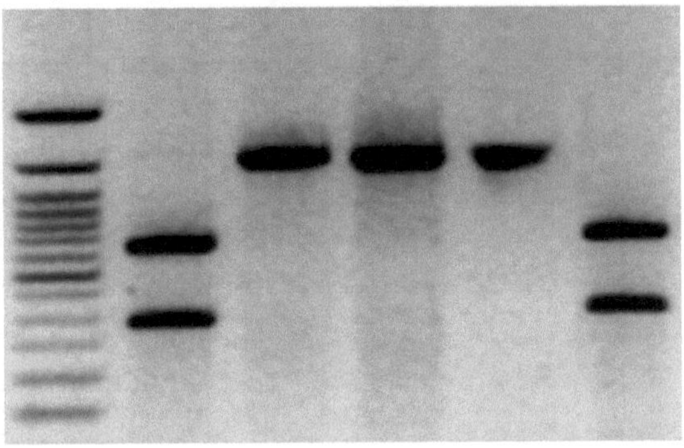

Fig. 6 The amplicon is generated using the RFLP primer pair. Cells transfected with mutated SDF represent our control (CTR, lane 2). All amplification products were digested with *BtsI*, except lane 5. Restriction patterns of sorted positive clone (lane 2) and of positive control cells (lane 6) were identical. No restriction bands were present in CTR (lane 3) and in sorted negative cells (lane 4) [31]

3.4.6 TaqMan Gene Expression Analysis

1. The presence of mRNA transcribed with endogenous wild-type sequence is assessed by PCR amplification from mRNA-derived cDNA.
2. First strand cDNA is made from 1 μg of RNA, following manufacturer's directions for high-capacity cDNA Archive kit.
3. Use 25 ng of cDNA.
4. TaqMan MGB Gene Expression Probe able to discriminate between wild-type and mutant transcript isoforms.
5. 7500 Fast Real-Time PCR System (Applied Biosystems).

4 Notes

1. All the solutions used are sterilized by filtering through 0.22-μm syringe filters.
2. Use a proof reading Taq polymerase to avoid nucleotide mis-incorporation errors.
3. Design therapeutic fragment with a length between 500 and 1,000 base pairs and bearing the wild-type nucleotide closer as possible to the center of the amplicon. The length of homologous DNA sequence incorporated in the amplicon significantly influences the frequency of homologous recombination. Source of the DNA must bear a sequence with wild-type nucleotide(s) with respect to the cells used for the targeting and could be genomic or plasmid DNA. As in other experiments, a restriction site could be incorporated into the SDF creating a silent mutation. The introduction of a unique restriction

enzyme cleavage site can be useful for further verification that the homologous replacement has occurred. In our experiment a glutamine (CAG) to stop codon (TAG) transition causes, at the same time, a fluorescence switch off and a *Bts*I restriction site disruption. Thus, *Bts*I site recovery highlights the correction of the eGFP gene [30].

4. Expose gel on UV transilluminator for no longer than 30–45 s in order to avoid nicking of DNA.

5. The quality and the concentration of DNA used for nucleofection in general play a central role for the efficiency of gene transfer. Use 50 °C pre-warmed endotoxin-free H_2O to increase yield. The amount of fragment used for SFHR transfections was based on the rate of homologous recombination in other systems. Currently, this rate is considered to generally occur with a frequency between 10^{-5} and 10^{-7}. Considering this low frequency, each cell is exposed to 12 × 106 dsDNA fragments and this corresponds to ~5–20 μg of fragment per transfection. The number of SDF per cell was calculated as follows:

$$\frac{(MW_{bp})(N)}{N_A} = \text{Y g}/\,SDF$$

where MW_{bp} = 660 amu/base pair (bp) = 660 g/mole bp, N = the number of bp/SDF (876-bp), $N_A = 6.022 \times 10^{23}$ molecules/mole.

Therefore:

$$\text{Y}=\frac{(660\text{g/mole bp})(876-\text{bp/SDF molecule})}{(6.022\ \ 10^{23}\text{molecules/mole})}=9.60\ \ 10^{-19}\text{g/SDF}$$

Thus, 12×10^6 SDF/cell is added to 1.7×10^6 cells = 20.4×10^{12} SDF or $(20.4 \times 10^{12}) \times (9.60 \times 10^{-19}$ g/SDF) = 0.0000195 g of SDF/1.7×10^6 cells corresponding to ~20 μg.

6. Approximately 50–70 μg of fragment will be produced per 96-plate reaction.

7. To determine whether the cell cycle phase might affect the efficiency of gene repair, we evaluated gene targeting in cell populations enriched in G_0/G_1, S, and G_2M phases. Cell synchronization was optimized in order to obtain a high cell cycle enrichment together with a high cell viability. Best synchronization conditions were evaluated by flow cytometry using propidium iodide (PI). After synchronization, cells were transfected with 12×10^6 molecules/cells of SDF, G_2M synchronized cells showed an increased correction efficiency with respect to G_0/G_1 and S synchronized cells [31].

8. Only cells with optimal confluency 60–70 % are used for nucleofection. Single cell suspension is mandatory; clumps lead to lower transfection efficiency and less reproducibility.

9. Once the Nucleofector supplement is added to the Nucleofector solution, it remains stable for 3 months at 4 °C. If few samples need to be nucleofected, consider the 4.5:1 ratio of Nucleofector solution to supplement. Thus for a single reaction, use 18 μL of supplement plus 82 μL of solution to make 100 μL total reaction volume. Final volume (buffer plus DNA) should not exceed 110 μL.
10. To ensure bubble elimination, tap the cuvette three times on the table.
11. It is recommended to test two Nucleofector programs when using MEF: A-23 and T-20 in parallel samples. In our scenario T-20 gave higher transfection efficiency and/or viability. Check datesheet and use program suggested.
12. Three drugs, potentially involved in SFHR mechanism, were tested to verify their effect on correction efficiency. Specifically KU55933, 1,5-Isoquinolinediol (1,5-ISQ), and α-Amanitin were added to transfected cells that are, respectively, inhibitor of ATM kinase, PARP-1, and RNA polymerase II. No statistically significant variations in modification efficiency were observed 3 days after transfection with respect to control sample, in which no drugs were added. Methylation is involved in hiding correction events, so 5-Aza-dC was added to all samples 24 h after transfection, resulting in an overall increase of fluorescence, statistically significant. When 5-Aza-dC is added to 1,5-ISQ treated cells, a statistically significant increase in correction efficiency was obtained with respect to cells untreated with 5-Aza-dC. These data indicated PARP-1 as a potential SFHR-efficiency modifier [31].
13. The isolation of a clonal population of cells that have undergone SFHR is mandatory. This can be accomplished in our system because of a selectable phenotypic marker in the cell, e.g., eGFP fluorescence restoration. Exception made for HPRT [26], to date no selection of corrected cells is possible. Therefore, limiting dilution must be performed to obtain cell clones bearing successfully SFHR-modified DNA.
14. As shown in Fig. 7, for SDF integration analysis, non-allele-specific oligonucleotides (RFLP F and RFLP R) should be designed upstream and downstream the SDF homology region to ensure that any randomly integrated or free-floating fragments will not be amplified. So, only replacement of endogenous sequence by the SDF will be detected.

    Allelic discrimination PCR-Real-Time can be performed as nested PCR with allele-specific probes to assay for homologous replacement.

    Another method of genomic analysis relies on genomic DNA gel purification [32] allowing direct use of allele-specific oligonucleotides to assess gene conversion.

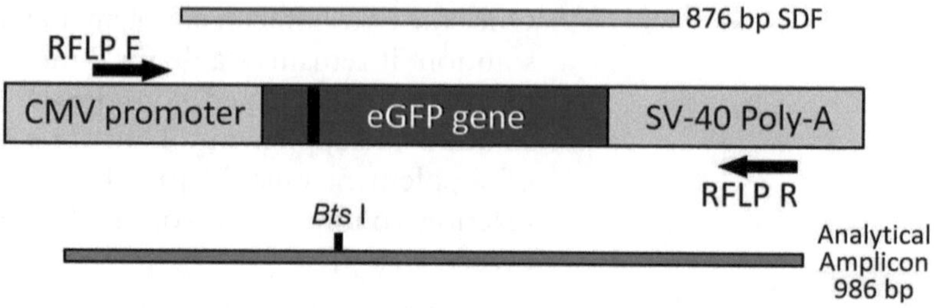

Fig. 7 Example of analytical amplicon design. Analytical oligonucleotides (RFLP F and RFLP R) must anneal outside the SDF homology sequence represented by the SDF generation primer pairs 1F/1R [31]

Acknowledgments

This work was supported by Fondazione Cenci Bolognetti and by Fondazione Roma.

References

1. Capecchi MR (1994) Targeted gene replacement. Sci Am 270:52–59
2. Sullenger B (2003) Targeted genetic repair: an emerging approach to genetic therapy. J Clin Invest 112:310–311
3. Horie K, Kuroiwa A, Ikawa M, Okabe M, Kondoh G, Matsuda Y, Takeda J (2001) Efficient chromosomal transposition of a Tc1/mariner-like transposon Sleeping Beauty in mice. Proc Natl Acad Sci U S A 98:9191–9196
4. Ivics Z, Plasterk RH, Izsvak Z (1997) Molecular reconstruction of Sleeping Beauty, a Tc1-like transposon from fish, and its transposition in human cells. Cell 91(4):501–510
5. Izsvak Z (2003) Sleeping beauty transposition: biology and applications for molecular therapy. Mol Ther 9(2):147–156
6. Yoon K, Cole-Strauss A, Kmiec EB (1996) Targeted gene correction of episomal DNA in mammalian cells mediated by a chimeric RNA-DNA oligonucleotide. Proc Natl Acad Sci U S A 93:2017–2076
7. Zhu T, Mettenburg K, Peterson DJ, Tagliani L, Baszczynski CL (2000) Engineering herbicide-resistant maize using chimeric RNA/DNA oligonucleotides. Nat Biotechnol 18:555–558
8. McManus MT, Sharp PA (2002) Gene silencing in mammals by small interfering RNAs. Nat Rev Genet 3(10):737–747
9. Urnov FD, Miller JC, Lee YL, Beausejour CM, Rock JM, Augustus S, Jamieson AC, Porteus MH, Gregory PD, Holmes MC (2005) Highly efficient endogenous human gene correction using designed zinc-finger nucleases. Nature 435(7042):646–651
10. Gruenert DC (1998) Gene correction with small DNA fragments. Curr Res Mol Therapeut 1:607–613
11. Gruenert DC (1999) Opportunities and challenges in targeting genes for therapy. Gene Ther 6:1347–1348
12. Yáñez RJ, Porter AC (1998) Therapeutic gene targeting. Gene Ther 5(2):149–159
13. Goncz KK, Gruenert DC (2000) Site-directed alteration of genomic DNA by small-fragment homologous replacement. Methods Mol Biol 133:85–99
14. Davis BR, Gruenert DC (2002) Application of SFHR to gene therapy of monogenic disorders. Gene Ther 9:691–694
15. Kunzelmann K, Legendre JY, Knoell DL, Escobar LC, Xu Z, Gruenert DC (1996) Gene targeting of CFTR DNA in CF epithelial cells. Gene Ther 3:859–867
16. Colosimo A, Goncz KK, Novelli G, Dallapiccola B, Gruenert DC (2001) Targeted correction of a defective selectable marker gene in human epithelial cells by small DNA fragments. Mol Ther 3:178–185
17. Sangiuolo F, Bruscia E, Serafino A, Nardone AM, Bonifazi E, Lais M, Gruenert DC, Novelli G (2002) In vitro correction of cystic fibrosis epithelial cell lines by small fragment homologous

replacement (SFHR) technique. BMC Med Genet 3:8

18. Maurisse R, Cheung J, Widdicombe J, Gruenert DC (2006) Modification of the pig CFTR gene mediated by small fragment homologous replacement. Ann N Y Acad Sci 1082:120–123
19. Sangiuolo F, Scaldaferri ML, Filareto A, Spitalieri P, Guerra L et al (2008) Cftr gene targeting in mouse embryonic stem cells mediated by Small Fragment Homologous Replacement (SFHR). Front Biosci 1:2989–2999
20. Quigley A, Lynch GS, Steeper K, Kornberg AJ, Gregorevic P, Austin L, Byrne E (2001) In vivo and in vitro correction of the mdx dystrophin gene nonsense mutation by short-fragment homologous replacement. Hum Gene Ther 12:629–642
21. Todaro M, Quigley A, Kita M, Chin J, Lowes K, Kornberg AJ, Cook MJ, Kapsa R (2007) Effective detection of corrected dystrophin loci in mdx mouse myogenic precursors. Hum Mutat 28:816–823
22. Sangiuolo F, Filareto A, Spitalieri P, Scaldaferri ML, Mango R et al (2005) In vitro restoration of functional SMN protein in human trophoblast cells affected by spinal muscular atrophy by small fragment homologous replacement. Hum Gene Ther 16:869–880
23. Spitalieri P, Cortese G, Pietropolli A, Filareto A, Dolci S et al (2009) Identification of multipotent cytotrophoblast cells from human first trimester chorionic villi. Cloning Stem Cells 11:535–546
24. Zayed H, McIvor RS, Wiest DL, Blazar BR (2006) In vitro functional correction of the mutation responsible for murine severe combined immune deficiency by small fragment homologous replacement. Hum Gene Ther 17:158–166
25. Goncz KK, Prokopishyn NL, Abdolmohammadi A, Bedayat B, Maurisse R, Davis BR, Gruenert DC (2006) Small fragment homologous replacement-mediated modification of genomic beta-globin sequences in human hematopoietic stem/progenitor cells. Oligonucleotides 16:213–224
26. Bedayat B, Abdolmohamadi A, Ye L, Maurisse R, Parsi H et al (2009) Sequence-specific correction of genomic hypoxanthine-guanine phosphoribosyl transferase mutations in lymphoblasts by small fragment homologous replacement. Oligonucleotides 20:7–16
27. Hu Y, Parekh-Olmedo H, Drury M, Skogen M, Kmiec EB (2005) Reaction parameters of targeted gene repair in mammalian cells. Mol Biotechnol 29:197–210
28. Engstrom JU, Kmiec EB (2008) DNA replication, cell cycle progression and the targeted gene repair reaction. Cell Cycle 7:1402–1414
29. Brachman EE, Kmiec EB (2005) Gene repair in mammalian cells is stimulated by the elongation of S phase and transient stalling of replication forks. DNA Repair 4:445–457
30. Gruenert DC, Bruscia E, Novelli G, Colosimo A, Dallapiccola B et al (2003) Sequence-specific modification of genomic DNA by small DNA fragments. J Clin Invest 112:637–641
31. Luchetti A, Filareto A, Sanchez M, Ferraguti G, Lucarelli M, Novelli G, Sangiuolo F, Malgieri A (2012) Small fragment homologous replacement: evaluation of factors influencing modification efficiency in an eukaryotic assay system. PLoS One 7(2):e30851
32. Maurisse R, Fichou Y, De Semir D, Cheung J, Ferec C, Gruenert DC (2006) Gel purification of genomic DNA removes contaminating small DNA fragments interfering with PCR analysis of SFHR. Oligonucleotides 16:375–386

Chapter 7

Preparation and Application of Triple Helix Forming Oligonucleotides and Single Strand Oligonucleotide Donors for Gene Correction

Md. Rowshon Alam, Arun Kalliat Thazhathveetil, Hong Li, and Michael M. Seidman

Abstract

Strategies for site-specific modulation of genomic sequences in mammalian cells require two components. One must be capable of recognizing and activating a specific target sequence in vivo, driving that site into an exploitable repair pathway. Information is transferred to the site via participation in the pathway by the second component, a donor nucleic acid, resulting in a permanent change in the target sequence. We have developed biologically active triple helix forming oligonucleotides (TFOs) as site-specific gene targeting reagents. These TFOs, linked to DNA reactive compounds (such as a cross-linking agent), activate pathways that can engage informational donors. We have used the combination of a psoralen-TFO and single strand oligonucleotide donors to generate novel cell lines with directed sequence changes at the target site.

Here we describe the synthesis and purification of bioactive psoralen-linked TFOs, their co-introduction into mammalian cells with donor nucleic acids, and the identification of cells with sequence conversion of the target site. We have emphasized details in the synthesis and purification of the oligonucleotides that are essential for preparation of reagents with optimal activity.

Key words Triple helix forming oligonucleotide, TFO, Gene targeting, Sequence conversion, Oligonucleotide modification, Oligonucleotide synthesis

1 Introduction

DNA triple helices were first described in 1957 [1] and are formed by a third strand of DNA in the major groove of an intact duplex. They are composed of polypyrimidine or polypurine third strands in complex with polypurine:polypyrimidine duplexes and are stabilized by sequence-specific hydrogen bonds between the third strand bases and the purines of the duplex. The recognition of a "triplex" binding code (T·A:T, C⁺·G:C for pyrimidine motif triplexes; A·A:T, G·G:C for purine motif triplexes), and the development of procedures and instruments for facile oligonucleotide synthesis, was the basis for the concept of triplex forming oligonucleotides (TFOs) as

Francesca Storici (ed.), *Gene Correction: Methods and Protocols*, Methods in Molecular Biology, vol. 1114,
DOI 10.1007/978-1-62703-761-7_7, © Springer Science+Business Media, LLC 2014

gene targeting reagents [2, 3]. These could introduce damage into specific sequences in the genome of living cells, either as carriers of DNA reactive moieties [4–7] or because the triplex structure might be recognized by cellular activities that would introduce nicks and breaks in an effort to "repair" the triplex [8]. Both scenarios could yield mutations at the site or, if accompanied by an informational donor that could enter an appropriate repair pathway, direct a change in the genomic sequence.

Bioactive TFOs must be resistant to nucleases and functional in a physiological environment. These issues have been addressed by introduction of base and sugar modifications in the oligonucleotides, usually in pyrimidine motif third strands [9–11]. For example, triplex stability is enhanced by RNA analogue sugars in the TFO [12]. The most extensively studied ribose derivatives are the 2′-O-methyl; the "locked (LNA)" or bridged (BNA) 2′-O-methyl-4′; and the positively charged 2′-O-aminoethyl (AE) [13–16]. Biochemical and biophysical characterization of triplexes formed by TFOs with these modifications demonstrated enhanced triplex stability and, particularly with the latter two, improved resistance to nucleases (reviewed in ref. 6). Our most active TFOs contain four contiguous 2′-aminoethoxy ribose residues, all other sugars being 2′-O-methyl, with 5-methylcytosine in place of cytosine [17].

In contrast to some purine motif third strands, the modified pyrimidine motif TFOs do not provoke mutagenesis at specific target sites in living cells. However, based on work with psoralen-linked TFOs by the Helene and Glazer groups [4, 5], we find that modified TFOs linked to psoralen can, dependent on photoactivation, induce mutations at genomic target sites [18]. The mutations are base substitutions and deletions, the latter identical to those formed by repair of double strand breaks by the nonhomologous end joining (NHEJ) pathway [19]. Double strand breaks are well-established inducers of recombinational repair, and the psoralen-linked TFOs can stimulate targeted sequence modulation by co-introduced duplex donors, thousand of bases long [20]. Single strand oligonucleotides are also effective informational donors. These can introduce small deletions and base substitutions at the target site at frequencies that are much higher than obtained with the duplex donors. Furthermore, the sequence conversion activity of the single strand oligonucleotide donors is not dependent on recombinational functions. Instead these donors enter an NHEJ pathway, indicating, contrary to conventional wisdom, that NHEJ can be templated [21].

2 Materials

2.1 Oligonucleotide Synthesis (See Note 1*)*

1. Controlled Pore Glass supports (500 Å).
2. Detritylation: 3 % trichloroacetic acid in dichloromethane.
3. {5′-*O*-(4, 4′-dimethoxytrityl)-5-methyluridine-2′-*O*-methyl-3′-*O*-(*β*-cyanoethyl-*N*, *N*-diisopropyl)} phosphoramidite

(2′-OMe-5Me U) was dissolved in 50 % THF/acetonitrile at a concentration of 0.1 M. Dry the solution over molecular sieves (4 Å) for at least 4 h prior to use (*see* **Note 2**).

4. Other amidites: 2′-OMe-5Me C, 2′-AE-5Me-U, 2′-AE-5ME-C, and psoralen are dissolved in 100 % anhydrous acetonitrile at a concentration of 0.1 M. Dry over molecular sieves as above.
5. Coupling: 0.45 M sublimed 1H-tetrazole in anhydrous acetonitrile.
6. Oxidation: 0.02 M iodine in THF/pyridine/H_2O (89.6/.4/10).
7. Capping mix A: 10 % acetic anhydride in tetrahydrofuran.
8. Capping mix B: 10 % *N*-methyl-imidazole in tetrahydrofuran/pyridine (8:1, v/v).
9. Deprotection: AMA, 28 % concentrated ammonium hydroxide and 40 % aqueous methyl amine (1:1, v/v).

2.2 TFO Purification and Characterization

1. Mobile phase A: 100 mM Tris–HCl (pH 7.8) containing 10 % acetonitrile.
2. Mobile phase B: 1 M NaCl in 100 mM Tris–HCl (pH 7.8) and 10 % acetonitrile.
3. Dionex DNAPac PA-100 column: 4.0 mm × 250 mm (analytical) and 9.0 mm × 250 mm (preparative).
4. Sep-Pak Plus C18 cartridge (Waters Corp).
5. 0.45 μm acrodisc filter.
6. 3-Hydroxypicolinic acid (50 mg/ml in 50 % aqueous acetonitrile, HPLC grade water).
7. Ammonium citrate (50 mg/ml, HPLC grade water).

2.3 Cell Culture and Electroporation Reagents

1. Culture medium: Dulbecco's modified Eagle medium supplemented with penicillin and streptomycin and 10 % fetal bovine serum.
2. PBS: phosphate buffered saline, 8 g of NaCl, 0.2 g of KCl, 1.44 g of Na_2HPO_4, 0.24 g of KH_2PO_4, adjusted to pH 7.4.
3. Nucleofection apparatus, cuvettes (Amaxa).
4. Nucleofection cell suspension solution (Amaxa).
5. UVA lamp.

2.4 DNA Purification and Analysis

1. Proteinase K/SDS: 100 μg/ml proteinase K, 0.5 % sodium dodecyl sulfate, in 10 mM Tris–HCl, pH 7.5, 1 mM EDTA.
2. NaCl–EtOH: 5 ml of 5 M NaCl mixed with 45 ml of freezer cold EtOH (will appear cloudy).
3. TE buffer: 10 mM Tris–HCl, pH 8.0, 1 mM EDTA.

2.5 PCR

1. 10× buffer: 200 mM Tris–HCl, pH 8.4; 500 mM NaCl.
2. 15 mM $MgCl_2$.
3. 2 mM each dNTP.
4. 5 μM forward and reverse primer.
5. Taq polymerase (5 U/μl).
6. PCR master mix: for 1 reaction (scale as appropriate). 2 μl each 10× buffer, 15 mM $MgCl_2$, 2 mM dNTP, Taq polymerase, and 7 μl H_2O.

3 Methods

3.1 Synthesis of Psoralen-Linked TFO

1. Detritylation: 5′-Detritylation of the DMTr group on the support bound oligonucleotide by treatment with 3 % trichloroacetic acid in dichloromethane for 1.5 min.
2. Coupling: 6 min are allocated for the 0.45 M tetrazole-activated reaction of the 2′-OMe-5Me-U and 2′-OMe-5Me-C phosphoramidites with the free 5′-hydroxyl group of the growing chain. The coupling time for the other amidites (2′-AE-5Me-U, 2′-AE-5ME-C, psoralen) is 15 min (*see* **Note 3**).
3. Oxidation: The phosphite triester linkage is oxidized to the corresponding phosphate linkage in 0.02 M iodine in THF/pyridine/H_2O for 30 s.
4. Capping: Residual uncoupled 5′-hydroxyl groups (typically <2 %) are blocked with Cap Mix A and Cap Mix B for 25 s to prevent formation of deletion sequences.

3.2 Deprotection of Oligonucleotides

1. After synthesis, dry the oligonucleotide bound supports by flushing with argon gas. Transfer the dry supports to a 3 ml glass vial and suspend in 1 ml of AMA solution, and tightly cap the vial. Shake the suspension for 90 min at room temperature (*see* **Note 4**). Remove the supports by filtration through a 0.45 μm acrodisc filter and wash the supports with HPLC grade water (0.5 ml ×2). Collect the filtrates and combine them with the deprotection filtrate. Immediately evaporate the combined solution to dryness in a speed vac. Resuspend the crude mass in HPLC grade water for anion exchange HPLC purification.

3.3 Purification of Oligonucleotides

1. The oligonucleotides are purified by ion exchange on a Dionex DNAPac PA-100 column on a Shimadzu HPLC system (LC-10ADvp) with a dual wavelength detector (SPD-10AVvp) and an autoinjector (SIL-10ADvp). Preparative (9.0 mm × 250 mm) columns are eluted with a linear gradient 0–50 % Buffer B in Buffer A for 45 min, followed by 100 %

Buffer B for 60 min, at 4 ml/min. UV monitors are 254 and 315 nm (λ max for psoralen) (*see* **Note 5**). Lyophilize the peak fractions and desalt two times using a Sep-Pak Plus C18 cartridge (Waters Corp.).

3.4 Desalting of the Purified TFOs

1. The flow rate of the solvents through the cartridge should be regulated at a rate of about 2 drops/s. Connect a 10 ml disposable syringe to the female end of the cartridge and have the male port terminate in a waste vessel. Flush the cartridge sequentially with 10 ml of acetonitrile, 10 ml of H_2O:acetonitrile (1:1), and 10–20 ml of 0.1 M Na Acetate in 2 % acetonitrile in water.
2. Resuspend the lyophilized oligonucleotides containing sodium salts in HPLC grade water (15–20 ml). Load the oligonucleotide solution onto the Sep-PAK cartridge. Discard the pass through. Flush the cartridge subsequently with 0.1 M NaCl solution (10 ml) and water (10 ml ×2). These can also be discarded. Elute the cartridge-bound desalted oligonucleotides by flushing the cartridge with 50 % aqueous acetonitrile (1 ml×3). Collect all the eluates (*see* **Note 6**).

3.5 Lyophilization of the TFO

1. After desalting, lyophilize the TFO solution containing 50 % aqueous acetonitrile to dryness. Dissolve the dry oligonucleotide in water and lyophilize to dryness three times to eliminate any trace of acetonitrile (*see* **Note 7**). Redissolve the TFO in sterile water, measure the OD_{260} and aliquot the oligonucleotides in stock concentrations (300–500 μM), and store at −80 °C.

3.6 Characterization of the TFO

1. Determine the chemical structure of the TFO by positive ion mode MALDI-TOF (matrix-assisted laser desorption ionization time of flight) mass spectroscopy on a Voyager Applied Biosystem instrument. The matrix for preparing the MALDI-TOF mass samples is a mixture of 3-hydroxypicolinic acid (50 mg/ml in 50 % aqueous acetonitrile) and ammonium citrate (50 mg/ml in HPLC grade water). The TFO (0.1 OD) is passed through a Millipore ZipTipC18 to remove any trace of salts in the sample. Then mix the oligonucleotide with matrix solution (5 μl) and place 1 μl of the mix solution on the plates for MALDI-TOF mass analysis.

3.7 Single Strand Oligonucleotide Donors

1. The 100 mers contain sequences with homology centered around the cross-link site in the genomic target sequence. They also contain variable nonhomologous sequence designed to modify the target site. They are obtained as HPLC-purified oligonucleotides from commercial suppliers and are used without further purification (but *see* **Note 7**).

3.8 Cell Culture

1. We have used a variety of rodent and human cell lines in targeting experiments. The procedure for Chinese hamster cells will be described here, although elements that must be tailored to the specific cell type are noted. Cells are grown in Dulbecco's modified Eagle medium supplemented with penicillin, streptomycin, glutamine, and 10 % fetal calf serum (this will vary according to the cell line or cell type). The best results are obtained if the cells are split the day before use. Cells should be plated so as to be subconfluent at the time of harvest. Cells are removed from the plates with trypsin and washed with PBS (*see* **Note 8**).

3.9 Electroporation

1. Suspend the cells in the nucleofection medium as specified by Amaxa/Lonza (solution T for CHO cells, this will vary depending on cell line or cell type) at a concentration of approximately $2 \times 10^6/95$ μl. Add the pso-TFO to a final concentration of 5 μM and 5 μg donor oligonucleotide (the combined volume of both oligonucleotide additions is no more than 5 μl), place the mixture in a cuvette (total volume 100 μl), and electroporate at settings described by the manufacturer for the cell line or cell type. Remove the cell suspension and place in a transparent 1.5 ml microfuge tube containing 0.5 ml complete medium. Cap the tube and hold at room temperature shielded from light for 3 h (*see* **Note 9**).

3.10 Photoactivation by Long-Wave UV Light (UVA)

1. Psoralen is photoactivated by 365 nm light. Place the cells, still in the tube, behind a glass shield and expose to UVA light for 3 min at 1.8 J/cm^2. Remove the cells and plate in a T75 flask. Incubate in a 37 °C/CO_2 incubator.

3.11 Determination of Targeting Frequencies by Small Pool PCR

1. The donor oligonucleotide contains information for the desired change in the target sequence, generally including the generation of a novel restriction site or the loss of an existing site. The frequency of the event can be ascertained relatively quickly by small pool PCR analysis of DNA harvested from treated cells. We allow cells to proceed through eight to ten population doublings, by which time both TFO and donor oligonucleotide are lost through dilution and degradation. No positive signals above background were detected in control experiments in which cells were treated with both donor and TFO without photoactivation. This is an important control as persistent donor oligonucleotides can enter the PCR reactions, giving rise to false positives. Extract DNA from cells using SDS/proteinase K/phenol extraction/ethanol precipitation procedures, and dissolve in TE buffer. There is no particular requirement for high-molecular weight DNA as the products of the PCR reactions will be less than a thousand nucleotides long. Adjust the genomic DNA concentration to 10 ng/ml (*see* **Note 10**).

2. Small pool PCR is performed on 50 pg aliquots of DNA (5 diploid genome equivalents). We find that this is more successful, and more efficient, for identifying positive signals than when single genome equivalents of DNA are examined. Our standard reactions contain 50 pg DNA, 0.2 μM primers, 1.5 mM $MgCl_2$, 0.2 mM dNTP, and 0.5 U Taq polymerase, in a final volume of 20 μl. Our analyses are performed in 96 well plates, but this can be adjusted according to the anticipated conversion frequency.
3. Prepare a reaction master mix scaled to the number of samples (*see* Subheading 2).
4. Place 5 μl of DNA solution in each well of a multiwall plate. Add 15 μl of master mix and mix thoroughly (*see* **Note 11**). Amplify in a thermal cycler (cycling conditions will vary with the specific primer set).
5. After amplification, remove 10 μl aliquots from each well and analyze by restriction digestion to identify wells in which a site was gained or lost, per the change directed by the donor oligonucleotide. From positive wells remove 1 μl and dilute 200-fold and perform nested PCR, with internal primers, in the reaction mix described above. Analyze 5 μl aliquots of these reactions by restriction digestion to verify the initial identification. Confirm identification of the PCR product by sequence analysis. Conversion frequencies are calculated by dividing the number of reactions with the desired change by the total number of successful amplification reactions multiplied by 5 (since each initial PCR reaction contains five genome equivalents) (*see* **Note 12**).

3.12 Isolation of Clones with Targeted Changes

1. Assuming that the small pool PCR analysis is encouraging (0.5–1 % conversion, at minimum), individual clones with the desired change can be isolated. Electroporate cells with the psoralen-TFO and donor oligonucleotide, followed by photoactivation as above. Plate the cells and culture for a few days to allow targeted conversion and clearance of the oligonucleotides. Then trypsinize and plate in multiple 100 mm petri dishes at 200 cells/plate. Allow colony formation and transfer the colonies to individual wells in 96 well culture plates, preloaded with 100 μl medium/well.

3.13 Colony Transfer

1. Place the tip of a 20 μl pipetting device on a colony and draw a small amount (5 μl) of medium and colony into the tip. Transfer to a well (*see* **Note 13**). When the cells reach near confluence, replicate the plates by drawing 10 μl medium up and down several times in a 20 μl pipet tip (we do this a row at a time with a multichannel pipetting device). Then transfer the aliquot, which contains cells and medium, to another multiwall

plate. Cells in the newly inoculated plate are allowed to grow. This plate serves as a reference plate. The first plate serves as a source of DNA for PCR analysis.

3.14 Purification of DNA from Multiwell Cultures

1. Remove medium from the wells, and wash 2× with PBS. Then add 50 μl of proteinase K/SDS solution. Wrap the plate in parafilm and enclose in a heat sealed bag and submerge the plate in a 50 °C water bath (*see* **Note 14**).

   After an overnight incubation, allow the plates to equilibrate at room temperature, then spin to collect all liquid in the wells, and add 150 μl NaCl/EtOH (Subheading 2). Mix by rocking and spin at 1156×*g* in a refrigerated centrifuge. Carefully remove the supernatant and wash 2× with 70 % EtOH and air-dry. Then dissolve the DNA in each well in 100 μl TE buffer and remove 5 μl for amplification as above and analysis by restriction digestion.

4 Notes

1. We synthesize our oligonucleotides on an Expedite 8909 synthesizer. Most labs will take advantage of commercial services for oligonucleotide synthesis. However, since it is likely that many suppliers will be unfamiliar with the variations in procedure required for the psoralen-linked TFOs with nonconventional base and sugar analogues, we have included specific details as to synthesis, deprotection, purification, etc.
2. Solutions of this amidite in 100 % acetonitrile usually precipitate at the concentration of 0.1 M over 24 h. To avoid precipitation the 2′-OMe-5Me U amidite is dissolved in 50 % THF/acetonitrile. The solution should be used within 5 days of preparation. All amidite solutions must be dry. We observe reduced yields, particularly during humid summer months, if these precautions are not taken.
3. Coupling efficiencies should be in excess of 98 %. If not, reagents may not be dry or may need to be replaced with fresh solutions.
4. Psoralen is inactivated by alkali. Thus psoralen-linked oligonucleotides cannot be deprotected using harsh alkaline conditions at elevated temperatures. Oligonucleotides with inactivated psoralen will compete for binding sites with those with active psoralen. Consequently, it is extremely important that the integrity of the psoralen be maintained. Many commercial oligonucleotide synthesis suppliers are unfamiliar with psoralen chemistry and must be carefully instructed.
5. When very high purity oligonucleotides are desired (these have the greatest biological activity), they are purified on an

analytical column (4.0 mm × 250 mm). A volume of 25 μl of the oligonucleotide is injected in multiple runs. The column is eluted with a gradient of 2 min 10–50 % B, 20 min 50–90 % B, and 21 min 90–100 % B, in Buffer A, with a flow rate of 1.5 ml/min.

6. A single cycle of desalting is adequate for oligonucleotides intended for biochemical experiments. Those employed in biological experiments are desalted twice.
7. We have found that acetonitrile is extremely toxic to cells. Thus, we employ multiple cycles of lyophilization out of water. It may be necessary to similarly treat donor oligonucleotides obtained from commercial sources if toxicity problems are observed.
8. Cell density has an effect on the efficiency of targeting. If the cells are too dense, or have been on the plates too long, they do not trypsinize cleanly and cannot be suspended as single cells in electroporation media. They clump and electroporation is inefficient.
9. The psoralen, linked to the TFO, must be photoactivated by 365 nm UVA. However, a few hours of incubation are required to permit equilibration of TFO binding at the target site. We have found that holding in medium in a transparent capped microfuge tube at room temperature keeps the cells at appropriate pH. Incubation at 37 °C results in lower targeting frequencies. We place the capped tubes on their sides as this eliminates cell packing during the incubation. They are shielded from light during this time.
10. Care must be taken to ensure that the DNA is completely dissolved prior to dilution. This is essential to avoid preparing a dilution with much less DNA than intended. Because there is no need for very high-molecular weight DNA, the solution can be pipetted which will both shear the DNA and aid solubilization.
11. Mix the DNA and master mix solutions extensively in each well, while taking care to avoid cross-well contamination. Thorough mixing is essential; otherwise many wells will fail to support a reaction.
12. Even with thorough mixing, not all wells will support amplification. These must be ignored in the frequency calculations.
13. If the cells form colonies, then these can be transferred as described. We find it helpful to illuminate the plates from below with a lamp. When held at the right angle, the colonies will be visible to the eye. Cultures of cells that do not form colonies may have to be screened by analysis of wells plated by limiting dilution.
14. Evaporation of sample during incubation at 50 °C can be a major problem. Submerging the sealed plates avoids the problem of having all the water in the sample condense on the surface

of the plate cover above the well. The plates must be weighted down during the incubation.

Acknowledgement

This research was supported entirely by the Intramural Research Program of the NIH, National Institute on Aging.

References

1. Felsenfeld G, Davies DR, Rich A (1957) Formation of a three stranded polynucleotide molecule. J Am Chem Soc 79:2023–2024
2. Thuong NT, Helene C (1993) Sequence specific recognition and modification of double helical DNA by oligonucleotides. Angew Chem Int Ed Engl 32:666–690
3. Sun JS, Garestier T, Helene C (1996) Oligonucleotide directed triple helix formation. Curr Opin Struct Biol 6:327–333
4. Takasugi M, Guendouz A, Chassignol M, Decout JL, Lhomme J, Thuong NT, Helene C (1991) Sequence-specific photo-induced crosslinking of the two strands of double-helical DNA by a psoralen covalently linked to a triple helix-forming oligonucleotide. Proc Natl Acad Sci U S A 88:5602–5606
5. Havre PA, Gunther EJ, Gasparro FP, Glazer PM (1993) Targeted mutagenesis of DNA using triple helix-forming oligonucleotides linked to psoralen. Proc Natl Acad Sci U S A 90:7879–7883
6. Seidman MM, Glazer PM (2003) The potential for gene repair via triple helix formation. J Clin Invest 112:487–494
7. Campbell MA, Miller PS (2009) Transplatin-conjugated triplex-forming oligonucleotides form adducts with both strands of DNA. Bioconjug Chem 20:2222–2230
8. Wang G, Seidman MM, Glazer PM (1996) Mutagenesis in mammalian cells induced by triple helix formation and transcription-coupled repair. Science 271:802–805
9. Lee JS, Woodsworth ML, Latimer LJ, Morgan AR (1984) Poly(pyrimidine).poly(purine) synthetic DNAs containing 5-methylcytosine form stable triplexes at neutral pH. Nucleic Acids Res 12:6603–6614
10. Ono A, Ts'o PO, Kan L (1991) Triple helix formation of oligonucleotides containing 2′-O-methylpseudoisocytidine in substitution for 2′-deoxycytidine. J Am Chem Soc 113: 4032–4033
11. Shahid KA, Majumdar A, Alam R, Liu ST, Kuan JY, Sui X, Cuenoud B, Glazer PM, Miller PS, Seidman MM (2006) Targeted crosslinking of the human beta-globin gene in living cells mediated by a triple helix forming oligonucleotide. Biochemistry 45:1970–1978
12. Roberts RW, Crothers DM (1992) Stability and properties of double and triple helices: dramatic effects of RNA or DNA backbone composition. Science 258:1463–1466
13. Escude C, Sun JS, Rougee M, Garestier T, Helene C (1992) Stable triple helices are formed upon binding of RNA oligonucleotides and their 2′-O-methyl derivatives to double-helical DNA. C R Acad Sci III 315:521–525
14. Kumar R, Singh SK, Koshkin AA, Rajwanshi VK, Meldgaard M, Wengel J (1998) The first analogues of LNA (locked nucleic acids): phosphorothioate-LNA and 2′-thio-LNA. Bioorg Med Chem Lett 8:2219–2222
15. Torigoe H, Obika S, Imanishi T (2000) Promotion of triplex formation by a fixed N-form sugar puckering: thermodynamic and kinetic studies. Nucleic Acids Symp Ser: 241–242
16. Blommers MJ, Natt F, Jahnke W, Cuenoud B (1998) Dual recognition of double-stranded DNA by 2′-aminoethoxy-modified oligonucleotides: the solution structure of an intramolecular triplex obtained by NMR spectroscopy. Biochemistry 37:17714–17725
17. Puri N, Majumdar A, Cuenoud B, Miller PS, Seidman MM (2004) Importance of clustered 2′-O-(2-aminoethyl) residues for the gene targeting activity of triple helix-forming oligonucleotides. Biochemistry 43:1343–1351
18. Majumdar A, Khorlin A, Dyatkina N, Lin FL, Powell J, Liu J, Fei Z, Khripine Y, Watanabe

KA, George J, Glazer PM, Seidman MM (1998) Targeted gene knockout mediated by triple helix forming oligonucleotides. Nat Genet 20:212–214

19. Puri N, Majumdar A, Cuenoud B, Natt F, Martin P, Boyd A, Miller PS, Seidman MM (2002) Minimum number of 2′-O-(2-aminoethyl) residues required for gene knockout activity by triple helix forming oligonucleotides. Biochemistry 41:7716–7724
20. Majumdar A, Muniandy PA, Liu J, Liu JL, Liu ST, Cuenoud B, Seidman MM (2008) Targeted gene knock in and sequence modulation mediated by a psoralen-linked triplex-forming oligonucleotide. J Biol Chem 283:11244–11252
21. Liu J, Majumdar A, Liu J, Thompson LH, Seidman MM (2010) Sequence conversion by single strand oligonucleotide donors via nonhomologous end joining in mammalian cells. J Biol Chem 285:23198–23207

Chapter 8

Triplex-Mediated Genome Targeting and Editing

Faisal Reza and Peter M. Glazer

Abstract

Genome targeting and editing in vitro and in vivo can be achieved through an interplay of exogenously introduced molecules and the induction of endogenous recombination machinery. The former includes a repertoire of sequence-specific binding molecules for targeted induction and appropriation of this machinery, such as by triplex-forming oligonucleotides (TFOs) or triplex-forming peptide nucleic acids (PNAs) and recombinagenic donor DNA, respectively. This versatile targeting and editing via recombination approach facilitates high-fidelity and low-off-target genome mutagenesis, repair, expression, and regulation.

Herein, we describe the current state-of-the-art in triplex-mediated genome targeting and editing with a perspective towards potential translational and therapeutic applications. We detail several materials and methods for the design, delivery, and use of triplex-forming and recombinagenic molecules for mediating and introducing specific, heritable, and safe genomic modifications. Furthermore we denote some guidelines for endogenous genome targeting and editing site identification and techniques to test targeting and editing efficiency.

Key words Genome targeting, Genome editing, Recombination, Triplex-forming oligonucleotide (TFO), Peptide nucleic acid (PNA), Recombinagenic donor DNA, Molecular delivery

1 Introduction

In 1953, the double stranded, or duplex, structure of deoxyribose nucleic acid (DNA) was described in a seminal work by Watson and Crick [1]. Four years later, a third strand of polyuracil was reported to bind to a duplex consisting of polyadenine and polyuracil to form a triplex structure by Felsenfeld, Davies, and Rich [2]. Then, in 1987, Dervan and coworkers as well as Helene and coworkers reported that not only could oligonucleotides bind the major groove of DNA to form a triplex structure, but that this also could, in turn, potentiate DNA metabolism, such as cleavage at a specific site [3, 4].

Since then, triplex-forming molecules have diversified in chemistries and content, as well as approaches to achieving their intended goals. Among these molecules are triplex-forming oligonucleotides (TFOs), which are composed of natural or unnatural nucleobases

Francesca Storici (ed.), *Gene Correction: Methods and Protocols*, Methods in Molecular Biology, vol. 1114, DOI 10.1007/978-1-62703-761-7_8, © Springer Science+Business Media, LLC 2014

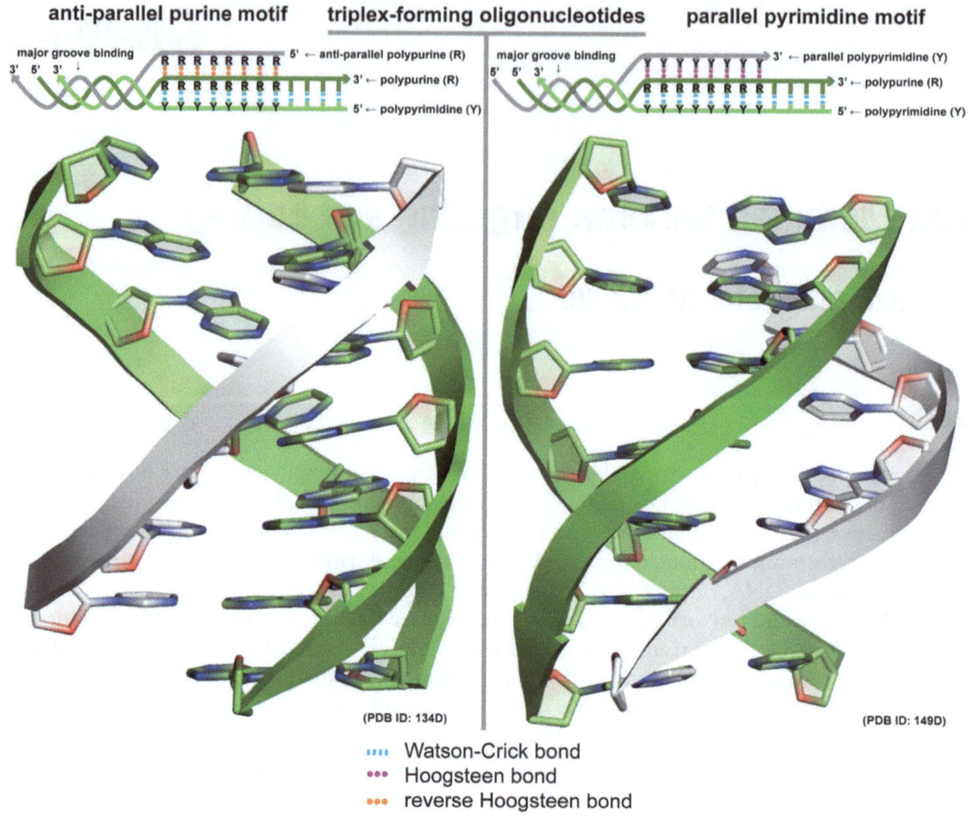

Fig. 1 Structures of triplex-forming molecules, such as triplex-forming oligonucleotides (TFOs). Third strand, or triplex, major groove binding by a TFO (*in grey*) to double-stranded, or duplex, genomic DNA (*in green*). TFO (*in grey*) bind in anti-parallel or parallel motifs relative to the polarity of the polypurine strand of genomic duplex DNA (*in green*). Anti-parallel triplex-forming oligonucleotide structure rendered from [58]; parallel triplex-forming oligonucleotide structure rendered from [59]

linked by phosphodiester backbones. A third strand of TFOs, parallel or anti-parallel in polarity to the polypurine strand of genomic DNA, is able to Hoogsteen or reverse Hoogsteen base pair its natural DNA nucleobases such as adenine, or A, guanine, or G, and thymine or T, with adenine, guanine, and adenine, respectively, of the Watson–Crick-base-paired DNA duplex to form the triplex (Fig. 1). Cytosine, or C, in this third strand requires protonation, so at physiologic pH, an unnatural nucleobase such as pseudoisocytidine, or J, a, C-nucleoside analogue that mimics the N3 protonation of cytosine, is preferred and is able to Hoogsteen base pair with the guanine of a Watson–Crick base pair in a pH independent manner (Fig. 2a).

Fig. 2 (continued) complementary to the genomic DNA target than an uncharged polyamide backbone in a triplex-forming PNA (*right*). Anti-parallel purine motifs rendered from [58] or model built from [58] and [59]; parallel pyrimidine motifs rendered from [59]. Triplex backbones for TFO and triplex-forming PNA rendered from [59] and [60], respectively

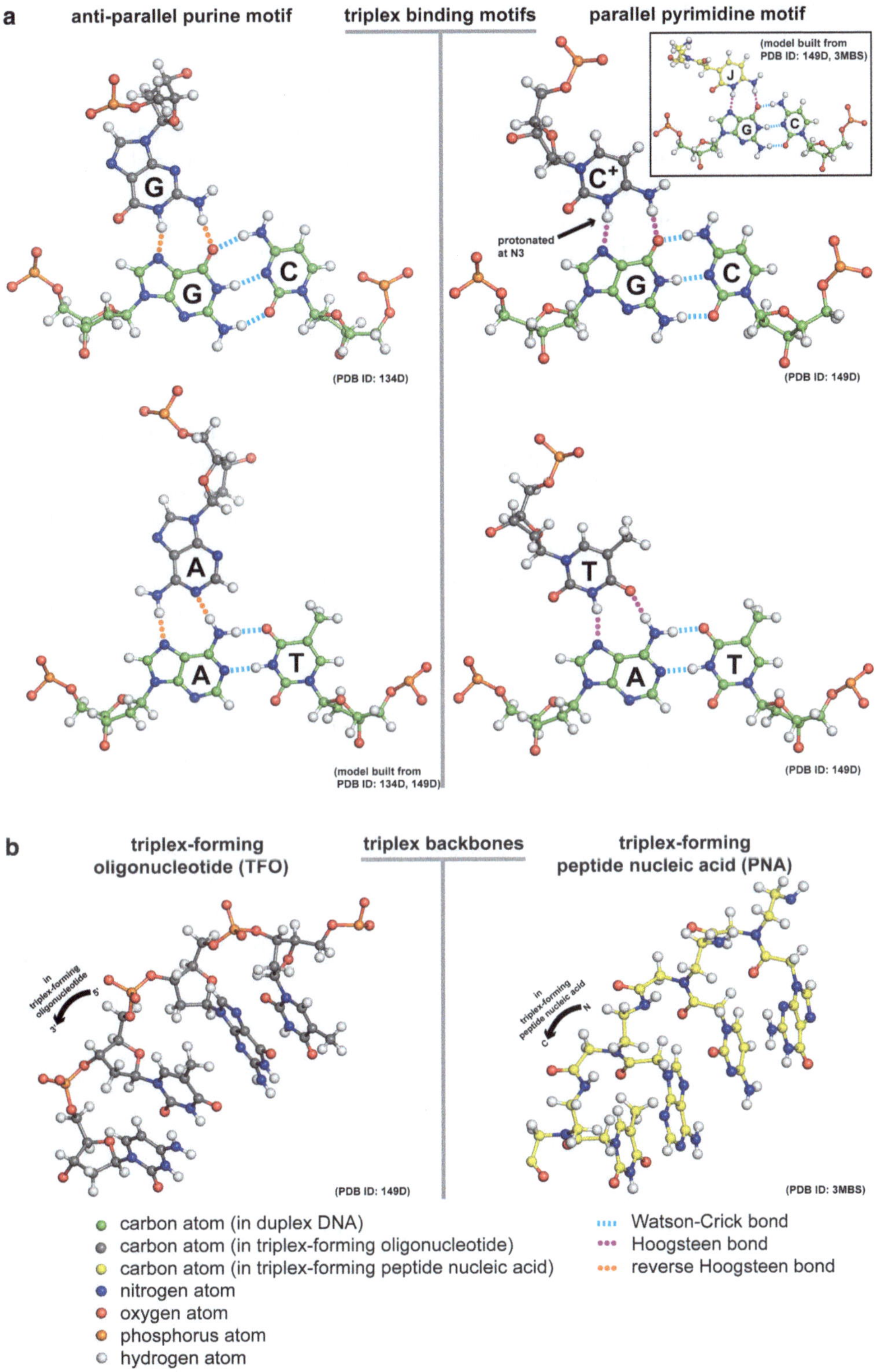

Fig. 2 Triplex binding motifs and triplex backbones of triplex-forming oligonucleotides (TFOs) and triplex-forming peptide nucleic acids (PNAs). (**a**) The triplex binding code permits TFO Hoogsteen binding to Watson–Crick bound base pairs. Genomic duplex DNA base pairs (*in green*) are bound by TFO bases (*in grey*) or triplex-forming PNA bases (*inset*, *in yellow*) using anti-parallel or parallel binding motifs. (**b**) A charged phosphodiester backbone in a DNA-based TFO (*left*) is more prone to degradation by nucleases and less electrostatically

TFOs bind into the major groove of specific sequences of duplex DNA to form the triplex molecule. The DNA sequences amenable to triplex formation typically consist of polypurine bases on one strand and polypyrimidine bases on the other. In the triplex binding code, the third strand can bind duplex DNA in a sequence-specific manner in either a parallel or anti-parallel motif. In the anti-parallel purine motif, a polypurine TFO binds to the purine strand of the DNA duplex through reverse Hoogsteen base pairing in an anti-parallel orientation; in the parallel pyrimidine motif, a polypyrimidine TFO binds to the purine strand of the DNA duplex, but now through regular Hoogsteen base pairing in a parallel orientation.

Another type of unnatural chemistry exists not in the nucleobases but in replacing the phosphodiester backbone with an alternative one, such as the uncharged polyamide backbone in triplex-forming peptide nucleic acids (PNAs) [5] (Fig. 2b). This hybrid molecule, being neither wholly nucleic nor amino acid, is resistant to nuclease and protease degradation, as well as more electrostatically favorable for interaction with the negatively charged genomic DNA backbone [6].

TFOs and triplex-forming PNAs retain the ability to bind sites on genomic and episomal DNA with high thermodynamic stability to form triplexes. Under acidic conditions, the natural base cytosine in a TFO can be protonated so that it can Hoogsteen bind to guanine bases in the genomic target (Fig. 3a). Similar to TFOs, triplex-forming PNAs can form triplexes in the major groove of DNA [7] under normal physiological conditions using the unnatural base pseudoisocytidine (Fig. 3b). Furthermore, by linking a triplex-forming PNA with a strand-invading PNA through a flexible linker, a triplex-forming bis-PNA can simultaneously Hoogsteen and Watson–Crick hydrogen bond sequence-specifically to a DNA target with an physiologically extremely high melting temperature [8] (Fig. 3c). By extending the length of the strand-invading PNA "tail" beyond that of the triplex PNA "clamp" a triplex-forming tail-clamp PNA garners greater affinity as well specificity for the target DNA [9] (Fig. 3d). This binding affinity of triplex-forming PNAs has been demonstrated to be considerable enough in vitro to block DNA polymerization, inhibit transcription initiation and elongation, and prevent sequence-specific protein binding. In addition, the conformational capacity of triplex-forming PNAs to potentially form D-loop open DNA conformations has been shown to act as artificial transcriptional promoters [10–14].

Recombinagenic DNA molecules, also containing varying chemistries and content, have been used to introduce edits to a recombination-potentiated genome locus. These donor DNA molecules are homologous to the region near the genome editing site, sans the nucleobases to be edited, whether by mutation, insertion, or deletion. Unlike the triplex-forming molecule, which has been found to potentiate the recombination machinery upon binding at, near, and as far away as 900 base pairs away from the editing site, the recombinagenic donor DNA molecule directly overlaps the editing

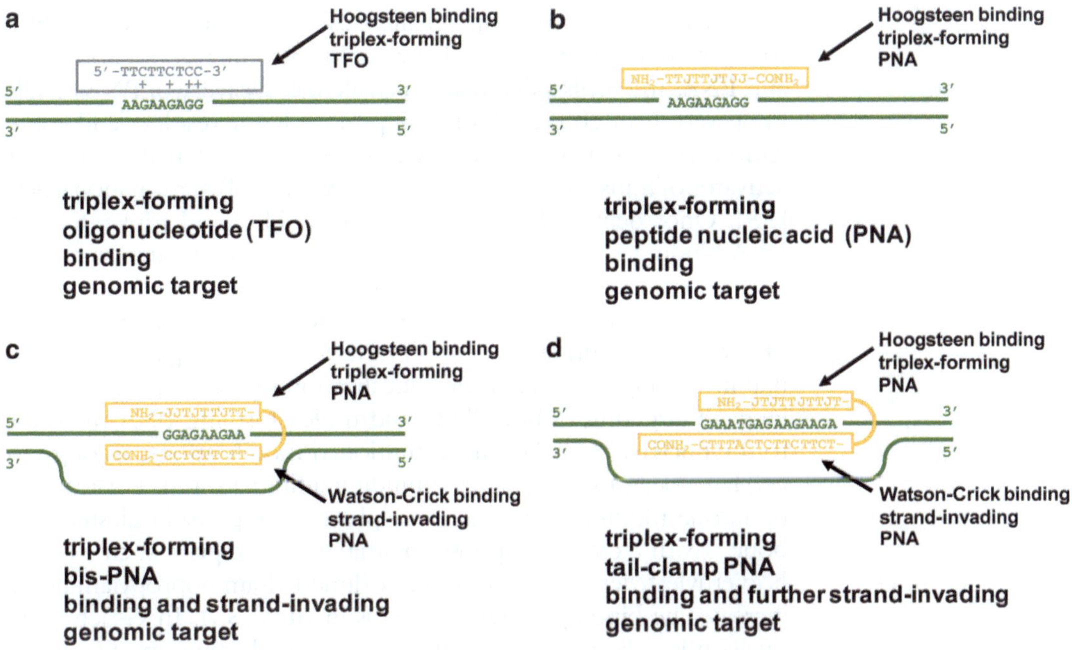

Fig. 3 Various compositions and configurations of TFO and triplex-forming PNA binding to their genomic targets. (**a**) A TFO with natural bases can bind its genomic target in the major groove under acidic conditions that satisfy protonation requirements for cytosine in order to Hoogsteen bond. (**b**) A triplex-forming PNA with unnatural bases can bind its genomic target in the major groove under physiological pH. (**c**) A triplex-forming bis-PNA can utilize both Hoogsteen major groove binding and Watson–Crick strand-invading binding to enhance its stability of binding the genomic target. (**d**) A triplex-forming tail-clamp PNA can extend the Watson–Crick strand-invading domain in order to increase stability and specificity from off-targeting

site itself, thus serving as a competing template to the site's endogenous complementary strand for homologous recombination. However, compositional variations still exist in the donor DNA molecule, such as its length, offset from centering the editing site, and homology to the strand to be edited, or the complementary strand.

The relative resistance of exogenously introduced molecules to cellular nucleic acid defenses can be extended to recombinagenic donor DNA as well. The three terminal nucleotides can be chemically modified so that the phosphodiester backbone linking the nucleobases is replaced with a phosphorothioate backbone. This modification renders the internucleotide backbone resistant to nuclease degradation, and has not been observed to alter the base pairing capabilities of the donor DNA or impact its recombinagenic potential.

By potentiating the cell's endogenous recombination machinery, triplex-forming molecules have been used to alter the behavior of the genome, such as by inducing [15] or inhibiting [16] DNA transcription, DNA replication [17, 18], and protein-DNA interactions [19–22], promoting site-specific DNA damage [23–26], mutagenesis [27, 28], or recombination on both chromosomal and episomal DNA [29–31].

There still exist many challenges in the implementation of triplex technology in vitro and in vivo. Efficient delivery of both TFOs and PNAs into cells still remains a challenge. Delivery of PNAs can be improved through the addition of positive lysine residues and conjugation to cell-penetrating peptides such as Antennapedia and trans-activator of transcription (TAT) [32] as well as polymeric nanoparticles [33]. Once delivered into the cytoplasm, TFO and triplex-forming PNA binding can be inhibited by cellular conditions, such as concentrations of potassium and magnesium. Within the nucleus, triplex-forming molecule interaction may also be limited by the accessibility of a target genomic site as a result of its location and packing in chromatin. Conjugation of a nuclear localization signal has been shown to increase targeting of both TFO and triplex-forming PNA molecules to the nucleus [34]. The aforementioned modifications, among others, have been used to increase binding affinity as well. For example, backbone modifications, such as replacing the phosphodiester backbone with cationic phosphoramidate linkages such as N, N-diethylethylene-diamine or N, N-dimethyl-aminopropylamine can increase the binding affinity of TFOs in vitro [35]. In G-rich oligonucleotides, the use of N, N-diethylethylene-diamine-modified bases may also mitigate G-quartet formation. Physiological pH necessitates the use of pseudoisocytosine rather than cytosine in TFOs and triplex-forming PNAs in order to Hoogsteen pair with the target DNA. Other nucleobase modifications, 5-methyl-2-deoxyuridine and 5-methyl-2-deoxycytidine, and sugar modifications, such as 2-O-(2-aminoethyl)-ribose, have all been shown to increase the binding affinity of TFOs in the pyrimidine motif [36, 37].

Targeted triplex formation leads to gene editing by provoking the cell's own DNA repair pathways, primarily the nucleotide excision repair machinery. TFO-induced editing can lead to heritable, lasting mutations in particular genes by, for example, directing a nonspecific mutagen such as psoralen (pso), which is conjugated to the TFO, near to a gene or directly inactivating a gene of interest. Pso-TFOs have been shown to induce mutations in a target site on plasmids in vitro, when transfected into mammalian cells, and in chromosomal targets [23, 27, 38]. Majumdar et al. [39] showed that by synchronizing and treating cells in late S-phase this targeted mutagenesis could be increased 5.5-fold over those in G1, and 2.5-fold over cells in early S-phase. By modulating the cell cycle or transcriptional state, it may be possible to increase the efficiency of triplex-induced mutagenesis.

In a case of TFO-induced mutagenesis on plasmid in vitro, mutation can be detected and quantified using a reporter system. This reporter system was composed of the supF reporter gene, which encodes an amber suppressor tyrosine tRNA and contains a TFO-binding site, cloned into an SV40 vector. When plated in the presence of 4-chloro-5-bromo-3-indolyl-β-D-galactopyranoside and isopropyl-β-D-1-thiogalactopyranoside, bacteria having an amber mutation in the *lacZ* gene but a functional copy of the *supF* gene produce blue colonies. However, a nonfunctional, or mutated, copy of the *supF* gene produces white colonies. This reporter system of blue and white colonies can be used to calculate the frequency of mutagenesis [23, 40]. Using this reporter system, pso-TFOs have been found to induce mutagenesis in mouse cells containing a chromosomally integrated copy of the *supF* reporter gene [41]. Whereas psoralen has been shown to induce mutations, triplex formation alone using either TFOs or triplex-forming PNAs has also been shown to be sufficient to stimulate mutagenesis in target genes [7, 41]. In addition, in transgenic mice containing the same reporter gene integrated in the chromosome, intraperitoneal injection of the TFO, AG30, also led to site-specific mutations [27].

In order to modify, correct, or replace genes another approach using triplex-forming molecules is to provoke homologous recombination. While it occurs at a low frequency, studies have indicated that DNA damage at the target site can upregulate the frequency of homologous recombination [42]. To exploit this phenomena, pso-TFOs have been used to create site-specific damage in order to sensitize a target site for homologous recombination [43, 44]. It was found that not only could pso-associated DNA damage increase intrachromosomal recombination [45], but also a TFO alone was sufficient to induce homologous recombination [30, 31, 46]. Triplex formation increased the level of recombination at a target site and led to correction of a specific mutation [30]. Using a plasmid with two tandem *supF* genes, each containing different point mutations, intramolecular recombination increased on binding of a TFO, resulting in gene correction of one copy of the gene [43]. When TFOs with or without pso were microinjected into the nuclei of mouse cells containing two mutant copies of the herpes simplex virus thymidine kinase gene in tandem, homologous recombination was stimulated at a frequency of 1 %, which is 2,000-fold over background [31]. Linking a TFO to a short donor fragment, it is thought, would act to position the donor over the target area whereas triplex formation would sensitize the site for homologous recombination [29]. The designed donor fragment would be completely homologous to the site adjacent to the triplex-binding site (TBS) except for the mutation or base correction desired in the target gene. Some studies have determined that antisense (homologous to the transcribed strand) donors are preferred over sense (matching the non-transcribed strand) donors at certain target sites [46] but this is not universally the case [8]. Further studies have shown that the donor does not need to be tethered to the TFO or be adjacent to the TFO-binding site; homologous recombination has been detected at sites up to 750 basepairs away from the triplex-binding site [47]. It has also been shown that TFOs can stimulate intermolecular recombination at a single-copy genomic locus in mammalian cells. In that finding, both pyrimidine and purine motif TFOs were able to induce site-specific recombination in a dose-dependent manner up to a frequency of 0.11 % [46].

Triplex formation has many applications in chromosomal gene targeting and editing. Herein, the use of TFOs or triplex-forming PNAs to induce site-specific therapeutic targeting and editing of a gene is discussed. To use this method of gene targeting and editing, one must first identify potential binding sites (polypurine stretch) within the gene of interest (*see* Subheading 3). A TFO or triplex-forming PNA, which can bind to the target, must be designed and evaluated for binding affinity to the target site. While these triplex-forming molecules are validated by showing a strong binding affinity in vitro and in vivo, further studies can be pursued to optimize delivery into the requisite cells.

There are a number of therapeutic gene targeting and editing capabilities of triplex-forming oligonucleotides and recombinagenic donor DNAs. Among these capabilities are gene correction, disruption, and induction with regulation, as discussed below.

In the capacity of gene correction, triplex-forming oligonucleotides provoke the endogenous homologous recombination machinery to specifically correct one or more bases in a particular gene that causes a deleterious state to the healthy state using recombinagenic donor DNA carrying the corrections. For example, a common beta-thalassemia mutation, that of a guanine to a adenine, has been successfully modified to a healthy state using triplex technology [8]. In a dual-pronged approach, using the aforementioned hamster GFP reporter system the correction modification of this particular splicing mutation was reported, as well as a demonstration of this to create the beta-thalassemia mutation modification in healthy human CD34+ cells. These modifications were specific and heritable, lasting in the progeny of the triplex and donor DNA molecule treated cells over the observation time of 21 days.

In a case of triplex-forming PNA-induced gene correction in vivo, the correction can be observed and measured also using a reporter system. A reporter system is composed of the green fluorescent protein (GFP) gene interrupted by the IVS2 intron of the beta-globin gene with a G→A mutation at position 1 [8]. This particular mutation is a common cause of many thalassemias by disrupting proper splicing of the beta-globin gene. If this mutation goes uncorrected, then the expressed transcript from this locus is aberrantly spliced and retains an additional 47 nucleotides of beta-globin intron in between the GFP transcript, causing the host cell (such as Chinese hamster ovary, or CHO) to not fluoresce (Fig. 4a). However, if this mutation is corrected using a triplex-forming bis-PNA at position 194 in conjunction with a recombinagenic donor DNA carrying the A→G correction at position 1, then the expressed transcript from

Fig. 4 (continued) chloroquine, which causes lysosomes to release molecules such as those transfected, further enhanced the gene correction percentages. **(f)** The particular cell cycle phase followed by uniform chloroquine treatment also affected percentage of gene correction for each of the co-transfected triplex-forming PNAs and donor DNAs

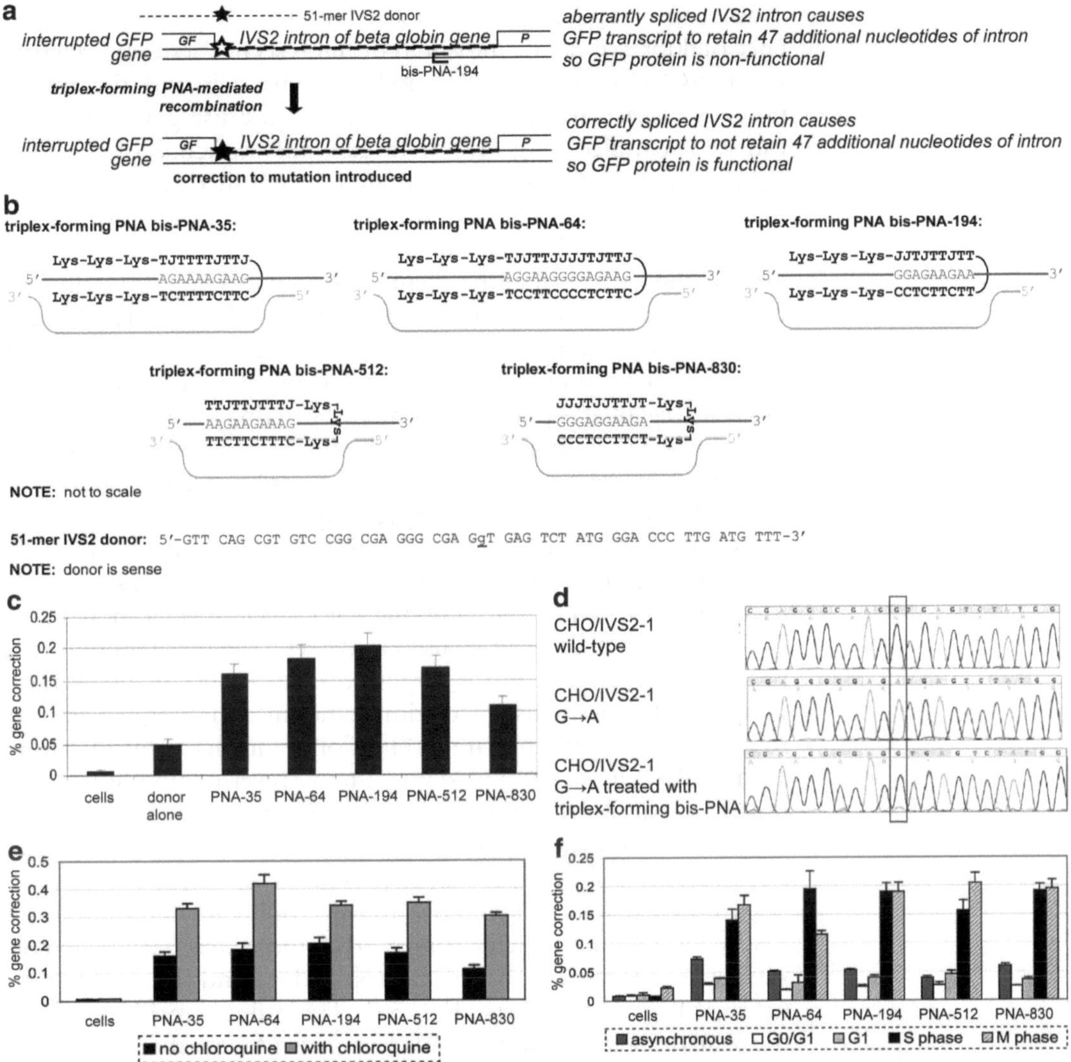

Fig. 4 Model gene correction with the IVS2-1 CHO reporter system using triplex-forming PNAs and recombinagenic donor DNAs. (**a**) Without gene correction of the first position of intron 2 in the human beta-globin gene (IVS2-1), aberrant splicing is unchecked, resulting in a GFP transcript that retains 47 additional nucleotides of the intron and thus a nonfunctional GFP protein and cells that do not fluoresce. This models the common intron 2 mutation that causes beta thalassemia. With gene correction, the aberrant splicing in some cells is corrected, resulting in a GFP transcript without these additional nucleotides and thus a functional GFP protein and cells that do fluoresce. (**b**) The triplex-forming bis-PNAs and recombinagenic donor DNAs that were designed to potentiate recombination in the IVS2 intron region of the beta-globin gene and to introduce the single base A → G correction at IVS2-1. (**c**) Cell studies in this Chinese Hamster Ovary (CHO) cell-based reporter system reveal that the gene correction by recombinagenic donor DNA is enhanced to varying percentages by co-introduction with the triplex-forming bis-PNAs [8]. (**d**) By DNA sequencing genomic DNA from CHO cells not carrying the mutation, i.e., wild-type (*top*), carrying the IVS2-1 G → A mutation (*middle*), and carrying the mutation and having been treated with bis-PNA-35 and the 51-mer IVS2 donor (*bottom*), it was shown that the one month cultured and then sorted GFP-expressing treated cells carried the A → G correction, and thus resembled the wild-type rather than the mutant genome sequence. (**e**) It was also shown that the addition of

this locus is correctly spliced so that the GFP transcript is whole, causing the CHO cell to fluoresce (Fig. 4b). The absence or presence of fluoresce can be measured by fluorescence activated cell sorting (FACS) of these CHO model system cells and the relative potency of each triplex-forming bis-PNA to cause gene correction can be quantified (Fig. 4c). The genomic sequence of the IVS2 region can be interrogated even after 1 month to reveal a stable, heritable correction (Fig. 4d). In addition to the triplex-forming bis-PNAs and donor DNAs themselves, the addition of chloroquine inhibits the lysosomal entrapment of these molecules, thus allowing these molecules to act on the genome and in turn resulting in increased correction percentage (Fig. 4e). The correction percentage can also be modulated by synchronizing the CHO cells at various cell cycle phases prior to treatment with triplex-forming bis-PNA and donor DNA molecules, with S and M phase synchronization permitting more favorable chromatic access and disintegration of the nuclear envelope barrier, respectively (Fig. 4f).

Gene correction can also be detected by various means. If the target is a reporter gene, then the relevant phenotypic assay can be used. If not, then at the genomic level, allele-specific polymerase chain reaction (PCR) can be used to detect a sequence modification in the genomic DNA of the target cells. Allele specific quantitative PCR (qPCR) can be used to estimate modification frequencies relative to samples with known modification frequencies. Furthermore, at the mRNA level, combined quantitative reverse transcriptase real-time PCR (qRT-PR) can be used to interrogate presence and level of expression from the modification of the genome. These materials and methods will be expounded upon in the subsequent sections.

In the capacity of gene disruption, triplex-forming tail-clamp PNAs and recombinagenic donor DNA molecules have been conversely used to create premature stop codons and thus create nonfunctional gene products (Fig. 5a). The tail-clamp PNAs, where the Watson–Crick binding "tails" are extended so that they strand invade further, are thought to confer greater binding propensity and specificity (Fig. 5b). These premature stop codons can thus disrupt a gene that would produce a gene product that conferred a particular vulnerability to the cell. For example, the *CCR5* gene encodes the CCR5 chemokine receptor protein that is vital for HIV-1 infection into human cells. A naturally occurring mutation, called *CCR5*-delta32, has been observed to prevent proper CCR5 chemokine trafficking and ultimately an inability of HIV-1 to enter cells [48]. Schleifman et al. have demonstrated disruption of the *CCR5* gene by introduction of premature stop codons using triplex-forming tail-clamp PNAs and recombinagenic donor DNAs in human CD34+ hematopoietic stem cells (HSCs) [9]. This is particularly relevant for clinical applications, since the sequence-specific with minimal off-target effects of co-transfected triplex-forming and recombinagenic molecules coupled with the self-renewal and heritable capacities of HSCs lends themselves to a

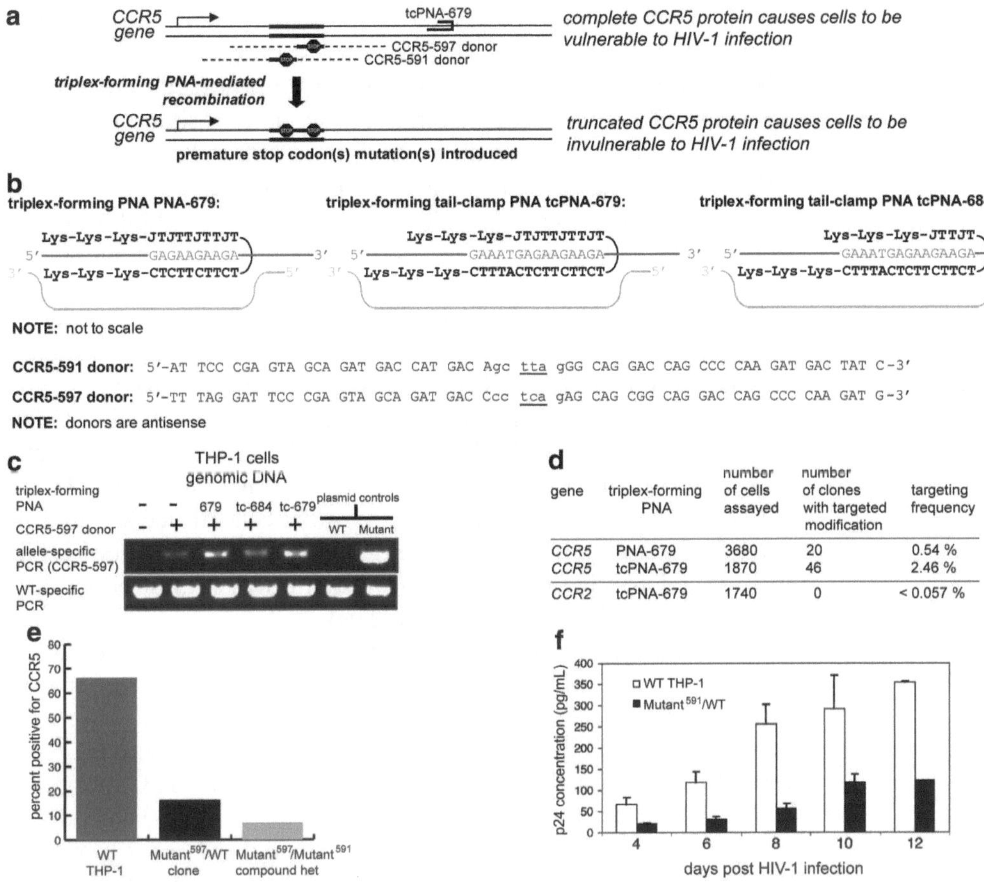

Fig. 5 Model gene disruption using triplex-forming PNAs and recombinagenic donor DNAs. (**a**) Without gene disruption, cells express a complete CCR5 protein that can localize to the cell surface and thus permit a means of infection by HIV-1. With gene disruption through introduction of premature stop codons in the *CCR5* gene, cells express a truncated chemokine receptor that fails to localize to the cell surface and thus abrogates infection by HIV-1. (**b**) The triplex-forming tail-clamp PNAs and recombinagenic donor DNAs that were designed to potentiate recombination in the *CCR5* gene to introduce multiple base modifications for the stop codon and silent mutations to enhance allele-specific PCR sensitivity. (**c**) THP-1 cells transfected with recombinagenic donor DNA and one of the various triplex-forming PNAs, from which genomic DNA was harvested after two days, shows varying levels of gene disruption modification as determined by allele specific PCR [9]. (**d**) Quantification of this gene disruption modification by enumeration of *CCR5* single cell clones and direct deep sequencing indicates that triplex technology has a high on-target frequency and negligible *CCR2* gene off-target frequency, the ratio of the latter to the former being 43-fold less despite the homology between these two gene sequences. (**e**) The cell surface of single and double mutant cells has significant reduced percentages of CCR5 expression, when compared to the parental cell line having a complete CCR5 protein. (**f**) This reduced expression of cell surface CCR5 conferred resistance to HIV-1 infection over the course of 12 days

targeted, specific therapeutic intervention that lasts and propagates further. Furthermore, human CD34+ HSCs are particularly attractive since they have been shown to remain unaffected even in HIV-1 afflicted patients [49].

This demonstration included allele-specific genomic investigation to reveal that the co-transfection of a triplex-forming tail-clamp PNA and donor DNA led to the greatest amount of gene conversion (Fig. 5c). Furthermore direct sequencing of single-cell clones showed that the designed triplex-forming PNA and recombinagenic DNA molecules led to an favorable safety profile, with specific on-target effects that were 43-fold greater than the negligible off-target effects (Fig. 5d). This demonstration also included the preservation of the differentiation capacity of THP-1 human acute monocytic leukemia cells into a macrophage-like state with resulting high levels of CCR5 expression on the cell surface. As such, it was observed using antibody staining and flow cytometry that, while the parental THP-1 could be differentiated to yield fivefold and 17-fold *CCR5* mRNA and CCR5 cell surface protein expression, respectively, the double mutant, i.e., carrying both 591 and 597 disruption modifications had minimal cell surface staining for CCR5 (Fig. 5e). Thus, the triplex-forming PNAs and donor DNAs introduced to create this double mutant was highly effective in disrupting *CCR5* and its gene product.

Ultimately, this disruption of *CCR5* and its gene product by triplex-forming PNAs and donor DNAs was shown to confer resistance to HIV-1 infection. Clones of THP-1 cells treated with these molecules were isolated, expanded, induced to express *CCR5* as described above and then challenged with live HIV-1. Core protein p24 antigen levels, which indicate the amount of HIV-1 infection in cells, were substantially decreased in the treated cells, as compared to the parental THP-1 (Fig. 5f). Thus, this suggests a potential translational opportunity of disrupting *CCR5* using triplex-forming technology to ameliorate or prevent HIV-1 infection.

In the capacity of gene induction and regulation disruption, triplex-forming oligonucleotides and recombinagenic donor DNA molecules have been used not to correct or create mutations, but rather induce functional but dormant genes via their promoters and perturb them by concurrently introducing a regulatory element (Fig. 6a). These induced gene products can rescue the cell from a state of disease [15]. For example, another more general cause of thalassemias, sickle cell disease, and other hemoglobinopathies is a defective beta-globin subunit in adult hemoglobin. Hereditary persistence of fetal hemoglobin (HPFH) is a benign condition in which the fetal gamma-globin subunit is expressed in adults. The targeted induction of this functional fetal gamma-globin subunit in cells carrying the nonfunctional adult beta-globin subunit can mimic HPFH and restore the working state of hemoglobin, thus lessening the severity of these diseases. A triplex-forming bis-PNA targeting the gamma-globin promoter was co-transfected with a HFPH donor at the −117 position that also introduced a hypoxia responsive element (HRE) to regulate the expression of the HPFH state (Fig. 6b). A non-related control

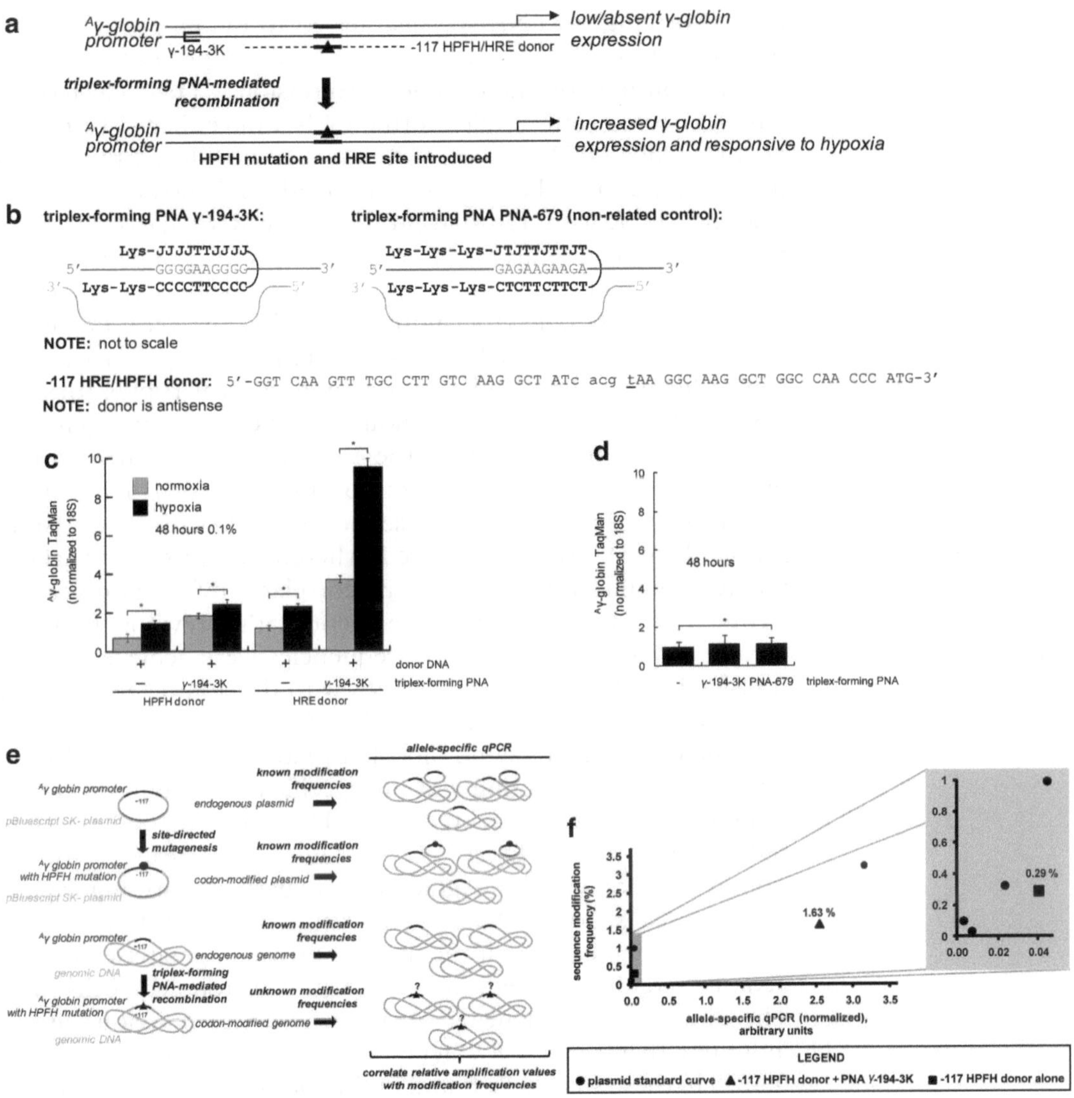

Fig. 6 Model gene induction and regulation using triplex-forming PNAs and recombinagenic donor DNAs. (**a**) Without gene induction and regulation, cells express a faulty adult beta-globin subunit which is the cause of hemoglobinopathies, such as thalassemia and sickle cell disease. With gene induction and regulation through modification of the gamma-globin promoter and introduction of a hypoxia responsive element (HRE), cells express the dormant fetal, yet functional, gamma-globin subunit in adults under exquisite regulation of hypoxia. (**b**) Prior to cell studies, the triplex-forming PNA were designed and tested for binding to their target DNAs. (reproduced from ref. 15). (**c**) The HRE recombinagenic donor, in combination with a triplex-forming PNA, leads to elevated expression of gamma-globin, whereas this induction is attenuated using a hereditary persistence of fetal hemoglobin (HPFH) donor. (**d**) This gene modification is due to the co-transfection of the recombinagenic donor and triplex-forming PNA, as the triplex-forming PNA alone is not acting as an artificial transcriptional promoter to the gamma-globin gene. (**e**) Quantification of the level of modification by allele specific qPCR of samples of known modification frequencies are correlated to those of unknown modification frequencies. (**f**) The modification of the gamma-globin promoter is greater with triplex-forming PNA and recombinagenic donor DNA, rather than with recombinagenic donor DNA alone

triplex-forming bis-PNA was also tested to examine whether the triplex-forming PNAs were acting as artificial promoters to and thus responsible for inducing gene expression. CD34-selected cell samples treated with a triplex-forming PNA and either −117 HRE/HPFH donor DNA or control donor that introduced the HPFH mutation but not HRE were subjected to hypoxia for 48 h. The qRT-PCR with TaqMan probes of the RNA harvested from these and an analogous set of normoxia cells for comparison indicated that gamma-globin expression was being regulated by hypoxia (Fig. 6c). Furthermore, qRT-PCR of RNA harvested from CD34-selected cell samples treated with a gamma-globin triplex-forming PNA and a non-related control triplex-forming PNA both showed little to no induction of expression of gamma-globin, thus suggesting that the mutation and element introduced by the donor DNA was responsible (Fig. 6d). To quantify the amount of gene promoter sequence modification, a plasmid standard curve was constructed whereby allele-specific qPCR relative amplification values were correlated with modification frequencies (Fig. 6e). In agreement with previous findings, greater sequence modification frequencies were achieved by co-transfection of triplex-forming PNA with the donor DNA, rather than with the donor DNA alone (Fig. 6f).

These gene therapeutic modalities have been translated to in vitro and in vivo targeting and site-specific editing in human cells using nanoparticles for delivery [33]. Human hematopoietic stem cells engrafted in NOD-*scid IL2r* γ^{null} mice were treated via intravenous injection with biodegradable nanoparticles carrying triplex-forming PNAs, donor DNAs, or a mixture of both. Of these, an in vitro screen indicated that the mixture provided greater efficacy of genome targeting and editing of human hematopoietic cells [50]. The direct deep sequencing of the genome from various tissues indicated that the *CCR5* gene was modified at frequencies of 0.43 % in these stem cells found in the spleen, and at 0.05 % in the bone marrow, with off-target effects two orders of magnitude lower than these frequencies. Furthermore, in vivo testing of an enhanced GFP-beta-globin reporter mouse further confirmed the potency of the triplex-forming PNA with donor DNA mixture over that of the donor DNA alone. This suggests that direct in vivo modification is not only possible, but effective, and possibly necessary for disease for which cells cannot be harvested, modified ex vivo and reintroduced.

Thus, an overall therapeutic goal for triplex-forming technology is to then harness homologous recombination and utilize it with a recombinagenic donor DNA to create a modified genome that is heritable and permanent (Fig. 7, inset). In the broader scheme of engrafting corrected cells and tissues, this technology may play a pivotal role in transforming and transplanting one's own once disease-carrying but now disease-free cells (Fig. 7). The modality of genetic disease treatment is using triplex-forming oligonucleotides

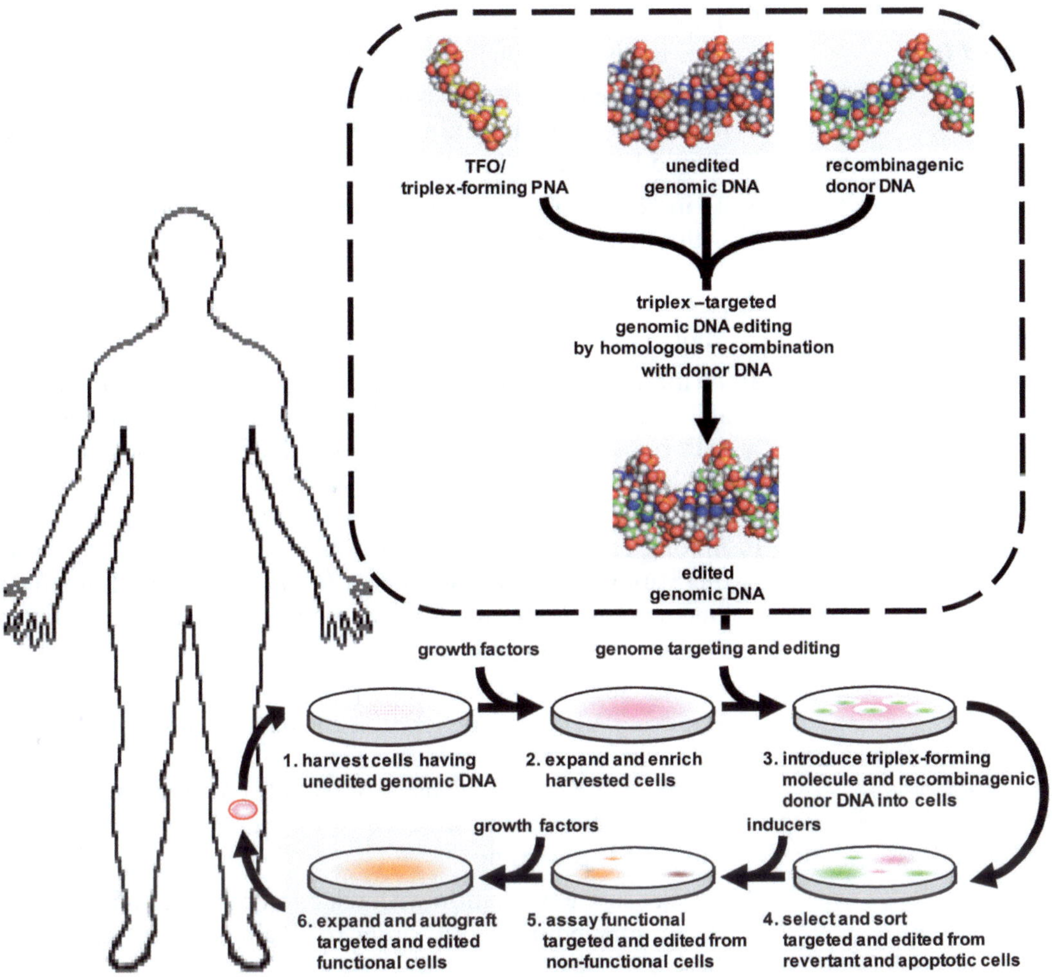

Fig. 7 Overall therapeutic application of triplex-mediated genome targeting and editing technology. Cells harvested from the affected individual can be expanded and readied, e.g., synchronized, for treatment with triplex-forming molecules and recombinagenic DNA molecules. Corrected cells can be selected and sorted from those that failed to reprogram or those that reverted or died. Prior to further expansion and autografting into the affected individual, assays can assess the satisfactory function of the reprogrammed cells from those that were reprogrammed but remain nonfunctional

versatile, as evidenced by the examples of gene correction, disruption, or induction and concomitant regulation and suitable for therapeutic genome targeting and editing in human beings.

2 Materials

Most oligonucleotides discussed here, including pso-conjugated TFOs, triplex-forming PNAs, and single-stranded donor DNAs, can be ordered from commercial vendors. All oligonucleotides should be purified using either high-pressure liquid chromatography or gel purification.

2.1 Triplex-Forming PNAs and TFOs

1. Triplex-forming PNAs can be ordered from Bio-Synthesis (Lewisville, TX), or Panagene (South Korea).
2. To increase electrostatic complementarity, and thus affinity, to negatively charged genomic DNA, the triplex-forming PNA should be modified with one or more terminal lysine residues.
3. To increase resistance to nucleases, the TFOs should be modified with one or more amine groups on the 3′ end (*see* **Note 1**).

2.2 Recombinagenic Donor DNAs

1. Recombinagenic donor DNAs can be ordered from The Midland Certified Reagent Company (Midland, TX).
2. To increase resistance to nucleases, and thus intracellular availability, the recombinagenic donor DNA should be modified with phosphorothioate linkages on the first 3 bases on 5′ end and last 3 bases on 3′ end (*see* **Note 2**).

2.3 Cells

1. Among cells successfully used for gene targeting and editing, there are Chinese hamster ovary (CHO) cells, and human cell lines, such as K562 and THP1, primary human CD34+ hematopoietic progenitor cells.
2. Vectors such as *FLuc+* from pGL3-Basic Vector (Promega, Madison, WI) or pBlueScript II-SK have been used.

2.3.1 Maintaining Chinese Hamster Ovary Cell Cultures

1. To maintain Chinese hamster ovary cells, the medium includes Ham's F12 medium with 10 % of fetal bovine serum, and 2 mM of L-glutamine.
2. Further supplementation with selection marker, such as hygromycin, may be needed in order to retain genomic cassettes in these cells (*see* **Note 3**).

2.3.2 Maintaining K562 Cell Cultures

1. For K562 cells the medium is composed of RPMI with 10 % of fetal bovine serum.
2. If necessary to maintain viable cell cultures, penicillin and streptomycin may be used (*see* **Note 4**).

2.3.3 Maintaining Primary Human CD34+ Hematopoietic Progenitor Cell Cultures

1. To maintain primary human CD34+ hematopoietic progenitor cells obtained and selected from the aphaeresis of granulocyte colony stimulating factor mobilized peripheral blood of healthy donors, StemSpan Serum-Free Expansion Media (StemCell Technologies Inc., Vancouver, Canada) should be used.
2. Media should be supplemented with StemSpan CC110 cytokine mixture.
3. The media contains bovine serum albumin, recombinant human insulin, human transferrin (iron-saturated), 2-mercaptoethanol, Iscove's Modified Dulbecco's Medium, and other supplements.

4. The cytokine mixture contains 100 ng/mL rh Flt-3 Ligand, 100 ng/mL rh Stem Cell Factor, 20 ng/mL rh IL-3, and 20 ng/mL rh IL-6 (*see* **Note 5**).

2.4 Solutions and Buffers

1. All solutions and buffers may be stored at room temperature unless otherwise noted.
2. In the TFO binding assay (*see* Subheading 3) the triplex-binding buffer is composed of:
   (a) 10 mM of Tris–HCl (pH 7.6).
   (b) 0.1 mM of $MgCl_2$.
   (c) 1 mM of spermine.
   (d) 10 % of glycerol (with or without 140 mM potassium) (*see* **Note 6**).
3. In the triplex-forming PNA binding assay (*see* Subheading 3) the silver stain solution is composed of:
   (a) Sodium borohydrate 0.1 % silver nitrate (Sigma Aldrich, St. Louis, MO), e.g. 1.0 g of $AgNO_3$ in 1 L of dH_2O (*see* **Note 7**).
4. In the triplex-forming PNA binding assay (*see* Subheading 3) the developer solution is composed of:
   (a) 15.0 g of NaOH.
   (b) 0.1 g of $NaBH_4$.
   (c) 5 mL of formaldehyde (*see* **Note 8**).
5. For restriction protection assay (*see* Subheading 3) the lysis buffer is composed of:
   (a) 50 mM of Tris–HCl (pH 7.5).
   (b) 20 mM of EDTA.
   (c) 100 nM of NaCl.
   (d) 0.1 % of sodium dodecyl sulfate.
   (e) TE pH 8.0.
   (f) 10 mM of Tris–HCl (pH 8.0).
   (g) 1 mM of EDTA (pH 8.0).

3 Methods

3.1 Identification of Target DNA Sites

TFOs and triplex-forming PNAs bind with high affinity to in the binding motifs described earlier. In the anti-parallel purine motif, for example, the target sequence should have a homopurine run; mixed sequences will decrease the triplex-forming molecule's binding affinity. By scanning the gene of interest, these homopurine stretches can be identified and TFOs that bind in either a parallel or anti-parallel orientation can be designed. TFO-binding sites should be typically 14–30 base pairs in length, while triplex-forming PNA-polypurine binding sites should be at least eight to ten base pairs in length.

Beyond adhering to the triplex binding code, it may be necessary to identify several possible target DNA sites. This is due to the aforementioned issue of accessibility in the chromatin structure and organization in the genome. In addition, the DNA target site should contain a 5′ TpA site at either the 5′- or 3′-end of the polypurine run, should psoralen be conjugated to the TFO. Psoralen is conjugated to site-specifically induce cross-links. DNA target sites may include regulatory regions of the gene of interest, which may be effective for modulating gene splicing and expression.

3.2 Design and Synthesis of Triplex-Forming Molecules

For each target DNA site under consideration, a variety of triplex-forming molecules can be designed and synthesized. Below are the guidelines that will help to determine which molecule will work best.

3.2.1 For the Design and Synthesis Triplex-Forming Oligonucleotides (TFOs)

For A-rich target sites, the pyrimidine motif is preferred and requires TFOs containing C and T or their analogs (C+ will form Hoogsteen bonds with a G in a G:C base pair, and T with the A in an A:T base pair). These TFOs will bind in an orientation parallel to the purine-rich strand of the genomic DNA target duplex in order to form the triplex.

For G-rich target sites, the purine motif is preferred and requires TFOs containing A (or T) and G or their analogs (G forms reverse Hoogsteen bonds with the G in a G:C base pair and A or T with the A in an A:T base pair). These oligonucleotides will bind in an anti-parallel orientation, forming reverse Hoogsteen bonds, with the polypurine strand of the duplex.

To induce targeted cross-links at the triplex site, psoralen can be conjugated on the 5′ or 3′ end of the TFO using psoralen phosphoramidites.

TFOs containing backbone or base modifications can also be synthesized to create a molecule with a high binding affinity to the target site (*see* **Note 9**).

3.2.2 For the Design and Synthesis of Triplex-Forming Peptide Nucleic Acids (PNAs)

Most often a triplex-forming bis-PNA containing two pyrimidine PNAs connected by a flexible linker (8-amino-3,6-dioxaoctanoic acid [O]) is used for targeting. This molecule oriented N-terminus to C-terminus parallels a 5′–3′orientation of a cognate DNA strand. As noted above, by extending the length of the strand-invading PNA "tail" beyond that of the triplex PNA "clamp" a triplex-forming tail-clamp PNA garners greater affinity as well specificity for the target DNA [9].

To aid in strand invasion, several lysine residues can be put on the triplex-forming bis-PNA. These positively charged residues have also been shown to facilitate uptake into cells [51]. Other conjugates such as the cell-penetrating peptides transactivator of transcription, the HIV transactivator of transcription, and Antennapedia, the third helix of the *Drosophila* homeodomain transcription factor, have also been shown to increase uptake of triplex-forming PNAs [52].

To overcome the pH dependence of N3-protonation of cytosine, pseudoisocytosine, J, can be used in place of cytosine in the Hoogsteen strand. This J base forms two Hoogsteen bonds with a G:C base pair [53].

3.2.3 For the Design and Synthesis of Recombinagenic Donor DNA

Single-stranded donors can be 30–2,000 bases in length and can be homologous to any region within 750 basepairs of the triplex site [46] (*see* **Note 10**).

Antisense (binding to the sense strand of the DNA) or sense (binding to the antisense strand of the DNA) donors can be designed. Since triplex-meditated editing by homologous recombination edits both strands of genomic DNA, evaluation with sense an antisense donors, and in co-introduction with various triplex-forming molecules, is recommended for assessing recombination efficiencies.

Donors can be synthesized by Midland Certified Reagent Company, or Panagene, or long double-stranded donors can be synthesized by PCR amplification of a plasmid. If donors are ordered from Midland or Panagene, the first and last 3 bases should be attached through phosphorothioate linkages to inhibit nuclease degradation.

When designing a donor, desired sequence changes should be kept toward the center of the oligo, and a sufficient number of bases must be on either side of the mismatch to allow for homologous recombination within the target.

3.3 Evaluation of Binding Under Physiological Conditions

3.3.1 TFO Binding Assay

To evaluate the binding of a TFO, a gel mobility shift assay is used. For this, a synthetic duplex containing the potential TFO-binding site should be designed. To do this, complementary oligomers containing the target sequence can be annealed to form duplexes.

The duplex DNA can then be 5′ end-labeled using T4 Polynucleotide kinase and (γ-32P) dATP in the reaction. Typical reactions contain 10^{-6} M duplex (final concentration) in a total volume of 20 μL.

Next electrophorese the duplex on a 15 % polyacrylamide gel to purify. Following electroelution of the duplex, add this purified duplex to a Centricon-3 column (Millipore, Bedford, MA) to concentrate (as per the manufacturer's instructions).

Incubate the duplex DNA at 37 °C for approx 12–24 h in a series of reactions containing increasing concentrations of TFO in triplex-binding buffer. A typical reaction is 20 μL, with 2 μL (10^{-6} M) of labeled duplex and 2 μL of tenfold dilutions of TFO (10^{-12}–10^{-7} M).

Next, run the binding reactions on a 12 % native (19:1 acrylamide:bis-acrylamide) gel at 60–70 V in 89 mM Tris, 89 mM of boric acid with 0.1 mM of MgCl to achieve separation of the triplex structures from the duplex.

The gel can then be imaged with a PhosphorImager (Amersham Biosciences, Piscataway, NJ) to quantify the band shift. This data can be used to calculate the dissociation constant of the TFO-binding to the target site, which represents the concentration at which binding is half-maximal.

3.3.2 Triplex-Forming PNA Binding Assay

In Eppendorf tubes add the triplex-forming PNA at concentrations ranging from 0 to 1 μM, 2 μg of plasmid DNA containing the target site flanked by known restriction enzymes, KCl to a final concentration of 10 μM, and TE to a final volume of 10 μL.

These tubes should be incubated overnight at 37 °C. On the following morning the entire reaction should restriction digested in 20 μL to recover the binding site. Note that the digest should progress for 1–2 h at a temperature suitable for all enzymes involved. To terminate the digest, restriction enzymes can be chosen that have heat inactivation properties. Ideally the digest should yield a single band during gel electrophoresis at a size of approximately 150–200 base pairs.

Add DNA loading dye to each sample and run them on a 10 % of polyacrylamide gel or 8 % native gel (19:1 in TBE) until bands are well separated. For example, the 8 % native gel is composed of 10 mL of 40 % 19:1 acrylamide–bisacrylamide, 10 mL of 5× TB, 200 μL of 0.5 M EDTA, 29.25 mL of dH_2O, 500 μL of 10 % ammonium persulfate, and 50 μL of TEMED, for a final working volume of 50 mL.

To visualize the resulting gel shift, first stain with silver stain solution for about 10 min, and then rinse the gel with dH_2O. Then develop with developer solution for 1 min, and again rinse the gel with dH_2O.

The expected result is that at 0 μM, only a single band representing the duplex fragment containing the target site should be seen (*see* **Note 11**). As concentrations of triplex-forming PNAs increase, there should be a shift in the band corresponding to binding.

3.4 Evaluation of Intracellular Binding

Several strategies exist to physically evaluate intracellular binding of a TFO or triplex-forming PNA to the target site. If psoralen is conjugated to the TFO or triplex-forming PNA, a restriction protection assay, a denaturation resistance assay or a PCR-based assay [54] can be used.

3.4.1 Restriction Protection Assay and Fragment Identification by Southern Blot

For this assay, a restriction enzyme (e.g., *Bam*HI) recognition site overlapping the target site is necessary. Cells containing your target site should be used. Design and synthesize a pso-TFO that will bind to the target site and direct a cross-link to a position within the restriction enzyme recognition site. Transfect 2–5 μg of this pso-TFO into cells containing the chromosomal target site (*see* **Note 12**).

Incubate the cells at 37 °C for 2–6 h post-transfection and then irradiate for a total dose of 1.8 J/cm^2 of ultraviolet A (UVA) irradiation (UVA light source centered at 365 nm; Southern New England Ultraviolet, Branford, CT). This will allow psoralen photoactivation to cross-link the pso-TFO to the duplex DNA.

Immediately following UVA irradiation, wash the cells with phosphate-buffered saline and resuspend in heated (60 °C) lysis buffer at a concentration of 5×10^6 cells/mL for 15 mins.

Treat with proteinase K at a concentration of 100 μg/mL overnight at 37 °C. Extract the genomic DNA by addition of phenol to the lysates once, followed by two extractions with chloroform–isoamyl alcohol (24:1). Ethanol precipitate the DNA and resuspend in TE pH 8.0 and 100 mM of KCl (~100 μL).

Incubate samples at 60 °C for 2 h to disrupt non-covalent triplexes and remove by filtration through a Centricon column (Millipore) (*see* **Note 13**). Digest the genomic DNA with the appropriate restriction enzyme (e.g., *Bam*HI) and two restriction enzymes that will cut out a defined fragment containing the target site.

Analyze the DNA by Southern blot using a designed probe to identify the fragment.

The expected result is that if cross-linking occurred at the target site, the psoralen adduct in the *Bam*HI site will block cleavage at this site. The product resulting from the restriction protection of the *Bam*HI site can be visualized and quantified in comparison with the extent of the *Bam*HI digestion products. This can then be used to quantify TFO directed pso adduct formation in the target.

3.5 Evaluation of Intracellular Binding of Pso-TFOs

Cells containing the endogenous gene or a chromosomally integrated exogenous gene can be used in the following assay [55].

3.5.1 Denaturation-Resistance Assay

Design and synthesize a pso-TFO, which binds to the target, and transfect 2–5 μg into cells.

Repeat steps as above for digesting genomic DNA. Digest 10–20 μg of genomic DNA with flanking restriction enzyme(s) that will cut out the target region.

Denature the digested DNA by heating in a solution of 90 % formamide (to a final concentration of ~80 %) to 80 °C for 15 mins.

Electrophorese the samples in a neutral 1.5 % TAE agarose gel.

Transfer the DNA from the gel to a nylon filter/membrane by Southern blot and hybridize the membrane with a 32P-labeled probe that contains the target region.

The expected results are that once denatured, un-cross-linked DNA fails to reanneal after it enters a neutral environment, whereas cross-linked DNA will "snap back" to reform a duplex. Denatured cross-linked DNA when run in a neutral gel will migrate at the same rate as nondenatured DNA, whereas un-cross-linked DNA will migrate faster. The relative amount of the denatured (fast migrating) and the snap back (slow migrating) bands can then be quantified as a measure of target site cross-linking.

3.6 Evaluation of Induced Recombination

A luciferase assay was created to evaluate triplex-induced gene correction by homologous recombination [40]. The *Fluc* reporter gene can be engineered to contain the target site from any gene (TBS) upstream of the coding region of luciferase. This wild-type construct can then be mutated through site-directed mutagenesis to insert a stop codon mutation within the *Fluc* coding region, which will prevent full translation of the gene. When a TFO or triplex-forming PNA binds to the TBS, it can stimulate recombination between an introduced DNA donor and the *luciferase* gene to correct the mutation. Site-specific recombination can then be evaluated based on the luciferase enzyme activity in the cell lysates when compared with a standard curve, which can be created by mixing cell populations containing the chromosomally integrated wild-type or mutant construct. This reporter system can be used as a model for recombination at any chromosomal gene of interest. By placing the TBS of the target gene upstream of the *luciferase* gene, a construct is created that can be chromosomally integrated to create stable cell lines.

These cell lines allow for systematic testing of various DNA-binding molecules for their effects on recombination at chromosomal targets. This assay can be used as an intermediate to evaluate induction of homologous recombination at a chromosomal target by a DNA-binding molecule at a specific target site and allows for quantification of this induction.

3.6.1 Luciferase Assay

A construct containing the TFO target site upstream of the start site of the wild-type firefly *luciferase* gene, *FLuc+*. A stop codon can then be created at varying distances from the target site (24–750 basepairs) [40] using site-directed mutagenesis such as the readily available QuikChange Site-Directed Mutagenesis Kit (Stratagene, LaJolla, CA). Vector containing the wild type *Fluc+* should be used as a control.

Donors are designed containing correct wild-type sequence and bind to the editing site, with some mismatches. Furthermore, TFO or triplex-forming PNA are designed to bind to the targeting site.

Wild-type and mutant cell lines can be stably created using the above constructs. Determine that the construct is integrated in a single copy by Southern blot.

Mix wild-type and mutant cells in fixed ratios to establish standard curves. Dilute mutant cells to a concentration of 1×10^5 cells/mL and seed 50,000 mutant cells per well into a 12-well plate. Dilute wild-type cells to 5×10^4 cells/mL, serially dilute cells and add to the wells containing the mutant cells.

Grow cells 48 h and then harvest and lyse. Analyze lysates as described earlier.

This standard curve can be used to normalize the experimental data and allow for quantification of recombination frequencies.

Then 24 h before transfection seed 12-well plates with wild-type or mutant cells so they will be approx 50 % confluent the day of transfection.

The following day cells are transfected using Geneporter 2 (Gene Therapy Systems, San Diego, CA) according to manufacturer's instructions delivering various concentrations of donor DNA alone, and donor DNA cotransfected with TFO to each well.

Incubate the cells for 24–48 h at 37 °C to allow for cellular repair and recombination processes to occur.

Then 48-h post-transfection, rinse the cells two times with phosphate-buffered saline (Gibco BRL, Carlsbad, CA) and then lyse cells with 250 μL of 1× passive lysis buffer (Promega).

Assay luciferase activity (gene correction) using the Promega Dual Luciferase Kit (Promega) and a luminometer (Berthold Technologies, Oak Ridge, TN), as per manufacturer's instructions.

The expected result is that if TFO-induced recombination between the donor and the gene occurred, correction of the stop codon would have taken place. This would restore translation of a fully functional *Luciferase* gene, resulting in luciferase activity above levels seen in mock transfected cells or cells treated with donor alone.

3.7 Evaluation of Target Gene Correction via Mutagenesis

Allele-specific PCR is a general technique to assess the amount of gene correction via mutagenesis using triplex-forming and recombinagenic donor DNA molecules. Different applications of this powerful technique can provide both qualitative and quantitative assessments of gene correction frequency.

3.7.1 Genotyping Analysis by Allele-Specific PCR

After treating cells containing the target gene with TFO, triplex-forming PNA, and/or donor DNA, gene modification can be detected by allele-specific PCR [56, 57]. Harvest genomic DNA from the treated cells using Wizard SV Genomic DNA Purification System (Promega) and dilute it to approx 50 μg/μL. A forward primer can be designed with its 3′ end containing the desired mutation, and the reverse primer should be similar in length and T_m. Add 50 ng of the genomic DNA to a 25 μL PCR reaction and a gradient run to determine the optimal annealing temperature of the primers. Plasmids containing the wild-type and mutant gene can be used as control templates (*see* **Note 14**).

The expected results are that a band should be present on the gel in lanes with mutant template and no bands for wild-type DNA (*see* **Note 15**). Alternatively, real-time PCR can be used for a more sensitive measure of PCR products.

3.8 Evaluation of Target Gene Disruption via Mutagenesis

To quantify the amount of gene modification in cells treated with TFO, triplex-forming PNA, and/or donor DNA, the genomes of single-cell clones can be directly deep sequenced [9, 50]. Cells can be treated in bulk and then diluted into multi-well dishes so that approximately a single cell on average occupies a single well. Each single-cell clone can proliferate within its dish to a density sufficient to extract the genomic DNA using a variant of the aforementioned System, Wizard SV 96 Genomic DNA Purification System (Promega).

3.8.1 Genotyping Analysis by Single-Cell Clone Direct Deep Sequencing

For direct deep sequencing, forward and reverse primers can be designed to insert barcode-tagging and to encapsulate the desired locus with the location of the mutation position centered, and be similar in length and T_m. Determine the optimal anneal temperature of the primers using a gradient run and 50 ng of the genomic DNA to a 25 μL PCR reaction. Gel electrophorese the PCR products and purify the desired bands using the QAquick gel purification kit (Qiagen). Ligate the direct deep sequencing PCR samples to adapters and sequence with 75 base-pair paired-end reads on an Illumina HiSeq platform (Illumina Inc., San Diego, CA).

The expected results are that a band should be present on the gel representing the amplified locus that was flanked by the primers. The ratio of the number of clones harboring the targeted modification to the total number of single-cell clones assayed represents the targeting frequency.

4 Notes

1. Note that triplex-forming PNAs and TFOs should be aliquoted and kept in the −20 °C freezer for storage. Also note that during usage for experiments, an aliquot of triplex-forming PNAs can be thawed but kept cold on ice.
2. Note that recombinagenic donor DNAs should be aliquoted and kept in the −20 °C freezer for storage. Also note that during usage for experiments, an aliquot of recombinagenic donor DNA can be thawed but kept cold on ice.
3. Note that an antibiotic such as hygromycin is light-sensitive. Thus, exercise caution when supplementing medium with hygromycin by shielding the medium flash with aluminum foil.
4. Note that antibiotics such as penicillin and streptomycin are not light-sensitive.
5. Note that CD34+ cells cannot be maintained indefinitely without differentiating. Thus it is advisable that they be obtained fresh or pre-frozen, and used immediately over a course of a few days.

6. Note that this buffer has been composed so that it is appropriate for evaluating TFO and triplex-forming PNA binding under physiological conditions.
7. Note that, for ideal gel staining, silver stain solution should be fresh and, if possible, during the time the loaded gel is running through the electrophoresis apparatus.
8. Note that, for ideal gel developing, developer solution should be fresh and, if possible, during the time the loaded gel is being stained.
9. Note that to prevent degradation by 3′ exonucleases it is recommended that TFOs have a 3′ end cap of an amine group chemical modification. Also note that triplex-forming PNAs do not require this chemical modification, since their peptide backbones make them resistant to degradation.
10. Note that the optimal donor length and offsets from the recombination center will vary from editing site to site. This often makes it necessary to try various donor candidates and at various concentrations.
11. Note that, rather than a single band, multiple bands may appear during the gel shift. This is not unexpected, since many triplex-forming PNAs have the ability to form multiple complex triplex structures with their target DNA.
12. Use transfection methods and parameters that are suitable for the cell type in use. Some of these transfection methods include digitonin permeabilization, electroporation, cationic lipids, or nanoparticles. It is advisable to optimize the transfection method and parameters that provide high delivery efficiency and low cell damage and/or death.
13. Note that the appropriate size Centricon column should be used based on the molecular weight of the duplex and/or triplex molecule to be filtered.
14. Note that first the wild-type gene can be cloned, to which mutations can be introduced with site-directed mutagenesis via the QuikChange Site-Directed Mutagenesis Kit (Stratagene, LaJolla, CA).
15. Note that, initially, 40 cycles of PCR can be applied during allele-specific PCR using genomic DNA. In general, this provides sufficient amplification over background. For allele-specific PCR using plasmid DNA, since it is much easier to PCR-amplify from the latter, it is preferable to reduce the concentration rather than modify the allele-specific PCR cycling program, if possible.

Acknowledgments

We gratefully acknowledge members of the Glazer Laboratory for helpful discussions. This work was supported by a National Institutes of Health (NIH) grant R01HL082655 and by a Doris Duke Innovations in Clinical Research Award (to P.M.G.). A National Institute of Diabetes and Digestive and Kidney Diseases Experimental and Human Pathobiology Postdoctoral Fellowship from NIH grant T32DK007556 also provided support (to F.R.). The authors declared no conflict of interest.

References

1. Watson JD, Crick FHC (1953) Molecular structure of nucleic acids—a structure for deoxyribose nucleic acid. Nature 171: 737–738
2. Felsenfeld G, Davies DR, Rich A (1957) Formation of a 3-stranded polynucleotide molecule. J Am Chem Soc 79:2023–2024
3. Moser HE, Dervan PB (1987) Sequence-specific cleavage of double helical DNA by triple helix formation. Science 238:645–650
4. Le Doan T et al (1987) Sequence-specific recognition, photocrosslinking and cleavage of the DNA double helix by an oligo-[alpha]-thymidylate covalently linked to an azidoproflavine derivative. Nucleic Acids Res 15:7749–7760
5. Seidman MM (2004) Oligonucleotide mediated gene targeting in mammalian cells. Curr Pharm Biotechnol 5:421–430
6. Demidov VV et al (1994) Stability of peptide nucleic-acids in human serum and cellular-extracts. Biochem Pharmacol 48:1310–1313
7. Faruqi AF, Egholm M, Glazer PM (1998) Peptide nucleic acid-targeted mutagenesis of a chromosomal gene in mouse cells. Proc Natl Acad Sci U S A 95:1398–1403
8. Chin JY et al (2008) Correction of a splice-site mutation in the beta-globin gene stimulated by triplex-forming peptide nucleic acids. Proc Natl Acad Sci U S A 105:13514–13519
9. Schleifman EB et al (2011) Targeted disruption of the *CCR5* gene in human hematopoietic stem cells stimulated by peptide nucleic acids. Chem Biol 18:1189–1198
10. Hanvey JC et al (1992) Antisense and antigene properties of peptide nucleic acids. Science 258:1481–1485
11. Koppelhus U, Zachar V, Nielsen PE, Liu XD, EugenOlsen J, Ebbesen P (1997) Efficient in vitro inhibition of HIV-1 gag reverse transcription by peptide nucleic acid (PNA) at minimal ratios of PNA/RNA. Nucleic Acids Res 25:2167–2173
12. Praseuth D et al (1996) Peptide nucleic acids directed to the promoter of the alpha-chain of the interleukin-2 receptor. Biochim Biophys Acta 1309:226–238
13. Nielsen PE, Egholm M, Berg RH, Buchardt O (1993) Sequence specific-inhibition of DNA restriction enzyme cleavage by PNA. Nucleic Acids Res 21:197–200
14. Mollegaard NE, Buchardt O, Egholm M, Nielsen PE (1994) Peptide nucleic acid·DNA strand displacement loops as artificial transcription promoters. Proc Natl Acad Sci U S A 91:3892–3895
15. Chin JY, Reza F, Glazer PM (2013) Triplex-forming peptide nucleic acids induce heritable elevations in gamma-globin expression in hematopoietic progenitor cells. Mol Ther 21:580–587
16. Faria M et al (2000) Targeted inhibition of transcription elongation in cells mediated by triplex-forming oligonucleotides. Proc Natl Acad Sci U S A 97:3862–3867
17. Birg F et al (1990) Inhibition of simian virus 40 DNA replication in CV-1 cells by an oligodeoxynucleotide covalently linked to an intercalating agent. Nucleic Acids Res 18:2901–2908
18. Volkmann S, Jendis J, Frauendorf A, Moelling K (1995) Inhibition of HIV-1 reverse transcription by triple-helix forming oligonucleotides with viral-RNA. Nucleic Acids Res 23:1204–1212
19. Maher LJ, Wold B, Dervan PB (1989) Inhibition of DNA-binding proteins by oligonucleotide-directed triple helix formation. Science 245:725–730
20. Francois JC, Saisonbehmoaras T, Thuong NT, Helene C (1989) Inhibition of restriction endonuclease cleavage via triple helix formation by homopyrimidine oligonucleotides. Biochemistry 28:9617–9619
21. Hanvey JC, Shimizu M, Wells RD (1990) Site-specific inhibition of EcoRI restriction modification enzymes by a DNA triple helix. Nucleic Acids Res 18:157–161

22. Mayfield C et al (1994) Triplex formation by the human Ha-ras promoter inhibits Sp1 binding and in vitro transcription. J Biol Chem 269:18232–18238
23. Havre PA, Gunther EJ, Gasparro FP, Glazer PM (1993) Targeted mutagenesis of DNA using triple helix-forming oligonucleotides linked to psoralen. Proc Natl Acad Sci U S A 90:7879–7883
24. Takasugi M et al (1991) Sequence-specific photo-induced cross-linking of the two strands of double-helical DNA by a psoralen covalently linked to a triple helix-forming oligonucleotide. Proc Natl Acad Sci U S A 88:5602–5606
25. Vasquez KM, Wensel TG, Hogan ME, Wilson JH (1996) High-efficiency triple-helix-mediated photo-cross-linking at a targeted site within a selectable mammalian gene. Biochemistry 35:10712–10719
26. Wang G, Seidman MM, Glazer PM (1996) Mutagenesis in mammalian cells induced by triple helix formation and transcription-coupled repair. Science 271:802–805
27. Vasquez KM, Narayanan L, Glazer PM (2000) Specific mutations induced by triplex-forming oligonucleotides in mice. Science 290:530–533
28. Wang X, Tolstonog G, Shoeman RL, Traub P (1996) Selective binding of specific mouse genomic DNA fragments by mouse vimentin filaments *in vitro*. DNA Cell Biol 15:209–225
29. Chan PP, Lin M, Faruqi AF, Powell J, Seidman MM, Glazer PM (1999) Targeted correction of an episomal gene in mammalian cells by a short DNA fragment tethered to a triplex-forming oligonucleotide. J Biol Chem 274:11541–11548
30. Datta HJ, Chan PP, Vasquez KM, Gupta RC, Glazer PM (2001) Triplex-induced recombination in human cell-free extracts—dependence on XPA and HsRad51. J Biol Chem 276:18018–18023
31. Luo ZJ, Macris MA, Faruqi AF, Glazer PM (2000) High-frequency intrachromosomal gene conversion induced by triplex-forming oligonucleotides microinjected into mouse cells. Proc Natl Acad Sci U S A 97:9003–9008
32. Rogers FA, Manoharan M, Rabinovitch P, Ward DC, Glazer PM (2004) Peptide conjugates for chromosomal gene targeting by triplex-forming oligonucleotides. Nucleic Acids Res 32:6595–6604
33. McNeer NA, Chin JY, Schleifman EB, Fields RJ, Glazer PM, Saltzman WM (2011) Nanoparticles deliver triplex-forming PNAs for site-specific genomic recombination in CD34+ human hematopoietic progenitors. Mol Ther 19:172–180
34. Branden LJ, Mohamed AJ, Smith CIE (1999) A peptide nucleic acid-nuclear localization signal fusion that mediates nuclear transport of DNA. Nat Biotechnol 17:784–787
35. Vasquez KM, Dagle JM, Weeks DL, Glazer PM (2001) Chromosome targeting at short polypurine sites by cationic triplex-forming oligonucleotides. J Biol Chem 276:38536–38541
36. Lacroix L et al (1999) Triplex formation by oligonucleotides containing 5-(1-propynyl)-2′-deoxyuridine: decreased magnesium dependence and improved intracellular gene targeting. Biochemistry 38:1893–1901
37. Puri N et al (2002) Minimum number of 2′-O-(2-aminoethyl) residues required for gene knockout activity by triple helix forming oligonucleotides. Biochemistry 41:7716–7724
38. Wang G, Glazer PM (1995) Altered repair of targeted psoralen photoadducts in the context of an oligonucleotide-mediated triple-helix. J Biol Chem 270:22595–22601
39. Majumdar A et al (2003) Cell cycle modulation of gene targeting by a triple helix-forming oligonucleotide. J Biol Chem 278:11072–11077
40. Macris MA, Glazer PM (2003) Transcription dependence of chromosomal gene targeting by triplex-forming oligonucleotides. J Biol Chem 278:3357–3362
41. Vasquez KM, Wang G, Havre PA, Glazer PM (1999) Chromosomal mutations induced by triplex-forming oligonucleotides in mammalian cells. Nucleic Acids Res 27:1176–1181
42. Sargent RG, Rolig RL, Kilburn AE, Adair GM, Wilson JH, Nairn RS (1997) Recombination-dependent deletion formation in mammalian cells deficient in the nucleotide excision repair gene *ERCC1*. Proc Natl Acad Sci U S A 94:13122–13127
43. Faruqi AF, Seidman MM, Segal DJ, Carroll D, Glazer PM (1996) Recombination induced by triple-helix-targeted DNA damage in mammalian cells. Mol Cell Biol 16:6820–6828
44. Sandor Z, Bredberg A (1995) Triple-helix directed psoralen adducts induce a low-frequency of recombination in an SV40 shuttle vector. Biochim Biophys Acta 1263:235–240
45. Faruqi AF, Datta HJ, Carroll D, Seidman MM, Glazer PM (2000) Triple-helix formation induces recombination in mammalian cells via a nucleotide excision repair-dependent pathway. Mol Cell Biol 20:990–1000
46. Knauert MP, Kalish JM, Hegan DC, Glazer PM (2006) Triplex-stimulated intermolecular recombination at a single-copy genomic target. Mol Ther 14:392–400

47. Knauert MP et al (2005) Distance and affinity dependence of triplex-induced recombination. Biochemistry 44:3856–3864
48. Samson M et al (1996) Resistance to HIV-1 infection in Caucasian individuals bearing mutant alleles of the CCR-5 chemokine receptor gene. Nature 382:722–725
49. Shen H et al (1999) Intrinsic human immunodeficiency virus type 1 resistance of hematopoietic stem cells despite coreceptor expression. J Virol 73:728–737
50. McNeer NA et al (2012) Systemic delivery of triplex-forming PNA and donor DNA by nanoparticles mediates site-specific genome editing of human hematopoietic cells in vivo. Gene Ther 20:658–669
51. Sazani P et al (2001) Nuclear antisense effects of neutral, anionic and cationic oligonucleotide analogs. Nucleic Acids Res 29:3965–3974
52. Koppelhus U, Awasthi SK, Zachar V, Holst HU, Ebbesen P, Nielsen PE (2002) Cell-dependent differential cellular uptake of PNA, peptides, and PNA-peptide conjugates. Antisense Nucleic Acid Drug Dev 12:51–63
53. Egholm M, Christensen L, Dueholm KL, Buchardt O, Coull J, Nielsen PE (1995) Efficient pH-independent sequence-specific DNA-binding by pseudoisocytosine-containing bis-PNA. Nucleic Acids Res 23:217–222
54. Diviacco S, Rapozzi V, Xodo L, Helene C, Quadrifoglio F, Giovannangeli C (2001) Site-directed inhibition of DNA replication by triple helix formation. FASEB J 15: 2660–2668
55. Shahid KA et al (2006) Targeted cross-linking of the human beta-globin gene in living cells mediated by a triple helix forming oligonucleotide. Biochemistry 45:1970–1978
56. Orou A, Fechner B, Utermann G, Menzel HJ (1995) Allele-specific competitive blocker PCR—a one-step method with applicability to pool screening. Hum Mutat 6:163–169
57. Parsons BL, McKinzie PB, Heflich RH (2005) Allele-specific competitive blocker—PCR detection of rare base substitution. Methods Mol Biol 291:235–245
58. Radhakrishnan I, Patel DJ (1993) Solution structure of a purine.purine.pyrimidine DNA triplex containing G.GC and T.AT triples. Structure 1:135–152
59. Radhakrishnan I, Patel DJ (1994) Solution structure of a pyrimidine.purine.pyrimidine DNA triplex containing T.AT, C+.GC and G.TA triples. Structure 2:17–32
60. Yeh JI et al (2010) The crystal structure of non-modified and bipyridine-modified PNA duplexes. Chemistry 16:11867–11875

Chapter 9

Targeting *piggyBac* Transposon Integrations in the Human Genome

Daniel L. Galvan, Claudia S. Kettlun, and Matthew H. Wilson

Abstract

DNA based transposon systems offer a technology for active and efficient delivery of genes into human cells. An emerging field is directed at manipulating such systems to achieve site-directed integration as compared to un-targeted integration which occurs with native or unmodified transposon systems. The naturally active *piggyBac* transposon system is derived from insects but has been shown to be very efficient in gene-modifying human cells. Recent efforts have utilized the fusion of DNA binding domains to the *piggyBac* transposase enzyme with the goal of targeting integration to specific locations in the human genome. In this chapter, we describe methodology for engineering and characterizing chimeric *piggyBac* transposase enzymes, including experimental approaches for evaluating activity and targeting specificity in the human genome.

Key words *piggyBac*, Site-directed integration, Transposon, Transposase, Zinc finger, Human cells

1 Introduction

Gene correction promises permanent repair of genetic deficiencies through manipulation of eukaryotic genomic DNA. Both viral and nonviral systems can be employed for insertion of new DNA into the human genome. These techniques, while powerful, necessitate an understanding of insertional locations within the genome. Undirected DNA insertional events can have undesired negative consequences by altering necessary gene products. For example, uncontrolled DNA modification of cells could result in a spectrum of results from precipitating cellular apoptosis to immortalization. Current research is focused on targeting insertional events to user-defined chromosomal locations which will not exert undesired pressures.

Our investigations have centered on an insect derived transposable system, *piggyBac* [1, 2]. This is a two component system in which the transposase enzyme, *piggyBac*, is codelivered with cognate terminal repeat elements flanking transposon cargo DNA.

Francesca Storici (ed.), *Gene Correction: Methods and Protocols*, Methods in Molecular Biology, vol. 1114,
DOI 10.1007/978-1-62703-761-7_9,

The enzyme is then capable of catalyzing the insertion of the transposon cargo into TTAA sequences in the genomic DNA. It is direct movement of DNA into DNA by a cut and paste mechanism and has advantages in that it is active in human cells [3], has the ability to deliver large DNA sequences [4, 5], and excisions leave no footprints [3, 6]. We and others have demonstrated that it is possible to direct or enrich for specific insertional site selection through joining of a DNA-binding protein to the transposase thereby creating a chimeric *piggyBac* enzyme [7–10].

We describe a methodology to create and test a DNA-sequence target-specific transposase using a zinc-finger DNA binding domain fused to *piggyBac*. First, the chimeric protein must be made and tested. Key features include localization of the zinc-finger to the amino-terminus of the *piggyBac* protein. In addition, an extraneous immunological-tag permits protein detection and an intervening flexible protein linker between the two subunits is described for optimal activity. Transposase expression and activity should be assessed in human cell lines.

Two complementary approaches can be used to verify transposition activity. The first is a colony count assay in which long-term acquisition of antibiotic resistance is utilized to approximate the ability of the chimeric protein to result in transposon integration. The second interplasmid assay verifies transposition between plasmid donors and acceptors within human cells. This assay in combination with DNA sequencing allows for recovery of DNA insertion points and intricate control of the DNA acceptor sequence.

Having verified the efficacy of the system, a chromatin immunoprecipitation assay may be used to verify actual binding of chimeric protein to its intended DNA sequence target. Finally, qPCR can be used to quantitatively access copy number, or the number of insertions, in selected cells lines. PCR may be used to recover the locations of inserted and stably integrated transposons in human cell lines to confirm site-directed integration.

2 Materials

2.1 Generation of a Chimeric ZFP-piggyBac Transposase

1. Oligonucleotide primer sequences for recombinant polymerase chain reaction (PCR) fusion of a ZFP binding domain to the amino-terminus of the *piggyBac* transposase (*see* Table 1).
2. pCMV-*piggyBac* transposase expression plasmid (*see* Fig. 1a). Plasmids may be obtained from the authors. The plasmid containing the ZFP utilized herein (subcloned from pcDNA3.1-CHK2-ZFP-KOX) was provided by and can be obtained from Sangamo Biosciences, Inc.
3. High fidelity thermostable DNA polymerase (e.g., Expand Taq, Roche Applied Biosciences).

Table 1
Primers for cloning the *piggyBac* transposase behind the CHK2-ZFP to create a chimeric transposase

ZFPpigbacF	CGCGGGATCCGGCGGCTCTGGAGGTTCCGGAGG CTCTGGTGGTTCTGGAACTAGTATGGGTAGTT CTTTAGACGATGAG
ZFPpigbacR	GATGCGTATGCCTACTCGAGTCAGAAACAACTTT GGCACATATC

All primers are 5′–3′, left to right

4. Standard reagents for PCR and cloning.
5. Calf alkaline phosphatase.
6. Restriction enzymes.

2.2 Western Blot of a Recombinant Chimeric ZFP-piggyBac Transposase

1. HEK293 cells transfected with expression plasmids.
2. RIPA Lysis Buffer: 50 mM Tris–HCl, pH 7.5, 150 mM NaCl, 1 % IPEGAL, 0.5 % deoxycholate, 0.1 % SDS, 1× protease inhibitor cocktail (Complete Mini, Roche Applied Science).
3. 10 % SDS-PAGE gels, PVDF membrane, and related equipment.
4. Mouse, anti-HA.11 antibody, 16B12 (Covance).
5. Rabbit anti-β-tubulin antibody (LiCOR), for detection of loading control.
6. IRDye800CW conjugated anti-mouse antibody (LICOR) and IRDye680 conjugated anti-rabbit antibody (LICOR) for secondary detection.
7. Tris buffered saline pH 7.6 (Invitrogen) with 0.1 % tween-20 (Sigma) (TBS-T) with 5 % nonfat dry milk (blocking agent).
8. Odyssey detection system (LICOR).

2.3 Transposition Assay

1. Human embryonic kidney (HEK293) cells (American Type Culture Collection).
2. HEK293 Medium: Minimum essential medium with 2 mM l-glutamine, 10 % fetal bovine serum, 50 U/ml penicillin, and 50 μg/ml streptomycin, sterile filtered.
3. FuGene-6 transfection reagent (Roche Applied Science).
4. Transposon plasmids: pTpB (confers neomycin resistance) or pT-eGFP (expresses eGFP and confers puromycin resistance) (*see* Fig. 1b).
5. Trypsin (0.25 % w/v) in 1 mM EDTA solution (Invitrogen).

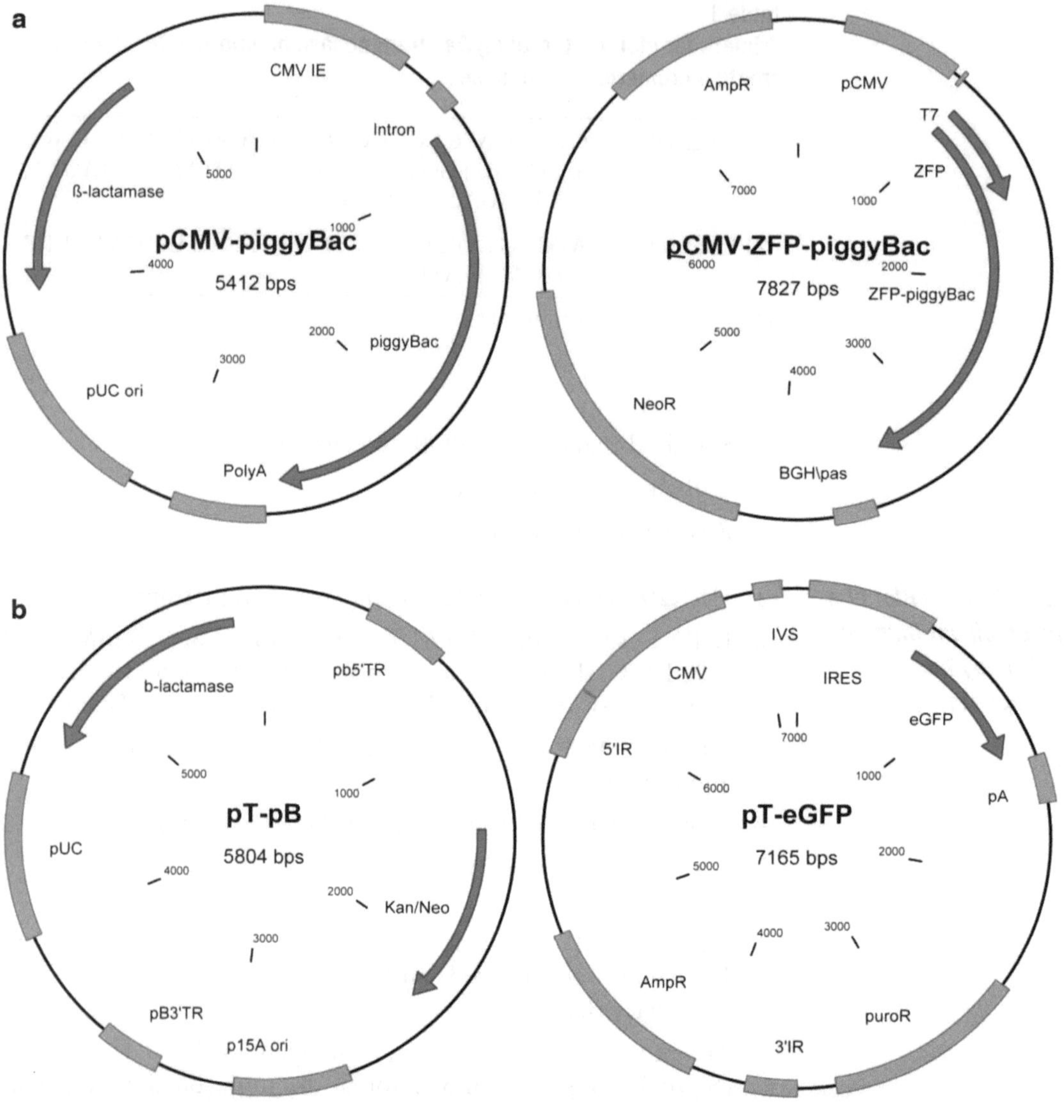

Fig. 1 Schematic plasmid maps for transposase (pCMV-*piggyBac*, *panel a*, *left*; and, pCMV-ZFP-*piggyBac*, *panel a*, *right*) and transposon (pTpB, *panel b*, *left*; and pT-eGFP, *panel b*, *right*). Abbreviations: *ZFP* zinc finger protein, *IR* inverted repeat element, *CMV* immediate early cytomegalovirus promoter, *IRES* internal ribosomal entry sequence, *IVS* intervening sequence, *intron* intron for improved expression, *eGFP* enhanced green fluorescent protein, *pA* polyadenylation sequence, *puromycinR* puromycin resistance gene (transcriptionally regulated by a SV40 promoter), β-*lactamase* confers ampicillin resistance (AmpR), *NeoR* neomycin (G418) resistance, *pUC* origin of replication, *p15A* origin of replication

6. G418: final concentration of 800 μg/ml in HEK293 Medium (Invitrogen).
7. 10 % Formalin Fixative (Azer Scientific, NBF-4-G).
8. Methylene blue (0.5 % w/v) in PBS, filtered.

Table 2
Primers for ChIP of the CHK2 site in human cells

ZFG11766F	GCAGATACAAACTCCACCCTCAG (native site)
ZFG12016R	CTATCTAGCCGTGGTCACTCG (native site)
pT3ChIPF	AGATGCCAGGAAGATACTTAACAG (engineered target site)
pT3ChIPR	GTCAATTGGAGGTGTACCTGTGGATG (engineered target site)

The top two primers allow PCR recovery of the endogenous CHK2 target site in the human genome. The bottom two primers allow PCR recovery from the engineered target sites described in Fig. 4

2.4 Chromatin Immunoprecipitation Assay

1. Oligonucleotide primer sequences for PCR recovery of ChIP assay products (*see* Table 2).
2. HEK293 cells transfected with transposases.
3. Sonic dismembrator (Fisher Scientific, Model 100).
4. Mouse, anti-HA.11 antibody, 16B12 (Covance).
5. Protein A/G Sepharose (Thermo).
6. BSA, Fraction V, Herring Sperm DNA, and Proteinase K (Sigma).
7. 1 % freshly prepared formaldehyde: 1 % paraformaldehyde, 36 mM sodium phosphate (dibasic), 14 mM sodium phosphate (monobasic), 1 mM EDTA.
8. 10× Tris–Glycine Buffer: 1.92 M glycine, 0.25 M Tris.
9. L1 buffer: 50 mM Tris HCl, pH 8.0, 10 % glycerol, 2 mM EDTA, 1 mM DTT, 0.1 % NP40.
10. Sonication Buffer: 50 mM Tris–HCl, pH 8.0, 150 mM NaCl, 5 mM EDTA, 1 % Triton-X-100, 0.1 % SDS, 0.1 mM DTT.
11. IP Buffer 1: 0.2 M NaCl, 10 mM Tris–HCl, pH 8.0, 10 % glycerol, 1 % Triton-X-100, 1 mM EDTA.
12. IP Buffer 2: 0.5 M NaCl, 10 mM Tris–HCl, pH 8.0, 1 % Triton-X-100, 0.1 % SDS, 1 mM EDTA.
13. IP Buffer 3: 0.3 M NaCl, 10 mM Tris–HCl, pH 8.0, 1 % Triton-X-100, 0.1 % SDS, 1 mM EDTA.
14. IP Buffer 4: 90 mM sodium bicarbonate, 1 % SDS, 10 mM DTT.
15. De-cross-linking solution: 2 M NaCl, 0.1 M EDTA, 0.4 M Tris pH 6.8, 20 Units/ml Proteinase-K.

2.5 Quantitative PCR for Copy Number Analysis

1. Stably transfected HEK293 cells with transposon integrations.
2. 1× iQ SYBR Green master mix (Bio-Rad).

Table 3
Primers for qPCR to determine integrated transposon copy number

qPB5IRF	CTAAATAGCGCGAATCCGTC (amplifies *piggyBac* 5′IR sequence)
QPB5IRR	TCATTTTGACTCACGCGG (amplifies *piggyBac* 5′IR sequence)
qKNF	CGGATGGAAGCCGGTCTTGTC (amplifies Kan/Neo cassette)
qKNR	AGAAGGCGATAGAAGGCGATG (amplifies Kan/Neo cassette)
RPF	AGATTTGGACCTGCGAGCG (amplifies *RNaseP*)
RPR	GAGCGGCTGTCTCCACAAGT (amplifies *RNaseP*)

3. Oligonucleotide primers for quantitative PCR (qPCR) of transposon integrations and *RNaseP* (the control gene) (*see* Table 3).
4. CFX96 Bio-Rad Real-Time System with C1000 Thermal Cycler, or equivalent.

2.6 Interplasmid Transposition Assay

1. HEK293 cells transfected with transposase plasmid (*ccdB*-pCMV-*piggyBac* or *ccdB*-pCMV-ZFP-*piggyBac*), transposon plasmid (*ccdB*-pTpB), and acceptor plasmid (pUC57-ZFS or pUC57-scrambled ZFS (control)) (*see* Subheading 3.6). These plasmids are available from the authors by request.
2. Hirt Buffer A: 10 mM Tris–HCl, pH 9.0, 10 mM EDTA, 10 mM EGTA.
3. Hirt Buffer B: 10 mM Tris–HCl, pH 9.0, 10 mM EDTA, 10 mM EGTA, 2 % SDS.
4. Top10 electrocompetent cells (Invitrogen).
5. Ampicillin/kanamycin-resistant bacterial agar plates.

2.7 PCR Recovery of Integration Sites

1. HEK293 cells stably transfected with transposon DNA.
2. DNeasy Blood and Tissue kit (Qiagen).
3. PCR primers for recovery of integration sites (*see* Table 4).
4. TOPO-TA cloning kit (Invitrogen).
5. DNA sequencing facility.

Table 4
Primers for PCR recovery of transposon integration sites

KN2171	ACCGCTTCCTCGTGCTTTAC (first round)
5/3PBF	CGGGATCCTATCTATAACAAGAAAATATATATATAA (first round)
3/5PBR	CGGAATTCTTTTGTTACTTTATAGAAGAAATTTTGA (first round)
KN2231	GCCTTCTTGACGAGTTCTTCTG (second round)
3nestPB92	CCTCGATATACAGACCGATAAAACAC (second round)
5nestPB121	TTGACTCACGCGGTCGTTATAG (second round)

These primers permit nested PCR recovery of integration sites. The first 3 are used in the first round of PCR and the last 3 are used in the second round of PCR

3 Methods

3.1 Generation of a Chimeric ZFP-piggyBac Transposase

1. Use PCR to create a chimeric transposases having the DNA binding domain of the CHK2-ZFP (or a ZFP of interest) fused to the N-terminus of the *piggyBac* transposase primer sequences provided in Table 1 and [8] (*see* **Notes 1–3**). The cDNA for the *piggyBac* transposase was PCR amplified from pCMV-piggyBac for cloning into the pCMV-ZFP plasmid [11]. The PCR product encoding the transposase was cloned into pCMV-ZFP to fuse it behind the CHK2-ZFP using *Bam*HI/*Xho*I sites.
2. The resultant pCMV-ZFP-*piggyBac* plasmid includes a hemagglutinin epitope tag (HA-tag, YPYDVPDYA) followed by an in-frame SV40 T-antigen nuclear localization signal (MPKKKRKV), the corresponding ZFP DNA binding domain, a 19 amino acid linker (GTGSGGSGGSGGSGTSGTG), and the transposase cDNA.
3. Verify all DNA constructs by DNA sequencing to exclude polymerase errors and to demonstrate an intact open reading frame.

3.2 Western Blot of Recombinant Chimeric ZF Transposases

1. Culture HEK293 cells in HEK293 Medium at 37 °C and 5 % CO_2. Transfect HEK293 cells at 30–50 % confluence in a 60 mm dish with FuGENE-6 and 3 μg of transposase plasmid.
2. Three days after transfection, lyse the cells in RIPA Lysis Buffer containing a 1× protease inhibitor cocktail.
3. Determine the protein concentrations (e.g., by Bicinchoninic Acid assay).

4. Electrophorese equivalent amounts of protein on a 10 % SDS-PAGE gel.
5. Transfer gel to a PVDF membrane and probe in TBS-T with blocking agent with an anti-HA antibody (1:2,000 dilution) followed by incubation with a secondary anti-mouse antibody conjugated to IRDye800 (1:5,000 dilution).
6. Detect epitope tagged proteins by Odyssey imaging detection. An example of expressed chimeric transposase enzymes is illustrated in Fig. 2a (*see* **Note 4**).

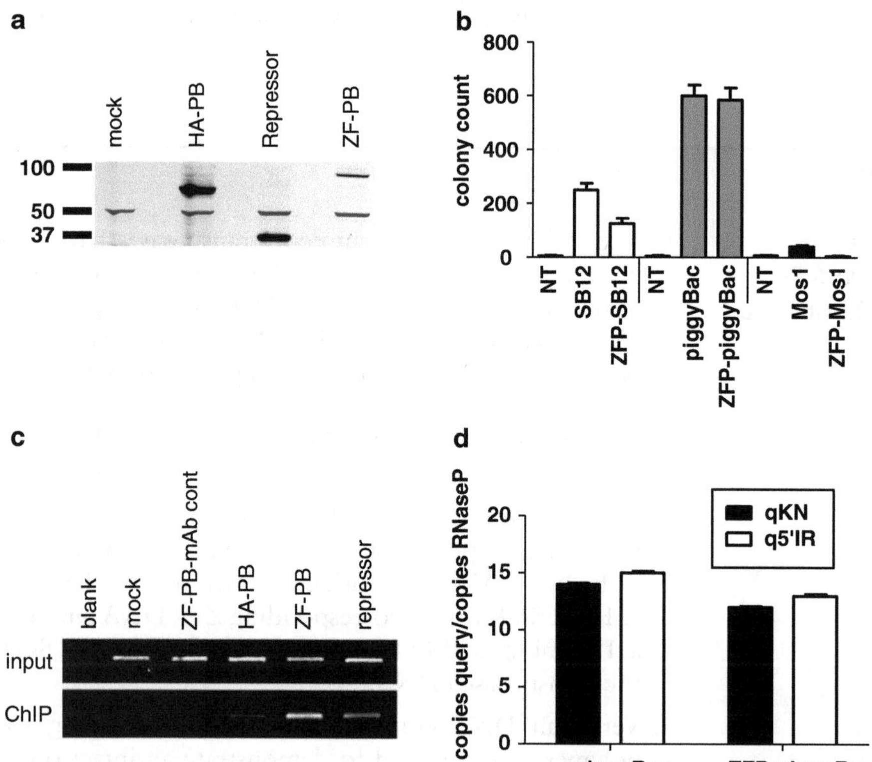

Fig. 2 (**a**) Immunoblot analysis showing the relative level of ZFP and *piggyBac* proteins expressed in HEK 293 cells. The 50 kDa band is β-tubulin, which serves as a loading control. (**b**) Comparison of gene transfer activity of native transposase to ZFP-transposase using SB12 (hyperactive *Sleeping Beauty* 12), *piggyBac*, or Mos1 in HEK-293 cells ($n=3\pm$SEM). NT corresponds to no transposase transfected in the presence of the respective transposon. (**c**) Chromatin immunoprecipitation (ChIP) of ZFP-*piggyBac* at the target ZFP site in the human genome. *ZFP-PB-mAb cont* monoclonal antibody control, *HA-PB* HA tagged native *piggyback* transposase, *repressor* ZFP-KOX protein (a control ZFP fusion protein with a KOX repressor domain). ChIP demonstrates that the target locus was recovered using ZFP-*piggyBac* and repressor (positive control protein) but not HA-tagged *piggyBac*. (**d**) Quantitative PCR (qPCR) was used to determine the number of transposons integrated into human cells. The copy number determined was normalized to the genomic copy number for *RNaseP* ($n=3\pm$SD). Adapted from [8] with permission

3.3 Transposition Assay

1. Transfect HEK293 cells in a 60 mm dish with both transposase and transposon (which contains a neomycin resistance gene) plasmids using FuGene-6.
2. Two days after transfection, split cells to varying densities (1:15 and 1:150 dilutions) and place in medium containing 800 μg/ml G418 for selection.
3. After 2 weeks of selection, fix colonies of cells for 10 min at room temperature (RT) using 10 % formalin, stain with methylene blue solution for 2 or more hours at RT, wash, dry, and count. A control plate transfected without the transposase is necessary to observe random events. The colonies present result from stable integration of the transposon DNA which confers G418 resistance. The number of colonies visualized is therefore a proxy for transposition activity (e.g., the higher the number of colonies means higher transposase activity; *see* **Notes 5** and **6**). At least three separate experiments should be performed to do a side-by-side comparison of native (i.e., unaltered) transposase to the designed chimeric ZFP-transposase to ensure that ZFP fusion has not altered transposase activity (*see* Fig. 2b). The colony count assay does not discriminate site-directed integration from random integration events.

3.4 ChIP Assay

1. Transfect 2×10^6 HEK293 cells in a 100 mm dish with transposase enzyme and Fugene 6.
2. Two days post-transfection the cells are washed twice in PBS and scraped into 1 ml of ice-cold PBS (optimal protein expression time is dependent on the recombinant protein expressed, *see* **Notes 1** and **4**).
3. Remove an aliquot that can be processed for western blotting later. Alternatively, if an expression check by blotting is not required the cells can be washed and fixed on the plate directly (*see* **Note 7**).
4. The cells are fixed with 1 % freshly prepared formaldehyde for 15 min at RT. If previously scraped from the plate the fixative can be added to dispersed cells or if fixed on the culture dish conduct the scraping to a separate tube while fixing.
5. The cells are pelleted at $500 \times g$, 5 min, RT.
6. Wash the cell pellet three times by resuspension in 1 ml of a 1:15 dilution of 10× Tris–Glycine Buffer with 1 mM EDTA and 1 mM DTT to quench the formaldehyde.
7. Prepare nuclei by resuspension in 1 ml of L1 buffer and incubation for 15 min at RT with periodically vortexing. Pellet nuclei at $1{,}000 \times g$, 5 min, RT followed by resuspension in 0.5 ml of Sonication Buffer on ice.

8. Fragment DNA by sonication until an approximate size of about 1,000 bp is achieved, observable by agarose gel electrophoresis.
9. After sonication, samples are clarified by 10 min centrifugation at 4 °C and max on a tabletop centrifuge and aliquot (*see* **Note 8**).
10. Immunoprecipitate by thawing cross-linked and sonicated DNA on ice. While the samples are thawing prepare the immunoprecipitating antibody dilution for use as well as the Protein-G/Sepharose.
11. Prepare a master mix of immunoprecipitating antibody for all the samples. Per sample, prepare 1 ml of a 1:500 dilution of antibody in IP Buffer #1 supplemented to 1 % BSA on ice. Aliquot 1 ml of cleared solution to a fresh eppendorf tube for each sample planned.
12. Remove 5 μl of the sonicated DNA sample to a fresh tube and set aside as starting material.
13. Remove an additional 50 μl of sonicated DNA for ChIP and add this to each sample tube already containing 1 ml of diluted antibody and incubate overnight at 4 °C with tumbling.
14. Block the Protein G sepharose Block in IP Buffer 1 supplemented to 1 % BSA and 0.1 mg/ml sheared herring sperm DNA overnight at 4 °C with tumbling (*see* **Note 9**).
15. The next morning wash the blocked sepharose with 3 with IP buffer 1. Following the final wash add ChIP sample to the wet beads and incubate sample and Protein-G at 4 °C for 2 h with tumbling.
16. Wash the sample 3×s with each IP Buffer changing the sample to a fresh eppendorf time each time the buffer changes (*see* **Note 10**).
17. Release the sample with addition of 0.25 ml of IP Buffer 4 and incubate at RT for 15 min, separate sepharose as before, but collect eluate in a fresh tube. Repeat this with an additional 0.25 ml and combine for 0.5 ml per sample. Add 0.5 ml of IP Buffer 4 to each input sample.
18. To all tubes (ChIP and Starting material) add 51 μl of De-cross-linking solution and incubate overnight at 65 °C.
19. Phenol–Chloroform-extract all samples twice, ethanol-precipitate, and resuspend in 50 μl of TE.
20. Perform PCR to observe ChIP DNA product (*see* Fig. 2c).

3.5 Copy Number Analysis

1. Stably transfect HEK293 cells and select them for a minimum of 4 weeks. Extract the gDNA from the cells as directed by the DNeasy Blood and Tissue kit instructions.
2. Copy number can be determined as an average for a population of selected cells. This is accomplished using transgene

specific primers against the gDNA in conjunction with plasmid DNA containing the transgenes to generate standard curves. Normalize the various gDNA samples use *RNaseP* gene specific primers [11] in conjunction with plasmid DNA containing the amplified sequence. *RNaseP* established human copy number of 1 copy per haploid genome. Set up PCR reactions according to the instructions given with the iQ SYBR Green PCR mix adding primers and DNA (*see* **Note 11**).

3. Create plasmid based standard curves for use with the gDNA samples. To do this determine the plasmid mass as (mass of plasmid) = (base pairs plasmid DNA)(1.096×10^{-21} g/bp). Next determine the mass of plasmid DNA required for a predetermined copy number. The previously determined mass is the number of g/copy so multiply by the number of copies desired. Prepare a dilution series containing 3×10^6, 3×10^5, 3×10^4, 3×10^3, 3×10^2, 3×10^1, 0 copies in a standardized volume in 0.1 % carboxymethylcellulose (as a carrier, *see* **Note 12**). For the transgene primer set it is important to include naïve or untransfected gDNA in each to account for interference of the gDNA samples. Plot $\log_{10}$ of the copy number versus C(t) values as the standard curve.
4. At the same time as running the standard curve set up a dilution series of gDNA. The exact amount will depend on the number of copies present and 6.6 pg human gDNA/genome is an established estimate. A reasonable starting point is 66 ng, 6.6 ng, 0.66 ng, 0.066 ng which will help establish both the optimal quantitative range and linearity of sample response if plotted as done for the standard curve (*see* **Note 13**).
5. Determine the number of copies from the plasmid generated standard curves and divide by the mass of gDNA added to get copies of transgene/ng of gDNA.
6. Repeat the whole process with the *RnaseP* standard to get copies of *RNaseP*/ng of gDNA.
7. Divide to get copies of transgene/copies of *RNaseP* bearing in mind that there are two copies of *RNaseP* per human genome (*see* Fig. 2d).

3.6 Interplasmid Transposition Assay

1. Cotransfect HEK293 cells with the necessary plasmids (Subheading 2.5; *see* **Note 14**; *see* Fig. 3a). The *ccdB* gene in the transposase and transposon plasmids inhibits *E. coli* DNA gyrase and eliminates recovery of these plasmids in the assay. Therefore, only the acceptor plasmid is recovered. The acceptor plasmid contains the target sequence or a scrambled target sequence as well as a different bacterial drug resistance gene (ampicillin resistance) from the transposon (kanamycin resistance).
2. Harvest episomal plasmid 2 days after transfection by the Hirt protocol [12]. Pellet the cells at 500 × *g*, 10 min and wash 1× in PBS.

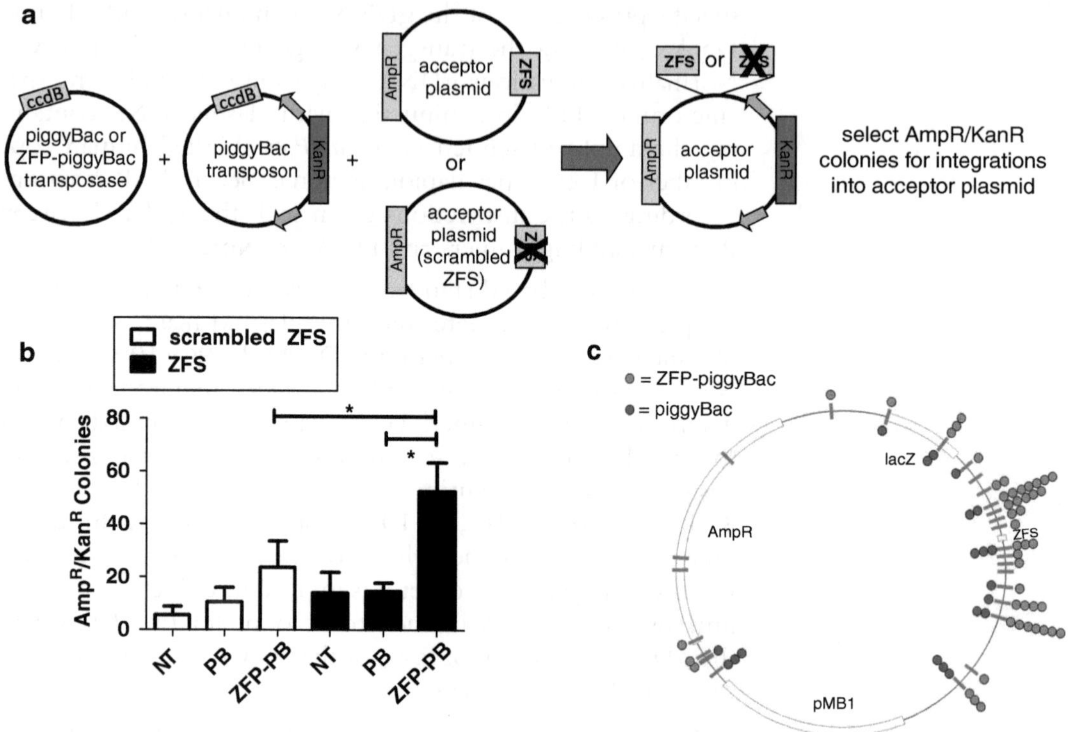

Fig. 3 (**a**) Schematic of plasmid-based transposition assay. The transposase and donor (transposon) plasmids contain the *ccdB* gene in their backbone to eliminate recovery of transposition events in these plasmids. The acceptor plasmid contains a single 18 bp ZFP site or a scrambled 18 bp sequence. (**b**) The number of Amp^R/ Kan^R colonies recovered from HEK293 cells after plasmid-based transposition comparing *piggyBac* to ZFP-*piggyBac* in the presence of an acceptor plasmid containing the 18 bp ZFP site (ZFS) or a scrambled 18 bp sequence (scrambled ZFS) ($n = 3 \pm SD$). ANOVA followed by Bonferroni post test comparison was used for statistical analysis. *$p < 0.05$ when comparing ZFP-*piggyBac* (with ZFS) to *piggyBac* (with ZFS) or ZFP-*piggyBac* (no ZFS). (**c**) Locations of sequenced integrations into the acceptor plasmid containing the 18 bp ZFP site comparing *piggyBac* to ZFP-*piggyBac*. The *green lines* indicate all of the tetranucleotide TTAA element locations throughout the acceptor plasmid. Adapted from [8] with permission

3. Resuspend the cell pellet in Hirt Buffer A, 70 μl.
4. To this add Hirt Buffer B, 80 μl.
5. After this addition and mixing add 40 μl of 5 M NaCl.
6. Place the cells on ice at 4 °C for 16 h.
7. Centrifuge the samples at 8,000 × *g*, 10 min to get a white cloudy pellet and remove the liquid.
8. Extract the aqueous layer 3×s with phenol–chloroform and precipitate the plasmid by addition of 2 volumes of ethanol.
9. Incubate the samples at −80 °C for 90 min.
10. Pellet the plasmid DNA and resuspend in TE.
11. Transform the plasmid DNA by electoporation into electrocompetent Top10 cells

12. Plate the bacteria onto selective media incorporating both the transposable drug resistance and the acceptor plasmid drug resistance antibiotics.
13. Count the colonies and compare using the analysis of variance followed by Bonferroni post-tests (*see* Fig. 3b).
14. Sequence recovered plasmid DNA to verify location of integration within the plasmid. Use Fisher's exact test for statistical comparison of the number of integrations within a window of the target integration sequence (an example is provided in Fig. 3c and Table 5).

3.7 Analysis of Targeting Integration into the Human Genome

1. One can introduce an artificial target into the genome with a different transposase or by random integration following endonuclease linearization of the transgene to evaluate targeting to such a site. The analysis method described would be applicable to an endogenous site as well. Stable expressing transgenic cells containing the artificial target sequence should be selected with G418 for at least 14 days after this procedure for use in targeting experiments (see **Note 15**).
2. Simultaneously transfect transgenic target cells with (1) plasmid containing the green fluorescent protein and puromycin resistance within a transposon sequence (*see* Fig. 1b) and (2) either no transposase, plasmid encoding the undirected transposase, or DNA-directed transposase.
3. Two days post-transfection, select these cells at low density with 3 μg/ml puromycin for at least 2 weeks.
4. Select and expand Individual colonies as clones to isolate gDNA using the DNeasy blood and tissue kit.
5. Use nested touchdown PCR to amplify joining region of gDNA with transposable element (*see* Fig. 4a). Evaluate the copy number of integrated transposons as in Subheading 3.5 (*see* Fig. 4b).

Table 5
Targeted inter-plasmid transposition

TTAA proximity to ZFP target	# of TTAAs	# of integrations		Ratio of integrations: TTAA	
		piggyBac	ZFP-*piggyBac*	*piggyBac*	ZFP-*piggyBac*
<250 bp	14	10	35	0.71	2.5
>250 bp	8	10	11	1.25	1.38

The target plasmid has 14 TTAAs within 500 bp (<250 bp in either direction) of the target ZFP site. Excluding the TTAAs in the ampicillin resistance gene (which would not be recovered in our assay) leaves eight TTAAs outside of this target window available for integration. ZFP-*piggyBac* was significantly different from *piggyBac* by the Fisher's exact test, $p<0.05$ demonstrating a bias toward the targeted window of DNA surrounding the ZFP site. Adapted from [8] with permission

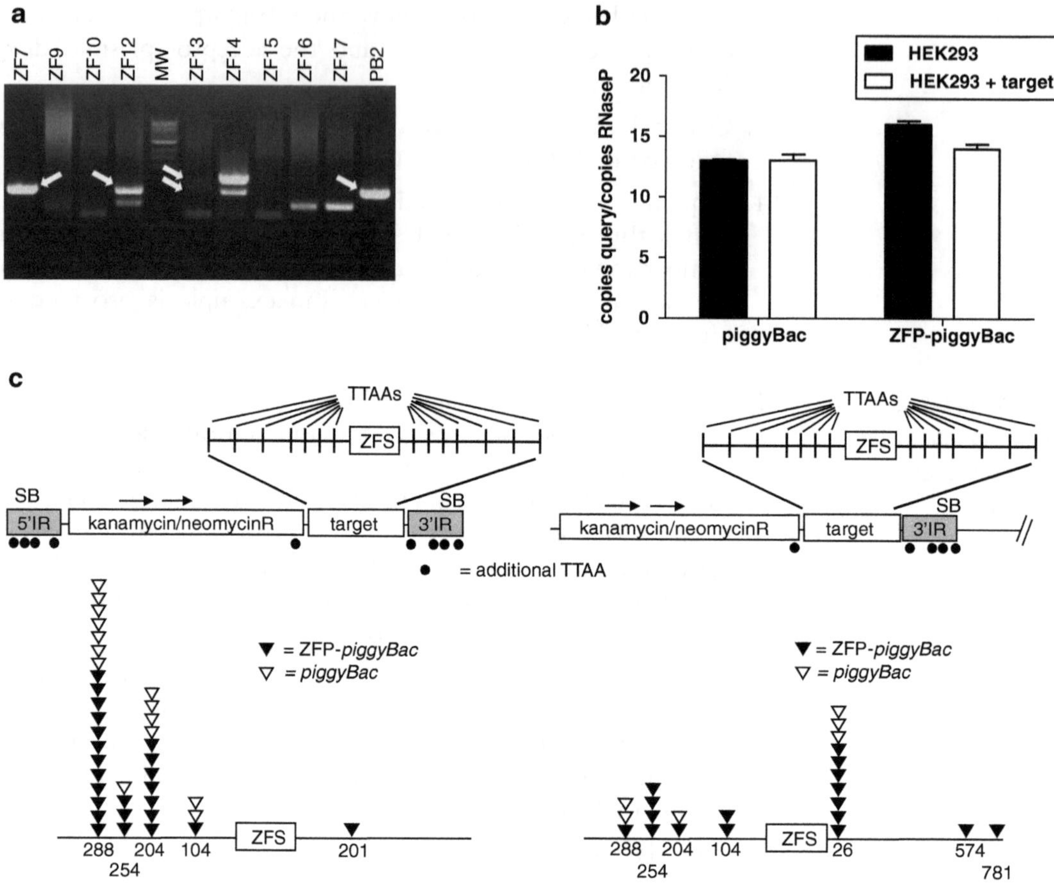

Fig. 4 (**a**) A representative experiment showing gel electrophoresis of PCR products after recovering putative site-directed integration events from HEK293 cell clones resultant from *piggyBac* or ZFP-*piggyBac* transposition. The *arrows* indicate site-directed events confirmed by DNA sequencing after TOPO cloning the corresponding PCR products. PCR products *without arrows* were either not integrations into the target site or not *piggyBac* derived integration events. The labels are for the ZFP-*piggyBac* (ZF#) or *piggyBac* (PB#) clones recovered after selection and genomic DNA isolation. ZF 13 was one of the clones that exhibited two independent site-directed events. MW corresponds to a 1 kb molecular weight ladder. (**b**) qPCR was used to determine the number of transposons integrated into human cells. Samples were derived from genomic DNA isolated from a population of HEK293 cells after transfection with *piggyBac* or ZFP-*piggyBac* and selected with puromycin for >4 weeks. HEK 293 cells lacking the engineered target site were used as a control to determine if the presence of the engineered target had an effect on the overall number of *piggyBac* integrations achieved. The target cell population sample was derived from genomic DNA isolated from a population of cells resultant from SB transposition of the engineered genomic target followed by selection with G418 as described in (**c**), *left panel*. For all samples, amplification of genomic *RNaseP* was used as a reference gene for analysis. Shown are the mean ± SD ($n = 3$). qPCR using the qKN primers revealed that the target cell population had 3 ± 0.358 (mean ± SD, $n = 3$) copies of the integrated SB target ((**c**), *left panel*) per copy of *RNaseP*. (**c**) Target sites integrated into the human genome. An identical target site was integrated using either a SB transposon (*left*) or by transfecting linearized plasmid DNA (*right*). A schematic of site-directed integration events recovered from cell clones isolated after *piggyBac* or ZFP-*piggyBac* transposition is below. The *numbers under the symbols* indicate the distance from the start of the TTAA site of integration from the end of the zinc finger site (ZFS). The *black dots* represent additional TTAAs outside of the engineered target area. Some of the clones recovered after transposition contained more than one site-directed event. Adapted from [8] with permission

Table 6
Targeted transposon integration into the genome of human cells

Method of stable transfection of the identical target site	*piggyBac*		ZFP-*piggyBac*		
	Clones analyzed	Clones with a targeted event	Clones analyzed	Clones with a targeted event	*p* value
Sleeping Beauty	45	12	41	19	0.046[a]
Linear DNA	34	6	41	17	0.023[a]
Total	79	18	82	36	0.003[a]
% of clones with a targeted event		22.5 %		43.9 %[a]	

Transposition mediated puromycin resistant clones (Fig. 4) were isolated and analyzed by PCR for site-directed events into the target site (Fig. 4). Clones were isolated from three separate experiments derived from *piggyBac* or ZFP-piggyBac transfections into cells containing the target
[a]When comparing the percentage of clones containing site-directed events for the total number of clones analyzed, ZFP-piggyBac was significantly different from *piggyBac* by the Fisher's exact test ($p < 0.05$). Reproduced from [8] with permission

TOPO-TA clone (Invitrogen) the PCR products containing the integration sites and sequence the plasmids (*see* Fig. 4c).

6. Use Fisher's exact *t*-test to compare the number of clones isolated containing directed events with each transposase. An example of analysis of the number of clones analyzed is presented in Table 6 (*see* **Note 16**).

4 Notes

1. Fusion of a ZFP DNA binding domain to a transposase may impair transposase activity. In previous studies, addition of a ZF element to native *SB* has resulted in no detectable activity [13–15]. In order to retain activity with *SB*, it would appear that hyperactive transposases (e.g., *SB12*) and transposons are necessary [13–15]. The native *piggyBac* transposase has been successfully modified using a ZFP element with detectable activity in mammalian cells [16]. A hyperactive *piggyBac* transposase has also recently been developed [17, 18].
2. Fusion of ZFPs to the amino-terminus preserves detectable transposase activity with both *SB12* and *piggyBac* [3, 13–16]. Carboxyl-terminal addition to *SB12* abolishes enzyme activity [13–15]. Although a ZFP has not been added to the C-terminus of *piggyBac*, one report has shown preserved *piggyBac* activity with C-terminal addition of another protein domain [19].

3. An optimal amino acid linker sequence separating the ZFP and transposase reading frames may need to be determined. Various linkers have been successfully used in creating chimeric *SB12* and *piggyBac* transposases [3, 13–16, 19]. The linker needs to be flexible and of a length to ensure activity of both the ZFP and transposase domains.
4. The addition of a ZFP element to a transposase can alter the expression level of the transposase enzyme [15]. Codon optimization can be used to increase the expression level of a transposase enzyme if necessary [17, 18].
5. The *SB* transposon system exhibits a phenomenon termed overproduction inhibition, in which increased transposase expression levels can inhibit transposition [3, 13, 20, 21]. The addition of a ZFP element to the *SB12* transposase may alter this phenomenon [13]. When using *SB* as a gene transfer system, it is advisable to do a titration of transposase and transposon plasmids to determine which ratio gives optimal gene transfer. The *piggyBac* transposon system has conflicting reports regarding overproduction inhibition and may or may not exhibit this phenomenon in different cell types under varying conditions [3, 16, 19, 22].
6. Transposition assays require dilution of the transfected cells to such an extent that plating in G418 will permit visualization of colony formation for quantification. Various dilutions are therefore required to ensure that colony formation can be observed. Colony formation will depend on the inherent activity of the transposon system evaluated.
7. It is better to fix the cells directly on the plate to preserve the cell architecture as quickly as possible. In this case, the relative protein expression levels can be determined independently in experiments dedicated to western blots and then inferred for the ChIP. However, removal of an aliquot of cells prior to fixation offers certain advantages. By removing a small aliquot of cells and then fixing the remainder the expression of the recombinant protein can be verified in the experiment directly and the amount of immunoprecipitable protein can be relatively compared between samples. This may aid data interpretation in the case of vastly differing protein expression levels between samples. It is less important if similar protein levels are observed between samples.
8. Fragmentation of the cross-linked DNA to a manageable size by sonication is the critical step for success. Small differences in experimental geometries, sonication powers, and probes make this a necessary step to determine empirically prior to

starting the experiments. When finished sonicating, de-cross-link the samples and run by agarose gel electrophoresis. Initially a slowly migrating band of approximately 12 kb should be apparent. With sonication this should disappear and a smear of fragmented DNA should replace it. The best results are when this smear migrates around the 1 kb marker. Use a sonicator power setting and space the sonication blasts such that the sample does not get too hot during the process. A 50 % power setting with 10 s blasts and 30–60 s rests on ice is a good starting point.

9. A 27.5 g tuberculin syringe excludes the beads and aids in fluid removal. Blocking the sepharose overnight reduces background and makes sure the beads are fully equilibrated.
10. Because of the sensitivity of the detection method, PCR, it is necessary to change tubes frequently. Even residual amounts of DNA adhering to the tube plastic can present a signal and interfere with getting reproducible results.
11. When using double stranded DNA binding dyes in quantitative PCR it is important to stringently test each primer set before use. Adjust PCR conditions such that the primers chosen amplify a single target when used against template and template mixed with genomic DNA. Make sure no products are observed when primers are used separately. If problems are evident it may be prudent to investigate alternative primer sets.
12. Because of the low concentration of plasmids employed in establishing standard cures it is advisable to use carboxymethylcellulose or another non-interfering DNA carrier to help prevent plasmid loss to sides of tubes etc.
13. It is important to run different dilutions of your genomic DNA sample for two reasons. The most obvious is to make sure that your sample stays within the limits of your plasmid standard curve. Secondly, it is important to establish that the sample has the expected linearity and free of interference or quenching.
14. The *ccdB* gene encodes a bacterial DNA gyrase which requires a special strain of *E. coli* for recovery. Because of this it aids the lack of recovery of the transposase and donor transposon plasmids in the procedure.
15. Creating transgenic cell lines which chromosomally incorporate the transgene by two independent methods adds valuable stringency to the experiments. This may be accomplished by using a different transposase, for example *SB*, to achieve target gene insertion. Alternatively the target gene can be integrated by standard means in which the DNA is linearized and transfected followed by selection for stable transfectants.

16. *PiggyBac* requires a TTAA nucleotide element for integration. If one creates an artificial target site with numerous TTAAs, it may be possible to observe native *piggyBac* integrations into the target site as well as ZFP-*piggyBac* integrations [8]. Statistical analysis should be used to confirm an enrichment of ZFP-*piggyBac* integrations compared to native *piggyBac* to ensure targeting at the artificial locus (*see* Table 6). This issue can potentially be obviated by attempting to target an endogenous locus in the human genome. The endogenous human CH2K site is TTAA deplete and required the generation of an artificial target to test for ZFP-*piggyBac* mediated targeted integration [8].

Acknowledgments

M.H.W. would like to acknowledge support from a Career Development Award from the Department of Veterans Affairs and NIH R01 DK093660.

References

1. Cary LC, Goebel M, Corsaro BG, Wang HG, Rosen E, Fraser MJ (1989) Transposon mutagenesis of baculoviruses: analysis of Trichoplusia ni transposon IFP2 insertions within the FP-locus of nuclear polyhedrosis viruses. Virology 172:156–169
2. Fraser MJ, Cary L, Boonvisudhi K, Wang HG (1995) Assay for movement of Lepidopteran transposon IFP2 in insect cells using a baculovirus genome as a target DNA. Virology 211:397–407
3. Wilson MH, Coates CJ, George AL Jr (2007) PiggyBac transposon-mediated gene transfer in human cells. Mol Ther 15:139–145
4. Li MA, Turner DJ, Ning Z et al (2011) Mobilization of giant piggyBac transposons in the mouse genome. Nucleic Acids Res 39:e148
5. Nakazawa Y, Huye LE, Dotti G et al (2009) Optimization of the PiggyBac transposon system for the sustained genetic modification of human T lymphocytes. J Immunother 32: 826–836
6. Fraser MJ, Ciszczon T, Elick T, Bauser C (1996) Precise excision of TTAA-specific lepidopteran transposons piggyBac (IFP2) and tagalong (TFP3) from the baculovirus genome in cell lines from two species of Lepidoptera. Insect Mol Biol 5:141–151
7. Maragathavally KJ, Kaminski JM, Coates CJ (2006) Chimeric Mos1 and piggyBac transposases result in site-directed integration. FASEB J 20:1880–1882
8. Kettlun C, Galvan DL, George AL, Kaja A, Wilson MH (2011) Manipulating piggyBac transposon chromosomal integration site selection in human cells. Mol Ther 19:1636–1644
9. Owens JB, Urschitz J, Stoytchev I et al (2012) Chimeric piggyBac transposases for genomic targeting in human cells. Nucleic Acids Res 40:6978–6991
10. Wang H, Mayhew D, Chen X, Johnston M, Mitra RD (2012) "Calling cards" for DNA-binding proteins in mammalian cells. Genetics 190:941–949
11. Tan SY, Guschin D, Davalos A et al (2003) Zinc-finger protein-targeted gene regulation: genomewide single-gene specificity. Proc Natl Acad Sci USA 100:11997–12002
12. Hirt B (1967) Selective extraction of polyoma DNA from infected mouse cell cultures. J Mol Biol 26:365–369
13. Wilson MH, Kaminski JM, George AL Jr (2005) Functional zinc finger/sleeping beauty transposase chimeras exhibit attenuated overproduction inhibition. FEBS Lett 579: 6205–6209

14. Ivics Z, Katzer A, Stuwe EE, Fiedler D, Knespel S, Izsvak Z (2007) Targeted Sleeping Beauty transposition in human cells. Mol Ther 15:1137–1144
15. Yant SR, Huang Y, Akache B, Kay MA (2007) Site-directed transposon integration in human cells. Nucleic Acids Res 35:e50
16. Wu SC, Meir YJ, Coates CJ et al (2006) piggyBac is a flexible and highly active transposon as compared to Sleeping Beauty, Tol2, and Mos1 in mammalian cells. Proc Natl Acad Sci USA 103:15008–15013
17. Yusa K, Zhou L, Li MA, Bradley A, Craig NL (2011) A hyperactive piggyBac transposase for mammalian applications. Proc Natl Acad Sci USA 108:1531–1536
18. Doherty JE, Huye LE, Yusa K, Zhou L, Craig NL, Wilson MH (2012) Hyperactive piggyBac gene transfer in human cells and in vivo. Hum Gene Ther 23:311–320
19. Cadinanos J, Bradley A (2007) Generation of an inducible and optimized piggyBac transposon system. Nucleic Acids Res 35:e87
20. Geurts AM, Yang Y, Clark KJ et al (2003) Gene transfer into genomes of human cells by the sleeping beauty transposon system. Mol Ther 8:108–117
21. Zayed H, Izsvak Z, Walisko O, Ivics Z (2004) Development of hyperactive Sleeping Beauty transposon vectors by mutational analysis. Mol Ther 9:292–304
22. Wang W, Lin C, Lu D et al (2008) Chromosomal transposition of PiggyBac in mouse embryonic stem cells. Proc Natl Acad Sci USA 105:9290–9295

Chapter 10

Gene Targeting in Human-Induced Pluripotent Stem Cells with Adenoviral Vectors

Kohnosuke Mitani

Abstract

Helper-dependent adenoviral vector (HDAdV), which is also called gutless AdV, has been used to deliver donor DNA for gene targeting in human pluripotent stem cells. Surprisingly, the targeting efficacies, both per chromosomal integration (drug-resistant colony) and per treated cells, are much higher than those by standard electroporation and equivalent to those by utilizing artificial nucleases, such as TAL effector nucleases (Aizawa et al., Mol Ther 20:424–431, 2012; Suzuki, Proc Natl Acad Sci U S A 105:13781–13786, 2008). Importantly, gene targeting with HDAdVs was equally efficient in transcriptionally inactive loci in human ES/iPS cells. Therefore, multiple gene-targeted clones can be obtained from human embryonic stem cells (hESCs) and human-induced pluripotent stem cells (hiPSCs) cultured in one 100-mm dish. For virus-mediated gene targeting, it is not required to introduce artificial double-strand breaks. By using electroporation for gene targeting, target cells should be expanded to 10^6–10^7 cells. In contrast, as an advantage of virus-mediated method, DNA delivery efficiency is high even in a smaller number of cells, resulting in minimizing the number of passages/cell divisions before performing gene targeting. The characteristics suggest that HDAdV-mediated gene targeting has potential advantages for manipulation of chromosomes of pluripotent stem cells for therapeutic applications.

Key words Adenoviral vector, Gene targeting, Pluripotent stem cells, Gene repair

1 Introduction

Low efficiencies of gene targeting via homologous recombination (HR) have limited basic research and applications using hESCs/hiPSCs. We showed highly and equally efficient gene knockout and knock-in at both transcriptionally active (*HPRT1*, *KU80*, *LIG1*, *LIG3*) and inactive (*HB9*) loci in these cells using high-capacity HDAdVs [1, 2]. Without the necessity of introducing artificial DNA double-strand breaks, 7–81 % of drug-resistant colonies were gene targeted by accurate HR, which were not accompanied with additional ectopic integrations. Even at the motor neuron-specific HB9 locus, the enhanced green fluorescent protein (EGFP) gene was accurately knocked in 23–57 % of drug-resistant colonies. In these clones, induced differentiation into the HB9-positive

Francesca Storici (ed.), *Gene Correction: Methods and Protocols*, Methods in Molecular Biology, vol. 1114,
DOI 10.1007/978-1-62703-761-7_10,

motor neuron correlated with EGFP expression. Furthermore, HDAdV infection had no detectable adverse effects on the undifferentiated state and pluripotency of hESCs and hiPSCs. These results suggest that HDAdV is one of the best methods for efficient and accurate gene targeting in hESCs and hiPSCs and might be especially useful for therapeutic applications.

2 Materials

Materials and methods for hiPSC culture have many varieties. Because of high gene targeting efficiencies with HDAdV, any culture method for hESCs/hiPSCs can be used.

2.1 Cell Culture of hiPSCs

1. SNL76/7 medium: DMEM (Sigma, #D5796), 7 % FCS, penicillin/streptomycin (Invitrogen).
2. hES/iPSC medium: DMEM/F12 (Sigma), 20 % KnockOut Serum Replacement (KSR) (Invitrogen), nonessential amino acids (Invitrogen, #11140-050), 0.1 mM 2-mercaptoethanol (Invitrogen), 5 ng/ml bFGF, penicillin/streptomycin (Invitrogen). Use the medium within 1 week.
3. hESC/iPSC dissociation solution (CTK solution): Add 5 ml of 2.5 % trypsin (Difco), 5 ml of 1 mg/ml collagenase IV(Invitrogen), 0.5 ml of 0.1 M $CaCl_2$, and 10 ml of KSR into 30 ml of distilled water. Filter sterilize and store at −20 °C. Avoid repeated freezing and thawing.
4. ROCK inhibitor, Y27632: Dissolve 5 mg of Y-27632 in 1.48 ml of distilled water (10 mM). Aliquot and store at −20 °C. Use at 10 μM of final concentration.
5. Recombinant basic fibroblast growth factor, human: To prepare PBS containing 0.1 % bovine serum albumin (BSA), add 50 μl of 10 % BSA into 5 ml of PBS. Dissolve 50 μg of bFGF in 5 ml of PBS containing 0.1 % BSA (10 μg/ml). Aliquot and store at −20 °C.
6. Cell dissociation solution: TrypLE (Invitrogen) and Accutase (Invitrogen).

3 Methods

Helper-dependent adenoviral vector for gene targeting is designed so that ~10 kb of homologous sequences are attached on the both ends of the marker gene. The targeting cassette is subcloned into helper-dependent adenoviral plasmids, such as pAMHDAdGT8-4 and pCIHDAdGT8-3, which encode the *E. coli* β-galactosidase gene for titration and the herpes simplex virus-1 thymidine kinase

(HSV-TK) gene for negative selection. The examples can be found in references [1] and [2]. These plasmids are available from Addgene. We use the recombineering technology to construct our gene targeting cassettes [3]. Helper-dependent adenoviral vectors can be propagated, as described previously [4].

3.1 Infection of Human ESCs/iPSCs with Adenoviral Vector

To prepare cells for infection, 50–70 % confluent culture in 10-cm dishes are dissociated with TrypLE. 50 % confluent culture in a 10-cm dish of hiPSCs should produce ~2×10^6 cells. Because the gene targeting efficiencies are usually 2×10^{-6}–5×10^{-5}, one to two 10-cm dishes of hESCs/iPSCs are enough.

1. The day before infection, plate 10-cm dishes of SNL76/7 cells with appropriate drug resistance.
2. Treat hiPSCs in a 10-cm dish with ROCK inhibitor for 1 h.
3. Prepare one 10-cm gelatinized dish, as described above.
4. Wash cells twice with 5 ml of PBS.
5. Add 1 ml of TrypLE and incubate for 3 min.
6. Add 5 ml of F12 gently and aspirate gently.
7. Tap the dish and destroy colonies.
8. Add 2 ml of hiPSC medium and pipet with P-1000.
9. Add 5 ml of hiPSC medium and transfer into a 15-ml tube.
10. Rinse the dish with 3 ml of F12 and transfer into the tube.
11. Centrifuge $120 \times g$ for 5 min at room temperature.
12. Suspend the cells with 10 ml of hiPSC medium with ROCK inhibitor.
13. Plate onto gelatinized dish and incubate for 1 h to remove contaminating STO cells.
14. Recover the floating cells by pipetting and transfer into a 15-ml tube.
15. Count the cell number.
16. Centrifuge $120 \times g$ for 5 min at room temperature.
17. Suspend the cells at 5×10^6/ml in hiPSC medium with ROCK inhibitor.
18. Transfer 0.4 ml (i.e., 2×10^6 cells) into siliconized 1.5-ml tubes for each infection.
19. Thaw an aliquot of purified HDAdV.
20. Add 2×10^8 genome copy (i.e., multiplicity of infection of 100) of HDAdV to the cells (*see* **Note 1**).
21. Incubate at RT for 1 h with occasional mixing.
22. Add hiPSC medium with ROCK inhibitor to a 10-cm dish of mitomycin C-treated SNL76/7 cells.

23. Add the infected human ES/iPS cells onto the SNL76/7 cells.
24. Change the medium to hiPSC medium (without ROCK inhibitor) the next day.
25. Start selection when the cells become ~50 % confluent. This is usually 1 or 2 days after infection.
26. Colonies usually start to appear after 10 days of infection.

3.2 Pickup and Passage of Colonies

Pick colonies before they become too big and start to differentiate. Once colonies start to appear, they grow rapidly. Check and pick colonies at least every other day. Because of the high average gene targeting efficiencies per integration (without negative selection with the HSV-TK gene), picking up 48 colonies is enough to obtain multiple gene-targeted hiPSC clones.

1. On the day before colony isolation, prepare MMC-treated SNL76/7 cells in 24-well dishes. The cell density should be $5–6 \times 10^4$ cells/well.
2. Treat dishes with the drug-resistant colonies with 10 μM of Y-27632 for 1 h.
3. Add 0.5 ml of hiPSC medium with ROCK inhibitor in each wells of 24-well STO feeder.
4. Add 100 μl of hiPSC medium with ROCK inhibitor into each well of round-well 96-well plates.
5. Under microscope, pick colonies using P-2 or P-10 Pipetman. Transfer them into medium in 96-well plates.
6. Every 8–16 colonies, pipet the colonies in 96-well plates 20–30 times with a multichannel pipettor. Transfer the disrupted colonies onto the SNL76/7 in 24-well dishes.
7. Change the medium to hiPSC medium with appropriate selection without ROCK inhibitor. Once the cells are visible in 24-well plates, start positive–negative selection with ganciclovir.
8. When the well is subconfluent (and before the cells start to differentiate), dissociate the cells with Accutase.
9. Prepare DNA from one third of the cells for PCR analyses. Transfer two thirds of the cells onto MMC-treated SNL76/7 on 6-well plates.
10. Once positive clones are identified by PCR, these clones are passaged for frozen stocks and for isolation of DNA for detailed DNA analyses by Southern hybridization.

4 Notes

1. If the gene targeting efficiency is low and no gene knockout clones are obtained, test lower and higher MOIs, such as 10, 30, 300, and 1,000.

Acknowledgments

This work was supported by national funds from Development of Technology to Create Research Model Cells Project of New Energy and Industrial Technology Development Organization (NEDO) and in part by a grant-in-aid for "Support Project of Strategic Research Center in Private Universities" from the Ministry of Education, Culture, Sports, Science and Technology (MEXT) to Research Center for Genomic Medicine, Saitama Medical University.

References

1. Aizawa E, Hirabayashi Y, Iwanaga Y, Suzuki K, Sakurai K, Shimoji M, Aiba K, Wada T, Tooi N, Kawase E, Suemori H, Nakatsuji N, Mitani K (2012) Efficient and accurate homologous recombination in hESCs and hiPSCs using helper-dependent adenoviral vectors. Mol Ther 20:424–431
2. Suzuki K, Mitsui K, Aizawa E, Hasegawa K, Kawase E, Yamagishi T, Shimizu Y, Suemori H, Nakatsuji N, Mitani K (2008) Highly efficient transient gene expression and gene targeting in primate embryonic stem cells with helper-dependent adenoviral vectors. Proc Natl Acad Sci USA 105:13781–13786
3. Skarnes WC, Rosen B, West AP, Koutsourakis M, Bushell W, Iyer V, Mujica AO, Thomas M, Harrow J, Cox T, Jackson D, Severin J, Biggs P, Fu J, Nefedov M, de Jong PJ, Stewart AF, Bradley A (2011) A conditional knockout resource for the genome-wide study of mouse gene function. Nature 474:337–342
4. Palmer DJ, Ng P (2011) Rescue, amplification, and large-scale production of helper-dependent adenoviral vectors. Cold Spring Harb Protoc 2011:857–866

Chapter 11

Enhanced Gene Targeting of Adult and Pluripotent Stem Cells Using Evolved Adeno-associated Virus

Melissa A. Bartel and David V. Schaffer

Abstract

Efficient approaches for the precise genetic engineering of stem cells can enhance both basic and applied stem cell research. Adeno-associated virus (AAV) vectors have demonstrated high-efficiency gene delivery and gene targeting to numerous cell types, and AAV vectors developed specifically for gene delivery to stem cells have further increased gene targeting frequency compared to plasmid construct techniques. This chapter details the production and purification techniques necessary to generate adeno-associated viral vectors for use in high-efficiency gene targeting of adult or pluripotent stem cell applications. Culture conditions used to achieve high gene targeting frequencies in rat neural stem cells and human pluripotent stem cells are also described.

Key words Adeno-associated virus, Gene targeting, Homologous recombination, Neural stem cells, Pluripotent stem cells

1 Introduction

Gene targeting manipulates an endogenous gene through the use of homologous recombination (HR) to swap a modified sequence in place of the host DNA. This engineering of a human stem cell's genome by exploiting the natural ability of cells to perform homologous recombination has broad applications and implications, including safe harbor integration of genes for basic or therapeutic application, creating in vitro models for investigating human development and disease, and high-throughput drug discovery and toxicity studies [1, 2]. Furthermore, high-efficiency gene targeting to stem cells can aid in the study of specific genes involved in stem cell self-renewal and differentiation into therapeutically relevant lineages.

Adeno-associated virus (AAV) is a nonpathogenic, nonenveloped virus containing a 4.7 kb single-stranded DNA genome. Recombinant versions of AAV can be created by replacing the viral genome with a gene or sequence of interest and providing the viral

Francesca Storici (ed.), *Gene Correction: Methods and Protocols*, Methods in Molecular Biology, vol. 1114,
DOI 10.1007/978-1-62703-761-7_11,

genome in trans along with helper viral genes during vector production [3]. The resulting recombinant vectors can efficiently deliver a transgene and safely mediate long-term gene expression in dividing and nondividing cells of numerous tissues [4]. Such AAV-based vectors have proven safe, efficient, and recently very effective for clinical applications [5–7], making them an attractive option for both basic and translational research. Importantly, Russell and colleagues have discovered that the secondary structure of AAV vector genomes carrying gene targeting constructs can mediate HR with target loci in a cellular genome at efficiencies 10^3- to 10^4-fold higher than genes of interest delivered using plasmid constructs [8]. This capacity would be further enabled through the development of vectors that can deliver these gene targeting constructs to the nuclei of target cells with high efficiency.

Using directed evolution, we have created AAV variants capable of enhanced gene delivery and gene targeting in both adult and pluripotent stem cells. In particular, Jang et al. developed an AAV2 variant that demonstrated enhanced gene delivery and targeting in rat neural stem cells (NSCs). Additionally, this variant exhibited increased transduction of murine NSCs, human fetal NSCs, and human embryonic stem cell-derived neural progenitor cells [9], which should additionally translate to enhanced gene targeting frequencies in these cell types. Furthermore, Asuri et al. developed an AAV variant that demonstrated enhanced gene targeting within several human embryonic stem cell (hESCs) and human induced pluripotent stem cell (hiPSCs) lines [10]. Here, we describe the techniques required for efficient gene targeting of neural stem cells or pluripotent stem cells using these AAV vectors.

The materials and methods presented in the following sections detail the production and purification of high titer AAV, as well as culture conditions for rat neural stem cells and human pluripotent stem cells that will enable high-efficiency gene targeting. Another important aspect of the high-efficiency gene targeting—the design of the gene targeting construct—will not be discussed. Research has shown that use of gene targeting constructs with longer homology arms and placement of a mutation in the center of the gene targeting construct, along with the secondary structure of the AAV inverted terminal repeats that facilitate the recruitment of host cell recombination or repair enzymes, can increase the frequency of gene targeting events during AAV-mediated gene targeting [11, 12]. The reader is referred to these valued references for advice in designing new gene targeting constructs for their specific applications.

When designing gene targeting constructs, one important consideration is the downstream selection of correctly targeted stem cells. That is, even high-efficiency gene targeting will require the use of some manner of selection to distinguish the modified cells from cells in which the construct has incorrectly integrated, and the options available for selection depend on the function of the gene being targeted. Some genes encode products whose loss can be readily detected. As one example, the *PIG-A* gene, located on the X chromosome, is required for several cell surface glycosyl-phosphatidyl-inositol-anchored proteins (GPI-APs) to attach to the cell membrane. Using gene targeting to knockout the function of this gene results in targeted cells that can be selected by fluorescence-activated cell sorting of cells deficient for GPI-APs, such as CD59 [13]. Genes whose targeting does not result in a straightforward molecular marker or phenotype require the insertion of genes to select the positively targeted cells, and the most common selection is the insertion of an antibiotic resistance gene, such as genes that confer resistance to puromycin or neomycin [14, 15]. These resistance genes can be inserted such that they replace one or more exons of the host target gene, such that the targeting construct offers the dual functionality of gene knockout and a positively selectable marker [14, 15]. Furthermore, a negative selection marker can be appended outside of the homology arms to eliminate cells in which the construct was inserted nonspecifically [14]. As a common example of a negative selectable marker, herpes simplex virus thymidine kinase (HSV-tk) converts the prodrug ganciclovir into a cytotoxic product [14]. The selected cell population can be further analyzed to determine both the on-target and off-target integration events. To confirm on-target integration in a cell population, Southern blot and polymerase chain reaction (PCR) are effective screening strategies [15]. To determine the frequency of off-target integration in a cell population, PCR can be used to amplify an end of the AAV donor construct between the homology arm and the AAV ITR. This region of the construct will not be inserted into the genome during homologous recombination and thus represents the presence of off-target integration. The amount of PCR product can be compared to a standard curve representing various frequencies of integration to quantify the frequency of off-target integration events, as demonstrated by Asuri et al. [10].

AAV thus represents an effective DNA donor, but gene targeting efficiency can be further enhanced by improving the properties of the acceptor locus. For example, introducing DNA damage such as a double-stranded breakage at the target site can greatly enhance homologous recombination, and the coadministration of site-specific DNA nucleases, such as zinc-finger nucleases (ZFNs) or transcription activator-like effector nucleases (TALENs), has been shown to substantially enhance the efficiency of plasmid DNA gene targeting constructs [10, 13, 16]. As we have shown, cDNA encoding ZFNs or TALENS also can be packaged into AAV capsids, and their delivery along with AAV genomes containing gene targeting constructs can lead to even higher gene targeting efficiencies [10]. Since ZFNs and TALENs must be designed to target a specific site within the host genome, the reader is referred to other references for more information in designing nucleases for their specific applications [13, 16].

2 Materials

2.1 Adeno-associated Virus Production

2.1.1 Viral Packaging

1. HEK 293T cells (ATCC, Manassas, VA).
2. Dulbecco's modified Eagle's medium (DMEM) supplemented with 10 % fetal bovine serum (Gibco, Carlsbad, CA) and 1 % penicillin/streptomycin (Invitrogen, Carlsbad, CA). Medium containing fetal bovine serum can be stored at 4 °C for up to 1 month.
3. 2× HEPES-buffered saline (HeBS) solution: 280 mM NaCl, 1.5 mM Na_2HPO_4, and 50 mM HEPES in water. Adjust pH to 7.10. Solution should be sterilized using a 0.22 μm filter in tissue culture hood and can be stored at room temperature.
4. 2.5 M $CaCl_2$ solution: Solution should be sterilized using a 0.22 μm filter in tissue culture hood and can be stored at room temperature.
5. Sterile tissue culture water.
6. Plasmids containing adenovirus helper genes, AAV rep and cap genes, and gene of interest between AAV ITRs. For example, plasmid pHelper (containing the adenovirus helper genes) [17], plasmid pXX2 (containing AAV2 rep and cap genes) [18], and plasmid t37GFP (containing a truncated GFP transgene-IRES-puromycin resistance cassette between AAV ITRs) were used by Asuri et al. to package AAV2 to evaluate the gene targeting efficiency of wild-type and novel AAV serotypes [10].
7. Lysis buffer: 50 mM Tris Base and 150 mM NaCl in water. Adjust pH to 8.2–8.5 using HCl. Solution should be sterilized using a 0.22 μm filter in tissue culture hood and can be stored at room temperature.
8. Benzonase nuclease: Dilute benzonase nuclease (Sigma Aldrich, St. Louis MO) to 1 U/μL in lysis buffer. Dilution can be stored at 4 °C.

2.1.2 Viral Purification: Iodixanol Density Gradient Centrifugation (See Note 1)

1. 10× PBS-MK: 10× PBS solution containing 10 mM $MgCl_2$ and 25 mM KCl. Solution should be sterilized using a 0.22 μm filter in tissue culture hood and can be stored at room temperature.
2. 1× PBS-MK with 2 M NaCl: Solution should be sterilized using a 0.22 μm filter in tissue culture hood and can be stored at room temperature.
3. 54 % iodixanol solution: 9 volumes of OptiPrep™ density gradient medium—60 % iodixanol solution (Sigma Aldrich)—and 1 volume of 10× PBS-MK. OptiPrep medium is sterile, so solution can be mixed in tissue culture hood. Solution can be stored at room temperature and should be protected from light.

4. 40 % iodixanol solution: 4 volumes of 54 % iodixanol solution and 1.4 volumes of 1× PBS-MK.
5. 25 % iodixanol solution: 2.5 volumes of 54 % iodixanol solution and 2.9 volumes of 1× PBS-MK.
6. 15 % iodixanol solution: 1.5 volumes of 54 % iodixanol solution, 2.7 volumes of 1× PBS-MK with 2 M NaCl, and 1.2 volumes of 1× PBS-MK.
7. Optional: Phenol red (0.01 μg/mL) can be added to 54 % and 25 % iodixanol solutions to enable easier identification of the layers.
8. OptiSeal polyallomer centrifuge tubes, capacity 4.9 mL (Beckman Coulter, Brea, CA).
9. 21G 1½ needles.
10. 1 mL sterile syringes.
11. Optima L-Series Preparative Ultracentrifuge (Beckman Coulter).
12. VTI 65.2 fixed-angle ultracentrifuge rotor (Beckman Coulter).
13. Sterile 15 mL conical tubes (BD, Franklin Lakes, NJ).

2.1.3 Viral Purification: Buffer Exchange and Concentration

1. PBS with 5 % Tween-20: Solution should be sterilized using a 0.22 μm filter in tissue culture hood and can be stored at room temperature.
2. PBS with 0.001 % Tween-20: Solution should be sterilized using a 0.22 μm filter in tissue culture hood and can be stored at room temperature.
3. Amicon Ultra-15 Centrifugal Filter Devices (Millipore, Billerica, MA).
4. J-Series high-speed centrifuge (Beckman Coulter).
5. JLA 16.250 fixed-angle high-speed centrifuge rotor (Beckman Coulter).
6. Sterile 1.7 mL microcentrifuge tubes (Eppendorf, Hauppauge, NY).

2.1.4 Viral Titering: Extraction of Viral Genome from Protein Capsid and qPCR Analysis of Viral Titer

1. 10× DNase buffer: 100 mM $MgCl_2$ and 250 mM Tris–HCl (pH 7.4) in water. Solution should be sterilized using a 0.22 μm filter in tissue culture hood and can be stored at room temperature.
2. DNase I: Dissolve DNase I (Roche, Indianapolis, IN) to a concentration of 10 U/μL in buffer containing 20 mM Tris–HCl (pH 7.5), 1 mM $MgCl_2$, and 50 % (w/v) glycerol. DNase solution can be stored at −20 °C.
3. 2× Proteinase K buffer: 10 mM Tris–HCl (pH 8.0), 20 mM EDTA (pH 8.0), and 20 mM NaCl and in water. Solution should be sterilized using a 0.22 μm filter in tissue culture hood and can be stored at room temperature.

4. Proteinase K (New England Biolabs, Ipswich, MA): Use in supplied concentration. Proteinase K can be stored at –20 °C.
5. 2× iCycler mix: 200 μL 10× PCR mix without Mg^{2+}, 120 μL 50 mM $MgCl_2$, 40 μL 10 mM dNTP mixture (*see* below), 640 μL molecular grade water.
6. 10 mM dNTP mixture: Dilute from 100 mM stock solutions of dATP, dCTP, dGTP, and dTTP (Invitrogen, Carlsbad, CA). Aliquots can be stored at –20 °C.
7. Primers against gene of interest.
8. 1 μM Fluorescein: Dilute stocks of 100 μM Fluorescein (Bio-Rad, Hercules, CA). Aliquots can be stored at –20 °C.
9. 40× SYBR: Dilute from 10,000× SYBRGreen stock (Invitrogen). Aliquots can be stored at –20 °C.
10. Taq/antibody mixture: Mix equal amounts of Taq DNA Polymerase (New England Biolabs) and JumpStart Taq Antibody (Sigma Aldrich). Incubate on ice for 20 min. Aliquots can be stored at –20 °C.
11. 20 ng/μL linearized plasmid standard containing gene of interest.
12. iQ5 Multicolor Real-Time PCR Detection System (Bio-Rad).

2.2 Rat Neural Stem Cell Gene Targeting

1. Tissue culture plates previously coated with 10 μg/mL poly-L-ornithine (Sigma Aldrich) and 5 μg/mL mouse laminin (Invitrogen). See Peltier et al. for a description of the plate coating protocol [19].
2. NSC proliferation medium: DMEM/F12 medium (Invitrogen) containing N-2 supplement (Invitrogen) and 20 ng/mL recombinant human fibroblast growth factor 2 (Peprotech, Rocky Hill, NJ). Medium containing supplements can be stored at 4 °C for up to 1 month.
3. Accutase (Innovative Cell Technologies, San Diego, CA).

2.3 Human Pluripotent Stem Cell (hPSC) Gene Targeting

1. Tissue culture plates previously coated with BD Matrigel™ hESC-qualified Matrix (BD). *See* Stover and Schwartz for a description of the plate coating protocol (*see* **Note 2**) [20].
2. hPSC proliferation medium: X-Vivo medium (Lonza, Norwalk, CT) supplemented with 80 ng/mL FGF-2 (PeproTech, Rocky Hill, NJ) and 0.5 ng/mL TGF-β1 (R&D Systems, Minneapolis, MN) or mTeSR1 maintenance medium (Stem Cell Technologies, Seattle, WA). Medium containing supplements can be stored at 4 °C for up to 1 week.
3. Collagenase, Type IV (Invitrogen).
4. Rho kinase inhibitor (ROCK inhibitor, Y-27632; CalBioChem, San Diego, CA).

3 Methods

3.1 Adeno-associated Virus Production

3.1.1 Viral Packaging

1. Plate HEK 293T cells in DMEM supplemented with 10 % FBS and 1 % penicillin/streptomycin on 10 cm or 15 cm plates 2 days prior to transfection. Cells should be 80–90 % confluent for transfection.
2. Change media 5–7 h prior to transfection.
3. Prepare HeBS and DNA/$CaCl_2$ solutions according to Table 1.
4. Add DNA/$CaCl_2$ solution dropwise to HeBS solution. Pipet entire volume two to three additional times to completely mix solutions.
5. Incubate transfection solution at room temperature for 2–3 min.
6. Add transfection solution dropwise to cells.
7. Change media 8–12 h post-transfection.
8. Virus can be harvested from cells 48–72 h post-transfection. To harvest virus, manually dissociate cells from tissue culture plate and pipet media and cells into conical tubes.
9. Centrifuge at 1,160 × *g* for 2.5 min to pellet cells.
10. Remove media and resuspend cells in 2 mL of lysis buffer for each 15 cm plate (or 1 mL of lysis buffer for each 10 cm plate).
11. Freeze/thaw samples three times using a dry ice/ethanol bath and a 37 °C water bath.
12. Add 10 U of benzonase nuclease per milliliter of cell lysate. Incubate samples at 37 °C for 30 min.
13. Centrifuge at 10,000 × *g* for 10 min.
14. Transfer supernatant to new tubes.
15. Optional: add 2 μL 5 % Tween-20 per milliliter of supernatant to decrease aggregation of viral particles.
16. Store supernatant at 4 °C until purification.

Table 1
AAV transfection solutions

15 cm plate		10 cm plate	
2× HeBS	2.5 mL	2× HeBS	1 mL
dH_2O	2.15 mL	dH_2O	860 μL
AAV *rep* and *cap* plasmid	15 μg	AAV *rep* and *cap* plasmid	9 μg
Adenovirus helper plasmid	15 μg	Adenovirus helper plasmid	7 μg
gene of interest plasmid	15 μg	Gene of interest plasmid	9 μg
2.5 M $CaCl_2$	300 μL	2.5 M $CaCl_2$	120 μL

3.1.2 Viral Purification: Iodixanol Density Gradient Centrifugation

1. Underlay 1,200 μL of 15 % iodixanol solution with 700 μL of 25 % iodixanol solution, 600 μL of 40 % iodixanol solution, and 600 μL of 54 % iodixanol solution.
2. Overlay each gradient with 1.6 mL of crude lysate from Subheading 3.1.1, **step 16** (*see* **Note 3**).
3. Weigh tubes to ensure that they are properly balanced. PBS can be added as necessary to balance tubes.
4. Seal tubes using the black caps provided with the tubes.
5. Centrifuge at 174,000 × g for 2 h at 18 °C. To prevent disruption of the iodixanol gradient, set the ultracentrifuge to slow acceleration and deceleration.
6. Puncture tube using 21G 1½ needle attached to 1 mL syringe at the interface between the 40 % and 54 % iodixanol solutions.
7. Collect the bottom 4/5 of the 40 % iodixanol solution and the top 1/5 of the 54 % iodixanol solution (*see* **Note 4**).
8. Store collected fractions in sterile 15 mL conical tubes at 4 °C until further purification.

3.1.3 Viral Purification: Buffer Exchange and Concentration

1. Incubate Amicon Ultra-15 centrifugal filter in PBS with 5 % Tween-20 at room temperature for 20 min. Following incubation, wash filter once using PBS with 0.001 % Tween-20.
2. Dilute collected iodixanol fractions to 15 mL in PBS with 0.001 % Tween-20 and apply diluted iodixanol solution to Amicon filter.
3. Centrifuge at 3,000 × g for 15–30 min (until solution has been concentrated to less than 2 mL).
4. Add PBS with 0.001 % Tween-20 to 15 mL and mix well.
5. Repeat **steps 3** and **4** until all iodixanol has been eliminated. This has occurred when solution is clear and viscosity is similar to PBS with 0.001 % Tween-20 solution.
6. On the final centrifugation step, concentrate virus solution to 200–300 μL.

3.1.4 Viral Titering: Extraction of Viral Genome from Protein Capsid and qPCR Analysis of Viral Titer

1. For each virus sample to be tested, combine 10 μL of virus stock solution with 34.5 μL of water, 5 μL of 10× DNase buffer, and 0.5 μL of DNase I to create a 50 μL total volume.
2. Incubate at 37 °C for 30 min.
3. Incubate at 56 °C for 10 min to deactivate DNase I.
4. Add 50 μL of 2× Proteinase K buffer and 10 μL of Proteinase K to each virus sample for a 110 μL total volume.
5. Incubate at 37 °C for 1 h.
6. Incubate at 95 °C for 20 min to deactivate Proteinase K.
7. Create tenfold serial dilutions of linearized plasmid standard between 0.2 ng/μL and 0.02 pg/μL for use as a standard curve.

Table 2
qPCR master mix (for ten samples)

2× iCycler mix	100 μL
10 μM sense primer	2 μL
10 μM antisense primer	2 μL
1 μM Fluorescein	2 μL
40× SYBR	2 μL
Molecular grade water	40 μL
Taq/antibody mixture	2 μL

Table 3
iCycler reaction protocol

Cycle	Step	Repeat	Temperature	Time (minutes)
1	1	1×	95 °C	5:00
2	1	40×	95 °C	0:30
	2		60 °C	0:30
	3		72 °C	0:10

8. Solutions resulting from viral purification are typically high titer (10^{11}–10^{12} viral genomes per mL), so the purified DNA solution should be diluted 1:10 before analysis in order to ensure that samples fall within the linear range of standards.
9. Prepare qPCR master mix according to Table 2.
10. Combine 15 μL of master mix and 5 μL of standard or diluted viral sample.
11. Run PCR protocol according to Table 3.
12. Threshold cycle values for standards and viral samples can be plotted to determine the genomic titer of each sample.

3.2 Rat Neural Stem Cell Gene Targeting (See Note 5)

1. Seed NSCs onto poly-L-ornithine/laminin-coated 24-well tissue culture plates at a density of 50,000 cells per well 24 h prior to AAV infection.
2. Add AAV gene targeting vector to NSCs at a genomic multiplicity of infection (MOI) of 5×10^5.
3. Change media 24 h postinfection.
4. Infected NSCs can be analyzed (or antibiotic selection, e.g., neomycin or puromycin, can begin) 72 h postinfection.

3.3 Human Pluripotent Stem Cell Gene Targeting (See Note 6)

1. Seed hPSCs onto Matrigel-coated 12-well tissue culture plates at a density of 10^5 cells per well 24 h prior to AAV infection. hPSCs should be seeded as small colonies to increase AAV transduction. 10 μM ROCK inhibitor can be added to hPSC culture to increase cell survival in small colony or single-cell culture conditions.
2. Add AAV gene targeting vector to hPSCs at a genomic MOI of 10^5.
3. Change media 24 h postinfection.
4. Infected hPSCs can be analyzed (or antibiotic selection can begin) 48 h postinfection.

4 Notes

1. The use of a viral purification protocol involving (1) iodixanol density gradient centrifugation and (2) buffer exchange and concentration results in a high titer virus solution that can be applied to sensitive in vitro cell culture systems (such as stem cell cultures) that will not result in cell toxicity. In addition, this virus solution is safe for in vivo administration.
2. The use of feeder layers may decrease the gene targeting frequency of the system, as a portion of the viral vectors may bind to and infect feeder cells instead of the hPSCs.
3. Tubes must be completely full to prevent deformation during centrifugation.
4. Contaminating proteins remaining from cell lysate will be concentrated in a band at the interface between the 25 % and 40 % iodixanol solutions. Do not collect these proteins.
5. General protocols for the maintenance of rat NSCs should be used to expand NSC cultures until the amount of cells needed for gene targeting has been obtained. See Peltier et al. for a complete protocol of rat NSC expansion and passaging [19]. Follow passaging protocol for cell seeding for gene targeting experiments.
6. General protocols for the maintenance of hPSCs should be used to expand hPSC cultures until the numbers of cells needed for gene targeting have been obtained. See the mTeSR technical manual for a complete protocol of hPSC maintenance and passaging for proliferation. Follow passaging protocol for cell seeding for gene targeting experiments.

Acknowledgments

This work was supported by California Institute for Regenerative Medicine Training Grants T1-00007 and RT1-01021 and an NSF Graduate Fellowship (to MAB).

References

1. Conrad C, Gupta R, Mohan H, Niess H, Bruns CJ, Kopp R, von Luettichau I, Guba M, Heeschen C, Jauch K-W, Huss R, Nelson PJ (2007) Genetically engineered stem cells for therapeutic gene delivery. Curr Gene Ther 7(4):249–260
2. Yates F, Daley GQ (2006) Progress and prospects: gene transfer into embryonic stem cells. Gene Ther 13:1431–1439
3. Flotte TR (2004) Gene therapy progress and prospects: recombinant adeno-associated virus (rAAV) vectors. Gene Ther 11:805–810
4. Schaffer DV, Koerber JT, K-i L (2008) Molecular engineering of viral gene delivery vehicles. Annu Rev Biomed Eng 10:169–194
5. Bainbridge JWB, Smith AJ, Barker SS, Robbie S, Henderson R, Balaggan K, Viswanathan A, Holder GE, Stockman A, Tyler N, Petersen-Jones S, Bhattacharya SS, Thrasher AJ, Fitzke FW, Carter BJ, Rubin GS, Moore AT, Ali RR (2008) Effect of gene therapy on visual function in Leber's congenital amaurosis. N Engl J Med 358:2231–2239
6. Maguire AM, High KA, Auricchio A, Wright JF, Pierce EA, Testa F, Mingozzi F, Bennicelli JL, G-s Y, Rossi S, Fulton A, Marshall KA, Banfi S, Chung DC, Morgan JIW, Hauck B, Zelenaia O, Zhu X, Raffini L, Coppieters F, Baere ED, Shindler KS, Volpe NJ, Surace EM, Acerra C, Lyubarsky A, Redmond TM, Stone E, Sun J, McDonnell JW, Leroy BP, Simonelli F, Bennett J (2009) Age-dependent effects of RPE65 gene therapy for Leber's congenital amaurosis: a phase 1 dose-escalation trial. Lancet 374(9701):1597–1605
7. Maguire AM, Simonelli F, Pierce EA, Pugh EN Jr, Mingozzi F, Bennicelli J, Banfi S, Marshall KA, Testa F, Surace EM, Rossi S, Lyubarsky A, Arruda VR, Konkle B, Stone E, Sun J, Jacobs J, Dell'Osso L, Hertle R, Ma J-X, Redmond TM, Zhu X, Hauck B, Zelenaia O, Shindler KS, Maguire MG, Wright JF, Volpe NJ, McDonnell JW, Auricchio A, High KA, Bennett J (2008) Safety and efficacy of gene transfer for Leber's congenital amaurosis. N Engl J Med 358:2240–2248
8. Russell DW, Hirata RK (1998) Human gene targeting by viral vectors. Nat Genet 18:325–330
9. Jang J-H, Koerber JT, Kim J-S, Asuri P, Vazin T, Bartel M, Keung A, Kwon I, Park KI, Schaffer DV (2011) An evolved adeno-associated viral variant enhances gene delivery and gene targeting in neural stem cells. Mol Ther 19(4):667–675
10. Asuri P, Bartel MA, Vazin T, Jang J-H, Wong TB, Schaffer DV (2012) Directed evolution of adeno-associated virus for enhanced gene delivery and gene targeting in human pluripotent stem cells. Mol Ther 20(2):329–338
11. Hendrie PC, Russell DW (2005) Gene targeting with viral vectors. Mol Ther 12(1):9–17
12. Hirata RK, Russell DW (2000) Design and packaging of adeno-associated virus gene targeting vectors. J Virol 74:4612–4620
13. Zou J, Maeder ML, Mali P, Pruett-Miller SM, Thibodeau-Beganny S, Chou B-K, Chen G, Ye Z, Park I-H, Daley GQ, Porteus MH, Joung JK, Cheng L (2009) Gene targeting of a disease-related gene in human induced pluripotent stem and embryonic stem cells. Cell Stem Cell 5:97–110
14. Mansour SL, Thomas KR, Capecchi MR (1988) Disruption of the proto-oncogene int-2 in mouse embryo-derived stem cells: a general strategy for targeting mutations to non-selectable genes. Nature 336(24):348–352
15. Wu S, Ying G, Wu Q, Capecchi MR (2008) A protocol for constructing gene targeting vectors: generating knockout mice for the cadherin family and beyond. Nat Protoc 3(6):1056–1076
16. Miller JC, Tan S, Qiao G, Barlow KA, Wang J, Xia DF, Meng X, Paschon DE, Leung E, Hinkley SJ, Dulay GP, Hua KL, Ankoudinova I, Cost GJ, Urnov FD, Zhang HS, Holmes MC, Zhang L, Gregory PD, Rebar EJ (2011) A TALE nuclease architecture for efficient genome editing. Nat Biotechnol 29:143–148
17. Matsushita T, Elliger S, Elliger C, Podsakoff G, Villarreal L, Kurtzman GJ, Iwaki Y, Colosi P (1998) Adeno-associated virus vectors can be efficiently produced without helper virus. Gene Ther 5(7):938–945
18. Xiao X, Li J, Samulski RJ (1998) Production of high-titer recombinant adeno-associated virus vectors in the absence of helper adenovirus. J Virol 72(3):2224–2232
19. Peltier J, Agrawal S, Robertson MJ, Schaffer DV (2010) In vitro culture and analysis of adult hippocampal neural progenitors. Methods Mol Biol 621:65–87
20. Stover AE, Schwartz PH (2011) Adaptation of human pluripotent stem cells to feeder-free conditions in chemically defined medium with enzymatic single-cell passaging. Methods Mol Biol 767:137–146

Chapter 12

Lentiviral Vectors Encoding Zinc-Finger Nucleases Specific for the Model Target Locus *HPRT1*

Laetitia P.L. Pelascini and Manuel A.F.V. Gonçalves

Abstract

Zinc-finger nucleases (ZFNs) are artificial proteins designed to induce double-stranded DNA breaks (DSBs) at predefined chromosomal positions. These site-specific genomic lesions facilitate the study of translocations and cellular DNA repair processes and serve as powerful stimuli for the editing of complex genomes. The delivery of ZFNs into a wide range of cell types is of utmost importance for the broad evaluation and deployment of the technology. Lentiviral vectors (LVs) are versatile gene delivery vehicles that transduce alike transformed and primary cells regardless of their division rate. In this chapter, we describe the generation of conventional and integrase-defective LVs encoding ZFNs targeting the human hypoxanthine phosphoribosyltransferase 1 (*HPRT1*) locus. Furthermore, we introduce a general LV titration method based on a cost-effective quantitative PCR protocol and implement a rapid and simple restriction enzyme site polymorphism assay to detected DSB formation at the *HPRT1* target sequence. Owing in part to the small molecule-based clone selection schemes conferred by *HPRT1* allelic knockouts, this X-linked gene has become a "classical" target model locus in mammalian cells. The reagents and techniques detailed herein yield LV preparations that induce *HPRT1*-specific DSBs. As a result, they should constitute a valuable resource to increase the robustness and decrease the timelines of the various protocols based on *HPRT1* gene disruption and targeting.

Key words Lentiviral vectors, Zinc-finger nucleases, Double-stranded DNA breaks, Gene correction, Gene targeting, DNA repair, Gene knockout, *HPRT1*

1 Introduction

Nucleases with customized DNA-binding specificities designed to catalyze the formation of double-stranded DNA breaks (DSBs) at predefined genomic sequences are opening a vast range of research possibilities [1–4]. These stem from the activation of the major DNA repair pathways of the cell and include the study of chromosomal translocations and cellular DNA repair processes [5] as well as the genetic manipulation of complex eukaryotic genomes [4]. Related to the latter aspect, the engagement of the error-free homologous recombination pathway at site-specific DSBs can lead to the chromosomal insertion of a transgene or to the correction

Francesca Storici (ed.), *Gene Correction: Methods and Protocols*, Methods in Molecular Biology, vol. 1114, DOI 10.1007/978-1-62703-761-7_12, © Springer Science+Business Media, LLC 2014

(or addition) of a mutation in the vicinity of the induced lesion. Alternatively, the recruitment of the error-prone nonhomologous end-joining DNA repair machinery at targeted DSBs made within coding sequences can yield gene knockouts due to the generation of frame-shifting insertions/deletions (indels). These genome editing principles are appealing from the fundamental and applied research standpoint. Indeed, gene knockouts are obviously more stringent than gene knockdowns and chromosomally targeted transgenes are less likely to suffer from positional effects on the one hand, and deregulate endogenous host gene expression on the other [4]. Of note, the latter aspects assume particular relevance in the context of gene therapy studies [6].

The strategies to assemble sequence-specific nucleases are evolving at a rapid pace. The technology undergoing development for longer is that of zinc-finger nucleases (ZFNs) [4, 7]. Related to this, ZFNs are the first and, as of yet, sole sequence-specific nuclease technology to enter clinical testing in the form of a candidate AIDS therapy. This approach is based on the knockout of the gene coding for the human immunodeficiency virus (HIV) co-receptor CCR5 in $CD4^+$ T cells [8]. ZFNs are modular proteins consisting of the nuclease domain of the type IIS restriction enzyme FokI, N-terminally fused to an array of typically 4–6 synthetic Cys_2-His_2 zinc-fingers whose individual units recognize a triplet on the double helix. ZFNs work in pairs and assemble in a tail-to-tail orientation at the target sequence on opposite DNA strands. The local dimerization of the tethered FokI motifs catalyzes the cleavage of the upper and lower DNA strands at the spacer sequence separating the two ZFN target sites [1]. Despite their broad appeal, designing ZFN pairs with high specificity and low cytotoxicity is not a trivial endeavor to most non-specialized laboratories. This has spurred industry and academic consortia to set up various platforms to assemble ZFNs based on bioinformatics and zinc-finger libraries (for a recent overview *see* ref. [4]). Newly generated ZFNs are most often evaluated through assays carried out in vitro or in transformed cell lines. ZFNs can be introduced into these readout systems either in the form of protein or as ZFN-encoding nucleic acids by using transfection agents or microinjection. In many settings, however, the goal is to test or to deploy them in hard-to-transfect populations of representative primary mammalian cells. In those instances, HIV-based lentiviral vectors (LVs) constitute a valuable gene delivery platform due to the infection mechanisms evolved by their parental virus [9]. These include the active transport, via the HIV pre-integration complex, of reverse transcribed genomes into the nuclei of non-dividing cells. Moreover, a working LV stock can be obtained within a week by using transient DNA transfection of human embryonic kidney 293T cells. This step is followed by ultracentrifugation-mediated concentration of the particles that bud off from the producer cell plasma membranes and are released into the medium. The 293T producer cells are co-transfected with split transfer and packaging plasmids and a vector particle-pseudotyping plasmid. Typically, the transfer plasmids harbor HIV *cis*-acting elements, namely, the 5′ and 3′ long terminal repeats (LTRs), the packaging signal, the Rev-responsive element (RRE) and the central polypurine tract (cPPT) plus the transgene(s) of interest, whereas the packaging constructs encode the *trans-acting* viral gene products. For increased biosafety, latest-generation LV production systems comprise packaging constructs expressing a minimal set of HIV genes (i.e., *gag*, *pol*, and *rev*), whilst the transfer plasmids have so-called self-inactivating (SIN) deletions within the 3′ LTR that remove viral-enhancer sequences [9]. Another feature that contributes to the versatility of LV technology is the ease with which vector particles can be pseudotyped by simply co-transfecting an expression plasmid encoding a heterologous envelope glycoprotein of choice that binds specific cellular receptors [10]. In most instances, to confer a broad host range and high stability to LV particles, the vesicular stomatitis virus glycoprotein-G (VSV-G) is used [10].

In this chapter, we describe reagents and techniques to produce integration-proficient and integration-defective lentiviral vectors (IDLVs) that code for ZFNs targeting the human hypoxanthine phosphoribosyltransferase 1 (*HPRT1*) gene whose genomic position is at Xq26.2-q26.3. The latter vectors are assembled by utilizing packaging constructs encoding HIV integrase (IN) moieties with so-called class I mutations [11]. Class I IN mutants have substitutions at one or more residue positions corresponding to the DDE triad present within the catalytic pocket domain of the enzyme (i.e., D64, D116, and E152). Importantly, these mutations are non-pleiotropic as they block specifically chromosomal cDNA integration and no other step of the transduction process [11]. Owing to their non-integrating phenotype, IDLVs are particularly suited for the synthesis of proteins, such as ZFNs, whose presence in target cells is only required for a short period of time.

Large-scale plasmid transfections, such as those involving the generation of LV and IDLV batches, become too expensive whenever using most commercially available formulations. Thus, it is sensible to deploy alternative cost-effective agents whose nucleic acid avidity yields macromolecular complexes that can be readily up taken by cells. Polyethyleneimine (PEI) is a low-cost cationic polymer produced in industrial amounts. Importantly, nucleic acid transfection is one among the various practical uses of PEI [12]. Indeed, PEI formulations have been utilized for the production of LV and IDLV preparations [13–16]. In this chapter, therefore, a PEI-based DNA transfection protocol is detailed to produce the ZFN-encoding LVs and IDLVs. Furthermore, we introduce a general quantitative PCR (qPCR) assay that makes use of SYBR® Green chemistry to titrate HIV-1-derived vectors and implement a rapid restriction enzyme-based genotyping method to detect target site cleavage by the *HPRT1*-specific ZFNs.

The *HPRT1* gene product hypoxanthine phosphoribosyltransferase (EC 2.4.2.8) plays a key role in the purine salvage pathway by catalyzing the conversion of guanine to guanosine monophosphate and hypoxanthine to inosine monophosphate. In its most severe forms, the deficiency of this metabolic enzyme leads to Lesch-Nyhan syndrome [17]. Historically, model systems based on *HPRT1* gene expression (or lack thereof) have been copiously utilized and were behind key proof-of-concepts such as those for hybridoma-based monoclonal antibody technology [18, 19], mammalian transgenesis [20, 21], and gene therapy [22, 23]. To a great extent, this overarching impact stems from the biochemical network of the enzyme and the availability of stringent cell selection regimens [18] that include the nucleotide analog 6-thioguanine and a cocktail of hypoxanthine, aminopterin, and thymidine (HAT medium). The protocols presented in this chapter expand the gamut of methodologies compatible with the experimental systems based on *HPRT1* gene disruption.

2 Materials

2.1 Production of LVs and IDLVs by DNA Transfection

1. Vertical laminar air flow cabinet for the handling of cells under aseptic conditions and incubator to keep the cells at 37 °C in a humidified atmosphere containing 5 or 10 % CO_2.
2. Pipetboy and disposable sterile pipettes: 1, 5, 10, and 25 ml.
3. Sterile pipette filter tips: 20, 200, and 1,000 μl.
4. Sterile 50-ml screw cap polypropylene centrifuge tube.
5. Sterile 50-ml syringes.
6. Disposable 0.22-μm cellulose filters.
7. 175-cm^2 cell culture flasks: CELLSTAR (Greiner Bio-One).
8. 25 mm-diameter Acrodisc syringe filters with 0.45-μm pore size HT Tuffryn® membranes (Pall Corporation).
9. Producer cells: 293T cells (ATCC number: CRL-11268).
10. Culture medium: Dulbecco's modified Eagle medium with 4.5 g/l glucose (DMEM) supplemented with 10 % fetal bovine serum (FBS).
11. Post-transfection medium: DMEM supplemented with 5 % FBS.
12. Phosphate-buffered saline (PBS) pH 7.4 and 0.05 % Trypsin–EDTA.
13. Transfer plasmids: pLV.ZFN-1^{HPRT} and pLV.ZFN-2^{HPRT} (*see* **Notes 1** and **2** and Fig. 1a). Plasmid sizes: 8,062 and 8,146 bp, respectively.

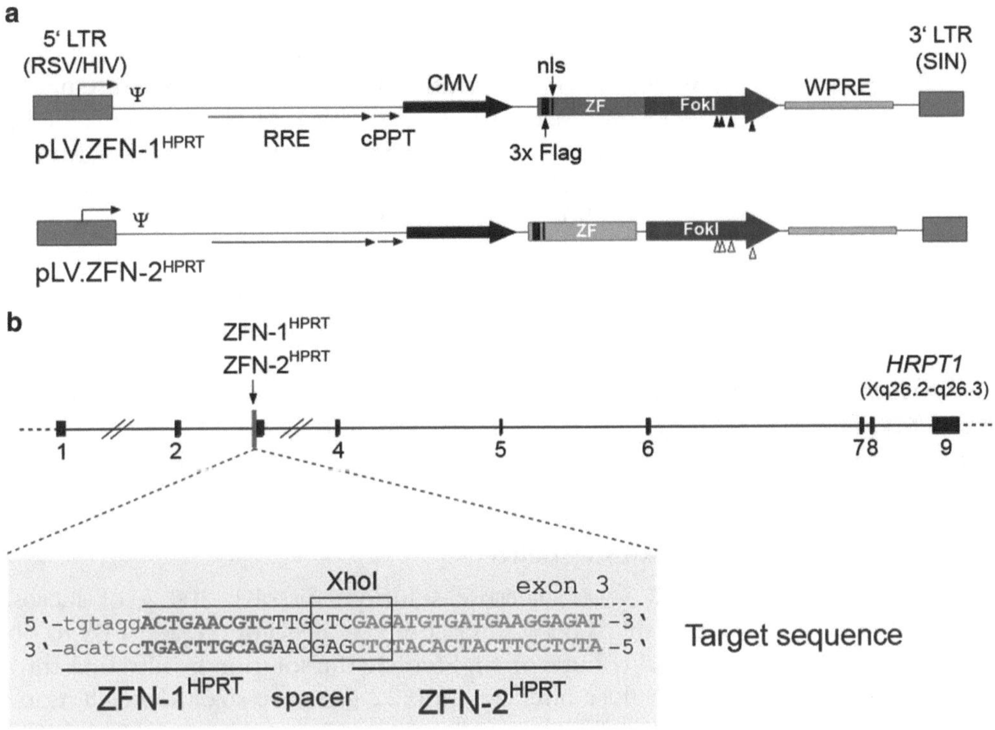

Fig. 1 Genetic structures of the LV transfer plasmids coding for *HPRT1*-specific ZFNs and their target sequences. (**a**) The constructs pLV.ZFN-1^{HPRT} and pLV.ZFN-2^{HPRT} code for the custom-made nucleases ZFN-1^{HPRT} and ZFN-2^{HPRT} whose synthetic DNA-binding zinc-finger arrays are directed at the upstream end of exon 3 of the human *HPRT1* gene. Close to their N-termini, the ZFN proteins are tagged by Flag epitopes and contain the simian virus 40 large T antigen nuclear localization signal (nls). *Boxes* with *broken arrow*, hybrid 5′ LTR composed of Rous sarcoma virus- and HIV-1-derived sequences; *boxes* without *broken arrow*, self-inactivating (SIN) 3′ LTR; ZF, DNA sequences encoding the zinc-finger arrays; FokI, DNA sequences encoding mutant FokI nuclease motifs with the engineered residues EELI and QKIK being indicated by *solid* and *open vertical arrowheads*, respectively. These mutations minimize ZFN-associated cytotoxicity by obligating assembly of target-specific heterodimers in detriment of nonspecific homodimers [25]. Ψ HIV-1 packaging signal, *RRE* Rev-responsive element, *cPPT* central polypurine tract, *WPRE* Woodchuck hepatitis virus posttranscriptional regulatory element. The ZFN open reading frames are under the transcriptional control of the *immediate-early cytomegalovirus* gene promoter (CMV) and the polyadenylation signal located within the SIN 3′ LTR. For the sake of simplicity the plasmid backbone containing the prokaryotic origin of replication and the β-*lactamase* gene conferring resistance to ampicillin, are not drawn. (**b**) *HPRT1* target locus. The nucleotide sequences of the two *HPRT1-specific* ZFN target sites as well as their position in the *HPRT1* locus is shown. *Solid boxes*, *HPRT1* exons 1 through 9. The nucleotide sequence of *HPRT1* corresponding to exon 3 is indicated in *upper case*, whereas the contiguous upstream flanking nucleotides belonging to the ZFN-1^{HPRT} recognition site, are depicted in *lower case*. The 6-bp intervening sequence framed by the two ZFN target sites (spacer) partially overlaps with that of the restriction enzyme XhoI

14. Packaging plasmids: psPAX2 (Addgene plasmid #12260) and psPAX2.IN^{D116N} to assemble LV and IDLV particles, respectively (*see* **Note 2**). Both have a size of 10,702 bp.
15. Pseudotyping plasmid: expression construct pLP/VSVG (Invitrogen) coding for the heterologous envelope protein VSV-G (*see* **Note 2**). Plasmid size: 5,821 bp.

16. PEI DNA transfection solution at 1 mg/ml: dissolve 45 mg of linear 25 kDa PEI in 40 ml of distilled H_2O pre-heated to 80 °C. Allow the solution to cool down to room temperature (RT). Adjust the pH to 7.4 with 1 M HCl. Add distilled H_2O to reach a total volume of 45 ml. Sterilize the solution by filtration through a cellulose filter with a 0.22-μm pore size, aliquot, and store at −80 °C (*see* **Note 3**).
17. 150 mM NaCl: dissolve 8.76 g of NaCl in 1 l of PBS. Sterilize the solution by filtration through cellulose filters with a 0.22-μm pore size. Prepare 50-ml aliquots and store at RT.

2.2 Purification of LVs and IDLVs by Ultracentrifugation

1. Sterile 50-ml screw cap polypropylene centrifuge tubes.
2. Sterile pipette filter tips: 20, 200, and 1,000 μl.
3. Autoclave-sterilized 0.5- and 2-ml Eppendorf tubes.
4. Sterilized 35.8-ml open-top polyallomer ultracentrifuge tubes (Beckman Coulter).
5. 20 % (w/v) sucrose solution: dissolve 200 g of sucrose in 800 ml of PBS. After complete dissolution, add PBS to obtain a final volume of 1 l. Sterilize the solution by filtration through a cellulose filter with a 0.22 μm pore size. Make aliquots and store at RT.
6. Culture medium: DMEM.
7. 100 % ethanol.
8. PBS–1 % BSA vector storage buffer: 1 % (w/v) bovine serum albumin (BSA) in PBS pH 7.4, filter-sterilized, aliquoted, and conserved at 4 °C.
9. Ultracentrifuge, such as Optima LE-80K with SW28 rotor (Beckman Coulter).
10. Weighing scale.
11. Platform shaker.

2.3 Titration of LVs and IDLVs by qPCR

1. Concentrated vector stocks: LV.ZFN-1^{HPRT}, LV.ZFN-2^{HPRT}, IDLV.ZFN-1^{HPRT}, and IDLV.ZFN-2^{HPRT}.
2. Target cells: Human cervix carcinoma HeLa cells (ATCC).
3. Culture medium: DMEM supplemented with 5 % FBS.
4. PBS pH 7.4 and 0.05 % Trypsin–EDTA.
5. Sterile pipette filter tips: 20, 200, and 1,000 μl.
6. 24-well cell culture plates CELLSTAR.
7. 5-ml polystyrene round-bottom FACS tubes (BD Biosciences).
8. DNeasy Blood and Tissue kit (Qiagen).
9. PCR reagent concentrations in 20-μl PCR mixtures: 2 U of GoTaq® Flexi DNA polymerase (Promega), 1× Colorless GoTaq® Flexi Buffer, $MgCl_2$ (Table 1), 400 nM deoxynucleoside

Table 1
Target templates, primer sequences, $MgCl_2$ concentrations, and amplicon sizes corresponding to the qPCR assay to measure vector genome copies/ml (vgc/ml)

Target sequences		Primer sequences (5′→3′)	$MgCl_2$ (mM)	Amplicon size (bp)
Primers used to quantify the amount of vector DNA molecules per sample				
HIV-1 5′ LTR U5—Ψ	Fwd	tagtgtgtgcccgtctgttgtg	1	242
	Rev	acccatctctctccttctagcctcc		
Primers used to quantify the amount of cellular genomes per sample				
Human *ALB* (intron 12)	Fwd	ctgtcatctcttgtgggctgtaatc	2	138
	Rev	actcatgggagctgctggtc		
Primers used to control for the carryover of transfer plasmid DNA into target cells				
E. coli β-lactamase	Fwd	agatgctgaagatcagttgggtgc	1	191
	Rev	aatagtgtatgcggcgaccgagtt		

triphosphates (dNTPs), 200 nM of each primer (Table 1), 0.1× SYBR® Green and DNase/RNase-free distilled H_2O.

10. qPCR standards: pCRII-TOPO cloning vector (Invitrogen) carrying either DNA encompassing the 5′ LTR U5 region plus the contiguous packaging element (Ψ) or a fragment of the human *ALB* gene (Table 1).
11. Restriction enzyme: NotI.
12. Spectrophotometer (e.g., NanoDrop).
13. Tris–EDTA (TE) buffer: 10 mM Tris–HCl (pH 8.0) and 1 mM EDTA.
14. The transfer plasmid(s) used to produce the vector stock(s) to control by qPCR for residual plasmid carryover into target cell cultures.
15. 96-well optical PCR plates, adhesive seals or optical caps.
16. Real-Time PCR machine, such as CFX Connect™ Real-Time PCR Detection System (Bio-Rad).

2.4 Detection of DSB formation at HPRT1 by XhoI Genotyping

1. Concentrated vector pair: LV.ZFN-1^{HPRT}/LV.ZFN-2^{HPRT} or IDLV.ZFN-1^{HPRT}/ILDV.ZFN-2^{HPRT}.
2. Target human cells of choice.
3. Culture medium: DMEM supplemented with 5 % FBS.
4. PBS and 0.05 % Trypsin–EDTA.
5. Sterile pipette filter tips: 20, 200, and 1,000 μl.
6. 24-well cell culture plates.
7. DNeasy Blood and Tissue kit.
8. NanoDrop.
9. PCR reagent concentrations in 25-μl PCR mixtures: 2 U of GoTaq® Flexi DNA polymerase, 1× Colorless GoTaq® Flexi

Buffer, 3 mM $MgCl_2$, 400 nM dNTPs, 200 nM of each primer and DNase/RNase-free distilled H_2O.

10. Primers designed to amplify a 379-bp fragment encompassing the ZFN pair cutting site located within exon 3 of *HPRT1* (Fig. 1b):Fwd.H1.Ex3: 5′-tcactgtattgcccaggttggtg-3′Rev. H1.Ex3: 5′-gaaagcaagtatggtttgcagagat-3′.
11. PCR strips and caps.
12. PCR machine allowing a touchdown PCR program, such as the DNA Engine Tetrad 2 Peltier Thermal Cycler (Bio-Rad).
13. Restriction enzyme: XhoI.
14. Agarose gel electrophoresis apparatus.
15. TAE buffer: to prepare a stock of 50× Tris–acetate–EDTA (TAE) buffer, dissolve 242 g Tris base with 57.1 ml glacial acetic acid and 100 ml 0.5 M EDTA (pH 8.0) in 800 ml of distilled H_2O. After complete dissolution, add H_2O to obtain a total volume of 1 l. Store at RT.

3 Methods

3.1 Production of LVs and IDLVs by DNA Transfection

The working activities for the production and handling of self-inactivating vectors based on HIV need to take place in a laboratory with a biosafety level II according to the pertinent national biosafety practices. Regular cell culturing of 293T cells and target cells should occur in a biosafety level I laboratory instead.

The following DNA transfection protocol for the generation of LV and IDLV stocks in 293T cells refers to the use of two 175-cm^2 cell culture flasks (*see* **Note 4**). Up scaling to twelve 175-cm^2 flasks is the maximum number of vessels compatible with a single round of ultracentrifugation-mediated vector concentration (*see* Subheading 3.2).

1. On day 0, seed 17×10^6 293T cells per 175-cm^2 cell culture flask in DMEM supplemented with 10 % FBS. Incubate the cells overnight at 37 °C in a humidified atmosphere containing 10 % CO_2. On the day of transfection, the cells should be homogenously distributed over the vessel surface and display a confluency of about 70–80 %.
2. On day 1, prepare the DNA transfection mixtures as follows: in a 50-ml screw cap polypropylene centrifuge tube add, to a 150 mM NaCl solution, 60 μg of a DNA cocktail consisting of transfer, packaging, and pseudotyping plasmids at a ratio of 2:1:1 (size-normalized for molecule copy numbers). The total volume of the DNA and NaCl solutions should be 1 ml. In a second 50-ml centrifuge tube add, to 820 μl of a 150 mM NaCl solution, 180 μl of PEI at 1 mg/ml (*see* **Note 5**).

3. Add drop wise 1 ml of the PEI–NaCl mixture to 1 ml of DNA–NaCl solution while gently vortexing the tube (i.e., 600 RPM). Immediately after, homogenize thoroughly the final mixture by vigorous vortexing (i.e., 1,400 RPM) for a few seconds (*see* **Note 6**).
4. Incubate for 15 min at RT to allow the formation of DNA/PEI complexes.
5. Add 38 ml of regular culture medium to the 2-ml colloidal solution of DNA/PEI complexes. After removing the conditioned medium, distribute 20 ml of the resulting transfection mixture per 175-cm^2 cell culture flask. Incubate the cells at 37 °C in a humidified atmosphere with 10 % CO_2.
6. On day 2, in the morning, remove the DNA transfection medium and replace it by 15 ml of DMEM supplemented with 5 % FBS. Incubate the cells at 37 °C in a humidified atmosphere containing 10 % CO_2.
7. On day 4, collect the 15 ml of the conditioned media from both 175-cm^2 cell culture flasks containing the cell-free vector particles and pool them in a sterile 50-ml screw cap polypropylene centrifuge tube.
8. Centrifuge the tubes for 10 min at 1,600 × *g* at RT in a MSE Mistral 2000 centrifuge or similar apparatus.
9. Transfer the circa 30 ml of vector supernatant into a syringe coupled to a HT Tuffryn® filter with 0.45-μm pore size membrane (Pall Corporation) and collect the filtrate in a new 50-ml screw cap polypropylene centrifuge tube.
10. Proceed directly to the subsequent purification steps or store the clarified vector particle-containing supernatant at −80 °C until further use.

3.2 Purification of LVs and IDLVs by Ultracentrifugation

For each round of ultracentrifugation, a maximum of six 30-ml clarified supernatant can be processed.

1. Sterilize the 35.8-ml polyallomer ultracentrifuge tubes and the ultracentrifuge buckets with 100 % ethanol and let them dry under UV light inside the flow cabinet. Make sure that all the ethanol has evaporated before proceeding with using the tubes.
2. Add 30 ml of clarified supernatant into a sterilized ultracentrifuge tube (*see* **Note 7**).
3. Carefully place the tip of a 5-ml pipet containing 7 ml of the sterile 20 % sucrose solution near the bottom of the ultracentrifugation tube. Slowly dispense 5 ml of the sucrose solution in order to create a block gradient consisting of the denser sucrose cushion underneath the lighter supernatant. Cautiously remove the pipette to avoid disturbing the interface between the two phases.

4. Transfer the loaded tubes into the ultracentrifuge buckets. All positions in the rotor must be occupied with the respective bucket. The weight of the loaded buckets facing each other in the rotor should, ideally, not vary by more than 0.02 g. Thus, place the weighing scale in the flow cabinet and use sterile plain DMEM to correct for weight differences between the buckets pairs.
5. Carefully place the loaded rotor in the ultracentrifuge to avoid disturbance of the interface between the two density phases of the block gradients. The ultracentrifugation step is carried out at 29,774 × *g* for 2 h at 10 °C with the rate of acceleration and breaking set at slow and off, respectively.
6. Discard the supernatants by carefully inverting the ultracentrifuge tubes and place them directly upside-down on a UV-sterilized tissue for about 3–4 min. Vector pellets are not always clearly visible on the bottom of the ultracentrifuge tubes (*see* **Note 8**).
7. Use a new sterilized tissue to remove supernatant remnants present on the edge of the ultracentrifuge tubes and transfer the ultracentrifuge tubes into 50-ml screw cap polypropylene centrifuge tubes. Cap these tubes well to avoid the over-drying of the vector pellets.
8. Add between 200 and 400 μl of PBS–1 % BSA to each of the vector pellets (*see* **Note 9**). Place the capped centrifuge tubes on a rocking platform and incubate them overnight at 4 °C.
9. On day 5, in the morning, place the 50-ml screw cap polypropylene tubes containing the ultracentrifuge tubes on ice. From now on, all work should be done on ice (*see* **Note 10**).
10. Pool and collect the vector particle suspensions in a pre-cooled 2-ml tube.
11. Add 100 μl of PBS–1 % BSA to wash each ultracentrifuge tube and transfer this "washing solution" into the corresponding 2-ml tube.
12. Divide the resuspended vector particles in aliquots of 20–50 μl using pre-cooled 0.5-ml tubes and store them at -80 °C until further usage (*see* **Note 10**).

3.3 Titration of LVs and IDLVs by qPCR

3.3.1 Cell Transduction and DNA Sample Preparation

1. On day 0, seed 1×10^5 HeLa cells per well of a 24-well plate in DMEM containing 5 % FBS. Eight wells are deployed for each vector stock titration. Incubate the cells overnight at 37 °C in a humidified atmosphere with 10 % CO_2.
2. On day 1, before initiating the transduction procedure, harvest the cells in one of the wells and count them. The number of cells/well value will be used in the vector titer calculations (*see* **step 17** of the current section).

3. Set up a series of dilutions of each vector stock in 5-ml FACS tubes as follows. After an initial 16.7-fold dilution, prepare a series of six threefold dilutions in regular HeLa cell culture medium using the volumes specified in the table below. Do not add vector material to the cells in one of the wells to provide for a negative control.

Vector stock dilutions	Vector stock (μl)	Medium (μl)
First dilution	60	940
Second to sixth dilution	300	600

4. Add 500 μl of each of the vector dilutions to each well and incubate the target cells overnight.
5. On day 2, wash the cells 4–5 times with 1 ml of PBS (*see* **Note 11**). Next, trypsinize and resuspend them in 700 μl of DMEM supplemented with 5 % FBS. Transfer the cells into 1.5-ml tubes, pellet them by centrifugation in a table top microcentrifuge (e.g., Eppendorf 5417C) at 1,200 RPM for 5 min, after which carefully remove the supernatants. Resuspend the cell pellets in 1 ml of PBS to wash the cells and repeat the 5-min centrifugation step. Carefully discard the supernatants.
6. Isolate total cellular DNA by using the DNeasy Blood and Tissue kit (Qiagen), following the manufacturer's instructions.
7. Proceed directly to the qPCR assay or store the DNA samples at −20 °C.

3.3.2 Preparation of the Standards for the qPCR Assay

8. The titers of LV and IDLV stocks are determined in terms of vector genome copies per ml (vgc/ml) on the basis of calibration curves using 8 tenfold serial dilutions of pCRII-TOPO cloning vectors containing the relevant target sequences. To quantify the amount of vector DNA molecules within a cell, the HIV-1 5′ LTR U5-Ψ segment common to most transfer plasmids is targeted. To normalize vector to chromosomal DNA copy numbers, a fragment of the human *ALB* gene coding for albumin, is used. Both fragments were cloned into pCRII-TOPO according to the manufacturer's instructions giving rise to pCRII-LTR/Ψ and pCRII-ALB (*see* Subheading 2.3 and Table 1).
9. Linearize pCRII-LTR/Ψ and pCRII-ALB with NotI following the manufacturer's recommendations. Precipitate the plasmids with isopropanol, wash the pellets with 70 % ethanol, and resuspend them in TE (*see* **Note 12**). Determine the concentration (ng/μl) of both plasmid solutions by using a NanoDrop spectrophotometer.

3.3.3 qPCR Set-up

10. Prepare two PCR “Master Mix” containing all the reagents but the DNA samples (*see* Subheading 2.3 and Table 1). The first “Master Mix” contains the primers for the amplification of the 5′ LTR U5-Ψ fragment, whereas the second has the primers for the amplification of the *ALB* target sequence. Take into account in the “Master Mix” volume calculations the negative controls (H_2O) and the eight dilution standards in duplicate.
11. Distribute 18 μl of each “Master Mix” into the wells of a 96-well optical PCR plate, after which add 2 μl of the various DNA samples, the plasmid standards and H_2O. Cover the plate with an adhesive seal and briefly centrifuge it at 1,200 RPM in a 5810R centrifuge (Eppendorf), or similar apparatus.
12. Place the 96-well plate in a real-time PCR thermal cycler machine and select the following cycling conditions:

1 cycle	95 °C	2 min
40 cycles	95 °C 60 °C 72 °C	20 s 20 s 20 s
Melting curves		

3.3.4 qPCR Data Analyses

13. Check all the melting curves. The melting temperatures (Tm) corresponding to 5′ LTR U5-Ψ and *ALB* fragments are 87.5 and 81.5 °C, respectively. Confirm the absence of nonspecific PCR products following the amplification of the 5′ LTR U5-Ψ target DNA.
14. The baseline thresholds are auto-calculated by the software of the real-time PCR machine. If preferred, they can be defined by the user instead. Export all Ct values into an Excel sheet.
15. *Standard analyses I:* Plasmid DNA copy number calculation to establish calibration curves:
    - Calculate the molecular weight (g/mol):Plasmid size (bp) + PCR insert (bp) × 330 Da × 2 nucleotides per bp
    - Calculate the molarity of your plasmid solution (mol/l):Molecular weight (g/mol)/DNA concentration (g/l)
    - Express in terms of copies/l:Molarity (mol/l) × Avogadro number (6.022E+23)
    - Express in terms of copies/μl:Molarity (mol/l) × Avogadro number (6.022E+23)/10E−06
    - Express as logarithm with base 10:Log_{10} (copies/μl)
16. *Standard analyses II:* For each plasmid dilution series, create the calibration curves with the Log_{10} (copies/μl) values and their corresponding Ct value. Using the Excel function, determine

Table 2
qPCR titration data (vgc/ml) corresponding to two independent LV and IDLV production rounds

	Vector batches (round #1)		Vector batches (round #2)	
	Mean	St. Dev.	Mean	St. Dev.
LV.ZFN-1^{HPRT}	3.70×10^8	3.61×10^7	2.04×10^8	1.21×10^7
LV.ZFN-2^{HPRT}	2.87×10^8	2.99×10^7	1.82×10^8	1.81×10^7
IDLV.ZFN-1^{HPRT}	3.11×10^8	1.81×10^7	1.89×10^8	2.65×10^7
IDLV.ZFN-2^{HPRT}	2.21×10^8	2.16×10^7	1.29×10^8	2.83×10^7

St. Dev.: standard deviation

the equation of the resulting regression curves and their coefficient of determination (R^2). This equation corresponds to the following model: $y = ax + b$, where y represents the Ct value, a corresponds to the slope, b is the y-axis interception value, and x represents the Log_{10} (copies/µl) (*see* **Note 13**).

17. *Sample analyses*, to calculate the vgc/ml values:
    - Calculate the copies/µl (in Log10): x = (y − b)/a
    - Calculate the copies/µl number: Power (10, Log10 [copies/µl])
    - Calculate the vector genome copies for each vector dilution:Copies/µl (5′ LTR U5-Ψ)/copies/µl (ALB)
    - Calculate the vgc/ml for each dilution [24]:Normalized vector genome copies × (target cell chromosomal ploidy) × number of transduced cells/volume of vector stock (ml)

    Calculate the vgc/ml mean and standard deviation values resulting from the six dilutions of each vector stock. qPCR titration results corresponding to two independent LV and IDLV production rounds are presented in Table 2.

3.3.5 Assessment of Residual Carryover of Transfer Plasmids

18. Prepare 8 tenfold dilutions of the transfer plasmid used to produce the relevant vector stock.
19. For qPCR preparation and data analysis: apply the protocol above-described and use the primers targeting the prokaryotic *β-lactamase* gene (*see* Subheading 2.3, Table 1 and **Note 14**).

3.4 Detection of DSB Formation at HPRT1 by XhoI Genotyping

The transduction protocol described below (**steps 1–4**) serves merely as an example to provide general guidelines. Clearly, human target cell type, cell amounts, tissue culture vessel formats and experimental design, including vector doses, are at the discretion of the researcher.

1. On day 0, seed 2×10^5 HeLa cells per well of a 24-well plate in DMEM supplemented with 5 % FBS. Incubate the cells overnight at 37 °C in a humidified atmosphere with 10 % CO_2.
2. On day 1, the cells are either mock-treated or are exposed to vector particles at the chosen multiplicities of infection (MOIs) in a total volume of 600 μl.
3. On day 2, remove the inocula and wash the cells once with a large volume of PBS. Next, add 1 ml of fresh culture medium onto the cells and incubate overnight under regular culture conditions.
4. On day 3, wash the cells once with a large volume of PBS. Subsequently, trypsinize the cells, add 700 μl of DMEM containing 5 % FBS and transfer the cell suspensions into a 1.5-ml tube. Collect the cells by a 5-min centrifugation at 1,200 RPM in a table top microcentrifuge (e.g., Eppendorf 5417C) and carefully remove the supernatants. Resuspend the cell pellets in 1 ml of PBS to wash the cells and harvest the cells again by repeating the centrifugation step.
5. Proceed to the total cellular DNA isolation step by using the DNeasy Blood and Tissue kit following the manufacturer's instructions. Determine the DNA concentrations with the aid of a spectrophotometer (e.g., NanoDrop) and dilute each DNA sample to a final concentration of 20 ng/μl.
6. Prepare a "Master Mix" containing all the PCR reagents except for the DNA template (*see* Subheading 2.4). Distribute 20 μl of this "Master Mix" into the wells of a strip, after which add 5 μl of the diluted DNA samples (20 ng/μl). Cap the strip of wells and briefly centrifuge it at 1,200 RPM in a regular table top centrifuge.
7. Place the samples in a PCR machine and select the following touchdown program:

1 cycle	95 °C	2 min
10 cycles	95 °C 63–58 °C[a] 72 °C	25 s 25 s 25 s
25 cycles	95 °C 58 °C 72 °C	25 s 25 s 25 s
1 cycle	72 °C	2 min
1 cycle	20 °C	10 s

[a]Annealing temperature slope: decrease by 0.5 °C per cycle.

8. Incubate a fraction of each PCR mixture with XhoI at 37 °C for 2 h (e.g., 8 μl). Take along a sample not treated with XhoI as a negative control. Subject the DNA to electrophoresis through a 2 % agarose gel in 1× TAE buffer and visualize

the DNA migration by staining with ethidium bromide or an alternative fluorescence-emitting nucleic acid intercalating agent. Figure 2b depicts an example of such an agarose gel corresponding to a transduction experiment deploying LV. ZFN-1HPRT and LV.ZFN-2^{HPRT}.

4 Notes

1. ZFNs can be assembled through a variety of proprietary and publicly available methods (for an overview *see* ref. [4]). The ZNFs targeting *HPRT1* have been designed by using the CompoZr® Custom ZFN platform (Sangamo Biosciences) and are available through Sigma-Aldrich. To reduce the chance for off-target chromosomal DNA cleavage, their FokI domains have been mutagenized at specific residue positions to favor the formation of target-specific heterodimers [25].
2. The protocols described in this chapter make use of psPAX2 (Addgene plasmid #12260) and its derivative psPAX2^{D116N} [16]. These constructs contain only four of the nine HIV-1 genes (i.e., *gag*, *pol*, *rev*, and *tat*) and yield, together with the transfer and pseudotyping plasmids, so-called second-generation vector particles. This 3-plasmid packaging system displays a high biosafety profile and routinely leads to high-titer vector preparations. This makes it suitable for most experimental purposes. For clinical trials, however, third-generation vectors are used instead. These vectors are produced by using split packaging constructs to further reduce the likelihood for the recombination-mediated assembly of replication-competent retroviruses (RCRs). One of these constructs expresses *gag* and *pol* (e.g., pLP1 [Invitrogen] and its derivative pLP1^{D116N}), whereas the other expresses *rev* (e.g., pLP2 [Invitrogen]). In addition, in this 4-plasmid packaging system, *tat* removal is rendered possible by deploying transfer plasmids with *tat-independent* hybrid 5′ LTRs composed of HIV-1 sequences fused to constitutively active transcriptional elements from other viruses (e.g., the Rous sarcoma virus [RSV] or the cytomegalovirus [CMV]). Since the vector genomic RNA-encoding transfer plasmids pLV.ZFN-1^{HPRT} and pLV.ZFN-2^{HPRT} harbor a RSV/HIV-1 hybrid 5′ LTR (Fig. 1a), they are compatible with third-generation packaging systems. As a result, the herein described protocols can be easily adapted to the production of third-generation LV and IDLV particles by simply substituting psPAX2 for pLP1 plus pLP2 and psPAX2^{D116N} for pLP1^{D116N} plus pLP2, respectively.
3. To reach the highest possible transfection efficiency the PEI solution should be frozen and thawed at least five times before its use in DNA transfection protocols.

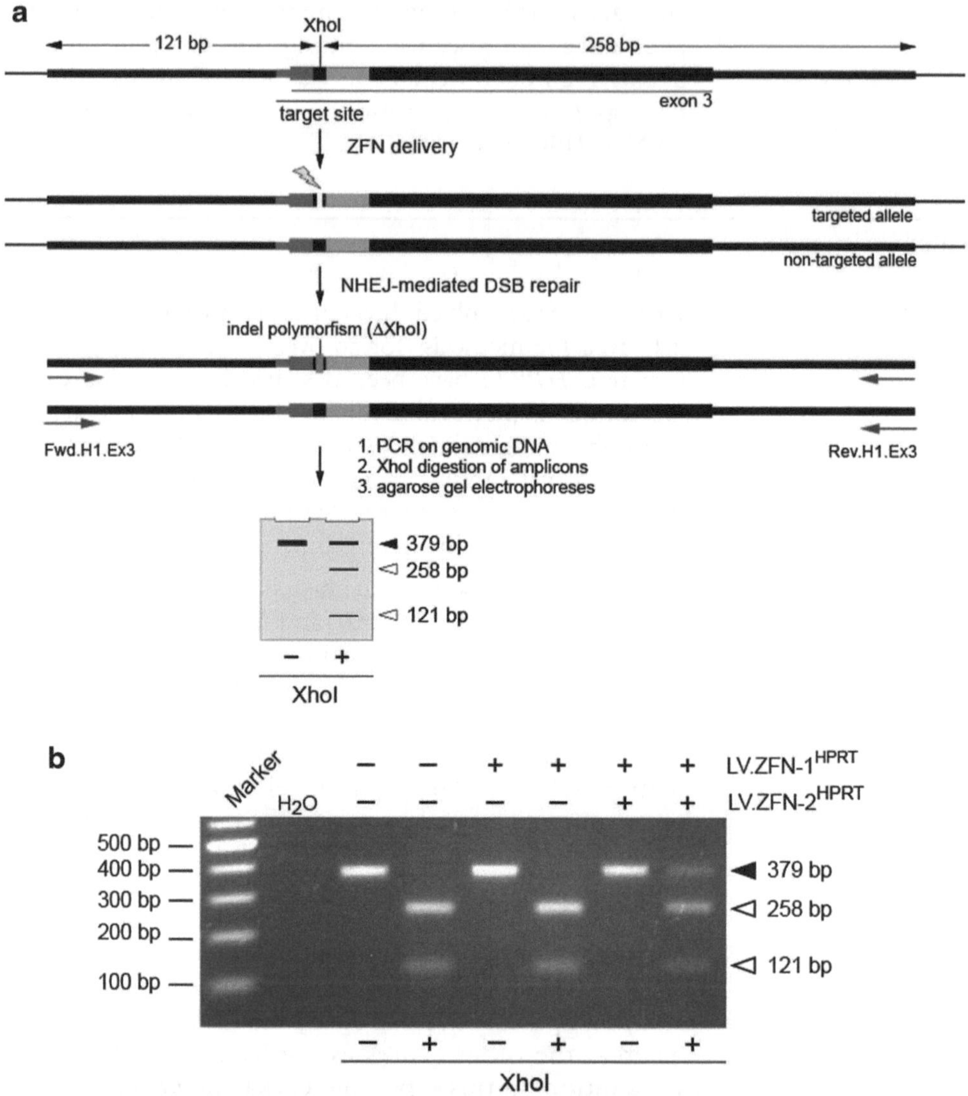

Fig. 2 Readout to detect targeted DSB formation at the exon 3 of the *HPRT1* locus. (**a**) Schematic representation of the XhoI-based genotyping assay. After the introduction into target cells of the ZFN-1^{HPRT} and ZFN-2^{HPRT} pair, DSB formation can ensue within the spacer sequence. The repair of this site-specific DNA lesion through the engagement of the error-prone NHEJ pathway results in the introduction of polymorphisms in the form of small insertions and deletions that disrupt the XhoI recognition sequence (Fig. 1b). PCR on genomic DNA extracted from cells exposed to the *HPRT1*-specific ZFNs should yield amplicons representing targeted and non-targeted alleles. These two types of amplicons can be discriminated through agarose gel electrophoreses of DNA fractions that are resistant to (*solid arrowhead*) or digested by (*open arrowheads*) XhoI. (**b**) Example of the XhoI-based genotyping assay. Human cervix carcinoma HeLa cells were co-transduced with a 1:1 mixture of LV.ZFN-1^{HPRT} and LV.ZFN-2^{HPRT} at a total dose of 3.20×10^7 vgc/ml or were transduced with LV.ZFN-1^{HPRT} alone (1.60×10^7 vgc/ml). Nuclease-free H_2O and total cellular DNA from mock-transduced HeLa cells provided for additional controls. PCR amplifications with the primer set depicted in Fig. 2a were carried out on total cellular DNA isolated at 48 h post-transduction. The resulting PCR products were treated or not treated with XhoI and were subjected to electrophoresis through a 2.0 % agarose gel in TAE 1× buffer. XhoI-resistant and XhoI-digested DNA fragments are indicated (*solid* and *open arrowheads*, respectively)

4. 293T cells are usually highly transfectable, but their good condition is a critical parameter for obtaining efficient DNA transfection. Thus, use cells at a low passage number and culture them twice a week with a constant split ratio of about 1:10 avoiding that they become overconfluent. Careful handling is required as 293T cells detach easily from their plastic support. It is advised to add medium or PBS down the side of the flask and not directly onto the cells.
5. High purity of plasmid DNA is critical for efficient transfection. An A_{260}/A_{280} ratio between 1.8 and 2.0 is desirable. If necessary, following DNA preparation, proceed with an extra purification step consisting of DNA precipitation in isopropanol followed by the washing of the resulting DNA pellet with 70 % ethanol. After removing the ethanol remnants and mildly drying the DNA, dissolve it in sterile TE.
6. It is critical to add the PEI into the DNA solution and not vice versa. It has been shown that adding drop wise the cationic polymer to a shaking plasmid solution leads to transfection levels that are approximately tenfold higher than those resulting from adding the plasmid into the polymer solution instead [12].
7. If necessary, add DMEM to reach a minimum volume of 25 ml per tube in order to avoid tube implosion during ultracentrifugation.
8. It is important avoiding that the vector pellets resulting from the ultracentrifugation step dry up for too long as it may cause significant losses in gene delivery activity.
9. Estimate the volume of the PBS–1 % BSA solution according to the size of the vector pellet. To a typically sized pellet of LV or IDLV particles in a SW28 rotor ultracentrifugate tube, add about 200–400 μl of PBS–1 % BSA. If the pellets are not clearly visible, resuspend them in smaller volumes.
10. It is important to keep lentiviral vectors at 4 °C or lower temperatures as their half-life dramatically declines at room or higher temperatures [26]. Moreover, frequent cycles of freezing and thawing rapidly decreases infectious vector titers as well [26]. Thus, it is recommended to divide vector stocks in small aliquots (20–50 μl) and store them at −80 °C.
11. We routinely isolate the total cellular DNA at 24 h post-transduction. This relates to the rapid loss of IDLV genomes in populations of dividing target cells [16, 27]. When exclusively working with integration-proficient LVs, target cells can be kept for longer periods (e.g., 4 days) in culture prior to total cellular DNA extraction [28].
12. As described herein, plasmids containing qPCR target sequences are often utilized for the generation of amplification standard curves. However, circular super-coiled plasmid DNA

is not amplified with the same efficiency as linear plasmid DNA resulting in overestimating standard curves [29]. Therefore, it is advised to linearize such plasmids with a restriction enzyme before using them as qPCR standards like in the above-described protocol in which a NotI digestion step has been incorporated.

13. The PCR efficiency (E) can be calculated by using the formula: $E = 10^{[-1/\text{slope}]}$. Acceptable values vary between 1.8 and 2.1.

14. Previous studies have established that the fortuitous carryover of plasmid DNA into target cell cultures is less of a concern when using concentrated vector stocks than when deploying clarified producer-cell supernatants. Indeed, it has been demonstrated that the ultracentrifugation step can substantially decrease the amount of co-transfected plasmid DNA (for a review *see* ref. [30]). In any case, it is important to establish the absence of significant amounts of plasmid DNA in target cell cultures so that qPCR measurements accurately reflect vector genome copy numbers utilized in transduction experiments. To this end, we describe a qPCR assay to detect prokaryotic *β-lactamase* sequences. If significant amounts of plasmid DNA are detected by qPCR, treatment with a DNA nuclease should be considered (e.g., RNase-free DNaseI or Benzonase®).

Acknowledgments

The authors thank Dr. Antoine A.F. de Vries (Department of Cardiology, LUMC, The Netherlands) for technical assistance in setting up the PEI transfection protocol and Ignazio Maggio (Department of Molecular Cell Biology, LUMC, The Netherlands) for generating the ZFN-encoding lentiviral vector transfer plasmids. We also thank Prof. Rob Hoeben (Department of Molecular Cell Biology, LUMC, The Netherlands) for his critical reading of the manuscript. This work was supported by the Prinses Beatrix Spierfonds (grant W.OR11-18) and by the European Community's 7th Framework Programme for Research and Technological Development: PERSIST-Persisting Transgenesis (grant agreement number 222878).

References

1. Klug A (2010) The discovery of zinc fingers and their applications in gene regulation and genome manipulation. Annu Rev Biochem 79:213–231
2. Silva G, Poirot L, Galetto R, Smith J, Montoya G, Duchateau P, Pâques F (2011) Meganucleases and other tools for targeted genome engineering: perspectives and challenges for gene therapy. Curr Gene Ther 11:11–27
3. Mussolino C, Cathomen T (2012) TALE nucleases: tailored genome engineering made easy. Curr Opin Biotechnol 23:644–650
4. Perez-Pinera P, Ousterout DG, Gersbach CA (2012) Advances in targeted genome editing. Curr Opin Chem Biol 16:268–277

5. Brunet E, Simsek D, Tomishima M, DeKelver R, Choi VM, Gregory P, Urnov F, Weinstock DM, Jasin M (2009) Chromosomal translocations induced at specified loci in human stem cells. Proc Natl Acad Sci U S A 106:10620–10625
6. Naldini L (2011) *Ex vivo* gene transfer and correction for cell-based therapies. Nat Rev Genet 12:301–315
7. Kim YG, Cha J, Chandrasegaran S (1996) Hybrid restriction enzymes: zinc finger fusions to Fok I cleavage domain. Proc Natl Acad Sci U S A 93:1156–1160
8. Perez EE, Wang J, Miller JC, Jouvenot Y, Kim KA, Liu O, Wang N, Lee G, Bartsevich VV, Lee YL, Guschin DY, Rupniewski I, Waite AJ, Carpenito C, Carroll RG, Orange JS, Urnov FD, Rebar EJ, Ando D, Gregory PD, Riley JL, Holmes MC, June CH (2008) Establishment of HIV-1 resistance in CD4+ T cells by genome editing using zinc-finger nucleases. Nat Biotechnol 26:808–816
9. Dropulić B (2011) Lentiviral vectors: their molecular design, safety, and use in laboratory and preclinical research. Hum Gene Ther 22:649–657
10. Cronin J, Zhang XY, Reiser J (2005) Altering the tropism of lentiviral vectors through pseudotyping. Curr Gene Ther 5:387–398, Erratum in: Curr Gene Ther 5:531, 2005
11. Philpott NJ, Thrasher AJ (2007) Use of nonintegrating lentiviral vectors for gene therapy. Hum Gene Ther 18:483–489
12. Boussif O, Lezoualc'h F, Zanta MA, Mergny MD, Scherman D, Demeneix B, Behr JP (1995) A versatile vector for gene and oligonucleotide transfer into cells in culture and in vivo: polyethylenimine. Proc Natl Acad Sci U S A 92:7297–7301
13. Kuroda H, Kutner RH, Bazan NG, Reiser J (2009) Simplified lentivirus vector production in protein-free media using polyethylenimine-mediated transfection. J Virol Methods 157:113–12
14. Segura MM, Garnier A, Durocher Y, Ansorge S, Kamen A (2010) New protocol for lentiviral vector mass production. Methods Mol Biol 614:39–52
15. Askar SF, Bingen BO, Swildens J, Ypey DL, van der Laarse A, Atsma DE, Zeppenfeld K, Schalij MJ, de Vries AAF, Pijnappels DA (2012) Connexin43 silencing in myofibroblasts prevents arrhythmias in myocardial cultures: role of maximal diastolic potential. Cardiovasc Res 93:434–444
16. Pelascini LP, Janssen JM, Gonçalves MAFV (2013) Histone deacetylase inhibition activates transgene expression from integration-defective lentiviral vectors in dividing and non-dividing cells. Hum Gene Ther 24:78–96
17. Torres RJ, Puig JG (2007) Hypoxanthine-guanine phosophoribosyltransferase (HPRT) deficiency: Lesch-Nyhan syndrome. Orphanet J Rare Dis 2:e48
18. Szybalski W (1992) Use of the HPRT gene and the HAT selection technique in DNA-mediated transformation of mammalian cells: First steps toward developing hybridoma techniques and gene therapy. Bioessays 14:495–500
19. Milstein C (1999) The hybridoma revolution: an offshoot of basic research. Bioessays 21:966–973
20. Doetschman T, Gregg RG, Maeda N, Hooper ML, Melton DW, Thompson S, Smithies O (1987) Targetted correction of a mutant HPRT gene in mouse embryonic stem cells. Nature 330:576–578
21. Thomas KR, Capecchi MR (1987) Site-directed mutagenesis by gene targeting in mouse embryo-derived stem cells. Cell 51:503–512
22. Miller AD, Jolly DJ, Friedmann T, Verma IM (1983) A transmissible retrovirus expressing human hypoxanthine phosphoribosyltransferase (HPRT): gene transfer into cells obtained from humans deficient in HPRT. Proc Natl Acad Sci U S A 80:4709–4713
23. Gonçalves MAFV (2005) A concise peer into the background, initial thoughts and practices of human gene therapy. Bioessays 17:506–517
24. Charrier S, Stockholm D, Seye K, Opolon P, Taveau M, Gross DA, Bucher-Laurent S, Delenda C, Vainchenker W, Danos O, Galy A (2005) A lentiviral vector encoding the human Wiskott-Aldrich syndrome protein corrects immune and cytoskeletal defects in WASP knockout mice. Gene Ther 12:597–606
25. Miller JC, Holmes MC, Wang J, Guschin DY, Lee YL, Rupniewski I, Beausejour CM, Waite AJ, Wang NS, Kim KA, Gregory PD, Pabo CO, Rebar EJ (2007) An improved zinc-finger nuclease architecture for highly specific genome editing. Nat Biotechnol 25:778–785
26. Higashikawa F, Chang L (2001) Kinetic analyses of stability of simple and complex retroviral vectors. Virology 280:124–131
27. Wanisch K, Yáñez-Muñoz RJ (2009) Integration-deficient lentiviral vectors: a slow coming of age. Mol Ther 17:1316–1332
28. Giry-Laterrière M, Verhoeyen E, Salmon P (2011) Lentiviral vectors. Methods Mol Biol 737:183–209
29. Hou Y, Zhang H, Miranda L, Lin S (2010) Serious overestimation in quantitative PCR by circular (supercoiled) plasmid standard: microalgal pcna as the model gene. PLoS One 5(3):e9545
30. Delenda C, Gaillard C (2005) Real-time quantitative PCR for the design of lentiviral vector analytical assays. Gene Ther 12:S36–S50

Part III

Breaking the DNA to Stimulate Gene Correction: Nuclease Design, In Vivo Test, Efficacy, and Off-Target Effect

Additional protocol described in this book for gene correction stimulated by DNA breaks

- zinc finger nucleases: Chapter 12

Chapter 13

Designing and Testing the Activities of TAL Effector Nucleases

Yanni Lin, Thomas J. Cradick, and Gang Bao

Abstract

Transcription activator-like effector nucleases (TALENs) have rapidly developed into a powerful tool for genome editing. To avoid labor-intensive and time-consuming experimental screening for active TALENs, a scoring system can help select optimal target sites. Here we describe a procedure to design active TALENs using a scoring system named Scoring Algorithm for Predicted TALEN Activity (SAPTA) and a method to test the activity of individual and pairs of TALENs.

Key words Transcription activator-like effector nuclease, TALEN, Scoring algorithm for predicted TALEN activity (SAPTA), Gene modification, Genome editing, Gene targeting, Gene therapy, Scoring function, Single-strand annealing (SSA) assay, T7 endonuclease I (T7E1) mutation detection assay

1 Introduction

Transcription activator-like (TAL) effectors are a family of DNA binding proteins, discovered in the plant pathogen Xanthomonas [1–3]. Each DNA-binding domain contains a variable number of 33–35 amino-acid repeats that specify the DNA-binding sequence primarily through their 12th and 13th repeat-variable di-residues (RVDs) [4]. Each RVD specifies one nucleotide with minimal context dependence [1, 3, 5]. TALENs target specific DNA sequences through a series of RVD-containing repeats, flanked by modified N- and C-termini [6, 7] that are linked to a FokI nuclease domain [8–10]. When a pair of TALENs binds to their specific half-sites with the correct orientation and spacing to allow the nuclease domains to dimerize, the intervening sequence is cleaved. TALENs have been used to edit genomic DNA sequences in a variety of biological systems, including human cells, rats, zebrafish, nematodes, and plants [5–7, 11–15].

TALENs are easy to design due to the modular nature of the DNA-binding domain. However, TALENs targeting different

Francesca Storici (ed.), *Gene Correction: Methods and Protocols*, Methods in Molecular Biology, vol. 1114,
DOI 10.1007/978-1-62703-761-7_13, © Springer Science+Business Media, LLC 2014

DNA sequences have cleavage activities varying over a wide range—sometimes the activity is below detection. Current design guidelines provide qualitative criteria for selecting TALEN target sites, but could not guarantee an efficient gene modification rate. In this chapter, we describe the use of a quantitative scoring system, SAPTA, to predict TALEN activities and select optimal TALEN target sites for robust gene modification [16]. Although SAPTA was optimized to design TALENs constructed using the NK RVD to target the nucleotide guanine, we found that it also yields highly active TALENs constructed using the NN RVD that also target guanine. The SAPTA scoring system was programmed into an online design tool to enable researchers to modify genes more effectively (http://baolab.bme.gatech.edu/Research/BioinformaticTools/TAL_targeter.html). We also describe assays that can be used to quantify the activities of TALENs individually, or in pairs. This assay may be important for quality control of TALEN constructs and for choosing optimal TALEN sets to study further. When there are several overlapping TALEN target sites, this assay can also be used to pick the optimal left and right monomers to test further.

2 Materials

2.1 Assembly of Homodimeric or Heterodimeric Target Plasmids

1. Backbone plasmids (*see* **Note 1**).
2. Oligonucleotides containing target sequences and spacers.
3. 10× T4 DNA ligase reaction buffer (New England Biolabs).
4. T4 polynucleotide kinase (New England Biolabs).
5. Two restriction enzymes unique to the backbone plasmid, e.g., AscI and SbfI.
6. Alkaline Phosphatase, Calf Intestinal (CIP) (New England Biolabs).
7. T4 DNA ligase (400,000 units/ml) (New England Biolabs).
8. Thermal cycler.
9. Agarose.
10. 10 mg/ml ethidium bromide solution.
11. 1× TAE buffer.
12. Microwave.
13. Agarose gel electrophoresis apparatus.
14. UV imaging station.
15. Gel extraction kit.
16. Tabletop microcentrifuge.
17. 37 °C oven.

18. Chemically competent *E. coli*.
19. LB plates containing appropriate antibiotic.
20. LB broth containing appropriate antibiotic.
21. Oligonucleotide primers for colony screen.
22. GoTaq Green Master Mix (Promega).
23. Plasmid miniprep kit.

2.2 Cell Transfection

1. 293T cells (ATCC, Manassas, VA).
2. 293T cell culture media: Dulbecco's Modified Eagle Medium (Sigma-Aldrich, D6429) is supplemented with 2 mM L-glutamine and 10 % FBS.
3. Any 24-well multiwell plate, sterile, tissue-culture treated, flat bottom with lid.
4. 0.1 % Gelatin.
5. Control plasmid for gauging transfection efficiency, e.g., EGFP plasmid.
6. Flow cytometer (e.g., BD Accuri C6).
7. Fugene HD Transfection Reagent (Promega). Store at 4 °C with cap closed tightly.
8. Filter sterilized 2 M $CaCl_2$. Store at room temperature.
9. Filter sterilized 2× HBS buffer: To make 100 ml buffer, dissolve 1.6 g NaCl, 1.2 g HEPES, 100 μl 1.5 M Na_2HPO_4 stock solution in 80 ml water. Adjust pH to 7.05 and bring the volume to 100 ml. Adjust the pH again to assure accurate pH. Make aliquots of the buffer and store at −20 °C. Once thawed, the buffer can stay stable at room temperature for several months. Do not refreeze the buffer.

2.3 Single Strand Annealing (SSA) Assay Using PCR

1. QuickExtract DNA extraction solution (Epicentre).
2. GoTaq Green Master Mix (Promega).
3. PCR tube strips with caps.
4. Oligonucleotide primers (*see* **Note 2**).
5. Vortex mixer.
6. Thermal cycler.
7. Agarose gel electrophoresis apparatus and reagents as in Subheading 2.1.

2.4 T7 Endonuclease I (T7E1) Mutation Detection Assay

1. Oligonucleotide primers (*see* **Note 3**).
2. QuickExtract DNA extraction solution (Epicentre).
3. AccuPrime Taq DNA polymerase high fidelity (Life Technologies, 12346-086 or 12346-094).
4. DMSO.

5. PCR tube strips with caps or 96-well PCR plates.
6. Thermal cycler.
7. Any PCR purification kit or gel extraction kit (*see* **Note 4**).
8. Agarose gel electrophoresis apparatus and reagents as in Subheading 2.1.
9. T7 Endonuclease I (New England Biolabs).
10. 6× xylene cyanol loading dye: 4 g of sucrose, 3 mg of xylene cyanol FF (Sigma-Aldrich), add sterile water to 10 ml. Store at 4 °C.
11. 0.5 M EDTA.

3 Methods

3.1 Design of TALENs Targeted to Genes of Interest

1. Obtain the genomic DNA sequence to be targeted and cleaved by TALENs. For example, if the disruption of a gene is desired, DNA sequences of the first several exons may be selected. To correct a specific mutation within a gene using the homologous-directed recombination (HDR) pathway, the sequence around the mutated site may be selected.
2. Direct an Internet browser to the SAPTA Home Page (http://baolab.bme.gatech.edu/Research/BioinformaticTools/TAL_targeter.html). Some instructions for using this tool are provided on the page, with links to further instructions and tutorials.
3. Paste the DNA sequence of interest into the input box below "Enter a DNA sequence" (*see* **Note 5**), and press "Submit." Typically a sequence range from 100 to 1,500 bp may be entered. Sequence shorter than 100 bp can be searched but may not yield high scoring TALEN pairs (*see* **Note 6**). To search a sequence longer than 1,500 bp, check "Override the 1,500 base pair limit?" below the sequence input box. The Web site also allows a variety of custom settings, such as specifying the lengths of spacer and TAL repeat array for the TALEN designs. Explanations of these custom settings are found in the online tutorials (click the "Tutorials" tab located on the top of the SAPTA Web page).
4. The output from SAPTA contains a table of TALEN pairs ranked by their composite scores. The table can be copied and pasted into Microsoft Excel to save the data. The columns of this table are explained as follows.

   Starting index: the position of the 5′ end of the left TALEN target sequence. If brackets where used on the DNA sequence entered, the numbering is relative to the bracketed nucleotides (*see* **Note 5**), otherwise the numbering is relative to the first nucleotide in the searched sequence.

Left and right TALEN sequences: the target sequences of left and right TALENs, respectively. The 5′ nucleotide is followed by a hyphen and the sequence targeted by the RVDs (e.g., T-TACTGAAGTAACCT).

Left TALEN size, spacer size, and right TALEN size: the nucleotide lengths (bp) of each target DNA sequences.

Left and right TALEN scores: the scores that predict the individual activities of left and right TALEN monomers, respectively. The score is calculated by the SAPTA function that was optimized based on a training set of intracellular TALEN activity data.

Composite score: predict the activity of a pair of TALENs by functionally combining the individual scores of left and right TALENs of each pair.

Restriction enzyme name and sequence: if there is a unique restriction enzyme site in the spacer of a TALEN pair, it will show in this column together with the sequence of this enzyme site. An enzyme site can be used for Restriction Fragment Length Polymorphism (RFLP) analysis to quantify the mutation rates resulted from TALEN cleavage and loss of the enzyme site.

5. The output also includes a separate table containing two columns: base and score. The column "base" indicates the starting positions of target sites as in the "starting index" discussed above. The column "score" shows the highest score of TALEN pairs targeting that particular base, when there are multiple pairs starting at the same position. This table can be copied into Microsoft Excel to generate a graph (*see* Fig. 1). If you use brackets in the input sequence to define your base of interest as position one (*see* **Note 5**), this graph can be very informative and help you visualize high-scoring TALEN target sites near your base of interest.

6. Choose pairs of TALENs to construct from the top scoring pairs on the output list (*see* **Notes 7** and **8**) with preference to TALEN target sites closer to your target base, if applicable. As SAPTA will evolve as more information become available, please check SAPTA online tutorials for the up-to-date recommended cutoff score for active TALEN pairs. The SAPTA algorithm was developed using experimental data of TALENs constructed with the NK RVD. However, we have used SAPTA to select target sites for TALENs with NN or NH RVDs and also observed effective cleavage with these NN- or NH-TALENs.

7. It is important to limit cleavage by designed TALENs at off-target sites. To determine if there are other sites with sequence homology to the intended target site, check for

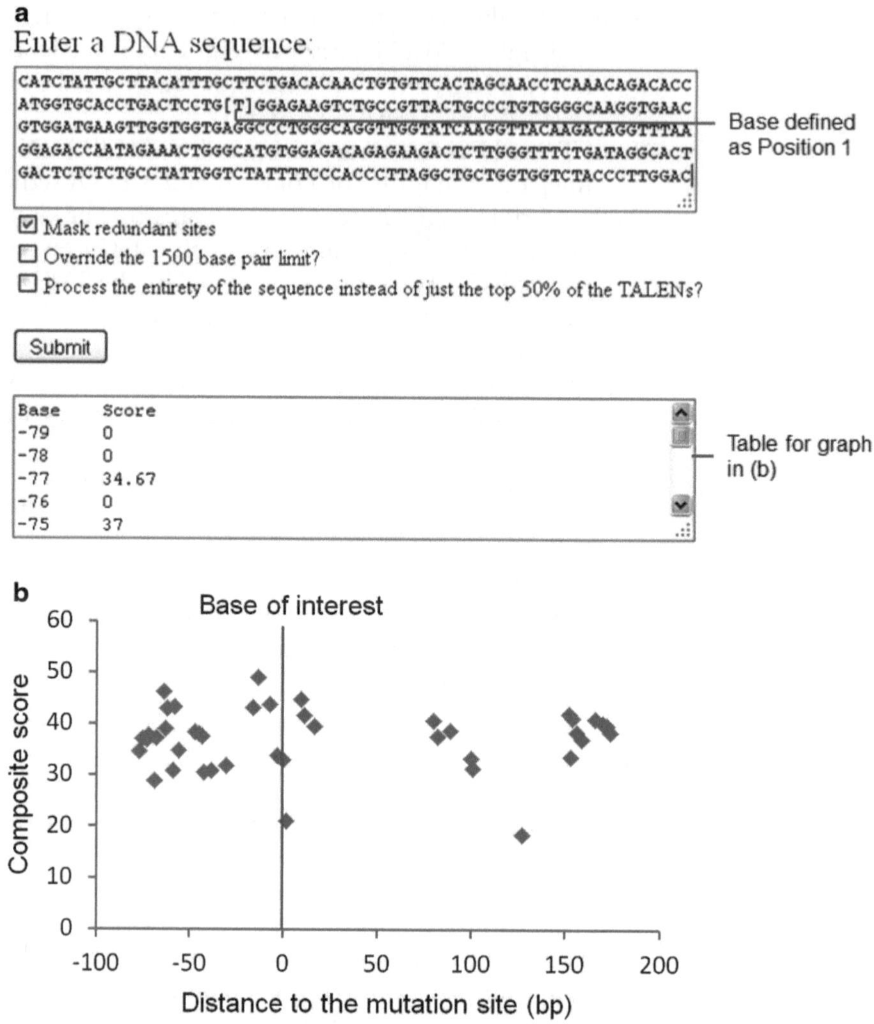

Fig. 1 Generating a graph showing the TALEN pair composite score versus target site position. (**a**) *Brackets* are placed around a base to define it as position one of the input sequence. The *table* below contains the numbering of bases next to the composite scores of TALEN pairs starting at the corresponding bases. This table can be copied into a spreadsheet to generate the graph shown in (**b**). (**b**) Composite scores plotted against target site position. Adapted from [16]

potential off-target sites using PROGNOS. This link (http://baolab.bme.gatech.edu/Research/BioinformaticTools/prognos.html) also contains tutorials. Chapter 24 by Eli J. Fine et al., this volume contains more details on the off-target analysis.

3.2 Assembly of Homodimeric or Heterodimeric Target Plasmids

1. Prepare an SSA target plasmid backbone (*see* **Note 1**). In our example, we use AscI and SbfI restriction sites for directional cloning of the TALEN target sequences.
2. Digest the SSA target plasmid backbone using the two chosen restriction enzymes. Here we use AscI and SbfI to clone

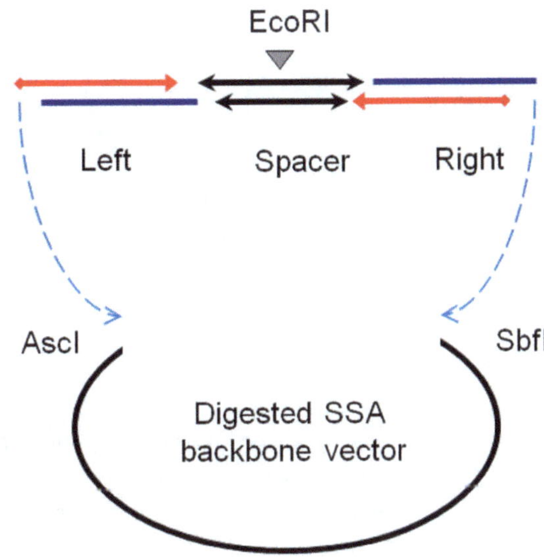

Fig. 2 Schematic of target plasmid assembly. Three pairs of oligonucleotides that contain the left TALEN half-site, a spacer with an EcoRI site, and the right TALEN half-site are ligated into the vector. The ends of each oligonucleotide pair are compatible with its adjacent oligonucleotide pair(s) or the backbone vector. Oligonucleotide pairs were ligated into the vector digested with AscI and SbfI resulting in a EGFP gene interrupted by a stop codon, zinc finger nuclease site and TALEN target site. Oligonucleotide pairs containing TALEN half-sites (red lines with arrows) can be used in assembling both homodimeric TALEN targets and heterodimeric TALEN targets. Various spacer lengths from 11 to 30 bp can be used in the ligation. Figure from [16]

our targets. To ensure complete digestion, allow the digestion to proceed for at least 3 h.

3. Load the digested vector on 0.7 % agarose gel to separate away from uncut DNA and the insert that is cut out. Gel-isolate the digested plasmid backbone and use a commercially available gel extraction kit to remove the agarose.
4. Order sense and antisense oligonucleotides that contain the sequences of TALEN target sites and a spacer. The ends of the oligonucleotide pairs should be compatible with each other and with the restriction-digested ends of the SSA plasmid backbone (*see* Fig. 2 and **Note 9**).
5. Resuspend these oligonucleotides using 10 mM pH 8.0 Tris–HCl solution to a final concentration of 10 μM.
6. Kinase the single-stranded oligonucleotides in the following reaction mixture: 0.7 μl 10 μM oligonucleotide, 2.5 μl 10× T4 DNA ligase reaction buffer (NEB) (*see* **Note 10**), 2 U of T4 polynucleotide kinase (NEB), and sterile water to 25 μl. Incubate this reaction mixture at 37 °C for 40 min, and 65 °C for 20 min to heat inactivate the enzyme.

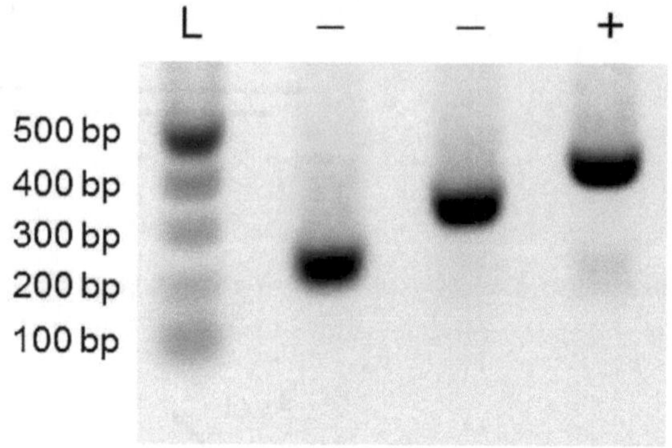

Fig. 3 Example gel image of colony PCR results for screening SSA target plasmid. Lane "L" is 100 bp DNA ladder (NEB). The two lanes marked by "–" are negative clones. The lane "+" contains the correct clone for an SSA target plasmid with TALEN site ligated into the vector

7. Mix the entire reaction mixture of the sense strand oligonucleotide with the antisense strand oligonucleotide. Heat the combined mixture to 95 °C for 10 min, and allow it to cool down at room temperature for at least 1 h. The sense and antisense oligonucleotides are now annealed and form double-stranded oligonucleotide pairs. Proceed to next step or stored the annealed pairs at –20 °C until use.
8. Ligate the left oligo pair, the spacer oligo pair, the right oligo pair (*see* **Note 9**), and the digested and gel-isolated plasmid backbone in the following reaction: 1 μl each of kinased and annealed left oligo pair, spacer oligo pair, and right oligo pair from **step** 7, respectively, 50 ng of digested and gel-isolated plasmid backbone from **step 3**, 2 μl of 10× T4 DNA ligase reaction buffer (NEB) (*see* **Note 10**), 200 U of T4 DNA ligase (NEB), and bring final volume to 20 μl with sterile water. Incubate at 16 °C overnight.
9. Transform the ligated products into chemically competent *E. coli*. Plate transformation onto LB plates with suitable antibiotic and incubate plates overnight at 37 °C.
10. Screen 1–3 colonies by colony PCR. Amplify using primers annealing to the plasmid backbone flanking the insert (*see* **Note 11** and Fig. 3).
11. Start overnight cultures of positive clones in LB broth with suitable antibiotic. Prepare plasmids using any commercially available miniprep kit for sequence confirmation and any downstream applications (*see* **Note 12**).

3.3 Cell Transfection for SSA Assay

1. Incubate 24-well cell culture plates with 0.1 % gelatin solution that covers bottom of the wells at 37 °C for 30 min to several hours. After incubation, aspirate all the gelatin solution.

2. Seed HEK293T cells into the gelatin-treated plates at 80,000 cells per well in 500 μl 293T cell culture media.
3. 3–6 h after seeding cells, prepare the following mixture per well for calcium phosphate transfection: 2.5 μl of 2 M $CaCl_2$, 100 ng of each TALEN monomer plasmid (total 200 ng for a pair of TALENs), 10 ng of an SSA target plasmid (*see* **Note 13**), and sterile H_2O to 20 μl.
4. Add 20 μl 2× HBS, pipet up and down gently for around ten times, and then completely depress plunger to blow bubbles several times.
5. Incubate the mixture for 2–5 min at room temperature. Incubation of more than 5 min may decrease the transfection efficiency.
6. Add 40 μl of the mixture drop-wise to each well of 24-well plates, and carefully return the plates to a cell culture incubator.
7. 16–20 h after transfection, replace the old media with 500 μl of 293T cell culture media.
8. 48 h after transfection, harvest the cells and pellet the transfected cells at 800 × *g* for 10 min. Gently remove the supernatant. Proceed to SSA assay or keep the cell pellets at −80 °C until use.

3.4 SSA Assay Using PCR

1. Add 70 μl of QuickExtract solution to the cell pellet from one well of a 24-well plate and pipette vigorously to completely resuspend the cells.
2. Transfer the solution to PCR tubes and run the following program in a thermal cycler to extract DNA: 68 °C for 15 min, 95 °C for 8 min, and then hold at 4 °C.
3. Use sterile water to make a fivefold dilution of the extracted DNA from **step 2** (*see* **Note 14**). Votex vigorously until a homogenous solution is obtained.
4. Set up a 25 μl PCR reaction using the 2× GoTaq Green Master Mix: 5 μl of diluted DNA from **step 3**, 1.25 μl of each primer at 10 μM (*see* **Note 2**), 12.5 μl GoTaq Green Master Mix, and 5 μl sterile water.
5. Run the PCR as follows: 95 °C for 5 min, 35× (95 °C for 30 s, 60 °C for 30 s, 72 °C for 1 min); 72 °C for 5 min; hold at 4 °C.
6. Run 5 μl of PCR reaction on a 2 % agarose gel.
7. Quantify the percentage of SSA-repaired band in the total PCR product using the free software ImageJ or other image-analysis softwares (*see* Fig. 4). This percentage of SSA is a measurement of TALEN activity.

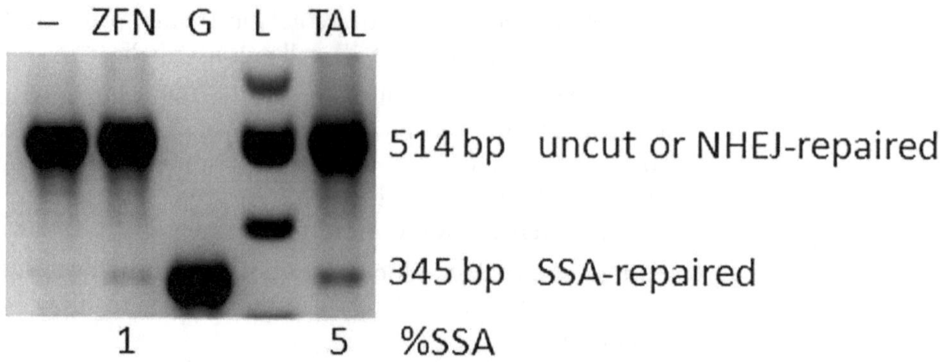

Fig. 4 Example SSA assay controls and results. Samples from a representative SSA assay used to quantitate the PCR products amplified from HEK293T cells. Agarose gels separated the 345-bp PCR fragments amplified from SSA-repaired target plasmids and the 514-bp PCR fragments amplified from uncut or NEHJ repaired plasmids. The percentage of the SSA-repaired products relative to the total PCR products was determined using ImageJ and shown below each lane. The negative controls cells transfected with an empty TALEN backbone and an SSA target plasmid are labeled "–". The positive control cells transfected with a pair of GFP-ZFNs [17] and an SSA target plasmid are labeled "ZFN". The positive control cells transfected with pEGFP plasmid and an empty TALEN backbone and are labeled "G". Lane "L" is 100 bp DNA ladder (NEB). The "TAL" lane contains the PCR from cells transfected with a pair of TALENs and the corresponding SSA target plasmid. Adapted from [16]

3.5 Cell Transfection for T7E1 Mutation Detection Assay

1. Incubate 24-well cell culture plates with 0.1 % gelatin solution that covers bottom of the wells at 37 °C for 30 min to several hours. After incubation, aspirate all the gelatin solution.
2. Seed HEK293T cells into the gelatin-treated plates at 40,000 cells per well in 500 μl 293T cell culture media with freshly added L-glutamine at 2 mM (*see* **Note 15**).
3. One day after seeding cells, prepare 24.1 μl of the plasmid mixture which contains 550 ng of each TALEN (total 1.1 μg TALEN) and 11 ng of EGFP plasmid. Bring this mixture to room temperature before next step.
4. Add 3.4 μl Fugene HD at room temperature to the plasmid mixture from **step 3**. Pipette up and down for more than 15 times or vortex briefly to mix (*see* **Note 16**).
5. Incubate the mixture for 10–15 min at room temperature.
6. Add 25 μl of the mixture drop-wise to each well of 24-well plates, and return the plates to a cell culture incubator.
7. 48 hours after transfection, replace the old media with 500 μl of 293T cell culture media.
8. 72 hours after transfection, harvest the cells and use flow cytometer to quantify the percentage of EGFP positive cells as an indication of transfection efficiency. Pellet the cells at 800 × *g* for 10 min. Gently remove the supernatant. Proceed to T7E1 mutation detection assay or keep the cell pellets at −80 °C until use.

3.6 T7E1 Mutation Detection Assay

1. Add 80 μl of QuickExtract solution to the cell pellet from one well of a 24-well plate and pipette vigorously to completely resuspend the cells.
2. Transfer the solution to PCR tubes and run the following program in a thermal cycler to extract DNA: 68 °C for 15 min, 95 °C for 8 min, and then hold at 4 °C.
3. Set up a 50 μl PCR reaction using the AccuPrime Taq DNA polymerase high fidelity kit: 2 μl of DNA from **step 2**, 2.5 μl of each primer at 10 μM, 5 μl of 10× AccuPrime buffer II, 0.2 μl of AccuPrime Taq DNA polymerase high fidelity, 2.5 μl DMSO, and sterile water to 50 μl. For every genomic locus tested, a negative control PCR reaction using cells not treated by TALENs need to be included. This negative PCR reaction will be treated the same way as the TALEN treated samples in the following steps.
4. Run the PCR as follows: 95 °C for 5 min, 35× (95 °C for 30 s, 60 °C for 30 s, 68 °C for 1 min); 68 °C for 5 min; hold at 4 °C.
5. To verify specific amplification, mix 5 μl of the PCR reaction with 1 μl of 6× xylene cyanol loading dye and load into a 2 % agarose gel (*see* **Note 17**).
6. Purify the PCR reaction using any commercially available PCR purification kit or other suitable methods (*see* **Note 4**).
7. Set up a reaction containing 200 ng of purified PCR product and 1.8 μl of 10× NEBuffer2 in 18 μl total volume.
8. Vortex and centrifuge briefly to mix.
9. Melt and re-anneal the DNA by placing in a thermal cycler: 95 °C for 10 min; decreasing at 0.1 °C/s down to 85 °C, hold at 85 °C for 2 min; decreasing at 0.1 °C/s down to 75 °C, hold at 75 °C for 3 min; decreasing at 0.1 °C/s down to 65 °C, hold at 65 °C for 3 min; decreasing at 0.1 °C/s down to 55 °C, hold at 55 °C for 3 min; decreasing at 0.1 °C/s down to 45 °C, hold at 45 °C for 3 min; decreasing at 0.1 °C/s down to 35 °C, hold at 35 °C for 3 min; decreasing at 0.1 °C/s down to 25 °C, hold at 25 °C for 3 min; hold at 4 °C.
10. The re-annealed DNA can be stored at −20 °C.
11. Vortex and centrifuge briefly to mix.
12. Make an enzyme master mix: every 2 μl contains 0.5 μl of T7 Endonuclease I, 0.2 μl of 10× NEBuffer2, and 1.3 μl of water.
13. Add 2 μl of the master mix to each reaction, vortex and centrifuge briefly to mix, and immediately place in a thermal cycler set for 37 °C for 60 min, with heated lid setting off.
14. After 60 min, immediately remove the reactions from the thermal cycler, centrifuge briefly, and quench by adding EDTA to a final concentration of 45 mM. This can be accomplished

by adding 6.94 μl of a mixture containing a ratio of 2.45:4.49 of 0.5 M EDTA to 6× xylene cyanol loading dye.

15. Vortex and centrifuge briefly to mix.
16. Quenched reactions can be stored at 4 °C for several hours or at −20 °C indefinitely.
17. Load as much of the reactions as possible (≥25 μl) into a 2 % agarose gel that was cast with wide wells to help visualize cleavage products clearly. We use combs with tooth width of 7 mm. For each genomic site tested, load the negative control sample mentioned in **step 3** next to the TALEN-treated sample for easy comparison. Run the gel until the bands are well separated.
18. When imaging the gel, make sure that the exposure time is properly adjusted so that none of the bands are saturated, as this will interfere with accurate quantification of the band intensities.
19. The free software ImageJ (NIH) can be used to quantify the bands on the gel. There are many online tutorials that detail this process. Chapter 24 by Eli J. Fine et al., this volume contains more details.
20. The percent of alleles that show evidence of nonhomologous end-joining (NHEJ) can be calculated from the band intensities according to the formulas [18] (*see* **Notes 18** and **19**):

$$f_{cut} = \frac{CleavageBand_1 + CleavageBand_2}{CleavageBand_1 + CleavageBand_2 + UncleavedBand}$$

$$\%NHEJ = 100 \times \left(1 - \sqrt{1 - f_{cut}}\right)$$

4 Notes

1. An SSA backbone plasmid should contain two repeat sequences separated by one or more stop codons, an optional control target site, and two unique restriction sites that are each present once in the plasmid. Our vector contains a ZFN target site that can be used as an activity control. The two restriction sites are used to insert TALEN target sites that include a left binding site, followed by a spacer and a right binding site.
   Our SSA backbone plasmid contains an EGFP gene, interrupted after 327 bp with a stop codon, the target site for a pair of GFP-targeted ZFNs [17], an AscI, and an SbfI cloning sites (plasmid kindly provided by Dr. Matthew Porteus, Stanford University). The downstream portion of the EGFP gene

includes a 42-bp region repeating the sequence of the EGFP gene before the stop codon. The effect of varying the repeat size was compared in the chapter 22. Part of the SSA backbone sequence is shown below with the 42-bp repeats shaded in black, the ZFN target sites shaded in gray, and the cloning restriction sites underlined.

```
GCCCGAAGGCTACGTCCAGGAGCGCACAATTTTTTTCAAGGATGATGGAAACTACAAGTAAG
                                       AscI             SbfI
GCGCGACCATCTTCTTCAAGGACGACGG^CGCGCCTGGGATCCTGCA^GGCAGGAGCGCACA

ATTTTTTTCAAGGATGATGGAAACTACAAGACCCGCGCCGAGGTGA
```

TALEN target sites can be cloned into the backbone plasmid using the AscI and SbfI sites, indicated above. The example SSA target plasmid, below, was cloned using these sites. The bold and underlined left and right TALEN binding sites are separated by a 17-bp spacer containing an EcoRI site.

```
GCCCGAAGGCTACGTCCAGGAGCGCACAATTTTTTTCAAGGATGATGGAAACTACAAGTAAG
                                       AscI
GCGCGACCATCTTCTTCAAGGACGACGG^CGCGCCTCATCCACGTTCACCTTGCCCCACAGG
          EcoRI                                             SbfI
GCAGTATCGATGAATTCTTAAGACTGCCCTGTGGGGCAAGGTGAACGTGGATGACCTGCA^G

GCAGGAGCGCACAATTTTTTTCAAGGATGATGGAAACTACAAGACCCGCGCCGAGGTGA
```

2. The forward primer for SSA PCR should be located upstream of the first 42-bp EGFP repeat (*see* **Note 1**), and the reverse primer should be located downstream of the second 42-bp repeat. The size of the PCR product should be in the range of 300–500 bp to better separate the PCR products from the original/uncut target plasmid and the SSA-repaired target plasmid on a 2 % agarose gel.

3. Primers for the T7E1 assay to amplify specific genomic loci can be designed using Primer-BLAST provided by NCBI, or similar programs. Ideally primers should have melting temperatures (Tm) around 60 °C. The PCR amplicon size should be between 300 and 600 bp because amplicons longer than 600 bp are prone to nonspecific degradation by the T7E1 enzyme. Primers should be carefully positioned so that the two cleaved bands generated by T7E1 digestion are both larger than 100 bp.

4. Any commercially available PCR purification kit can be used to clean up the PCR amplifications for the T7E1 assay. If a large number of samples need to be processed, you can alternatively use a high-throughput, magnetic-bead-based method to purify your PCR reactions (*see* Chapter 24 by Eli J. Fine et al., this volume for details). If nonspecific PCR products are present, gel isolation and gel extraction kit can be used to purify the band of desired size.

5. When you paste the sequence of interest into the input box on the SAPTA Web site, you may place brackets around the nucleotide you hope to target (e.g., ACGT[T]GTA). The nucleotide in brackets will be defined as position one in the output tables of the search, which allows users to more easily identify the distance of each scored TALEN target sites from the specific base of interest.
6. The frequency of high-scoring TALEN pairs in the human genome can be found in the study by Lin et al. [16]. Target sites with SAPTA scores of 35 and higher were found on average within the first 29 bp of open reading frames of 48 human genes, with a standard deviation of 32 bp.
7. Various methods can be used to construct TALENs, including those published by Zhang et al. [19], Cermak et al. [5], Briggs et al. [20], Reyon et al. [21], Schmid-Burgk et al. [22], and Kim et al. [23].
8. To ensure efficient targeting of the gene of interest, you may construct 2–3 top-ranking pairs per site. TALENs with less than 21 repeat arrays are preferred [21]. TALEN binding can be impaired by cytosine methylation in CpG islands [23–25]. When you choose target sites, you should avoid CpG dinucleotide if possible.
9. Examples of oligonucleotide pairs for making the SSA target in **Note 1** are shown below with each sense and antisense oligonucleotide paired. The overhangs of these oligonucleotide pairs were designed to be ligated into the AscI and SbfI digested backbone in the following order: backbone AscI site—left oligo pair—spacer oligo pair—right oligo pair—backbone SbfI site.

   Left oligo pair:

   5′-CGCGCCTCATCCACGTTCACCTTGCCCCACAGGGCAGT-3′

   3′-GGAGTAGGTGCAAGTGGAACGGGGTGTCCCGTCATAG-5′

   Spacer oligo pair (for 17 bp spacer):

   5′-ATCGATGAATTCTTAAG-3′

   3′-CTACTTAAGAA-5′

   Right oligo pair:

   5′-ACTGCCCTGTGGGGCAAGGTGAACGTGGATGACCTGCA-3′

   3′-TTCTGACGGGACACCCCGTTCCACTTGCACCTACTGG-5′.

10. T4 DNA ligase reaction buffer contains ATP which is not stable when subjected to many freeze–thaw cycles. It is recommended to aliquot this buffer into small volumes (20–100 μl depending on the volume to be used each time) and store at −20 °C.

11. To screen for positive clones of the SSA target plasmid, we amplified with a forward primer annealing about 175 bp upstream of the insert (TALEN target site) and a reverse primer annealing 130 bp downstream of the insert. Using our primers, the expected PCR product of a positive clone is ~400 bp. Since the insert is approximately 80 bp, the PCR product should be in the size range that allows discerning an 80-bp difference.
12. Sequencing the homodimeric SSA target plasmids (the left and the right TALEN binding sites are identical) is complicated by the palindromic sequence. Reads were more successful using the Power Read DNA sequencing service provided by Eurofins MWG Operon. Heterodimeric target plasmids do not have this problem and can be sequenced by any regular sequencing service.
13. To estimate transfection efficiency for the SSA assay, we transfect 200 ng of a stuffer plasmid and 10 ng of an EGFP plasmid, and use flow cytometry to quantify the percentage of GFP positive cells when cells are harvested 48 h after transfection.
14. It is difficult to pipette and dispense a small volume of the DNA extraction prepared by the QuickExtract solution due to its stickiness. We found that diluting the DNA extraction can reduce pipetting artifact and lead to more consistent results.
15. L-Glutamine can degrade over time in liquid cell culture medium. We found that adding fresh L-glutamine when cells are seeded is required for high TALEN transfection efficiency for the T7E1 assay.
16. Fugene HD reagent can only be kept in air-tight, non-silanized glass containers until mixed with plasmid solution. Prior to the mixing step, do not dilute Fugene HD or aliquot this reagent into regular microcentrifuge tubes or strip tubes made of polypropylene or other non-glass materials.
17. T7E1 assay involves DNA fragments in the 100–600 bp range. Many loading dyes contain components that run in this range, which would obscure the bands that need to be visualized. Make sure you choose a dye that does not have anything that will co-migrate with DNA fragments of that size. The xylene cyanol dye we used migrates at 800 bp to 1 kb on a 2 % agarose gel and thus will not interfere with the visualization of DNA bands in the T7E1 assay.
18. The T7E1 assay has a detection limit of ~1 % gene modification [18]. For higher sensitivity of detection, please *see* Chapter 24 by Eli J. Fine et al., this volume.
19. The gene-modifying efficiency of TALENs is influenced by genomic context (e.g., binding of other cellular proteins, how

accessible the genomic loci are, the methylation status) at the target sites. SSA assay, on the other hand, is a method to measure the activity of TALENs without affected by genomic context. If SSA assay shows high activities of TALENs, but T7E1 assay fails to show any activity at the endogenous loci, it may indicate that some genomic context is limiting the cleavage by TALENs.

Acknowledgments

This work was supported by the National Institutes of Health as an NIH Nanomedicine Development Center Award (PN2EY018244 to G.B.).

References

1. Boch J, Scholze H, Schornack S, Landgraf A, Hahn S, Kay S, Lahaye T, Nickstadt A, Bonas U (2009) Breaking the code of DNA binding specificity of TAL-type III effectors. Science 326(5959):1509–1512. doi:10.1126/science.1178811, 1178811 [pii]
2. Boch J, Bonas U (2009) Xanthomonas AvrBs3 family-type III effectors: discovery and function. Annu Rev Phytopathol 48:419–436. doi:10.1146/annurev-phyto-080508-081936
3. Moscou MJ, Bogdanove AJ (2009) A simple cipher governs DNA recognition by TAL effectors. Science 326(5959):1501. doi:10.1126/science.1178817, 1178817 [pii]
4. Schornack S, Meyer A, Römer P, Jordan T, Lahaye T (2006) Gene-for-gene-mediated recognition of nuclear-targeted AvrBs3-like bacterial effector proteins. J Plant Physiol 163(3):256–272. doi:10.1016/j.jplph.2005.12.001, S0176-1617(05)00462-1 [pii]
5. Cermak T, Doyle EL, Christian M, Wang L, Zhang Y, Schmidt C, Baller JA, Somia NV, Bogdanove AJ, Voytas DF (2011) Efficient design and assembly of custom TALEN and other TAL effector-based constructs for DNA targeting. Nucleic Acids Res. doi:10.1093/nar/gkr218, gkr218 [pii]
6. Miller JC, Tan S, Qiao G, Barlow KA, Wang J, Xia DF, Meng X, Paschon DE, Leung E, Hinkley SJ, Dulay GP, Hua KL, Ankoudinova I, Cost GJ, Urnov FD, Zhang HS, Holmes MC, Zhang L, Gregory PD, Rebar EJ (2010) A TALE nuclease architecture for efficient genome editing. Nat Biotechnol 29(2):143–148. doi:10.1038/nbt.1755, nbt.1755 [pii]
7. Tesson L, Usal C, Ménoret S, Leung E, Niles BJ, Remy S, Santiago Y, Vincent AI, Meng X, Zhang L, Gregory PD, Anegon I, Cost GJ (2011) Knockout rats generated by embryo microinjection of TALENs. Nat Biotechnol 29(8):695–696. doi:10.1038/nbt.1940, nbt.1940 [pii]
8. Christian M, Cermak T, Doyle EL, Schmidt C, Zhang F, Hummel A, Bogdanove AJ, Voytas DF (2010) Targeting DNA double-strand breaks with TAL effector nucleases. Genetics 186(2):757–761. doi:10.1534/genetics.110.120717, genetics.110.120717 [pii]
9. Li T, Huang S, Jiang WZ, Wright D, Spalding MH, Weeks DP, Yang B (2010) TAL nucleases (TALNs): hybrid proteins composed of TAL effectors and FokI DNA-cleavage domain. Nucleic Acids Res 39(1):359–372. doi:10.1093/nar/gkq704, gkq704 [pii]
10. Mahfouz MM, Li L, Shamimuzzaman M, Wibowo A, Fang X, Zhu JK (2011) De novo-engineered transcription activator-like effector (TALE) hybrid nuclease with novel DNA binding specificity creates double-strand breaks. Proc Natl Acad Sci USA 108(6):2623–2628. doi:10.1073/pnas.1019533108, 1019533108 [pii]
11. Wood AJ, Lo TW, Zeitler B, Pickle CS, Ralston EJ, Lee AH, Amora R, Miller JC, Leung E, Meng X, Zhang L, Rebar EJ, Gregory PD, Urnov FD, Meyer BJ (2011) Targeted genome editing across species using ZFNs and TALENs. Science 333(6040):307. doi:10.1126/science.1207773, science.1207773 [pii]
12. Sander JD, Cade L, Khayter C, Reyon D, Peterson RT, Joung JK, Yeh JR (2011)

Targeted gene disruption in somatic zebrafish cells using engineered TALENs. Nat Biotechnol 29(8):697–698. doi:10.1038/nbt.1934, nbt.1934 [pii]
13. Huang P, Xiao A, Zhou M, Zhu Z, Lin S, Zhang B (2011) Heritable gene targeting in zebrafish using customized TALENs. Nat Biotechnol 29(8):699–700. doi:10.1038/nbt.1939, nbt.1939 [pii]
14. Mussolino C, Morbitzer R, Lütge F, Dannemann N, Lahaye T, Cathomen T (2011) A novel TALE nuclease scaffold enables high genome editing activity in combination with low toxicity. Nucleic Acids Res 39(21):9283–9293. doi:10.1093/nar/gkr597, gkr597 [pii]
15. Hockemeyer D, Wang H, Kiani S, Lai CS, Gao Q, Cassady JP, Cost GJ, Zhang L, Santiago Y, Miller JC, Zeitler B, Cherone JM, Meng X, Hinkley SJ, Rebar EJ, Gregory PD, Urnov FD, Jaenisch R (2011) Genetic engineering of human pluripotent cells using TALE nucleases. Nat Biotechnol 29(8):731–734. doi:10.1038/nbt.1927, nbt.1927 [pii]
16. Lin Y, Fine EJ, Zheng Z, Antico CJ, Voit RA, Porteus MH, Cradick TJ, Bao G (in press) SAPTA: a new design tool for improving TALE nuclease activity. Nucleic Acids Research
17. Porteus MH (2006) Mammalian gene targeting with designed zinc finger nucleases. Mol Ther 13(2):438–446. doi:10.1016/j.ymthe.2005.08.003
18. Guschin DY, Waite AJ, Katibah GE, Miller JC, Holmes MC, Rebar EJ (2010) A rapid and general assay for monitoring endogenous gene modification. Methods Mol Biol 649:247–256. doi:10.1007/978-1-60761-753-2_15
19. Zhang F, Cong L, Lodato S, Kosuri S, Church GM, Arlotta P (2011) Efficient construction of sequence-specific TAL effectors for modulating mammalian transcription. Nat Biotechnol 29(2):149–153. doi:10.1038/nbt.1775, nbt.1775 [pii]
20. Briggs AW, Rios X, Chari R, Yang L, Zhang F, Mali P, Church GM (2012) Iterative capped assembly: rapid and scalable synthesis of repeat-module DNA such as TAL effectors from individual monomers. Nucleic Acids Res 40(15):e117. doi:10.1093/nar/gks624, gks624 [pii]
21. Reyon D, Tsai SQ, Khayter C, Foden JA, Sander JD, Joung JK (2012) FLASH assembly of TALENs for high-throughput genome editing. Nat Biotechnol 30(5):460–465. doi:10.1038/nbt.2170, nbt.2170 [pii]
22. Schmid-Burgk JL, Schmidt T, Kaiser V, Höning K, Hornung V (2013) A ligation-independent cloning technique for high-throughput assembly of transcription activator–like effector genes. Nat Biotechnol 31(1):76–81. doi:10.1038/nbt.2460
23. Kim Y, Kweon J, Kim A, Chon JK, Yoo JY, Kim HJ, Kim S, Lee C, Jeong E, Chung E, Kim D, Lee MS, Go EM, Song HJ, Kim H, Cho N, Bang D, Kim JS (2013) A library of TAL effector nucleases spanning the human genome. Nat Biotechnol 31(3):251–258. doi:10.1038/nbt.2517
24. Deng D, Yin P, Yan C, Pan X, Gong X, Qi S, Xie T, Mahfouz M, Zhu JK, Yan N, Shi Y (2012) Recognition of methylated DNA by TAL effectors. Cell Res. doi:10.1038/cr.2012.127, cr2012127 [pii]
25. Valton J, Dupuy A, Daboussi F, Thomas S, Maréchal A, Macmaster R, Melliand K, Juillerat A, Duchateau P (2012) Overcoming transcription activator-like effector (TALE) DNA binding domain sensitivity to cytosine methylation. The Journal of biological chemistry 287(46):38427–38432. doi:10.1074/jbc.C112.408864

Chapter 14

A Bacterial One-Hybrid System to Isolate Homing Endonuclease Variants with Altered DNA Target Specificities

Rakesh Joshi and Frederick S. Gimble

Abstract

Chromosomal cleavage near the site of mutations that cause disease can facilitate the targeted repair of the locus. Gene therapy protocols therefore require the engineering of DNA endonucleases that target specific genomic loci. Here, we describe a bacterial one-hybrid selection system that has been used to isolate derivatives of the I-SceI homing endonuclease from combinatorial libraries that display altered DNA recognition specificities. The construction of plasmid expression libraries, the development of reporter strains, and the utilization of these components in the bacterial one-hybrid system are detailed.

Key words Bacterial one-hybrid selection system, Homing endonucleases, I-SceI, Gene targeting, Altered specificity

1 Introduction

Homing endonucleases are site-specific enzymes encoded by selfish DNA elements that bind and cleave long recognition sequences [1]. Their ability to introduce double-strand breaks at defined loci in complex genomes in vivo has been utilized in targeted gene therapy methods as a means to stimulate DNA recombination and repair using donor template DNAs [2]. The ability to redesign homing enzyme recognition site specificities is an important prerequisite for targeting double-strand cleavage to specific genes of interest. High-throughput screens and genetic selections from combinatorial libraries have been effective in redesigning specificities of several naturally occurring homing endonucleases of the LAGLIDADG family [3–7]. These approaches have been complemented by computational redesign strategies to obtain enzymes with significantly altered specificities [8, 9].

Francesca Storici (ed.), *Gene Correction: Methods and Protocols*, Methods in Molecular Biology, vol. 1114,
DOI 10.1007/978-1-62703-761-7_14, © Springer Science+Business Media, LLC 2014

Success has been achieved in rapidly engineering proteins with a variety of desired functions by using "semi-rational" methodologies that take advantage of both directed evolution and rational design [10]. This approach begins with the creation of a combinatorial library of protein derivatives containing random residues at positions chosen from a high-resolution structure where mutations are likely to result in altered phenotypes. Variants are subsequently selected from the library that express proteins with the desired phenotype. A major advantage of the directed evolution nature of the approach is that it permits the rapid creation and testing of multiple permutations of mutations. The coupling of rational design to directed evolution is especially effective when structural data are only available that describe the wild-type protein bound to its canonical molecular target. Semi-rational approaches that change the recognition site specificity of homing enzymes have selected for (1) binding to altered DNA sequences, (2) cleavage of altered DNA sequences, or (3) increased DNA recombination at altered sequences. Variations of one-hybrid or two-hybrid systems have been used to select variants of the PI-SceI [11], I-PpoI [12], and I-SceI [7] homing enzymes with altered DNA binding activities. An alternative approach employed phage display to select I-CreI variants that recognize different targets [13]. The changes in DNA binding specificity of selected variants usually parallel the changes in DNA cleavage site specificity. Selection systems to isolate homing enzymes with altered DNA cleavage activities involved the isolation of variants that confer cell survival by cleaving plasmids that express toxic gene products [3, 4, 14]. Selections based on recombination activity required that the homing enzyme derivative cleave an altered recognition sequence in yeast, which resulted in the reassembly of selectable genes by recombination [13].

In a previous report, a two-hybrid selection system was used to "biopan" protein derivatives of the PI-SceI enzyme with desired target binding profiles [11, 15]. Here, we describe the construction and application of an *E. coli*-based one-hybrid selection system (Fig. 1) that was used to isolate derivatives of the I-SceI homing endonuclease that display altered DNA recognition specificities.

2 Materials

2.1 Plasmid Library Construction and Transformation

1. Plasmid pACLαGal4 [16, 18] containing the yeast *GAL4* gene fused to the *E. coli* RNA polymerase α subunit and controlled by the lacUV5 promoter.
2. Restriction enzymes (New England Biolabs, Inc.) including NotI, AvrII.
3. Restriction enzyme buffers used according to manufacturer's protocols.
4. 0.8 % (w/v) agarose gel in TBE buffer.

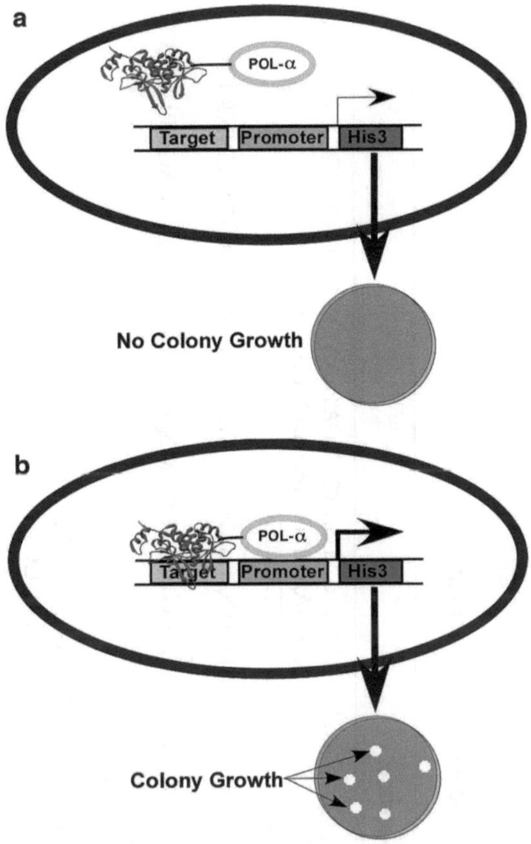

Fig. 1 Overview of bacterial one-hybrid selection for homing endonucleases with altered DNA binding specificity. (**a**) A homing endonuclease-RNA polymerase α subunit fusion protein that does not bind to the target sequence does not stimulate transcription of the *HIS3* gene, thereby slowing the growth of bacterial colonies on histidine-selective media. (**b**) Fusion protein variants expressed within the randomized libraries that have high affinity for the target sequence of interest stimulate transcription of *HIS3* and result in bacterial colonies that grow rapidly

5. TBE buffer: 89 mM Tris–HCl, 89 mM boric acid, 2 mM EDTA, pH 8.3 at 25 °C.
6. Commercial agarose gel extraction kit.
7. Homing endonuclease gene-specific DNA oligonucleotide primers for introducing a homing endonuclease gene into pACLαGal4 (Table 1).
8. A noncoding ~1 kb plasmid DNA sequence that can be introduced into pACLαGal4 containing the homing endonuclease gene that will function as a "stuffer" sequence.
9. Vent high-fidelity DNA polymerase, ThermoPol buffer (New England Biolabs, Inc.), and dNTPs.
10. T4 DNA ligase and T4 DNA ligase buffer (New England Biolabs, Inc.).

Table 1
Examples of oligonucleotides designed for one-hybrid selection of I-SceI homing enzyme variants

Oligonucleotide function	Oligonucleotide sequence
Cloning I-SceI into pACLαGAL4 plasmid[a]	5′-CCAGAGGCGGCCGCAGATTATAAAGATGACGATGATAAATTTCGTACCGGCAGCAAAACCCCGCCGCATCGTAGC**ATGAAAAACATCAAAAAAAACCAG**-3′ 5′-GCTCCCTAGGCTGCAGGTCGACTTAGT**GGT**-3′
Randomization of five I-SceI amino acid codons[b]	5′-TGGGTGATGCATACATCAGATCTCGTGATGAAGGTAAAA CCTACTGTATG**NNK**TTC**NNK**TGGAAAAACAAAGCATAC-3′ 5′-TGTTGAAAGCTTGGTGTTTGAAAGTCTGGGC**MNN**CCAGGTGATTACCAGGTTACCCAGGTGGTTAAC**MNN**TTC**MNN**TTTGTGCGGCGGGGACAG-3′
Wild-type I-SceI DNA target cassettes[c]	5′-AATTACGCTAGGGATAA**CA**GGGTAATAC-3′ 5′-GGCCGTATTACCC**TG**TTATCCCTAGCGT-3′
Confirming double-crossover recombination of DNA targets in reporter strains by detection of the *lacZ-aadA* and *LacI*-kan junctions	5′-AACATGGAAATCGCTGATT-3′ 5′-CTGCCGGCCCAGTATCAG-3′ 5′-GATTGGCGTTGCCACCTCCAG-3′ 5′-AACGGGTGCGCATAGAAA-3′

[a]The underlined nucleotides denote the NotI and AvrII sites used for cloning into pACLαGal4. The sequences in bold encode the I-SceI amino-terminal residues MKNIKKNQ (preceded by a AAADYKDDDDKFRTGSKTPPHRS linker) in one oligonucleotide and the carboxyl-terminal histidine of a his_6 tag in the second

[b]Codons that were randomized for creating an I-SceI protein library are in bold

[c]Two oligonucleotides used to produce a cassette containing the 18-bp I-SceI recognition sequence. The positions shown in bold were mutagenized to produce binding substrates in the one-hybrid selection

11. *E. coli* XL1 Blue MRF′ Supercompetent cells (Agilent Technologies).
12. LB agar plates containing tetracycline (Tc, 15 μg/mL) and chloramphenicol (Cm, 30 μg/mL).
13. Commercial DNA miniprep kit.
14. Degenerate mutagenic primers specific for the homing endonuclease that is mutagenized (Table 1).
15. Commercial PCR DNA fragment purification kit.
16. Electroporation cuvettes (1 mm).
17. SOC medium: 0.5 % yeast extract, 2 % tryptone, 10 mM NaCl, 2.5 mM KCl, 10 mM $MgCl_2$, 10 mM $MgSO_4$, 20 mM glucose containing 15 % glycerol and supplemented with tetracycline (Tc, 15 μg/mL) and chloramphenicol (Cm, 30 μg/mL).
18. M13K07 bacteriophage (New England Biolabs, Inc.).

2.2 Reporter Strain Construction

1. Complementary DNA oligonucleotides containing the designed target sequences with EcoRI and NotI-compatible overhangs (Table 1).
2. Plasmid pF11-stuffer-HIS3-aadA-PheS [11].
3. Restriction enzymes (New England Biolabs, Inc.): including EcoRI and NotI.
4. Commercial agarose gel extraction kit.
5. T4 DNA ligase and T4 DNA ligase buffer (New England Biolabs, Inc.).
6. Electroporation cuvettes (1 mm).
7. Bacterial strains: electrocompetent CSH100 (*F′ lacproA+, B+ (lacI^q lacPL8)/araD (gpt-lac)*5) and KJ1C (F^-, *ΔhisB463, Δ(gpt-proAB-arg-lac)XIII, zaj::Tn10*) were provided by Drs. Keith Joung and Carl Pabo.
8. SOC medium: 0.5 % yeast extract, 2 % tryptone, 10 mM NaCl, 2.5 mM KCl, 10 mM $MgCl_2$, 10 mM $MgSO_4$, and 20 mM glucose supplemented with kanamycin (Kn, 50 μg/mL).
9. LB agar plates supplemented with kanamycin (Kn, 50 μg/mL).
10. LB media supplemented with kanamycin (Kn, 50 μg/mL) or tetracycline (Tc, 15 μg/mL).
11. YEG agar plates supplemented with kanamycin (Kn, 50 μg/mL), tetracycline (15 μg/mL), and D,L-*p*-Cl-phenylalanine (2 mg/mL, Sigma).
12. Gene-specific DNA oligonucleotides used in PCR to verify a double-crossover recombination event (Table 1).
13. Commercial DNA miniprep kit.

2.3 Bacterial One-Hybrid Selection

1. NM liquid medium [15] supplemented with kanamycin (Kn, 50 μg/mL), tetracycline (Tc, 15 μg/mL), chloramphenicol (Cm, 30 μg/mL), and isopropyl β-D-1-thiogalactopyranoside (IPTG, 250 μM, Indofine Chemical Co.).
2. NM agar plates supplemented with kanamycin (Kn, 50 μg/mL), tetracycline (Tc, 15 μg/mL), chloramphenicol (Cm, 30 μg/mL), isopropyl β-D-1-thiogalactopyranoside (IPTG, 250 μM), and appropriate amounts of 3-aminotriazole (3-AT, Acros Organics, Morris Plains, NJ).
3. LB medium and LB agar plates supplemented with 20 mM 3-AT, antibiotics, and IPTG as described [7, 11, 15].
4. Electrocompetent KJ1C cells containing appropriate reporter episomes.
5. SOC medium: 0.5 % yeast extract, 2 % tryptone, 10 mM NaCl, 2.5 mM KCl, 10 mM $MgCl_2$, and 10 mM $MgSO_4$.

3 Methods

If detailed structural information is available for a homing endonuclease, rational design can be used to randomize specific amino acids that are predicted to play roles in DNA binding. Alternatively, if little is known about the DNA binding determinants, mutations can be introduced throughout the protein sequence using error-prone PCR. However, in the latter case, it will not be possible to sample the entire sequence space of all possible variants. It is important to screen randomly chosen candidates taken from the library under nonselective conditions in order to assess the library diversity [17]. An important feature of the one-hybrid selection method is that the stringency can be fine-tuned by varying the concentration of 3-aminotriazole, which is a competitive inhibitor of the *HIS3* gene product. Since the intracellular concentrations of homing endonuclease fusion proteins are dependent on the particular protein being expressed and are difficult to determine relative to the target DNA concentrations and since the binding affinities of the protein variants that arise in the library differ, an empirical approach is recommended in order to identify conditions for selecting high-affinity variants.

3.1 Construction of a Plasmid That Expresses a RNA Polymerase-Homing Endonuclease Fusion Protein and Randomized Plasmid Expression Libraries

3.1.1 Constructing a Plasmid Vector (pACLα-HE) That Expresses the RNA Polymerase-Homing Endonuclease Fusion Protein

1. The pACLαGal4 plasmid DNA [16, 18] is digested for 2 h at 37 °C using a fivefold excess of NotI and AvrII according to the manufacturer's procedures. The 4 kb DNA fragment containing the RNA polymerase gene is separated from the *GAL4*-containing DNA fragment by electrophoresis on a 0.8 % agarose gel and is purified using a DNA gel extraction kit (*see* **Note 1**). The DNA fragment is recovered in 100 μL buffer and stored at −20 °C.
2. Use gene-specific oligonucleotide primers (examples shown in Table 1) in a polymerase chain reaction (PCR) to generate a DNA fragment containing the homing endonuclease with flanking NotI and AvrII sites at either end (*see* **Note 2**). The template gene for the PCR should contain an allele of the homing endonuclease gene that encodes an inactivating mutation at the active site (*see* **Note 3**). The PCR mixture includes the following: 1× ThermoPol™ Reaction Buffer, 20 ng homing endonuclease template plasmid DNA, 1 μL Vent DNA polymerase (New England Biolabs, Inc.), 2 mM $MgSO_4$, 200 μM of each dNTP, 0.4 μM primer oligonucleotides, and H_2O (50 μL total volume).
3. The PCR fragment is digested with NotI and AvrII, resolved on a 1 % agarose gel and recovered using a gel extraction kit.
4. Ligate the prepared AvrII/NotI-digested pACLαGal4 and PCR fragments overnight at 16 °C. The ligation mixture contains 200 units T4 DNA ligase, 1× T4 NEB ligation buffer, 250 ng AvrII/NotI-digested pACLαGal4, and 200 ng AvrII/NotI-digested PCR insert (2:1 plasmid- insert molar ratio; 20 μL total reaction volume).
5. Desalt the ligation product using a spin column kit.
6. Electroporate the ligated product into electrocompetent *E. coli* XL1 Blue MRF′ Supercompetent cells (50 μL; Agilent Technologies, Inc.) as described in the product manual (*see* **Note 4**).
7. Plate the electroporated XL1 Blue cells on LB Tc/Cm agar plates.
8. Use a DNA miniprep kit to extract DNA from transformants and confirm the cloning of the pACLα-HE plasmid by DNA sequencing.
9. Clone a ~1 kb DNA fragment into the pACLα-HE plasmid to generate a pACLα-HE-stuffer plasmid (*see* **Note 5**). The described protocols for enzyme digestion, PCR, DNA ligation, DNA transformation, and preparation are followed according to manufacturer's recommendations.

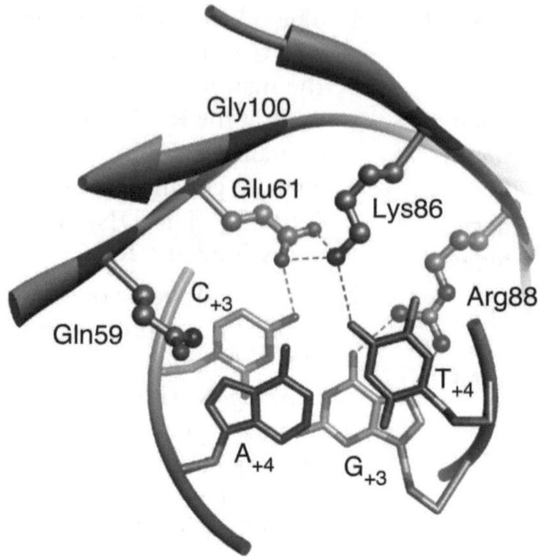

Fig. 2 Detailed view of I-SceI residues and recognition sequence base pairs targeted for mutagenesis in altered binding specificity studies. The positions of the five amino acids in wild-type I-SceI, Gln59, Glu61, Lys86, Arg88, and Gly100, whose codons were randomized in one plasmid library, are depicted in relation to the A:T_{+4} base pair of the wild-type recognition sequence that was substituted for the selection. Hydrogen bonds between amino acid side chains and DNA base pairs are indicated by *dashed lines*

3.1.2 Purification of pACLα-HE-Stuffer DNA for Cloning

1. Electroporate pACLα-HE-stuffer plasmid DNA into electrocompetent *E. coli* XL1 Blue MRF′ cells and pick a single colony for culturing and DNA extraction using a commercial DNA preparation kit as described (*see* **Note 6**).
2. Store the pACLα-HE-stuffer DNA at −20 °C.

3.1.3 Generating DNA Fragments Encoding Homing Endonucleases That Are Randomized at Specific Positions

1. Choose the amino acid residues that will be randomized in the HE (Fig. 2, *see* **Note 7**).
2. Use homing endonuclease gene-specific degenerate oligonucleotides (*see* Table 1) in PCR to create DNA fragments encoding all or a part of the homing endonuclease that can be inserted into pACLα-HE stuffer (*see* **Note 8**). These oligonucleotides contain NNK mixed-base codons to randomly encode each amino acid at the desired positions. Use 1× ThermoPol Reaction Buffer, 20 ng pACLα-HE plasmid DNA containing the inactivated HE gene as template, 1 μL Vent™ DNA polymerase, 2 mM $MgSO_4$, 200 μM of each dNTP, 0.4 μM of degenerate oligonucleotides, and H_2O (50 μL total PCR reaction mix; scale up the reaction tenfold).
3. Gel purify the randomized DNA fragment using a DNA extraction kit and digest with restriction enzymes that will permit its insertion into pACLα-HE stuffer (*see* **Note 8**).
4. Purify the digested insert DNA using a spin column kit and elute in a final volume of 100 μL of deionized H_2O.

3.1.4 Directional Cloning of the Randomized Homing Endonuclease PCR Product into Plasmid pACLα-HE Stuffer

1. Digest 5 μg pACLα-HE-stuffer DNA using the same two restriction enzymes that were used to create the overhangs on the PCR fragment containing the HE gene (*see* **Note 9**).
2. Resolve the ~4 kb DNA on a 0.8 % agarose gel and extract the DNA from the gel slice using a gel extraction kit.
3. Use T4 DNA ligase to ligate the digested PCR product to the gel-purified digested pACLα-HE-stuffer fragment in a 10 μL total volume.
4. Electroporate all of the desalted ligation product into electro-competent E. coli XL1 Blue MRF′ cells (1 mL) and grow for 1 h at 37 °C in SOC media.
5. Plate a small aliquot (5 μL) of the transformation mixture on LB plates supplemented with tetracycline or tetracycline and chloramphenicol. The total number of cells is determined from plates supplemented with tetracycline, while the number of transformants is determined from plates that are also supplemented with chloramphenicol.
6. Amplify the remainder of the XL1 Blue MRF′ transformants harboring the randomized library by growing for 1 h at 37 °C and then dilute into SOC/Tn/Cm media containing 15 % glycerol (total volume 50 mL).
7. Store the amplified E. coli XL1 cells at −80 °C.
8. Culture 20–30 transformants produced using the ligation mixture and extract DNA for sequencing using described protocols to confirm the library diversity.

3.2 Construction of Reporter Strains

3.2.1 Construction of DNA Target Cassettes

1. Design complementary oligonucleotides (~28-mers, *see* Table 1) containing the desired DNA recognition sequence (*see* **Note 10**). These should include cassettes containing the wild-type recognition sequence, to use as a positive control, and mutant targets. EcoRI- and NotI-compatible overhangs should be incorporated at the two ends of the DNA duplex.
2. Anneal the complementary oligonucleotides (28 mers) harboring the WT or mutant targets by heating them together at 95 °C for 2 min and then allowing them to cool slowly to room temperature.

3.2.2 Subcloning of DNA Target Cassettes into pF11-Stuffer-HIS3-aadA-PheS Plasmid

1. Digest the pF11-stuffer-HIS3-aadA-PheS plasmid [11] with EcoRI and NotI restriction enzymes using manufacturer's protocols.
2. Resolve the ~7.8 kb plasmid DNA on a 0.8 % agarose gel and purify using a gel extraction kit.
3. Ligate the DNA cassettes containing the wild-type and mutant homing endonuclease target sequences to the gel-purified pF11-stuffer-HIS3-aadA-PheS plasmid fragment in a final volume of 10 μL (see Subheading 3.1.1 for DNA processing protocols).

4. Electroporate a 2 μL aliquot of the ligation mixtures into 50 μL electrocompetent CSH100 cells and grow for 1 h in SOC/Kn media.
5. Grow the transformants on LB plates supplemented with kanamycin.
6. Extract DNA from colonies growing on experimental plates using a DNA preparation kit and perform DNA sequencing to confirm the correct clones of the pF11-target-HIS3-aadA-PheS constructs.
7. Grow single colonies of the CSH100 cultures that harbor the DNA targets overnight in LB medium at 37 °C and cryofreeze a 1.5 mL sample in 15 % glycerol at –80 °C.

3.2.3 Transfer of the Homing Endonuclease Target Sequences to F′ Episomes by Homologous Recombination

1. Grow 5 mL cultures of the CSH100 strain that harbors the pF11-target-HIS3-aadA-PheS plasmid containing the homing endonuclease target and the KJIC strain in LB/Kn and LB/Tc, respectively, for 16 h at 37 °C.
2. Inoculate 4 mL LB medium either with 0.2 mL of the CSH100 cell cultures or with 0.2 mL of the KJIC culture and allow each to grow at 37 °C for 2 h without agitation.
3. Slowly agitate (100 rpm) the KJ1C culture for the final 20 min.
4. Combine 0.5 mL of a CSH100 culture and 0.5 mL of the KJIC culture and incubate for 1 h without agitation at 37 °C.
5. Slowly agitate (100 rpm) the cultures at 37 °C for 30 min and then for an additional 1 h at normal speed (220 rpm).
6. Plate aliquots of the cell culture on YEG/Kn/Tc plates, to determine the total number of cells present, and on YEG/Kn/Tc plates supplemented with p-Cl-phe, to select for cells containing F′ episomes that have undergone a double-crossover event (*see* **Note 11**).
7. Use plasmid DNA miniprep kits to prepare DNA from Kn/Tc/p-Cl-phe KJIC cultures. Save aliquots of the cultures to prepare cryofrozen stocks.
8. Verify that a double-crossover event occurred during the bacterial conjugation by using oligonucleotide primers in PCR (see Table 1) to detect the two new junctions created between the F′ episome and a segment of the pF11-target-HIS3-aadA-PheS plasmid (*see* Subheading 3.1.1 for PCR protocols and *see* **Note 12**).

3.3 Application of the Bacterial One-Hybrid Selection System

3.3.1 Testing for In Vivo Binding of the RNA Polymerase-Homing Endonuclease Fusion Protein to Its Recognition Sequence

1. Prepare electrocompetent cells of a KJIC strain containing the F′ episome harboring the wild-type HE recognition sequence.
2. Electroporate 50 μL aliquots of these bacteria with 1 ng of pACLα-HE or with 1 ng of pACLαGal4 [18] and grow each culture with agitation in 1 mL SOC medium for 1 h at 37 °C.
3. Prepare tenfold serial dilutions and plate 5 μL aliquots of cells of each dilution on LB/Tc/Kn/Cm plates to determine the total number of transformants (*see* **Note 13**).
4. Harvest the remainder of the transformed cells 15 min by centrifugation and remove the SOC media by washing the transformants 3 times with 5 mL NM/Tc/Kn/Cm/IPTG media, centrifuging after each wash. Resuspend the cells in 0.5 mL NM/Tc/Kn/Cm/IPTG media.
5. Grow the washed cells on NM/Tc/Kn/Cm/IPTG plates supplemented with 20 or 40 mM 3-AT for 60 h at 37 °C to determine whether the HE fusion protein confers growth on the selective medium (*see* **Note 14**).

3.3.2 Isolation of Plasmid Variants that Confer Growth on Histidine-Selective Medium

1. Prepare electrocompetent KJIC reporter strains according to published procedures.
2. Thaw E. coli XL1 Blue cells harboring the pACLα-HE plasmid library.
3. Extract the pACLα-HE library DNA using a midi-prep DNA kit according to manufacturer's protocols. Resuspend the purified plasmid DNA into 500 μL deionized water.
4. Electroporate 50 μL aliquots of each KJIC strain that harbors the F′ episomes containing the different HE recognition sequences with DNA and grow with agitation in 5 mL SOC medium for 1 h at 37 °C.
5. Prepare tenfold serial dilutions and grow 2 μL aliquots of cells of each dilution on LB/Tc/Kn/Cm plates to determine the total number of transformants. Allow the plates to incubate overnight at 37 °C (*see* **Note 15**).
6. Harvest the remainder of the transformed cells by centrifugation and remove the SOC media by washing the transformants 3 times with 5 mL NM/Tc/Kn/Cm/IPTG media, centrifuging after each wash. Resuspend the cells in 0.5 mL NM/Tc/Kn/Cm/IPTG media.
7. Grow the washed cells on NM/Tc/Kn/Cm/IPTG plates supplemented with 0, 20, or 40 mM 3-AT for 60 h at 37 °C to determine whether the HE fusion protein confers growth on the selective medium (*see* **Note 16**).

3.3.3 Confirming and Characterizing Isolates Selected on the Histidine-Selective Media by Retransformation and Sequencing

1. Choose individual colonies from NM selective media plates and colony purify them on agarose plates containing the NM/Tc/Kn/Cm/IPTG medium (*see* **Note 17**).
2. Prepare 5 mL cultures in LB Tc/Kn/Cm medium of all re-streaked KJIC isolates that grow on the selective plates.
3. Extract DNA of isolates using a mini-DNA prep kit.
4. Retransform the prepared DNA into electrocompetent KJ1C cells containing the same reporter episome present during the original selection. Grow for 1 h in SOC medium at 37 °C.
5. Wash the cells 3 times in NM/Tc/Kn/Cm/IPTG media. Harvest the cells by centrifugation and resuspend the cells in 0.1 mL of the wash media. Grow for 60 h at 37 °C on NM/Tc/Kn/Cm/IPTG plates supplemented with 20 and 40 mM 3-AT.
6. Sequence the HE DNA of isolated colonies that grow on a specific reporter strain to identify the mutation(s) responsible for the phenotype (*see* **Note 18**).

4 Notes

1. Resolution of the linearized vector DNA fragment from uncut vector is critical to being able to construct the protein fusion vector. This can be accomplished by resolving the DNAs for ~2 h at 100 V.
2. PCR primers are designed such that one introduces a 5′ NotI site and encodes a 23-amino acid linker (AAADYKDDDDKFRTGSKTPPHRS) before encoding part of the amino-terminal region of the homing endonuclease. The second primer encodes the carboxyl-terminal region and stop codon of the homing endonuclease and includes a 5′ AvrII site at its end. Insertion of the synthesized PCR fragment into NotI/AvrII-digested pACLαGal4 DNA extends the open reading frame that encodes the RNA polymerase such that it also encodes a carboxyl-terminal homing endonuclease.
3. In order to prevent the RNA polymerase-homing endonuclease fusion protein from cleaving and eliminating the reporter episome that contains its recognition sequence during the selection, it is necessary to introduce an inactivating mutation at its active site. The acidic residues within each of the two LAGLIDADG signature motifs are conserved and are essential for DNA cleavage activity, but not for DNA binding activity [19]. Standard sequence alignment analysis can be used to identify these residues to permit their substitution.
4. Non-ligated AvrII/NotI-digested pACLαGal4 vector (250 ng) was used as a negative control for ligation. The number

of transformants that grow on LB/Tc/Cm plates indicates the background growth levels due to contamination by undigested plasmid DNA.

5. Insertion of a large "stuffer" fragment into pACLα-HE facilitates the subsequent construction of the randomized library since it permits the doubly digested plasmid fragment used for cloning to be resolved on agarose gels from singly cut DNA. The source and sequence of the inserted DNA fragment is not critical since it will be excised during the construction of the plasmid library. The only requirement for the insertion site of the "stuffer" fragment is that it must be situated between the restriction sites that will be used to generate the vector DNA fragment required for the library construction.
6. Extract and purify >20 μg of plasmid DNA to acquire sufficient DNA to construct randomized libraries.
7. Codons in the plasmid libraries chosen for randomization are those that encode residues within the DNA-bound wild-type HE structure where substitutions have the potential to establish new DNA contacts. Any of several molecular visualization software applications, including Swiss-PdbViewer ([20], http://spdbv.vital-it.ch) or UCSF Chimera ([21], http://www.cgl.ucsf.edu/chimera), can be used to perform modeling of the mutant proteins.
8. The design of the homing endonuclease gene-specific degenerate oligonucleotides will reflect the available restriction sites that are present in the homing enzyme that is being studied. If the region of the gene that will be randomized is flanked by two different restriction sites that are unique in pACLα-HE-stuffer, then degenerate oligonucleotides can be designed that will introduce the restriction sites at the two ends of the randomized PCR product so that directional cloning can be performed. However, if unique restriction sites are not available, then overlapping PCR can be used to generate two randomized PCR products that can then be joined using flanking primers in a second round of PCR. AvrII and NotI are potential restriction sites that could be used to introduce the gene into the vector.
9. Triple digesting the plasmid with restriction enzymes will decrease the risk of contamination of the plasmid libraries by undigested or singly digested pACLα-HE-stuffer plasmid.
10. It is difficult to predict whether HE derivatives can be isolated from a library that bind to a particular mutant recognition site. Therefore, it is necessary to perform the selection using a variety of different target sites, each with different base pair substitutions at particular positions. We designed sets of target sequence derivatives in which each of the three different alternative base pairs was substituted at particular positions.

11. Para-chlorophenylalanine is toxic to cells that express the PheS gene. Isolation of *p*-Cl-Phe-resistant, kanamycin-resistant cells selects for F′ episomes in which the homing endonuclease target, the *HIS3* gene, and the neighboring *aadA* and kanamycin-resistance genes have become integrated via a double-crossover event. Episomes that have integrated the entire pF11-stuffer-HIS3-aadA-PheS plasmid by a single crossover event, which are not desired, do not survive the selection.
12. The first junction is detected by the presence of an 818 bp product when using oligonucleotides complementary to part of the F′ episome *lacZ* sequence (5′-AACATGGAAATCGCTGATT-3′) and the *aadA* sequence of the plasmid (5′-CTGCCGGCCCAGTATCAG-3′). The second junction is detected by the presence of a 1,012-bp product generated using oligonucleotides complementary to a region of the *LacI* sequence of the F′ episome (5′-GATTGGCGTTGCCACCTCCAG-3′) and a region of the kanamycin-resistance gene of the pF11-HIS3-aadA-PheS plasmid (5′-AACGGGTGCGCATAGAAA-3′).
13. Titration of the cells is required in order to obtain accurate counts of the number of colonies.
14. It is expected that expression of the homing endonuclease from pACLα-HE will confer growth on the selective media. However, since it is not known how much of the HE is expressed in the bacteria nor how tightly the protein binds to the recognition sequence, it is necessary to adjust the stringency of the selection empirically by using two different concentrations of 3-AT in the experiment.
15. Plasmid libraries typically yielded 6.2×10^6 Kn/Tc/Cm resistant colonies.
16. The number of colonies that appear on NM plates lacking 3-AT and on LB plates should be similar because selection does not occur in the absence of the competitive inhibitor. If the HE fusion protein binds to its recognition sequence, it is expected that colonies should grow on plates containing 20 mM 3-AT or on both the 20 and 40 mM 3-AT plates.
17. Linkage between the growth phenotype and the plasmid DNA is established by (1) re-streaking the colonies on plates containing the same 3-AT concentration used for selection, and (2) by retransformation of the plasmid DNA obtained from selected isolates into naive KJ1C bacteria that contain the same recognition sequence on the F′ episome. This reduces the number of false positives picked for sequencing and subsequent biochemical characterization. Growth of the colonies on selective media containing different concentrations of 3-AT can be used to qualitatively assess the DNA binding affinity to the recognition sequence.

18. DNA sequence analysis of isolates can provide a measure of the frequencies with which amino acids occur at each position. A sequence logo can be generated from these data using web servers that indicate the information content at each randomized amino acid position [22] Assessing the plasmid library diversity can be accomplished by using the wild-type recognition sequence to select HEs, each of which contains a set of the various possible codons that encode the wild-type amino acids at the randomized positions. If several possible codon combinations are permitted and all are obtained multiple times in the sequenced set of candidates, there is assurance that the library is diverse.

Acknowledgments

We thank Weijia Luo, Jennifer L. Meyers, and Kristen Tenney for the technical assistance. This work was supported by grants from the National Institutes of Health (GM 070553) and the National Science Foundation (MCB-0321550) to F.S.G. Figures 1 and 2 reprinted from [7] with permission from Elsevier.

References

1. Gimble FS (2000) Invasion of a multitude of genetic niches by homing endonuclease genes. FEMS Microbiol Lett 185:99–107
2. Redondo P, Prieto J, Munoz IG et al (2008) Molecular basis of xeroderma pigmentosum group C DNA recognition by engineered meganucleases. Nature 456(7218):107–111
3. Chen Z, Zhao H (2005) A highly sensitive selection method for directed evolution of homing endonucleases. Nucleic Acids Res 33(18):e154
4. Doyon JB, Pattanayak V, Meyer CB et al (2006) Directed evolution and substrate specificity profile of homing endonuclease I-SceI. J Am Chem Soc 128(7):2477–2484
5. Scalley-Kim M, McConnell-Smith A, Stoddard BL (2007) Coevolution of a homing endonuclease and its host target sequence. J Mol Biol 372(5):1305–1319
6. Jarjour J, West-Foyle H, Certo MT et al (2009) High-resolution profiling of homing endonuclease binding and catalytic specificity using yeast surface display. Nucleic Acids Res 37(20):6871–6880
7. Joshi R, Ho KK, Tenney K et al (2011) Evolution of I-SceI homing endonucleases with increased DNA recognition site specificity. J Mol Biol 405(1):185–200
8. Ashworth J, Havranek JJ, Duarte CM et al (2006) Computational redesign of endonuclease DNA binding and cleavage specificity. Nature 441(7093):656–659
9. Ashworth J, Taylor GK, Havranek JJ et al (2010) Computational reprogramming of homing endonuclease specificity at multiple adjacent base pairs. Nucleic Acids Res 38(16):5601–5608
10. Chica RA, Doucet N, Pelletier JN (2005) Semi-rational approaches to engineering enzyme activity: combining the benefits of directed evolution and rational design. Curr Opin Biotechnol 16(4):378–384
11. Gimble FS, Moure CM, Posey KL (2003) Assessing the plasticity of DNA target site recognition of the PI-SceI homing endonuclease using a bacterial two-hybrid selection system. J Mol Biol 334(5):993–1008
12. Eklund JL, Ulge UY, Eastberg J et al (2007) Altered target site specificity variants of the I-PpoI His-Cys box homing endonuclease. Nucleic Acids Res 35(17):5839–5850
13. Chames P, Epinat JC, Guillier S et al (2005) In vivo selection of engineered homing endonucleases using double-strand break induced homologous recombination. Nucleic Acids Res 33(20):e178

14. Chen Z, Wen F, Sun N et al (2009) Directed evolution of homing endonuclease I-SceI with altered sequence specificity. Protein Eng Des Sel 22:249–256
15. Serebriiskii IG, Fang R, Latypova E et al (2005) A combined yeast/bacteria two-hybrid system: development and evaluation. Mol Cell Proteomics 4(6):819–826
16. Serebriiskii IG, Joung JK (2002) Yeast and bacterial two-hybrid selection systems for studying protein-protein interactions. In: Golemis E (ed) Protein-protein interactions: a molecular cloning manual. Cold Spring Harbor Laboratory Press, Cold Spring Harbor, NY, pp 93–142
17. Volles MJ, Lansbury PTJ (2005) A computer program for the estimation of protein and nucleic acid sequence diversity in random point mutagenesis libraries. Nucleic Acids Res 33:3667–3677
18. Joung JK, Ramm EI, Pabo CO (2000) A bacterial two-hybrid selection system for studying protein-DNA and protein-protein interactions. Proc Natl Acad Sci U S A 97(13):7382–7387
19. Gimble FS, Stephens BW (1995) Substitutions in conserved dodecapeptide motifs that uncouple the DNA binding and DNA cleavage activities of PI-*Sce*I endonuclease. J Biol Chem 270:5849–5856
20. Guex N, Peitsch MC (1997) SWISS-MODEL and the Swiss-PdbViewer: an environment for comparative protein modeling. Electrophoresis 18(15):2714–2723
21. Pettersen EF, Goddard TD, Huang CC et al (2004) UCSF Chimera - visualization system for exploratory research and analysis. J Comput Chem 25:1605–1612
22. Schneider TD, Stephens RM (1990) Sequence logos: a new way to display consensus sequences. Nucleic Acids Res 18(20):6097–6100

Chapter 15

Design and Analysis of Site-Specific Single-Strand Nicking Endonucleases for Gene Correction

Michael J. Metzger and Michael T. Certo

Abstract

Single-strand nicking endonucleases ("nickases") have been shown to induce homology-mediated gene correction with reduced toxicity of DNA double-strand break-producing enzymes, and nickases have been engineered from both homing endonuclease and *Fok*I-based scaffolds. We describe the strategies used to engineer these site-specific nickases as well as the in vitro methods used to confirm their activity and specificity. Additionally, we describe the Traffic Light Reporter system, which uses a flow cytometric assay to simultaneously detect both gene repair and mutagenic nonhomologous end-joining outcomes at a single targeted site in mammalian cells. With these methods, novel nickases can be designed and tested for use in gene correction with novel target sites.

Key words Gene correction, Genome engineering, Homologous recombination, Nickase, Single-strand DNA break, Nicking endonuclease, Zinc-finger nuclease, TALE nuclease, Homing endonuclease, Gene editing, CRISPR/Cas9

1 Introduction

The use of site-specific DNA double-strand break (DSB)-inducing enzymes for genome editing is a promising strategy for gene and cell therapy, but the high frequency and toxicity of mutagenic nonhomologous end-joining (NHEJ) of DSBs in mammalian cells are significant problems for the application of this technology. Recently, single-strand DNA nicks have been shown to induce homologous recombination-mediated gene correction with reduced production of mutagenic outcomes [1–5] or 2 nicks have been created in proximity to introduce a DSB with reduced off target effects [6, 7]. Site-specific nickases have been engineered from homing endonucleases (I-*Ani*I [3] and I-*Sce*I [8]), the *Fok*I catalytic core used in zinc-finger nucleases [5, 9, 10], and the RNA-guided CRISPR/Cas9 system [11, 12]. Additionally, targeted nickases have been generated by the fusion of MutH and I-*Tev*I domains with DNA-binding TALE repeats [13, 14]. The following strategies can be used to design nickases for new chromosomal targets.

Francesca Storici (ed.), *Gene Correction: Methods and Protocols*, Methods in Molecular Biology, vol. 1114, DOI 10.1007/978-1-62703-761-7_15, © Springer Science+Business Media, LLC 2014

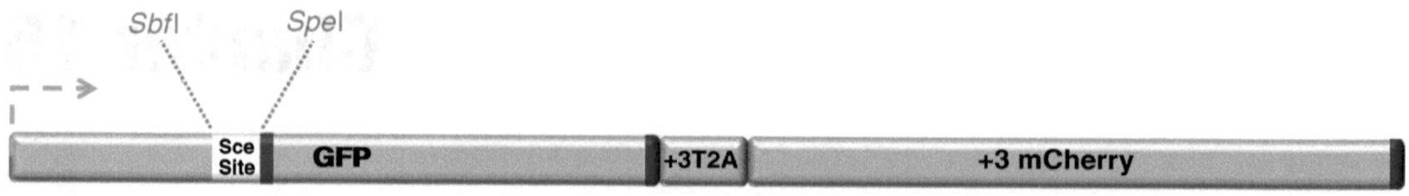

Fig. 1 Schematic of the traffic light reporter. *Arrow* indicates promoter, *SbfI* and *SpeI* restriction enzyme sites are shown flanking an example I-*SceI* target site. The two genes which can express fluorescent proteins, green fluorescent protein (GFP) and mCherry, are located as shown, separated by the translational pausing linker T2A. *Red bars* indicate stop codons, and reading frames are indicated via "+1" or "+3" relative to the initial start codon (color figure online)

A new nickase can be generated by the fusion of site-specific zinc finger or TALE DNA-binding domains to the *Fok*I nickase heterodimer (*see* other chapters in the volume for general zinc-finger nuclease methods and **Note 1** for nickase mutations). Alternatively, the I-*Ani*I and I-*Sce*I nickases can be engineered to recognize novel target sites, or other LAGLIDADG homing endonucleases can be engineered to generate novel nickases by mutation of one of the two catalytic sites in the pseudodimeric enzyme (*see* other chapters in the volume for general homing endonuclease methods and **Note 2** for engineering of nickase mutations). The catalytic activity of novel nickases can be determined by several in vitro methods described here, including nicking of a plasmid target and nicking of fluorescently labeled PCR products. Additionally, the Traffic Light Reporter system [1] uses a flow cytometric assay to simultaneously detect both gene repair and mutagenic nonhomologous end-joining outcomes at a single targeted site in mammalian cells. In this assay, the target site interrupts an integrated green fluorescent protein (GFP) coding sequence with a downstream out-of-frame mCherry fluorescent protein sequence. Repair of the target site using an exogenously delivered donor template will repair the GFP coding sequence, and a proportion of insertions and deletions at the target site introduced by mutagenic NHEJ will generate an in-frame mCherry downstream (Fig. 1). The ratio of GFP to mCherry can thus be used to determine the nickase with the greatest activity and optimal repair profile for use in downstream applications with a variety of delivery systems (as described in other chapters in this volume).

2 Materials

2.1 *In Vitro* Plasmid Nicking

1. TLR vector plasmid: pTLR (addgene #31482), digested with *Sbf*I and *Spe*I.
2. Forward and reverse oligonucleotides for insertion of the desired target into the *Sbf*I/*Spe*I site (*see* **Note 3** for oligo design).

3. 10× ZFN reaction buffer: 500 mM potassium acetate, 200 mM Tris-acetate, 100 mM magnesium acetate, 10 mM dithiothreitol, pH 7.9 (Buffer 4, New England Biolabs). Store at −20 °C.
4. 10× HE reaction buffer: 100 mM $MgCl_2$, 1 M NaCl, 500 mM Tris–HCl, pH 7.5. Store at −20 °C.
5. 2× Stop Solution: 1 % SDS, 100 mM EDTA, 20 % glycerol, and 0.2 % bromophenol blue or xylene cyanol FF. Store at room temperature.

2.2 In Vitro Fluorescently Labeled PCR Product Nicking

1. Synthesize oligonucleotides flanking the desired target site with a 6-carboxyfluorescein (6-FAM) modification on the 5′ end of primers. If the target site is present in a pTLR-based plasmid, the primers TLRqF2b (CGATCCTCGAGCGCCACCAT) and TLRqR2b (GCGCGGGTCTTGTAGTTGCC) can be used, generating an amplicon of 334 bp (plus the size of the inserted target site).
2. Plasmid containing target site: TLR plasmid containing target site (*see* Subheading 2.1 and **Note 3**) or any plasmid containing the target site can be used. Appropriate primers must be designed and synthesized if an alternative plasmid containing the target is used.

2.3 Traffic Light Reporter

1. Tissue culture media: Dulbecco's Modified Eagle Medium (DMEM), with 10 % fetal bovine serum (FBS), 2 mM glutamine, and 1 % penicillin/streptomycin (Pen/Strep).
2. 1× trypsin: (0.05 % trypsin-EDTA).
3. Flow cytometry buffer: phosphate buffered saline (PBS), with 2 mM EDTA and 2 % FBS.
4. 1,000× puromycin: 1 mg/mL in sterile water. Filter (0.2 μm) and store at −20 °C.
5. TLR plasmid (addgene #31482) containing desired target site (*see* Subheading 2.1 and **Note 3**).
6. TLDonor plasmid (addgene #31475).
7. Nickase-expression plasmid.
8. Lentiviral production plasmids [15]: pCMVΔR8.2 (addgene #12263) and pMD2.G (addgene #12259).

3 Methods

3.1 In Vitro Plasmid Nicking

1. Digest pTLR with *Sbf*I and *Spe*I, and ligate annealed oligos containing the desired target sequence. Confirm insertion by sequencing; primer TLRqF2b (Subheading 2.2, **item 1**) can be used.

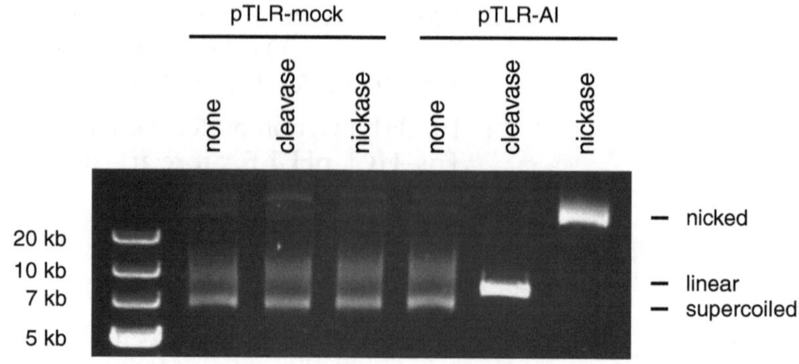

Fig. 2 Detection of plasmid nicking. TLR plasmids (9.0 kb) without an I-*Ani*Y2 target (pTLR-mock) and with an I-*Ani*Y2 target (pTLR-AI) were incubated for 1 h at 37 °C as in Subheading 3.1, with no enzyme, 1 μM I-*Ani*Y2 cleavase, and 1 μM I-*Ani*Y2 nickase. The nicked, linear, and supercoiled products are marked on the agarose gel next to the GeneRuler 1 kb Plus DNA Ladder (Thermo Scientific, Waltham, MA, USA)

2. Generate midi- or maxi-preps of either the TLR plasmid or any other plasmid containing the desired target site using standard protocols.
3. Combine 100 ng of pTLR, purified nickase enzyme, and 1 μL of 10× ZFN buffer or 10× HE buffer for a 10 μL reaction. Use 1 nM to 1 mM enzyme depending on the activity (a representative reaction using 1 μM I-*Ani*IY2 is shown in Fig. 2).
4. Incubate at 37 °C for 1 h.
5. Add 10 μL 2× stop buffer (*see* **Note 4**). Separate supercoiled, nicked, and linear plasmids using agarose gel electrophoresis and image using a DNA intercalating dye (*see* Fig. 2 and **Note 5**).

3.2 *In Vitro* Fluorescently Labeled PCR Product Nicking

1. Generate 6-FAM-labeled PCR product containing the target site using Expand High Fidelity PCR System (Roche Applied Science) (*see* **Note 6**). Purify product using MinElute PCR Purification Kit (Qiagen).
2. Digest PCR product as in Subheading 3.1, **step 3**, at 37 °C for 1 h, using opaque tubes to protect 6-FAM-labeled product from light.
3. Purify digestions using MinElute PCR Purification Kit.
4. Detect the size of the fluorescent products using denaturing capillary electrophoresis. Analyze using Peak Scanner Software v1.0 (Applied Biosystems). The cleaved top strand will be approximately 30 bp longer than the cleaved bottom strand when using primers TLRqF2b and TLRqR2b with the TLR plasmid.

3.3 Generating Traffic Light Reporter

All cells are grown in DMEM with 10 % FBS, 2 mM glutamine, and 1 % Pen/Strep at 5 % CO_2.

1. Transfect HEK293T cells with plasmids for TLR lentivirus production: 0.9 μg pCMVΔR8.2, 0.3 μg pMD2.G, 0.6 μg pTLR containing the desired target site.
2. Replace medium 24 h after transfection.
3. Collect medium containing virus 24 h post media change (48 h post transfection). Filter medium (0.45 μm) and save at −80 °C. Virus can be collected every 24 h for up to 3 days.
4. Plate HEK293 or HEK293T cells in three 6-cm dishes at 1×10^6 cells/dish. Two will be used for titration of virus, and an extra 6-cm dish will be an untransduced control of puromycin selection.
5. One day later, add filtered medium containing virus to each dish of cells (add 10 μL, 1 μL, and nothing to the three dishes).
6. Two days after transduction, trypsinize cells and plate all cells from each 6-cm dish into either a 10-cm dish or a T75 flask in media containing 1 μg/mL puromycin for selection.
7. Change selective medium every 2–3 days for 10 days. Check that control dish has no cells remaining.
8. Stain the colonies with Coomassie and count to determine titer of lentivirus (colony forming units per mL).
9. Transduce new HEK293 or HEK293T cells (**steps 4–7**) with 100 colony forming units of TLR lentivirus (*see* **Note 7**). Alternatively, if one of the two originally transduced dishes has a polyclonal population of approximately 100–500 puromycin-resistant colonies, this can be used as the TLR cell population.
10. Use fluorescence-activated cell sorting to remove rare mCherry+ cells which likely result from frameshift mutations during reverse transcription (*see* **Note 8**).

3.4 Traffic Light Reporter Assay (See Notes 9 and 10)

1. Plate 293/TLR cells at 3.0×10^5 cells per well of a 12-well plate 1 day prior to transfection.
2. Transfect each well with 0.5 μg of nickase-expression plasmid and 0.5 μg pTLDonor, using X-tremeGENE 9 (Roche, Indianapolis, IN, USA) according to manufacturer's standard protocol.
3. Three days after transfection, remove media from cells and add 250 μL 1× trypsin to each well.
4. After cells have detached, add 1 mL flow cytometry buffer and analyze by flow cytometry for GFP and mCherry expression. Gate the live population of cells using Forward Scatter/Side Scatter, and determine the percentage of this live gate that are GFP+ and mCherry+.

4 Notes

1. The nickase catalytic domain is heterodimeric, with a D450A inactivating mutation in one of the monomers [5, 9, 10], and each zinc-finger pair can generate two zinc-finger nickases that cleave opposite strands of DNA by using both orientations of the *Fok*I heterodimer. Some zinc-finger pairs have been shown to function better with the nickase *Fok*I heterodimer than others. Thus, it is recommended that several zinc-finger pairs and both orientations of the heterodimer be created and tested to determine the optimal enzyme for further applications.
2. When converting a new homing endonuclease into a nickase, catalytic mutations should be made and tested in both halves of the enzyme. In the engineering of nickases from I-*Sce*I and I-*Ani*IY2 (a variant of I-*Ani*I with two point mutations, F13Y and S111Y, that enhance enzyme cleavage activity [16]), mutations in the lysine residues in the active sites were introduced, and in each case, one mutation generated a nickase with greater activity and specificity than the other (K227M for I-*Ani*IY2 [3] and K223I for I-*Sce*I [6]). An alternative strategy is to engineer new DNA sequence specificity to the I-*Ani*IY2 scaffold and use the K227M nickase mutation.
3. Forward oligo must contain the desired nickase recognition site followed by a stop codon in the +1 reading frame (e.g., TAA). The reverse oligo contains the reverse complement of the forward oligo with the flanking overhang sequences 5′-CTAG and TGCA-3′ for cloning into the *Sbf*I/*Spe*I sites. The target oligo should render the downstream mCherry in the +2 or +3 reading frame relative to the initial GFP start codon (+1 reading frame), so that the alternative reading frame in GFP leading to mCherry can only be active in response to a frameshift mutation occurring as a result of NHEJ. This is achieved when the number of base pairs inserted by the forward oligo is a multiple of 3, or one less than a multiple of 3 (e.g., 24 or 23 bp). Additionally, the target oligo must be designed to avoid any stop codons in the chosen mCherry reading frame downstream of the intended break site.
4. The 2× Stop Solution serves as a loading buffer and contains SDS in order to prevent protein–DNA interactions from forming aggregates which can impede DNA migration into the agarose gel. Additionally, the reaction time and enzyme concentrations can be modified based on the specific enzyme activity.
5. Supercoiled DNA migrates the fastest through an agarose gel, followed by linear DNA, and nicked plasmid DNA runs the slowest. The electrophoretic mobility of supercoiled plasmid

DNA is affected by the presence of ethidium bromide in the gel, as ethidium bromide intercalates into DNA, reducing negative supercoiling and introducing positive supercoiling, depending upon concentration. With some plasmids, the separation between supercoiled and linear may be increased by preparing and running the gel without ethidium bromide and staining with ethidium bromide or other methods such as SYBR Green afterwards. An uncut sample of plasmid should always be included as a control. A fraction of most plasmid preparations will be nicked, and a small amount may also be linear. Additionally, greater separation between supercoiled and linear bands can be achieved with smaller target plasmids.

6. Other polymerases can be used with 6-FAM-labeled primers, but the use of a non-proofreading polymerase, such as Taq, will result in a doublet peak for the full-length product due to the addition of an untemplated adenine to the 3′ end of some, but not all, of the PCR products. Additionally, 6-FAM-modified primers may be incompatible with some polymerases. In our experience, the Expand High Fidelity PCR System (Roche Applied Science) and Phusion Polymerase (New England Biolabs) work well with these primers.
7. The Traffic Light Reporter system uses lentivirus to integrate the reporter into chromosomal DNA. A polyclonal population of cells transduced at a low MOI will contain primarily cells with a single copy of the reporter. Single clones can be isolated, but different clones will likely have different rates of repair due to insertional position effects [17].
8. The Traffic Light Reporter cells can be used immediately after puromycin selection; however, a small fraction (0.1–2 %) will express mCherry due to frameshifts during reverse transcription of the lentiviral vector. This background can be removed from the polyclonal population by FACS sorting, saving the mCherry-negative cells and discarding the mCherry-positive cells.
9. Alternative delivery strategies for the Traffic Light Reporter have also been developed. Adeno-associated virus, integration-deficient lentiviral vectors, as well as RNA transfection can be used to deliver the endonuclease and donor template.
10. The ratio of HR (GFP) to mutNHEJ (mCherry) is affected by the time before flow cytometric analysis as well as the amounts of endonuclease expression construct and donor template. It is essential to keep these conditions the same if comparisons are to be made between experiments.

Acknowledgments

This work was supported by fellowships from National Institute of Health to MJM (CA009657 and CA009503) and MTC (GM007270). We thank Ryo Takeuchi for the purified I-*Ani*IY2.

References

1. Certo MT, Ryu BY, Annis JE et al (2011) Tracking genome engineering outcome at individual DNA breakpoints. Nat Methods 8(8): 671–676
2. Davis L, Maizels N (2011) DNA nicks promote efficient and safe targeted gene correction. PLoS One 6(9):e23981
3. McConnell SA, Takeuchi R, Pellenz S et al (2009) Generation of a nicking enzyme that stimulates site-specific gene conversion from the I-AniI LAGLIDADG homing endonuclease. Proc Natl Acad Sci USA 106(13):5099–5104
4. Metzger MJ, McConnell-Smith A, Stoddard BL et al (2011) Single-strand nicks induce homologous recombination with less toxicity than double-strand breaks using an AAV vector template. Nucleic Acids Res 39(3):926–935
5. Ramirez CL, Certo MT, Mussolino C et al (2012) Engineered zinc finger nickases induce homology-directed repair with reduced mutagenic effects. Nucleic Acids Res 40(12):5560–5568
6. Mali P, Aach J, Stranges PB et al (2013) CAS9 transcriptional activators for target specificity screening and paired nickases for cooperative genome engineering. Nature Biotechnol 31(9): 833–838
7. Ran FA, Hsu PD, Lin CY et al (2013) Double nicking by RNA-guided CRISPR Cas9 for enhanced genome editing specificity. Cell 154(6):1380–1389
8. Niu Y, Tenney K, Li H et al (2008) Engineering variants of the I-SceI homing endonuclease with strand-specific and site-specific DNA-nicking activity. J Mol Biol 382(1):188–202
9. Kim E, Kim S, Kim DH et al (2012) Precision genome engineering with programmable DNA-nicking enzymes. Genome Res 22(7): 1327–1333
10. Wang J, Friedman G, Doyon Y et al (2012) Targeted gene addition to a predetermined site in the human genome using a ZFN-based nicking enzyme. Genome Res 22(7):1316–1326
11. Cong L, Ran FA, Cox D et al (2013) Multiplex genome engineering using CRISPR/Cas systems. Science 339(6121):819–823
12. Mali P, Yang L, Esvelt KM et al (2013) RNA-guided human genome engineering via Cas9. Science 339(6121):823-826
13. Gabsalilow L, Schierling B, Friedhoff P et al (2013) Site- and strand-specific nicking of DNA by fusion proteins derived from MutH and I-SceI or TALE repeats. Nucleic Acids Res 41(7):e83
14. Beurdeley M, Bietz F, Li J et al (2013) Compact designer TALENs for efficient genome engineering. Nat Commun 4:1762
15. Naldini L, Blomer U, Gage FH et al (1996) Efficient transfer, integration, and sustained long-term expression of the transgene in adult rat brains injected with a lentiviral vector. Proc Natl Acad Sci U S A 93(21):11382–11388
16. Takeuchi R, Certo M, Caprara MG et al (2009) Optimization of in vivo activity of a bifunctional homing endonuclease and maturase reverses evolutionary degradation. Nucleic Acids Res 37(3):877–890
17. Cornea AM, Russell DW (2010) Chromosomal position effects on AAV-mediated gene targeting. Nucleic Acids Res 38(11):3582–3594

Chapter 16

CRISPR-Cas-Mediated Targeted Genome Editing in Human Cells

Luhan Yang, Prashant Mali, Caroline Kim-Kiselak, and George Church

Abstract

The clustered regularly interspaced short palindromic repeats (CRISPR) and CRISPR-associated (Cas) systems have evolved as an adaptive surveillance and defense mechanism in bacteria and archaea that uses short RNAs to direct degradation of foreign genetic elements. Here, we present our protocol for utilizing the *S. pyogenes* type II bacterial CRISPR system to achieve sequence-specific genome alterations in human cells. In principle, any genomic sequence of the form $N_{19}NGG$ can be targeted with the generation of custom guide RNA (gRNA) which functions to direct the Cas9 protein to genomic targets and induce DNA cleavage. Here, we describe our methods for designing and generating gRNA expression constructs either singly or in a multiplexed manner, as well as optimized protocols for the delivery of Cas9-gRNA components into human cells. Genomic alterations at the target site are then introduced either through nonhomologous end joining (NHEJ) or through homologous recombination (HR) in the presence of an appropriate donor sequence. This RNA-guided editing tool offers greater ease of customization and synthesis in comparison to existing sequence-specific endonucleases and promises to become a highly versatile and multiplexable human genome engineering platform.

Key words CRISPR, Cas9, Human genome engineering, hiPSCs

1 Introduction

The CRISPR-Cas system, found in most bacteria and archaea, provides immunity against invading genetic elements [1–3]. In the immunization phase, upon exposure to viral DNA or foreign plasmids, short fragments of exogenous DNA are integrated into the CRISPR locus in the prokaryote genome as spacers between repeat sequences [4]. The locus consists of CRISPR-associated (Cas) genes in operons in addition to the spacer-repeat array. In the immunity phase, the spacer-repeat array is expressed as a pre-CRISPR RNA (pre-crRNA), and the transcript is processed into individual spacer-repeat units as small guiding crRNAs. These crRNAs then form a complex with Cas proteins to induce cleavage of foreign nucleic acids complementary to the spacer sequence [5].

Francesca Storici (ed.), *Gene Correction: Methods and Protocols*, Methods in Molecular Biology, vol. 1114,
DOI 10.1007/978-1-62703-761-7_16, © Springer Science+Business Media, LLC 2014

The CRISPR-Cas systems described thus far fall into three major classes, each with distinct mechanisms of action and composed of different Cas gene family [6]. While the type I and III systems utilize multiple distinct effector proteins to direct endonuclease activity, the type II system relies primarily on a single protein Cas9 to both generate crRNA and cleave corresponding target DNA and thus represents the most convenient format for transferring this machinery into eukaryotic cells. Specifically, in type II system in *S.* pyogenes, transactivating crRNA (tracrRNA) first hybridizes to the repeat regions of pre-crRNA. This duplex is then recognized and cleaved by Cas9 and RNase III, and the resulting mature crRNA remains in complex with tracrRNA and Cas9 to form the functional unit that ultimately targets and cleaves foreign DNA homologous to the crRNA spacer sequence.

Jinek et al. [7] recently demonstrated that a single tracrRNA-crRNA chimera in complex with the Cas9 protein was capable of introducing sequence-specific double-stranded DNA cleavage in vitro. We engineered this two-component system for function in eukaryotic cells [8] by (1) inducing direct expression of an optimized tracrRNA-crRNA chimera construct, which we have termed guide RNA (gRNA), and (2) expressing the human codon-optimized Cas9 (hCas9) protein (Fig. 1a).

Cas9 detects genomic targets by unwinding the DNA duplex and scanning for complementarity between the genomic DNA and the spacer sequence in the gRNA. However, Cas9 will only cleave DNA if the correct protospacer-adjacent motif (PAM) is detected [7, 9]. While each type II system has individual PAM requirements, the *S. pyogenes* system described in this protocol requires an NGG sequence immediately downstream of the protospacer, with N being any nucleotide (Fig. 1b).

The Cas9 protein contains two nuclease domains, an HNH domain and a RuvC-like domain, which together generate a double-stranded break at the 3′ end of the protospacer following target recognition. If one of the two nuclease domains is inactivated, Cas9 will function as a nickase in vitro [7] as well as in human cells [8]. Cas9-induced double-stranded breaks or single-stranded nicks are subsequently detected by host DNA repair mechanisms, most often through the nonhomologous end-joining (NHEJ) pathway in mammalian cells, an error-prone process which can result mutations at targeted genomic sites.

In contrast to other recent technologies such as zinc finger nucleases and TALENs [10, 11], the CRISPR system relies upon easily engineered RNA molecules for sequence specificity versus DNA-binding proteins that must be reconfigured for each targeted region. Here, we describe a robust platform for gRNA synthesis and expression which can be used to generate single gRNAs as well as large pooled libraries synthesized from custom DNA arrays. Towards this, as a ready reference, a genome-wide resource of ~200,000 unique gRNAs targeting ~40 % of human exons featuring

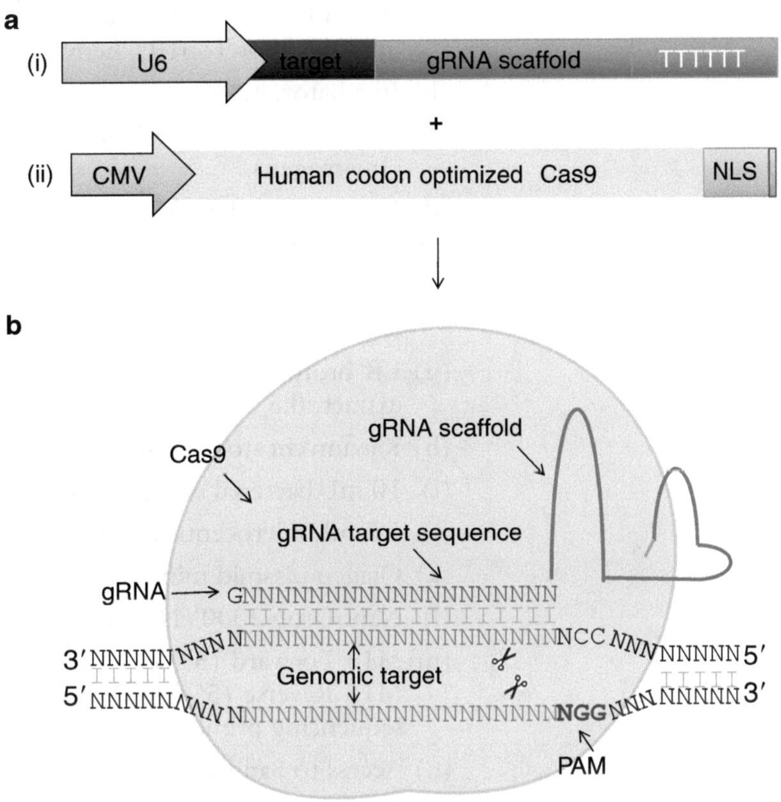

Fig. 1 Schematic outline of Cas9-gRNA-mediated genomic targeting

the NGG PAM sequence, at a multiplicity of ~2.4 sites per targeted exonic region as described in Mali and Yang et al. [8], is available for reference at http://arep.med.harvard.edu/human_crispr/. Finally, we describe our protocols for targeted genome editing in human cells using this engineered system. With its simplicity of design and construction, the Cas9-gRNA system represents an appealing and facile approach for human genome engineering with broad applications in basic and applied medical research.

2 Materials

2.1 Determination of gRNA Target Sequence

1. Sequence analysis software.
2. Sequence alignment search software such as NCBI BLAST or UCSC Genome Browser BLAT.

2.2 gRNA Synthesis

2.2.1 Option A

1. gBlock (IDT) containing desired gRNA target sequence.
2. PCR-grade sterile deionized water.
3. Option 3a:

(a) PCR-Blunt II-TOPO kit (Invitrogen, K2800-20) including One Shot Top10 Chemically Competent *E. coli* cells.
(b) Incubator, 42 °C.
(c) SOC medium, provided with One Shot Top10 cells (Invitrogen).
(d) Incubator shaker, 37 °C.
(e) Sterilized glass beads.
(f) LB agar: 1 % (w/v) Bactotryptone, 0.5 % (w/v) yeast extract, 0.5 % (w/v) sodium chloride, 1.5 % (w/v) agar.
(g) LB broth: 1 % (w/v) Bactotryptone, 0.5 % (w/v) yeast extract, 0.5 % (w/v) sodium chloride.
(h) Kanamycin stock at 50 mg/mL in water.
(i) 10 mL bacterial culture tubes.
(j) 1.5 mL microcentrifuge tubes.
(k) Qiagen plasmid miniprep kit (Qiagen).
(l) NanoDrop 2000 (NanoDrop).
(m) M13 Forward (5′-GTTTTCCCAGTCACGACG-3′) and M13 Reverse (5′-AACAGCTATGACCATG-3′) universal sequencing primers.
(n) Access to Sanger sequencing facility.
(o) HiSpeed Plasmid Maxi Kit (Qiagen).

Option 3b:

(a) PCR primers gRNA_F (5′-TGTACAAAAAAGCAGGC TTTAAAG-3′) and gRNA_R (5′-TAATGCCAACTTTG TACAAGAAAG-3′).
(b) 2× KAPA HiFi HotStart ReadyMix (Kapa, KK2601).
(c) PCR tubes.
(d) Thermocycler.
(e) E-Gel EX_Gel 2 % (Invitrogen).
(f) E-Gel agarose gel electrophoresis system (Invitrogen).
(g) 2-log DNA ladder (NEB).
(h) PCR purification kit (Qiagen).
(i) 1.5 mL microcentrifuge tubes.
(j) NanoDrop 2000 (NanoDrop).
(k) Speedvac.

2.2.2 Option B

1. PCR-grade sterile deionized water.
2. 60mer oligonucleotides Insert_F and Insert_R, sequence in Subheading 3.2.2, **step 1**.
3. 1.5 mL microcentrifuge tubes.

4. Water.
5. 250 mL glass beaker.
6. Phusion polymerase kit (NEB, M0530S).
7. PCR tubes.
8. Thermocycler.
9. gRNA cloning vector bacterial glycerol stock (Addgene, plasmid ID 41824).
10. LB agar.
11. Incubator.
12. LB broth.
13. Kanamycin stock at 50 mg/mL in water.
14. Incubator shaker (37 °C).
15. HiSpeed Plasmid Maxi Kit (Qiagen).
16. AflII restriction enzyme (NEB, R0520).
17. E-Gel EX_Gel 2 % (Invitrogen).
18. E-Gel agarose gel electrophoresis system (Invitrogen).
19. 2-log DNA ladder (NEB).
20. QIAquick Gel Extraction Kit (Qiagen).
21. Gibson Assembly Master Mix (NEB, E2611).
22. One Shot Top10 Chemically Competent *E. coli* cells (Invitrogen, C4040-03).
23. SOC medium, provided with One Shot Top10 cells (Invitrogen).
24. 10 mL bacterial culture tubes.
25. Qiagen plasmid miniprep kit (Qiagen).
26. M13 Forward (5′-GTTTTCCCAGTCACGACG-3′) and M13 Reverse (5′-AACAGCTATGACCATG-3′) universal sequencing primers.
27. Access to Sanger sequencing facility.
28. HiSpeed Plasmid Maxi Kit (Qiagen).

2.3 gRNA Preparation from Microarrays for Multiplexible Targeting

1. 200 bp oligonucleotide pool (Custom Array Inc.).
2. PCR-grade sterile deionized water.
3. 2× KAPA HiFi HotStart ReadyMix (Kapa, KK2601).
4. PCR primers.
   (a) gRNA pool F1: TATGAGGACGAATCTCCCGCTTATA
   (b) gRNA pool R1: GGTCTTGACAAACGTGTGCTTGTAC
   (c) Target-specific barcode forward primer (25 bp).
   (d) Target-specific barcode reverse primer (25 bp).
   (e) gRNA pool F2: tttcttggctttatatatcttgtggaaaggac

   (f) gRNA pool R2: GACTAGCCTTATTTTAACTTG-CTATTTCTAGCT
5. Thermocycler.
6. QIAquick PCR purification kit (Qiagen).
7. 1.5 mL microcentrifuge tubes.
8. AflII restriction enzyme (NEB, R0520).
9. E-Gel EX_Gel 2 % (Invitrogen).
10. E-Gel agarose gel electrophoresis system (Invitrogen).
11. 2-log DNA ladder (NEB).
12. QIAquick Gel Extraction Kit (Qiagen).
13. Gibson Assembly Master Mix (NEB, E2611).
14. One Shot Top10 Chemically Competent *E. coli* cells (Invitrogen, C4040-03).
15. SOC medium, provided with One Shot Top10 cells (Invitrogen).
16. LB agar.
17. Kanamycin stock at 50 mg/mL in water.
18. Incubator.
19. LB broth.
20. Incubator shaker (37 °C).
21. 10 mL bacterial culture tubes.
22. Qiagen plasmid miniprep kit (Qiagen).
23. M13 Forward and M13 Reverse universal sequencing primers.
24. Access to Sanger sequencing facility.
25. HiSpeed Plasmid Maxi Kit (Qiagen).

2.4 Prepare hCas9 Plasmid

1. hCas9 bacterial glycerol stock (Addgene, plasmid ID 41824).
2. LB agar.
3. Ampicillin stock at 50 mg/mL in water.
4. Incubator(37 °C).
5. LB broth.
6. Incubator shaker.
7. HiSpeed Plasmid Maxi Kit (Qiagen).
8. PCR-grade sterile deionized water.

2.5 Transfection of 293 HEK Cells

2.5.1 Day 0: Plate 293 Cells for Transfection

1. 293 human embryonic kidney cells (Invitrogen).
2. 6 well tissue culture-treated plates.
3. Humidified incubator, with cells maintained at 37 °C and 5 % CO_2.
4. High-glucose DMEM media (Invitrogen).

5. Fetal bovine serum (FBS).
6. Penicillin/streptomycin solution (P/S).
7. Nonessential amino acids (NEAA).
8. TrypLE Express (Invitrogen, 12604-013).
9. Countess cell counter.
10. 15 mL centrifuge tubes.
11. Tabletop centrifuge.

2.5.2 Day 1: Transfection

1. Complete DMEM media (10 % FBS, 1× P/S, and 1× NEAA).
2. Lipofectamine 20000 (Invitrogen, 11668027).
3. Opti-MEM Medium (Invitrogen, 31985062).
4. 1.5 mL microcentrifuge tube.
5. hCas9 plasmid DNA generated in Subheading 3.3.
6. gRNA expression vector generated in Subheading 3.2.
7. 1.5 mL microcentrifuge tubes.
8. Humidified incubator, with cells maintained at 37 °C and 5 % CO_2.

2.6 Transfection of PGP1 iPS Cells

2.6.1 Matrigel Preparation

1. Matrigel (hESC-qualified) (BD Biosciences, 354277).

2.6.2 Transfection

1. PGP1 iPS cells adapted for growth on Matrigel.
2. Humidified incubator, with cells maintained at 37 °C and 5 % CO_2.
3. Matrigel aliquot prepared in Subheading 3.6.1.
4. DMEM/F12 media (Invitrogen).
5. 48 well tissue culture-treated plates (BD Biosciences).
6. mTeSR1 medium (Stemcell Technologies, 05850).
7. In Solution Rho kinase (ROCK) inhibitor (Calbiochem, Y-27632).
8. PBS (Invitrogen).
9. TrypLE Express (Invitrogen, 12604-013).
10. Countess automated cell counter (Invitrogen).
11. 15 mL centrifuge tubes.
12. Tabletop centrifuge.

13. P3 Primary Cell 4D-Nucleofector X kit containing P3 and Supplement 1 solutions in addition to 16-well Nucleocuvette strips (Lonza, V4XP-4032).
14. Amaxa 4D-Nucleofector System (Lonza, CD-MN025).
15. hCas9 plasmid DNA generated in Subheading 3.3.
16. gRNA expression vector generated in Subheading 3.2.
17. 15 mL centrifuge tubes.

3 Methods

The following is a general protocol (Fig. 2) for performing CRISPR-mediated genome editing through the identification of gRNA target sites (Fig. 3), generation of gRNA expression vectors either singly or in multiplexed fashion (Figs. 4 and 5), and the transfection of CRISPR elements into 293 HEK and PGP1 iPS cell lines (Fig. 6).

In this system, we have utilized the human U6 promoter to directly drive gRNA transcription, due to its well-defined transcriptional start and end points [12, 13]. However, this approach requires that gRNA transcription initiate with G, which is then followed by the remaining 19 nt sequence target in genomic sequence preceding the NGG PAM sequence (Fig. 1). The inclusion of this initial G in the final gRNA, which may not be present in the genomic sequence, does not affect sequence specificity due to its presence at the 5′ end of the spacer region, where mismatches are tolerated [7, 9].

This protocol also specifically describes our method for generating a large library of gRNA sequences targeting the human exome, designed in a 200 bp format compatible with multiplex synthesis on custom DNA arrays and efficient cloning into a common expression vector [8]. This permits the targeted retrieval of individual gRNA sequences as each sequence is associated with a

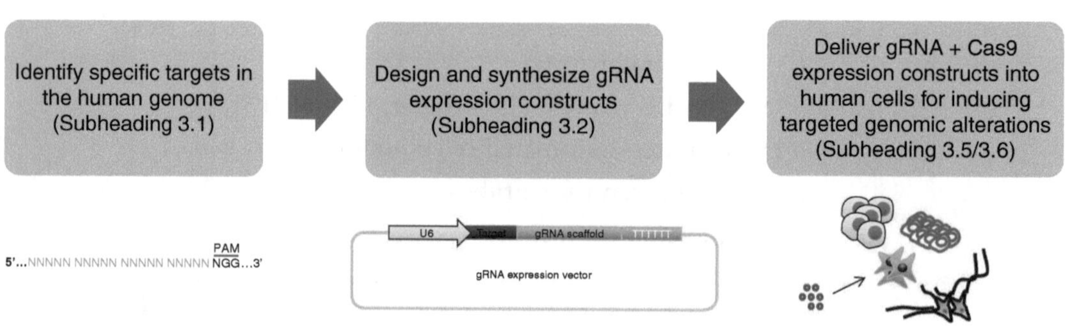

Fig. 2 Flow chart of using Cas9-gRNA for introducing human genomic targeting

Step 1: Find all 23bp genomic sites of the form 5'-N_{19}NGG-3' near your intended target site (ideally ±50bp). These may reside on the + or –strand. Representative example:

Target sequence PAM

N_{19} NGG

5′-TAATACTTTTATCTGTCCCCTCCACCCCACAGTGGGGCCAC**TAGGGACAGGATTGG**TGACAGAAAAGCCCC -3′

3′-ATTATGAAAATAGACAGGGGAGGTGGGGTGTCACCCCGGTGATCCCTGTCCTAACCACTGTCTTTTCGGGG -5′

↓

Step 2 : Using NCBI blast, select sequences for which none or very few sequences of the form 5'-NNNN NNSSS SSSSS SSSSS NGG-3' exist at any other location in the human genome (here the S's represent the actual sequence bases at the target genomic location). Representative example:

TAGGGACAGGAT**NGG**	BLAST hits in human genome
TAGGGACAGGAT**AGG**	0
TAGGGACAGGAT**TGG**	1
TAGGGACAGGAT**CGG**	0
TAGGGACAGGAT**GGG**	0

Fig. 3 Design of Cas9-gRNA targeting site (the sequence and the BLAST results are hypothetical for the purpose of demonstration)

Step 3 (option A) : Incorporate 19bp of the selected target sequence as highlighted here: 5'-NNNN NNNNN NNNNN NNNNN NGG-3' into the DNA fragment as indicated below:

TGTACAAAAAAGCAGGCTTTAAAGGAACCAATTCAGTCGACTGGATCCGGTAC
CAAGGTCGGGCAGGAAGAGGGCCTATTTCCCATGATTCCTTCATATTTGCATA
TACGATACAAGGCTGTTAGAGAGATAATTAGAATTAATTTGACTGTAAACACA
AAGATATTAGTACAAAATACGTGACGTAGAAAGTAATAATTTCTTGGGTAGTT
TGCAGTTTTAAAATTATGTTTTAAAATGGACTATCATATGCTTACCGTAACTT
GAAAGTATTTCGATTTCTTGGCTTTATATATCTTGTGGAAAGGACGAAACACC
G*NNNNNNNNNNNNNNNNNNN* GTTTTAGAGCTAGAAATAGCAAGTTAAAATAAG
GCTAGTCCGTTATCAACTTGAAAAAGTGGCACCGAGTCGGTGCT TTTTTT CTA
GACCCAGCTTTCTTGTACAAAGTTGGCATTA

→

TGTACAAAAAAGCAGGCTTTAAAGGAACCAATTCAGTCGACTGGATCCGGTAC
CAAGGTCGGGCAGGAAGAGGGCCTATTTCCCATGATTCCTTCATATTTGCATA
TACGATACAAGGCTGTTAGAGAGATAATTAGAATTAATTTGACTGTAAACACA
AAGATATTAGTACAAAATACGTGACGTAGAAAGTAATAATTTCTTGGGTAGTT
TGCAGTTTTAAAATTATGTTTTAAAATGGACTATCATATGCTTACCGTAACTT
GAAAGTATTTCGATTTCTTGGCTTTATATATCTTGTGGAAAGGACGAAACACC
G*GGCCAC TAGGGACAGGAT* GTTTTAGAGCTAGAAATAGCAAGTTAAAATAAGG
CTAGTCCGTTATCAACTTGAAAAAGTGGCACCGAGTCGGTGCT TTTTTT CTAG
ACCCAGCTTTCTTGTACAAAGTTGGCATTA

↓

This 455bp fragment bears all components necessary for gRNA expression, namely: U6 promoter + target sequence + guide RNA scaffold + termination signal.

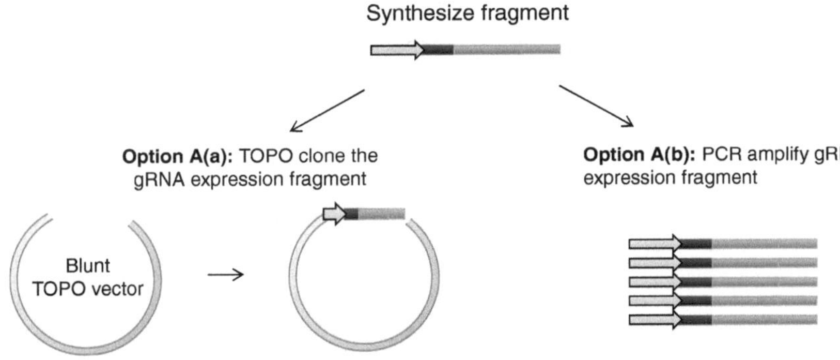

Fig. 4 Schematic outline of gRNA expression constructs design and synthesis approaches A (Subheading 3.2.1)

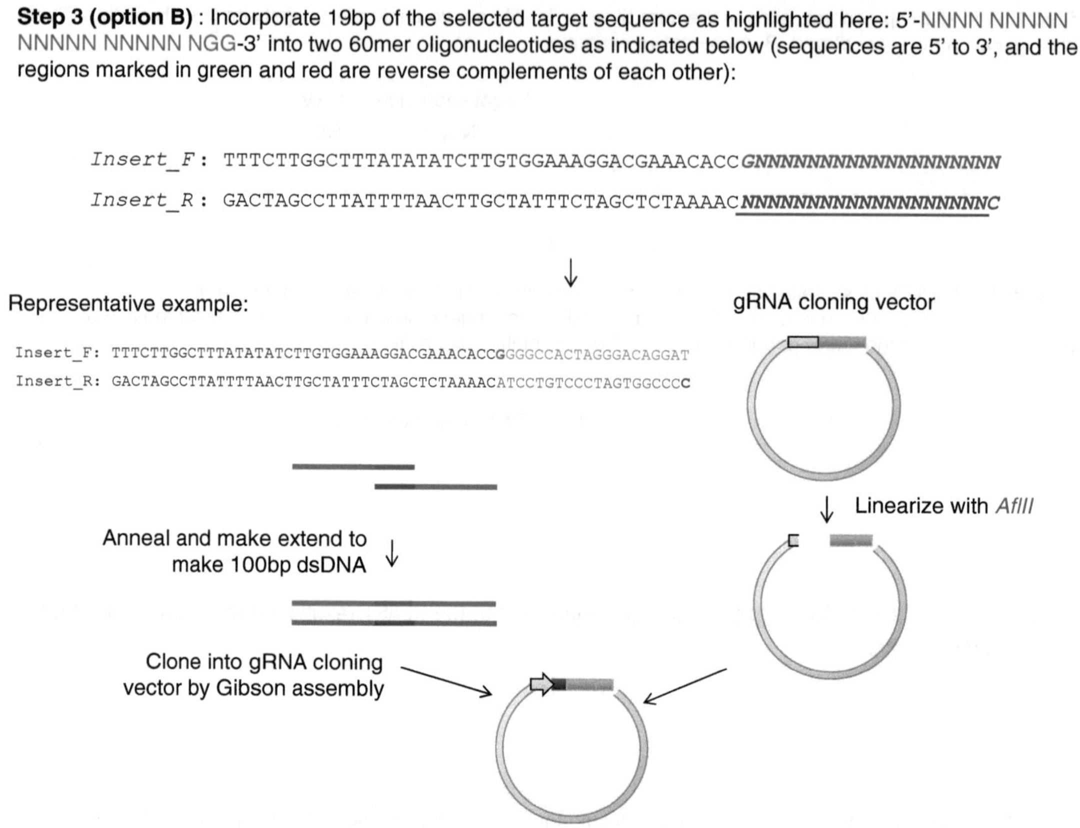

Fig. 5 Schematic outline of gRNA expression constructs design and synthesis approaches B (Subheading 3.2.2)

unique barcode. We anticipate that users will develop many alternative approaches to this basic design, such as using barcodes associated with several gRNA sequences for pooled retrieval or generating cheaper 100 bp arrays as a more cost-effective approach for certain applications.

While we have had success with these methods thus far as presented, several aspects of this protocol may require additional optimization, especially when applied to other cell lines. We discuss this topic as well as other potential limitations of this system in Subheading 4 of the manuscript.

3.1 Determination of Target Sequence

1. Using sequence analysis software, identify all 22 bp regions within 50 bp of your intended genomic target in the form of **5′**-N_{19}-NGG-**3′**. These 22 bp regions may be located on either strand and should ideally overlap the target sequence (Fig. 3).
2. Use an alignment search tool to check whether the selected target sequences are unique in the genome to prevent any undesired off-target effects. Exact correspondence in the 13 bp

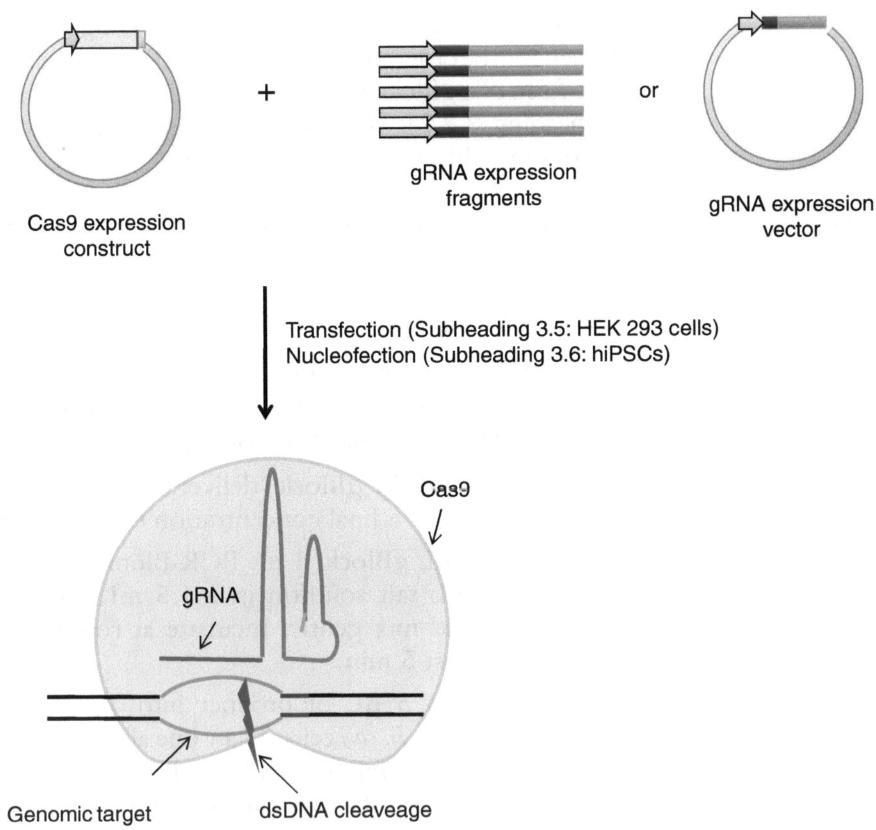

Fig. 6 Schematic representation of introducing Cas9/gRNA into HEK293 cells (Subheading 3.5) and hiPSCs (Subheading 3.6) for specific genomic targeting

region immediately proximal to the NGG PAM sequence (Fig. 3) is required for Cas9-gRNA specific recognition, while mismatches in the 5′ region are better tolerated. Thus, to determine whether Cas9-gRNA cleavage sequence is unique to the genome, one needs to check the existence of other sites in the genome in the form of $S_{13}NGG$ (S, N = A, T, C, G) (Fig. 3). If multiple hits besides the desired genomic target are observed, other alternative sites need to be chosen and checked. The specificity of gRNA target sequence design is discussed below (*see* **Note 1**).

3.2 gRNA Synthesis

1. Design gRNA expression constructs according to the target sequence 5′-N_{19}-NGG-3′. Incorporate the customized N_{19} sequence into the gRNA expression fragment design. This final fragment contains all of the components necessary for gRNA expression, including the U6 promoter, customized target sequence, the gRNA scaffold, and termination signal (Fig. 4).

2. Synthesize the gRNA expression construct. The gRNA expression fragment can be ordered directly through IDT (Subheading 3.2.1) in the form of a gBlock or synthesized with an in-house protocol (Subheading 3.2.2). We discussed the application of different methods in Subheading 4 (*see* **Note 2**).

3.2.1 Option A

1. Synthesize the final gRNA expression fragment (455 bp) as a standard gBlock without any 5′ modifications from IDT (https://www.idtdna.com/pages/products/genes/gblocks-gene-fragments).
2. The synthesized gBlock can be amplified in two ways (Fig. 4):
   (a) Clone the synthesized gBlock into the PCR-Blunt II-TOPO vector from Invitrogen.
      - Resuspend the gBlock (delivered at 200 ng) in 20 μL of water for a final concentration of 10 ng/μL.
      - Pipet 1 μL gBlock, 1 μL PCR-BluntII-TOPO vector, and 4 μL salt solution in a 1.5 mL microcentrifuge tube, and mix gently. Incubate at room temperature for at least 5 min.
      - Transform 5 μL of product into Top10 Chemically Competent *E. coli* cells. Thaw one aliquot of Top10 cells in ice for 10 min, then add 5 μL of the TOPO cloning reaction from the previous step and incubate on ice for 30 min. Heat-shock the cells at 42 °C, then return to ice for 2 min. Add 250 μL of room temperature S.O.C. medium and incubate in a 37 °C shaker for 1 h.
      - Spread 100 μL of the transformation using sterilized glass beads onto a prewarmed LB agar-kanamycin (50 μg/mL) plate and incubate overnight at 37 °C. Expect 10–100 colonies, with the majority containing the desired insert.
      - Pick 1–5 colonies the next day. Culture each colony in 5 mL LB medium containing 50 μg/mL kanamycin overnight at 37 °C with shaking at 200 rpm.
      - Use a plasmid miniprep kit to isolate DNA from the majority of each 5 mL culture. Elute DNA from each column with 30 μL water. Save the remainder of each culture by storing at 4 °C.
      - Determine DNA concentration for each sample with a NanoDrop, and submit samples for Sanger sequencing using M13 Forward and M13 Reverse universal sequencing primers to determine correct transformants.
      - Grow a maxiprep culture of the correct transformant by inoculating 200 mL of LB medium containing 50 μg/mL kanamycin with 100 μL of the original

miniprep culture. Grow overnight at 37 °C with shaking at 200 rpm.

- Isolate plasmid DNA using a plasmid maxiprep kit. Resuspend plasmid DNA at ~1 μg/mL in water. Use this product for transfection.

(b) Alternatively, directly PCR amplify the fragment for transfection and expression.

- Resuspend the gBlock (delivered at 200 ng) in 20 μL of water for a final concentration of 10 ng/μL; resuspend gRNA F and gRNA R primers at 100 mM in water.
- Prepare 100 μL PCR reaction as following and aliquot 25 μL reaction into each PCR tube.
  - 4 μL gBlock.
  - 4 μL gRNA F primer (10 μM).
  - 4 μL gRNA R primer (10 μM).
  - 50 μL 2× KAPA HiFi HotStart ReadyMix.
  - 38 μL PCR-grade water.
  - 100 μL total volume.
- Place tubes in thermocycler, and run the following program:

Temperature	Time	Number of cycles
95 °C	5 min	1
98 °C 65 °C 72 °C	20 s 15 s 30 s	30
72 °C	3 min	1
4–10 °C	∞	1

- Dilute 2 μL PCR product into 18 μL water and run alongside 2-log ladder on 2 % EX_Gel to confirm whether a specific band of 455 bp is obtained.
- Purify PCR product using a PCR cleanup kit, and elute the DNA from the column in 30 μL water.
- Measure the purified PCR product concentration with NanoDrop, then speedvac and resuspend the DNA at 1 μg/μL in water. Use this product for transfection.

3.2.2 Option B

1. Incorporate 19 bp of the selected target sequence (i.e., 5′-N_{19} of 5′-N_{19}-NGG-3′) into two 60mer oligonucleotides which have 20 nt complementary region with each other at 3′ ends (bold).

Underlined region in Insert_R is the reverse complement of 19 bp (N_{19}) (Fig. 5).

(a) Insert_F: TTTCTTGGCTTTATATATCTTG TGGA-AAGGACGAAACACC**G$\mathbf{N_{19}}$**

(b) Insert_R: GACTAGCCTTATTTTAAC TTGCTATTT-CTAGCTCTAAAACN_{19}**C**

2. Order the two 60mer oligonucleotides (25 nmol) from a commercial source such as IDT.
3. Suspend each 60mer oligonucleotide at 100 mM in water.
4. Anneal the two oligos to create a DNA fragment containing 20 bp double-stranded regions flanked by 40 nt 5′ overhangs at each end.

(a) Mix 10 μL Insert_F (100 μM) and 10 μL Insert_R (100 μM) in a PCR tube.

(b) Place the tube in a thermocycler with the following protocol:

Temperature	Time
95 °C	5 min
Ramp to 4 °C	0.1 °C/s
4–10 °C	∞

(c) The resulting product can be stored at 4 °C.

5. Extend the annealed DNA fragment using the Phusion polymerase to create a 100 bp fully double-stranded DNA product.

(a) Create the following reaction mixture:

- 2 μL annealed DNA product (50 μM).
- 4 μL 5× Phusion HF buffer.
- 1 μL 10 mM dNTPs.
- 1 μL Phusion DNA polymerase.
- 12 μL nuclease-free H_2O.
- 20 μL total volume.

(b) Place tubes in thermocycler, and run the following program:

Temperature	Time
72 °C	15 min
4–10 °C	∞

(c) Purify the reaction with PCR cleanup kit and elute the product in 30 μL water.

(d) The final product is a 100 bp dsDNA fragment. Measure the concentration of purified product with a NanoDrop.

6. Linearize the gRNA cloning vector using AflII.

(a) Obtain the gRNA cloning vector from Addgene (plasmid ID 41824) as a bacterial stab in agar.

(b) Use the tip of a sterile pipet tip or toothpick to scrape the frozen bacterial stock, then streak onto an LB agar-kanamycin plate (50 μg/mL). Incubate plate at 37 °C overnight.

(c) Pick a single colony from the plate the next day to inoculate 200 mL of LB-kanamycin (50 μg/mL), and grow overnight at 37 °C with shaking at 200 rpm.

(d) Use a plasmid maxiprep kit to purify the gRNA cloning vectors per manufacturer's instruction.

(e) Measure the concentration with a NanoDrop.

(f) Digest the gRNA cloning vector:

- 1 μg gRNA cloning vector.
- 1 μL AflII restriction enzyme.
- 2 μL 10× NEBuffer 4.
- 2 μL BSA (1×).
- H_2O to 20 μL.
- 20 μL total volume.

(g) Incubate digest reaction at 37 °C for 1 h.

(h) Dilute 20 μL digestion reaction with 40 μL water, load 20 μL diluted reaction on 1 % EX Gel alongside a 2-log DNA ladder. Excise the 3519 bp band and purify the linearized vector using a gel extraction kit. Elute the DNA from the column in 30 μL water. Measure the concentration with a NanoDrop.

7. Incorporate the 100 bp DNA fragment using isothermal assembly to create the final gRNA expression vector.

(a) In a PCR tube, combine the following:

- 100 ng double-stranded DNA product from **step 3**.
- 100 ng digested gRNA cloning vector from **step 4**.
- 10 μL Gibson Assembly Master Mix (2×).
- H_2O to 20 μL.
- 20 μL total volume.

(b) Place tubes in thermocycler, and run the following program:

Temperature	Time
50 °C	60 min
4–10 °C	∞

(c) Transform 5 μL of product into Top10 Chemically Competent *E. coli* cells. Thaw one aliquot of Top10 cells in ice for 10 min, and then add 5 μL of the isothermal-assembly reaction from the previous step and incubate on ice for 30 min. Heat-shock the cells at 42 °C, then return to ice for 2 min. Add 250 μL of room temperature S.O.C. medium and incubate in a 37 °C shaker for 1 h.

(d) Spread 100 μL of the transformation using sterilized glass beads onto a prewarmed LB agar-kanamycin (50 μg/mL) plate and incubate overnight at 37 °C.

(e) Pick ~10 colonies the next day. Culture each colony in 5 mL LB medium containing 50 μg/mL kanamycin overnight at 37 °C with shaking at 200 rpm.

(f) Use a plasmid miniprep kit to isolate DNA from the majority of each 5 mL culture. Elute DNA from each column with 30 μL water. Save the remainder of each culture by storing at 4 °C.

(g) Determine DNA concentration for each sample, and submit samples for Sanger sequencing using M13 Forward and M13 Reverse universal sequencing primers to determine transformants that have successfully incorporated the 100 bp DNA containing the target sequence.

(h) Grow a maxiprep culture of the correct transformant by inoculating 200 mL of LB medium containing 50 μg/mL kanamycin with 100 μL of the original miniprep culture. Grow overnight at 37 °C with shaking at 200 rpm.

(i) Isolate plasmid DNA using a plasmid maxi prep kit, resuspend DNA at ~1 μg/μL in water. Use this product for transfection.

3.3 gRNA Preparation from a Microarray for Multiplexible Genome Engineering

1. Design oligonucleotide library of a 200 nt oligonucleotide pool from Custom Array Inc. (Design provided in Supplementary Table 2: http://arep.med.harvard.edu/human_crispr/.)
2. Perform the first round of PCR to amplify the full oligonucleotide library.

3. Resuspend gRNA pool F1 and gRNA pool R1 primers at 100 mM in water.
4. In a PCR tube, add the following:
   (a) 10 ng pooled oligonucleotide library template.
   (b) 2.5 μL gRNA pool F1 primer (10 μM).
   (c) 2.5 μL gRNA pool R1 primer (10 μM).
   (d) 1.5 μL DMSO.
   (e) 25 μL 2× Phusion GC Buffer Master Mix.
   (f) Add H_2O to 50 μL.
   (g) 50 μL total volume.
5. Place tubes in thermocycler, and run the following program:

Temperature	Time	Number of cycles
98 °C	30 s	1
98 °C 70 °C 72 °C	10 s 30 s 10 s	25
72 °C	1 min	1
4 °C	∞	1

6. Purify the product using a PCR cleanup kit. Elute the DNA from each column with 30 μL water. Measure the concentration with a NanoDrop.
7. Perform the second round of PCR to specifically amplify the desired subpopulation of the oligonucleotide library.
   (a) The PCR primers for this step will be specific to the desired target(s) and consist of the 25 bp barcodes flanking the target sequence(s) (barcodes available in Supplementary Table 2 at http://arep.med.harvard.edu/human_crispr/). Resuspend primers at 100 mM in water before use.
   (b) In a PCR tube, add the following:
      - 10 ng PCR product from **step 2**.
      - 1 μL target-specific barcode forward primer (10 μM).
      - 1 μL target-specific barcode reverse primer (10 μM).
      - 12.5 μL 2× KAPA HiFi HotStart ReadyMix.
      - Add H_2O to 25 μL.
      - 25 μL total volume.

(c) Place tubes in thermocycler, and run the following program:

Temperature	Time	Number of cycles
95 °C	5 min	1
98 °C 65 °C 72 °C	20 s 15 s 30 s	30
72 °C	3 min	1
4–10 °C	∞	1

(d) Purify the product using a PCR cleanup kit. Elute the DNA from each column with 30 μL water. Measure the concentration with a NanoDrop.

8. Perform the final round of PCR to subtract the identifying barcode from each gRNA target sequence.

(a) Resuspend gRNA pool F2 and gRNA pool R2 primers at 100 mM in water.

(b) In a PCR tube, add the following:

- 10 ng pooled oligonucleotide library template.
- 2.5 μL gRNA pool F1 primer (10 μM).
- 2.5 μL gRNA pool R1 primer (10 μM).
- 1.5 μL DMSO.
- 25 μL 2× Phusion GC Buffer Master Mix.
- Add H_2O to 50 μL.
- 50 μL total volume.

(c) Place tubes in thermocycler, and run the following program:

Temperature	Time	Number of cycles
98 °C	30 s	1
98 °C 68 °C 72 °C	10 s 30 s 10 s	25
72 °C	1 min	1
4 °C	∞	1

Purify the product using a PCR purification kit. Elute the DNA from each column with 30 μL water.

(d) Linearize the gRNA cloning vector using AflII as described in Subheading 3.2.2, **step 4**.

(e) Incorporate the target-containing fragment from **step 4** into the linearized gRNA cloning vector via isothermal assembly and perform transformation as described in Subheading 3.2.2, **step 5**. Use this product for transfection.

3.4 Prepare hCas9 Plasmid

1. Obtain the hCas9 plasmid from Addgene (plasmid ID 41815) as a bacterial stab in agar.
2. Use the tip of a sterile pipet tip or toothpick to scrape the bacterial stock, streak onto an LB agar-ampicillin plate (100 μg/mL), and then incubate plate at 37 °C overnight.
3. Pick a single colony from the plate the next day to inoculate 200 mL of LB-ampicillin (100 μg/mL), then grow overnight in at 37 °C with shaking at 200 rpm.
4. Isolate plasmid DNA using a plasmid maxiprep kit, resuspend DNA at ~1 μg/μL in water. Use this product for transfection.

3.5 Transfection of 293 HEK Cells

3.5.1 Day 0: Plating 293 Cells for Transfection

1. Culture cells in 6 well plates until cells are healthy and at ~70 % confluency.
2. Gently wash cells with 2 mL room temperature PBS.
3. Aspirate PBS, add 1 mL TrypLE Express, and incubate at 37 °C for 2 min.
4. Resuspend cells with 3 mL prewarmed DMEM containing 10 % FBS, 1× P/S, and 1× NEAA (hereafter referred to as "complete DMEM media").
5. Count cells and calculate volume required for 0.1×10^6 cells (single transfection, scale as needed).
6. Place desired volume of cells into 15 mL centrifuge tube.
7. Spin at $200 \times g$ for 5 min at room temperature and aspirate supernatant.
8. Resuspend in 1 mL media.
9. Plate cells in a 12-well tissue culture plate, return to incubator.

3.5.2 Day 1: Transfection

1. Change media on cells to 1 mL fresh prewarmed complete DMEM media, return to incubator while preparing DNA mix.
2. Add 5 μL Lipofectamine 2000 to 50 μL Opti-MEM in a 1.5 mL microcentrifuge tube, flip several times to mix. Incubate the mixture in the room temperature for 5 min.
3. Add 1 μg Cas9 plasmid and 1 μg gRNA to 50 μL Opti-MEM in a 1.5 mL microcentrifuge tube.
4. Add diluted DNA to diluted Lipofectamine mixture from **step 2**, flick several times to mix.
5. Incubate 15 min at room temperature.
6. Add 100 μL mixture dropwise to cells.

7. Change media after 24 h to fresh prewarmed complete DMEM media.
8. Harvest cells 3 days after transfection, use desired method to assess targeting efficiency.

3.6 Transfection of PGP1 iPS Cells

3.6.1 Matrigel Preparation

1. Store frozen Matrigel at −80 until use.
2. Thaw Matrigel in the cold room over night before using.
3. Quickly aliquot 300 μL barely thawed Matrigel into cold 1.5 mL microcentrifuge tubes on ice. Of note, Matrigel will solidify if allowed to come to room temperature and will no longer dissolve in media.
4. Immediately store at −20 °C until use (good for 2 months).

3.6.2 Transfection

1. Culture PGP1 iPS cells in 6 well plates until healthy and at ~80 % confluency.
2. On day of transfection, prepare wells for transfected cells by coating with Matrigel.
   (a) Thaw a vial of Matrigel on ice.
   (b) Add 300 μL of cold Matrigel to 24 mL cold DMEM/F12 on ice.
   (c) Invert to mix.
   (d) Add 500 μL per well of 48-well plate (1 well per transfection), then leave the plate at room temperature for 1 h.
   (e) Aspirate Matrigel and replace with prewarmed 500 μL mTESR media with ROCK inhibitor (2 μL/mL) before plating transfected cells.
3. Change media to 2 mL prewarmed mTeSR media with ROCK inhibitor (2 μL/mL), 2 h before nucleofection.
4. Aspirate the media, wash cells with 2 mL room temperature PBS.
5. Aspirate PBS, add 1 mL TrypLE Express, incubate at 37 °C for 5 min.
6. Resuspend cells by gently pipetting up and down five times, place into 15 mL centrifuge tube, and add 10 mL mTeSR-1 followed by several rounds of gentle pipetting.
7. Count cells with cell counter and calculate volume required for 1×10^6 cells (single transfection, scale as needed).
8. Place desired quantity of cells in 15 mL centrifuge tube, spin at $200 \times g$ for 5 min at room temperature and aspirate supernatant.
9. Resuspend 1×10^6 cells in master mix of 21 μL, including the following:

(a) 16.4 μL P3.
(b) 3.6 μL Supplement 1.
(c) 1 μL Cas9 plasmid (1 μg/μL).
(d) ~21 μL total.

10. Add 1 μL of gRNA (1 μg/μL) to cell suspension mixture.
11. Quickly transfer cells into central chamber of one well of a Nucleocuvette strip, and place strip into the 4D-Nucleofector device.
12. Nucleofect cells using program CB150.
13. Quickly add 80 μL of prewarmed mTESR media containing ROCK inhibitor to each well of electroporated cells, pipet up and down 1–2× to mix.
14. Transfer cells from the strip to Matrigel-coated plate in mTeSR1 media with ROCK inhibitor (prepared in Subheading 3.6.2, **step 2**).
15. Centrifuge plate at 70 × *g* for 3 min at room temperature.
16. Place cells into incubator.
17. After 24 h, change to fresh mTeSR1 media without ROCK inhibitor.
18. Harvest cells 3 days after nucleofection and use desired method to assess targeting efficiency (*see* **Note 3**).

4 Notes

1. gRNA Design and Specificity.
Here, we describe a method for gRNA design for conducting human genomic targeting against sites with the pattern of 5′-N_{19}-NGG-3′. Despite the simplicity of the gRNA design and its broad applications, we emphasize the following caveats when one attempts to use this system to conduct genome engineering. The works of Jinek et al. [7] and Cong et al. [14] have indicated that the 13 nt immediately upstream of the PAM in the targeted sequence (N_{13} of 5′-N_{13}-NGG-3′) is primarily responsible for the specificity of Cas9-gRNA. Thus, the requirement of an NGG PAM motif adjacent to the targeting site and the limited numbers of nucleotides (13 + 2 = 15 bp) in gRNA which define specificity represent the primary constraints to choosing unique Cas9-gRNA targeting sites in the human genome. Additionally, a deeper understanding of the specificity of Cas9-gRNA is under current investigation [15]. We expect these rules to evolve as we further evaluate the specificity of Cas9-gRNA. It is important to note that because the nature of off-target events with this technology remains an open question, we strongly recommend that users screen for

potential off-target sites in the genome after obtaining edited monoclonal cells with the desired targeting event.

2. Different Assembly Methods.
   We present three different gRNA assembly methods. We suggest ordering synthesized gRNA expression constructs from a commercialized source (https://www.idtdna.com/pages/products/genes/gblocks-gene-fragments) as a convenient method, while Gibson Assembly starting with either 60mer ssODNs or 200-mer oligonucleotide pools is amenable to construction of multiple gRNA expression constructs.
3. Expected Targeting Efficiency and Toxicity in hiPSCs.
   The expected cell confluency of PGP1 hiPSCs is approximately 30 % on the second day after transfection. Additionally, the rate of mutation at the targeted site via NHEJ is approximately 1 % using this protocol [8]. However, the toxicity of transfection and targeting efficiency may vary from cell line to cell line. Thus, optimization of the amount of transfected DNA is recommended when applying this protocol to other human iPS cell lines.

Acknowledgements

We thank Church lab members and Alejandro De Los Angeles (George Daley's group, Harvard) for providing suggestions and feedbacks on the protocols. This work was supported by NIH grant P50 HG005550.

References

1. Wiedenheft B, Sternberg SH, Doudna JA (2012) RNA-guided genetic silencing systems in bacteria and archaea. Nature 482:331–338
2. Horvath P, Barrangou R (2010) CRISPR/Cas, the immune system of bacteria and archaea. Science (New York, NY) 327:167–170
3. Bhaya D, Davison M, Barrangou R (2011) CRISPR-Cas systems in bacteria and archaea: versatile small RNAs for adaptive defense and regulation. Annu Rev Genet 45:273–297
4. Barrangou R, Fremaux C, Deveau H et al (2007) CRISPR provides acquired resistance against viruses in prokaryotes. Science (New York, NY) 315:1709–1712
5. Grissa I, Vergnaud G, Pourcel C (2007) The CRISPRdb database and tools to display CRISPRs and to generate dictionaries of spacers and repeats. BMC Bioinformatics 8:172
6. Makarova KS, Haft DH, Barrangou R et al (2011) Evolution and classification of the CRISPR-Cas systems. Nat Rev Microbiol 9:467–477
7. Jinek M, Chylinski K, Fonfara I et al (2012) A programmable dual-RNA-guided DNA endonuclease in adaptive bacterial immunity. Science (New York, NY) 337:816–821
8. Mali P, Yang L, Esvelt KM et al (2013) RNA-guided human genome engineering via Cas9. Science (New York, NY) 339:823–826

9. Sapranauskas R, Gasiunas G, Fremaux C et al (2011) The Streptococcus thermophilus CRISPR/Cas system provides immunity in Escherichia coli. Nucleic Acids Res 39: 9275–9282
10. Urnov FD, Rebar EJ, Holmes MC et al (2010) Genome editing with engineered zinc finger nucleases. Nat Rev Genet 11:636–646
11. Bogdanove AJ, Voytas DF (2011) TAL effectors: customizable proteins for DNA targeting. Science (New York, NY) 333:1843–1846
12. Miyagishi M, Taira K (2002) U6 promoter-driven siRNAs with four uridine 3′ overhangs efficiently suppress targeted gene expression in mammalian cells. Nat Biotechnol 20:497–500
13. Brummelkamp TR, Bernards R, Agami R (2002) A system for stable expression of short interfering RNAs in mammalian cells. Science (New York, NY) 296:550–553
14. Cong L, Ran FA, Cox D et al (2013) Multiplex genome engineering using CRISPR/Cas systems. Science (New York, NY) 339:819–823
15. Jiang W, Bikard D, Cox D et al (2013) RNA-guided editing of bacterial genomes using CRISPR-Cas systems. Nat Biotechnol 31: 233–239

Chapter 17

RNA-Guided Genome Editing of Mammalian Cells

Neena K. Pyzocha, F. Ann Ran, Patrick D. Hsu, and Feng Zhang

Abstract

The microbial CRISPR-Cas adaptive immune system can be harnessed to facilitate genome editing in eukaryotic cells (Cong L et al., Science 339, 819–823, 2013; Mali P et al., Science 339, 823–826, 2013). Here we describe a protocol for the use of the RNA-guided Cas9 nuclease from the *Streptococcus pyogenes* type II CRISPR system to achieve specific, scalable, and cost-efficient genome editing in mammalian cells.

Key words CRISPR-Cas Genome editing DNA cleavage Cas9 Guide RNA PAM sequence NHEJ Gene knockout

1 Introduction

The ability to introduce targeted modifications into genomes and engineer model organisms holds enormous promise for biomedical and biotechnological applications. The development of programmable nucleases [1–9] has allowed targeting of specific genomic loci to introduce double-strand breaks (DSBs) in the DNA. These DSBs are subsequently repaired through either the error-prone nonhomologous end-joining (NHEJ) pathway or the homology-directed repair (HDR) pathway, allowing formation of indels or precise editing of the genome, respectively [10]. These endonucleases can be used for studies in basic biology, biotechnology, and medicine, including the development of reporter cell lines [11], transgenic organisms [12], disease models [13], and gene therapy [14], among others. Although ZFNs and TALENs can be reprogrammed to target specific DNA sequences, these tools still require time-consuming engineering of proteins de novo for each target, and there remains a deficit for technologies that are easily customizable, multiplexable, and affordable.

Francesca Storici (ed.), *Gene Correction: Methods and Protocols*, Methods in Molecular Biology, vol. 1114,
DOI 10.1007/978-1-62703-761-7_17,

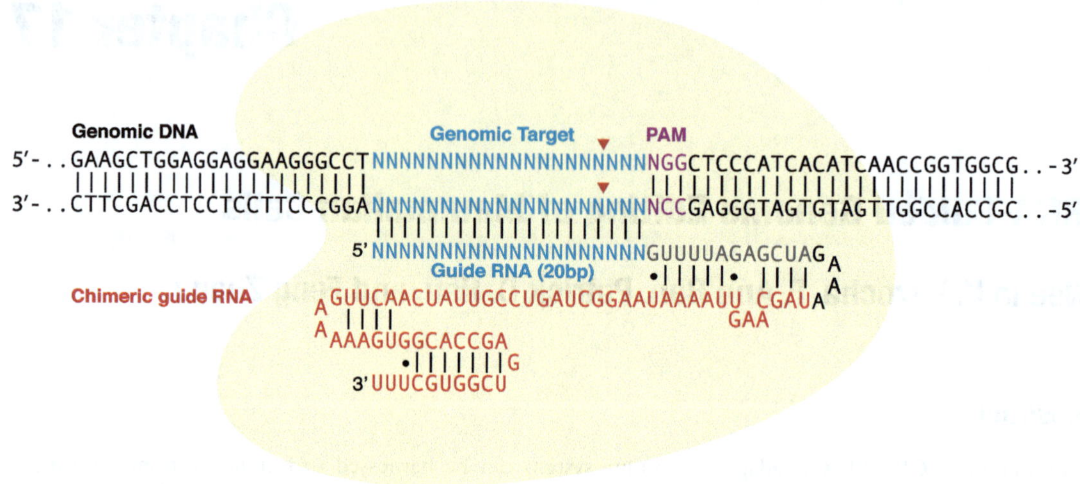

Fig. 1 Targeted DNA cleavage by SpCas9 in the human *EMX1* locus. The SpCas9 enzyme (*yellow*) interacts with its genomic target (*blue*) with the help of a guide RNA. The genomic target is directly 5′ to the PAM sequence, which is -NGG- for SpCas9. The guide RNA is composed of the guide sequence (*blue*), which anneals with the genomic target via Watson–Crick base pairing and a chimeric guide RNA scaffold consisting of a fusion between the crRNA (*gray*) and the tracrRNA (*red*)

The microbial adaptive immune system CRISPR (Clustered Regularly Interspaced Short Palindromic Repeats) consists of a set of enzymes and noncoding RNA elements [15–17]. Among the three types of CRISPR systems in bacteria and archaea [15, 16], type II requires only a single protein, Cas9 (formerly Csn1), to mediate DNA cleavage [18]. Cas9 is targeted to specific DNA sequences by a pair of noncoding RNA elements: the CRISPR RNA (crRNA), which carries the target-specifying guide sequence via Watson–Crick base pairing (Fig. 1), and the trans-activating crRNA (tracrRNA), which hybridizes with crRNA and is required for loading onto Cas9 [19, 20].

The type II CRISPR system of *Streptococcus pyogenes* can be reconstituted in mammalian cells to mediate DNA cleavage with three minimal components: Cas9, crRNA, and tracrRNA. The latter two components can further be truncated and fused into a single chimeric guide RNA scaffold (Fig. 1) [18] for a target sequence selected from any genomic locus with its 3' end followed by a NGG trinucleotide motif [19]. This protospacer-adjacent motif (PAM) is specific to each CRISPR system [21]. Generation of specific guide RNAs for targeted genome editing only requires the purchase of two short oligos and simple cloning that can take as little as two days.

The wild-type *S. pyogenes* Cas9 (SpCas9) enzyme has multiple endonuclease domains, two of which cleave DNA in a strand-specific manner. Two catalytic residues, D10 or H840 [18], can be mutated to convert the wild-type SpCas9 into a DNA-nicking enzyme (SpCas9n) [1, 18]. Given that single-stranded nicks in the

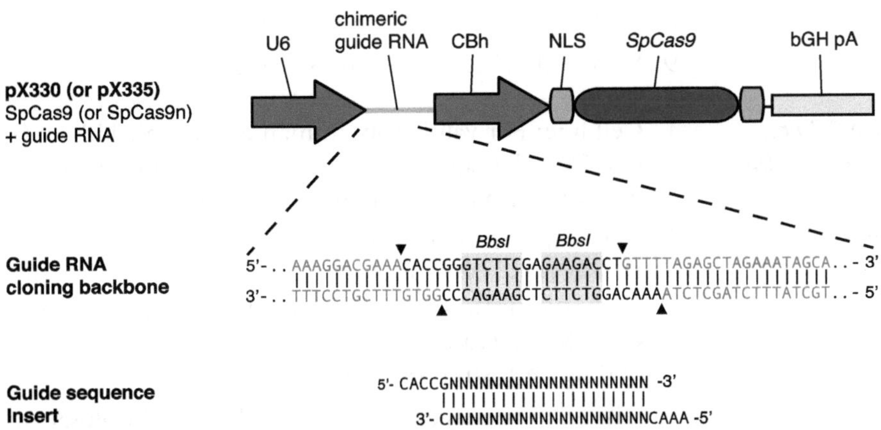

Fig. 2 Bicistronic expression vector for guide RNA and SpCas9 (or SpCas9n). A genomic target directly upstream to the PAM sequence can be cloned into the expression vector. After a target is selected, two DNA oligos can be designed based on the schematic showing the guide sequence insert. One oligo (*top* strand, written 5′–3′) contains ligation adapter sequences for cloning into the expression vector and $G(N)_{19}$, which is the selected genomic target sequence. The other oligo (*bottom* strand, written 3′–5′) also contains ligation adapter sequences for cloning into the expression vector and the complementary bases to the genomic target sequence. Once annealed and phosphorylated, the oligos can be inserted into the vector digested with *BbsI*

target DNA can also stimulate HDR, SpCas9n reduces the likelihood of error-prone repair by NHEJ. Furthermore, both catalytic domains of SpCas9 can be mutated to convert SpCas9 into a RNA-guided DNA-binding protein [18, 22]. This chapter describes a set of protocols for using the SpCas9 system for genome editing in mammalian cells.

2 Materials

2.1 Molecular Cloning Components

1. Cloning plasmids: *pX330* (CBh::SpCas9 + U6::chimeric guide RNA) (Addgene) or *pX335* (CBh::SpCas9n (D10A) + U6:: chimeric guide RNA) (Addgene) (Fig. 2).
2. Oligos for target sequence. *See* Subheading 3.1 for discussion regarding locus selection and Subheading 3.2 on oligo design (Integrated DNA Technologies).
3. Restriction enzymes and phosphatase: FastDigest *Bbs*I (Fermentas), FastAP (Fermentas), 10× FastDigest Buffer (Fermentas) (*see* **Note 1**).
4. QIAquick Gel Extraction Kit (QIAGEN).
5. Phosphorylation, annealing, and ligation reagents: 10× T4 Ligation Buffer (NEB), T4 Polynucleotide Kinase (NEB), 2× Quick Ligation Reaction Buffer (NEB), Quick Ligase (NEB).
6. Plasmid-Safe exonuclease (Epicentre Biotechnologies).
7. Competent cells and bacterial growth reagents.

8. QIAGEN Plasmid Midi Kit (QIAGEN).
9. Standard gel electrophoresis reagents.

2.2 Tissue Culture, Transfection, and DNA Extraction Components

1. Cell line: For validation, human embryonic kidney (HEK) cell line 293FT (Life Technologies). For additional discussions on working with other cell lines, *see* **Note 2**.
2. Cell culture reagents for maintenance of 293FT cells: Dulbecco's Modified Eagle's Medium (DMEM) (Life Technologies), 10 % fetal bovine serum (HyClone), 2 mM GlutaMAX (Life Technologies), 100 U/mL penicillin, and 100 μg/mL streptomycin.
3. Dissociation reagent: TrypLE™ (Life Technologies).
4. Transfection reagent: Lipofectamine 2000 (Life Technologies) for HEK293FT or Neuro-2a cells (Sigma Aldrich) (*see* **Note 3**).
5. 24-well tissue culture plates (Corning).
6. Transfection Control Plasmid: pMaxGFP (Lonza).
7. QuickExtract™ DNA extraction kit (Epicentre Biotechnologies).

2.3 Components for the Analysis of Genome Modification

1. SURVEYOR Mutation Detection Kit (Transgenomic).
2. 4–20 % Novex TBE polyacrylamide gels (Life Technologies).
3. Amplification primers specific to the targeted locus (Integrated DNA Technologies).
4. Herculase II High Fidelity Polymerase (Agilent).

3 Methods

3.1 Target Selection

For use with the SpCas9 system, target sites must be followed by a NGG trinucleotide motif on the 3′ end (*see* **Notes 4** and **5**).

3.2 Construct Design

We designed cloning vectors (pX330 for SpCas9 or pX335 for SpCas9n, a D10A nickase) to aid co-expression of SpCas9 and guide RNA in mammalian cells (Fig. 2). In this vector, SpCas9 is driven by the CBh promoter [23], and the guide RNA is driven by the human PolIII promoter U6. Phosphorylated and annealed oligos (design indicated in Fig. 2) can be cloned into the *BbsI* digested plasmid containing the entire guide RNA scaffold. The oligos are designed based on the target site sequence (20 bp sequence corresponding to the target site). The $G(N)_{19}$ refers to the sequence selected upstream of the PAM sequence in the genomic DNA (*see* **Note 6**). Create oligos using the schematic in Fig. 2.

3.3 Molecular Cloning: Oligo Annealing and Cloning into Backbone Vectors

1. Digest 1 μg of pX330 or pX335 with *Bbs*I for 30 min at 37 °C:

1 μg	pX330 or pX335
1 μL	FastDigest *Bbs*I (Fermentas)
1 μL	FastAP (Fermentas)
2 μL	10× FastDigest Buffer
X μL	ddH_2O
20 μL	Total

2. Gel purify digested pX330 or pX335 using QIAquick Gel Extraction Kit and elute in EB.
3. Phosphorylate and anneal each pair of oligos for the insert piece:

1 μL	Oligo 1 (100 mM)
1 μL	Oligo 2 (100 mM)
1 μL	10× T4 Ligation Buffer (NEB)
6.5 μL	ddH_2O
0.5 μL	T4 PNK (NEB)
10 μL	Total

Anneal in a thermocycler using the following parameters:

37 °C	30 min (for addition of 5′ phosphates)
95 °C	5 min and then ramp down to 25 °C at 5 °C/min

4. Set up ligation reaction and the negative control. Incubate at room temperature for 10 min:

X μL	*Bbs*I digested pX330 or pX335 from **step 2** (50 ng)
1 μL	Phosphorylated and annealed oligo duplex from **step 3** (1:200 dilution)
5 μL	2× Quick Ligation Buffer (NEB)
X μL	ddH_2O
10 μL	Subtotal
1 μL	Quick Ligase (NEB)
11 μL	Total

5. (Optional but highly recommended) Treat ligation reaction with Plasmid-Safe exonuclease (*see* **Note 7**):

11 μL	Ligation reaction from **step 4**
1.5 μL	10× Plasmid-Safe buffer
1.5 μL	10 mM ATP
0.7 μL	ddH_2O
0.3 μL	Plasmid-Safe DNAse
15 μL	Total

Incubate reaction at 37 °C for 30 min.

6. Transform 2 μL of reaction from **step 5** into competent cells and plate on ampicillin selection plates.
7. Pick two colonies the following day and analyze for correct insertion of the target sequence oligos.

3.4 Cell Culture and Transfection

The CRISPR-Cas DNA cleavage system has been validated for use in a variety of mammalian cell lines [1, 2, 24] (*see* **Note 2**). The protocol below is for HEK239FT cells.

1. HEK293FT cells are maintained in DMEM supplemented with 10 % fetal bovine serum and passaged before reaching 70 % confluency. Cells are maintained in an incubator set at 37 °C supplemented with 5 % CO_2.
2. HEK293FT cells can be transfected using Lipofectamine 2000 according to the manufacturer's protocol.
3. For each well of a 24-well plate, a total of 500 ng of plasmid is transfected. One well should be a control to see the relative transfection efficiency using a plasmid such as pmaxGFP.
4. After 12 h of transfection, replace the medium with pre-warmed maintenance medium. After 72 h, genomic DNA can be isolated using the QuickExtract DNA extraction kit following the manufacturer's protocol. Briefly, cells are resuspended in QuickExtract solution (50 μL per 24 well) and incubated at 65 °C for 15 min followed by 98 °C for 10 min.

3.5 Analysis of Genomic Modification: SURVEYOR and Sequencing

1. The efficiency of cleavage can be detected by assessing the percentage of cells containing indels in the target region (*see* **Note 8**). In order to detect indels in the DNA, follow the instructions provided in the SURVEYOR Mutation Detection Kit manual (*see* **Note 9**).
2. It is recommended that the SURVEYOR Nuclease digestion products are analyzed on a PAGE gel.

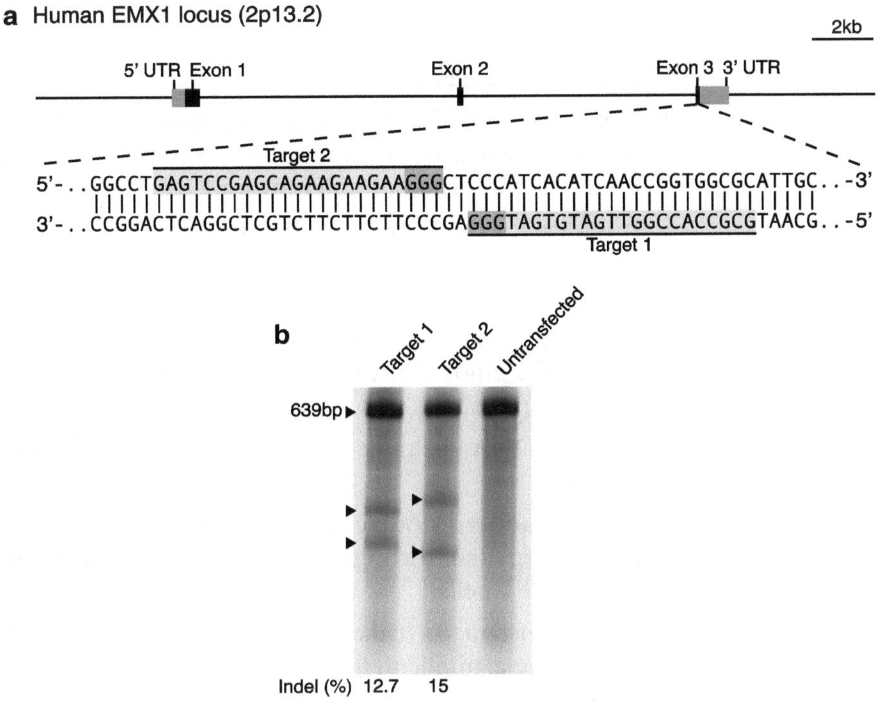

Fig. 3 SURVEYOR assay comparing SpCas9-mediated DNA cleavage at two different targets in the same gene. (**a**) The third exon of the human *EMX1* locus was targeted using guide RNAs at two unique sites. (**b**) A representative SURVEYOR assay gel image comparing the targeted cleavage efficiency by SpCas9 at the two targets in the human *EMX1* locus

3. To calculate the percent cutting efficiency of a CRISPR locus, use the following formula: $\%indel = \left(1 - \sqrt{1 - \frac{(a+b)}{(a+b+c)}}\right)100$

   where a and b refer to the relative concentrations of the cut bands and c equals the relative concentration of the full-length PCR template. A representative SURVEYOR gel image and quantitation is shown in Fig. 3.

4 Notes

1. Conventional restriction enzymes can be substituted for FastDigest restriction enzymes. In this case, adjust digestion reagents and digestion times according to manufacturer's protocol.
2. Experimental conditions may need to be optimized for each cell line.

3. For other cell lines, we suggest doing an initial comparison of different transfection reagents (e.g., FuGENE HD, nucleofection, and TransIT).
4. A free computational resource maintained by the Zhang lab (http://www.genome-engineering.org) contains the most up-to-date information relevant for Cas9 systems.
5. It is ideal for these targets to be unique within the genome. We also recommend testing multiple target sites for each gene and selecting the most effective target.
6. Selecting a target site with a 5′ G allows for efficient transcription of the guide RNA from the U6 promoter.
7. Plasmid-Safe treatment is recommended because it degrades linear dsDNA, helping to prevent unwanted recombination products.
8. SpCas9-induced double-strand breaks in the target DNA are usually repaired through the error-prone NHEJ process in HEK293FT cells.
9. It is important to make sure that the genomic PCR primers yield a single amplicon for reliable quantification of the percent cutting efficiency. In the case that primers do not yield a single amplicon, the PCR product needs to be gel purified or new primers should be designed.

Acknowledgments

We thank Randall Platt for comments and members of the Zhang Lab for discussion, support, and advice. N.P. is supported by the National Science Foundation Graduate Research Fellowship, Primary Award #1122374. P.D.H. is a James Mills Pierce Fellow. F.Z. is supported by the NIH Transformative R01 Award (R01-NS073124); the NIH Director's Pioneer Award (DP1-MH100706); the Keck, McKnight, Gates, Damon Runyon, Searle Scholars, Merkin, Klingenstein, and Simons Foundations; Bob Metcalfe; Mike Boylan; and Jane Pauley. Sequence and reagent information are available through http://www.genome-engineering.org.

References

1. Cong L et al (2013) Multiplex genome engineering using CRISPR/Cas systems. Science 339:819–823
2. Mali P et al (2013) RNA-guided human genome engineering via Cas9. Science 339:823–826
3. Urnov FD et al (2005) Highly efficient endogenous human gene correction using designed zinc-finger nucleases. Nature 435:646–651
4. Porteus MH, Baltimore D (2003) Chimeric nucleases stimulate gene targeting in human cells. Science 300:763

5. Miller JC et al (2007) An improved zinc-finger nuclease architecture for highly specific genome editing. Nat Biotechnol 25:778–785
6. Christian M et al (2010) Targeting DNA double-strand breaks with TAL effector nucleases. Genetics 186:757–761
7. Miller JC et al (2011) A TALE nuclease architecture for efficient genome editing. Nat Biotechnol 29:143–148
8. Sanjana NE et al (2012) A transcription activator-like effector toolbox for genome engineering. Nat Protoc 7:171–192
9. Stoddard BL (2005) Homing endonuclease structure and function. Q Rev Biophys 38: 49–95
10. Rouet P, Smih F, Jasin M (1994) Expression of a site-specific endonuclease stimulates homologous recombination in mammalian cells. Proc Natl Acad Sci U S A 91:6064–6068
11. Hockemeyer D et al (2011) Genetic engineering of human pluripotent cells using TALE nucleases. Nat Biotechnol 29:731–734
12. Cui X et al (2011) Targeted integration in rat and mouse embryos with zinc-finger nucleases. Nat Biotechnol 29:64–67
13. Ding Q et al (2013) A TALEN genome-editing system for generating human stem cell-based disease models. Cell Stem Cell 12:238–251
14. Maier D et al (2013) Efficient clinical scale gene modification via zinc finger nuclease targeted disruption of the HIV co-receptor CCR5. Hum Gene Ther 24:245–258
15. Deveau H, Garneau JE, Moineau S (2010) CRISPR/Cas system and its role in phage-bacteria interactions. Annu Rev Microbiol 64:475–493
16. Garneau JE et al (2010) The CRISPR/Cas bacterial immune system cleaves bacteriophage and plasmid DNA. Nature 468:67–71
17. Horvath P, Barrangou R (2010) CRISPR/Cas, the immune system of bacteria and archaea. Science 327:167–170
18. Jinek M et al (2012) A programmable dual-RNA-guided DNA endonuclease in adaptive bacterial immunity. Science 337:816–821
19. Deltcheva E et al (2011) CRISPR RNA maturation by trans-encoded small RNA and host factor RNase III. Nature 471:602–607
20. Jore MM et al (2011) Structural basis for CRISPR RNA-guided DNA recognition by Cascade. Nat Struct Mol Biol 18:529–536
21. Mojica FJ, Diez-Villasenor C, Garcia-Martinez J, Almendros C (2009) Short motif sequences determine the targets of the prokaryotic CRISPR defence system. Microbiology 155: 733–740
22. Qi LS et al (2013) Repurposing CRISPR as an RNA-guided platform for sequence-specific control of gene expression. Cell 152:1173–1183
23. Gray SJ et al (2011) Optimizing promoters for recombinant adeno-associated virus-mediated gene expression in the peripheral and central nervous system using self-complementary vectors. Hum Gene Ther 22:1143–1153
24. Cho SW, Kim S, Kim JM, Kim JS (2013) Targeted genome engineering in human cells with the Cas9 RNA-guided endonuclease. Nat Biotechnol 31:230–232

Chapter 18

Nuclease-Mediated Double-Strand Break (DSB) Enhancement of Small Fragment Homologous Recombination (SFHR) Gene Modification in Human-Induced Pluripotent Stem Cells (hiPSCs)

R. Geoffrey Sargent, Shingo Suzuki, and Dieter C. Gruenert

Abstract

Recent developments in methods to specifically modify genomic DNA using sequence-specific endonucleases and donor DNA have opened the door to a new therapeutic paradigm for cell and gene therapy of inherited diseases. Sequence-specific endonucleases, in particular transcription activator-like (TAL) effector nucleases (TALENs), have been coupled with polynucleotide small/short DNA fragments (SDFs) to correct the most common mutation in the cystic fibrosis (CF) transmembrane conductance regulator (CFTR) gene, a 3-base-pair deletion at codon 508 (delF508), in induced pluripotent stem (iPS) cells. The studies presented here describe the generation of candidate TALENs and their co-transfection with wild-type (wt) CFTR-SDFs into CF-iPS cells homozygous for the delF508 mutation. Using an allele-specific PCR (AS-PCR)-based cyclic enrichment protocol, clonal populations of corrected CF-iPS cells were isolated and expanded.

Key words Sequence-specific endonucleases, TALENs, SFHR, Gene targeting, Cystic fibrosis, iPS cells, Clonal isolation, Nucleofection, Polynucleotides, SDFs

1 Introduction

Sequence-specific modification of genomic DNA in human pluripotent and multipotent stem cells is a strategy that would promote the development of autologous cell and gene therapies for inherited diseases, aging associated diseases, and regenerative medicine. While classical gene-targeting protocols, utilizing cellular DNA repair pathways, have been successfully used for engineering pluripotent mouse embryonic stem cells (mESCs) and immortalized human cell lines, there has been only limited success with human pluripotent stem cell (hPSC) systems for sequence-specific genomic modifications. Classical gene-targeting protocols typically involve the transfection of cells with a plasmid-base gene-targeting vector that is

Francesca Storici (ed.), *Gene Correction: Methods and Protocols*, Methods in Molecular Biology, vol. 1114, DOI 10.1007/978-1-62703-761-7_18,

homologous to a genomic target sequence and uses drug selection enrichment strategies to identify and isolate cells that have the desired genomic modification. This approach is, however, highly inefficient where approximately 10^{-6}–10^{-7} of the surviving cells contain the desired genomic modification. In addition, the classical gene-targeting vectors contain drug-selectable marker genes and other nonnative DNA sequences that are therapeutically undesirable.

Small fragment homologous replacement (SFHR) is one alternative approach that avoids some of the limitations associated with classical gene targeting [1–3]. Polynucleotide small DNA fragments (SDFs) used as donor DNA for the SFHR-mediated, sequence-specific homologous exchange are typically about 200–1,000 base pairs and can be prepared in large quantities by polymerase chain reaction (PCR) amplification. SFHR does not rely on drug selection and thereby avoids the introduction of nonnative DNA sequences. SFHR has been previously shown to successfully modify genomic DNA in mouse ESCs [4] (Gruenert and Emamekhoo, unpublished data), immortalized and primary human epithelial cell line [1, 2, 5–7], hematopoietic stem cells [8, 9], peripheral blood monocytes [10], human lymphoblasts [11], mouse T cells [12], and myoblasts [13–15] of a number of target genes, at efficiencies approaching 1–10 %.

Homologous recombination (HR) in hPSCs, and other mammalian cells, can be dramatically improved by introducing double-strand breaks (DSBs) in the region of the targeted genomic DNA sequences. This was initially demonstrated using a model system based on the yeast I-SceI homing endonuclease to introduce DSBs at ectopic sites in rodent and human cellular genomic DNA [16–18]. However, the I-SceI endonuclease recognition sequences are not naturally found in human genomic DNA and have, therefore, not proven to be a useful therapeutic tool that can be readily implemented to create site-specific double-strand breaks at any desired DNA sequence.

Recent advances in the development of sequence-specific chimeric endonucleases have demonstrated that fusion of the nuclease domain of the FokI endonuclease to DNA sequence-specific zinc-finger or transcription activator-like effector (TALE) binding motifs can generate hybrid zinc-finger nucleases (ZFNs) or transcription activator-like effector nucleases (TALENs) that introduce DSBs in specific DNA regions [19–24]. The ZF or TALE amino acid motifs are designed as pairs to bind specific target DNA sequences 5′ and 3′ to the proposed FokI cut site. Since the FokI endonuclease is only active as a dimer, the inactive monomeric FokI nuclease domains fused to the 3′ end of each ZF or TALE binding pair will generate DSBs in mammalian cells at specific DNA target sequences where the nuclease domains are in close proximity to dimerize. In this chapter, we describe TALEN enhancement of SDF-potentiated correction of the delF508 mutation in exon 11 of the CFTR gene in human-induced pluripotent stem cells (hiPSCs).

2 Materials

2.1 Tissue Culture

1. *Pluripotent cell lines*: CF-induced pluripotent stem (CF-iPS) cell line, CF1-iPS4, homozygous for the delF508 mutation (Suzuki, Sargent, and Gruenert, unpublished data).
2. *Media*: TeSR1 (StemCell, Inc, Vancouver, BC, Canada).
3. *Growth substrates*: Matrigel (BD Biosciences, San Jose, CA).
4. *Cell dissociation*: Accutase (Life Technologies, Grand Island, NY), Dispase (StemCell, Inc).
5. *Culture reagents*: Rho-associated kinase (ROCK) inhibitor Y27632 (Sigma, St Louis, MO).
6. *Tissue culture plates*: 24-well plate (Corning/Costar, Corning, NY), 60 mm tissue culture dish (Falcon/Becton Dickenson Labware, Franklin Lakes, NJ).

2.2 Preparation of Polynucleotide Small DNA Fragments (SDFs)

1. *DNA sequence*: CFTR genomic sequences [25, 26].
2. *Genomic and plasmid DNA*: Human genomic DNA or plasmid containing human wild-type (wt) CFTR DNA sequence [6, 7].
3. *PCR oligonucleotide primers*: Table 1.
4. *PCR buffer and polymerase*: MyTaq HS Mix, 2× (Bioline, Taunton, MA).

Table 1
PCR primers used for generating analytical or preparative products from genomic and genomic DNA-derived PCR templates are indicated

Primer	Sequence (5′–3′)	Product (bp)
CF1 (f) CF5 (r)	GCAGAGTACCTGAAACAGGA CATTCACAGTAGCTTACCCA	491 (wt)/488 (delF508)
CF1B (f)	CCTTCTCTGTGAACCTCTATCA	
CF7C (r)	ATAGGAAACACCAAAGATGA	392
CF8C (r)	ATAGGAAACACCAATGATAT	389
CF17 (f)	GAGGGATTTGGGGAATTATTTG	
CF7C (r)	ATAGGAAACACCAAAGATGA	330
CF8C (r)	ATAGGAAACACCAATGATAT	327

Primer pair CF1/CF5 was used to generate preparative quantities of SDF by PCR. Analysis of the wtCFTR and delF508-CFTR alleles in genomic DNA was performed by AS-PCR with primer pairs CF1B/CF7C (wt) and CF1B/CF8C (delF508). Primers CF7C and CF8C are allele specific and common to both DNA and RNA analyses; however, when paired with the non-allele-specific primer CF17 in exon10, amplification is specific for RNA-derived cDNA (AS-RT-PCR) crosses intron/exon boundaries. Forward primer = f, Reverse primer = r

5. *Agarose gel electrophoresis*: Agarose—Genetic Analysis Grade (Thermo Fisher Scientific, Waltham, MA), ethidium bromide 1 % (10 mg/mL) (Thermo Fisher Scientific).

2.3 TALEN Preparation

1. Golden Gate TALEN plasmid kit (Addgene, Cambridge, MA).
2. Plasmid backbone (MR015, Porteus and Rahdar, unpublished data).

2.4 Electroporation/Nucleofection

1. Amaxa 4D-Nucleofector X apparatus (Lonza, Allendale, NJ).
2. P3 Primary Cell 4D-Nucleofector Kit L with pmaxGFP (Lonza).

3 Methods

3.1 Tissue Culture

1. The CF1-iPS4 cell line and corrected clones are cultured on Matrigel-coated tissue culture plates using TeSR1 medium (StemCell, Inc) with daily medium changes.
2. Cultures of iPS cells are routinely subcultured for maintenance and expansion, or for clonal purification, by enzymatic dissociation of individual iPS cell colonies with Dispase (StemCell, Inc) or by mechanical isolation. Mechanical isolation is typically used to enrich clonal populations for undifferentiated hiPSC colonies and/or pick candidate corrected clones (*see* **Notes 1** and **2**).
3. Cells are replated onto Matrigel-coated culture plates grown in fresh TeSR1 medium as described above.
4. The passage number for the CF1-iPS4 hiPS cell line and corrected clones are designated as $Pn_1 \times n_2 \times n_3 - n_z$, where n_1 = the number of passages as a primary culture before reprogramming, n_2 = the number of passages the CF1-iPS4 cell line is grown after reprogramming, and n_3 = the number of passages clones are grown after transfection/nucleofection with SDF, where the z in n_z = the number of manipulations/treatments/modifications that the cells have undergone since their isolation [27] (*see* **Notes 3** and **4**).

3.2 Small DNA Fragment (SDF) Preparation

1. The wtCFTR SDF (491z-SDF), capable of correcting the delF508 mutation, is generated by PCR amplification with primer pair CF1/CF5 (Table 1) using the p491z-plasmid DNA as template [2, 6, 7].
2. The PCR amplification conditions for generating the wtCFTR 491z-SDF are as follows: A 50 μL reaction solution containing 1.0 μM of each primer, MyTaq HS Mix (Bioline), and 0.02 ng p491z-plasmid DNA is amplified with an initial denaturation for 2 min at 95 °C, followed by denaturation at 95 °C for 30 s; annealing at 55 °C for 30 s; and extension at 72 °C for 1 min for 35 cycles, with a final extension of 3 min at 72 °C (Table 2).

Table 2
PCR amplification conditions for DNA and RNA indicating the denaturation, annealing, and extension temperatures and times as well as amplification cycle number

Step	Temperature (°C)	Time	Cycles
SDF and non-AS-PCR amplification conditions (CF1/CF5)			
Initial denaturation	95	2 min	1
Denaturation	95	30 s	35
Annealing	55	30 s	
Extension	72	1 min	
Final extension	72	3 min	1
AS-PCR amplification conditions (CF1B/CF7C or CF8C)			
Initial denaturation	95	2 min	1
Denaturation	95	30 s	35
Annealing	60	30 s	
Extension	72	1 min	
Final extension	72	8 min	1
AS-RT-PCR amplification conditions (CF17/CF7C or CF8C)			
Reverse transcription	42	1 h	
Initial denaturation	95	2 min	1
Denaturation	95	30 s	35
Annealing	52	30 s	
Extension	72	1 min	
Final extension	72	8 min	1

3. The 491z-SDF is separated from the p491z plasmid template by agarose gel electrophoresis and purified using a silica-based DNA purification protocol [28, 29] (*see* **Note 5**).
4. A second round of PCR amplification is carried out with 2 pg of the 491z-SDF as template. The amplified 491z-SDF is then purified using silica-based purification as indicated above.

3.3 TALEN Preparation

1. CFTR exon 11-specific TALENs are designed using the Web-based software, TALE-NT 2.0, https://boglab.plp.iastate.edu/ [23], and the following sequences were selected: TALEN pairs CFTAL-1B, 5′-T CTCAGTTTTCCTGGATTAT; spacer, gcctggcaccattaaagaa; CFTAL-2B, AATATCATTGGTGTTT CCT A-3′ (*see* **Note 6**).
2. Plasmids for expression of the CFTR-B TALENs are constructed using the Golden Gate TALEN assembly method [21] with the Golden Gate TALEN plasmid kit (Addgene) (*see* **Notes 7–10**).
3. A novel plasmid backbone (MR015, Porteus and Rahdar, unpublished data) can be used for optimal expression of the CFTR-B TALENs in mammalian cells.

3.4 Enhancement of SFHR-Mediated CFTR Correction by TALENs

1. The CF-iPS cells are pretreated with 10 μM of the Rho-associated kinase inhibitor, Y27632 (Sigma), for at least 2 h and harvested as a single cell suspension by treatment with Accutase (Life Technologies).
2. The wild-type 491z-SDFs alone or in the presence of 1 μg of each CFTAL-1B and CFTAL-2B TALEN expression vector are introduced into CF1-iPS4 cells by Amaxa nucleofection (electroporation) (Lonza) [30] to correct the genomic delF508 mutation.
3. Two doses of 491z-SDFs, 10^7 SDFs/cell (4.32 μg) and 2×10^7 SDFs/cell (8.64 μg), are introduced into ~8×10^5 cells with the Amaxa 4D-Nucleofector X apparatus using Solution P3 (solution:supplement = 82:18) and Program CB150 (Lonza). Duplicate nucleofections are carried out for each SDF amount.
4. Cells from duplicate electroporations are mixed and then plated into two wells of a 24-well plate (Corning/Costar) coated with Matrigel, with TeSR1 medium containing 10 μM Y27632 (*see* **Note 11**).
5. A sample of cells is also nucleofected with a GFP expression plasmid to monitor nucleofection efficiency. The GFP nucleofection control is evaluated by fluorescence microscopy 24 and 48 h post-electroporation to determine the approximate nucleofection efficiencies.
6. At 3 days post-nucleofection cells in one well of a nucleofection duplicate are dissociated with Accutase to assess the presence of wtCFTR sequences to determine whether the correction is successful. If successful, the other well of the duplicate is dissociated with Dispase and distributed in approximately equal numbers into 12 wells of a 24-well plate coated with Matrigel (*see* **Note 12**).
7. The enrichment process (see below) is initiated by isolating genomic DNA from nucleofecte cells 7–9 days post-nucleofection from the each well of the 12 wells of the 24-well plate containing cells and assaying for the relative amounts of wtCFTR by allele specific (AS)-PCR amplification with primer pairs CF1B/CF7C (wt) or CF1B/CF8C (delF508) [6, 7]. PCR products are assessed by electrophoretic banding on a 2 % agarose gel and visualized after staining with ethidium bromide (*see* **Note 13**).

3.5 Isolation of Corrected CF-iPS Cells

1. The isolation of clonal populations of corrected cells requires initial enrichment by a cyclic assessment protocol that involves redistribution of approximately equal numbers of the cells with the highest wtCFTR signal into multiple wells of a 24-well plate (Fig. 1).

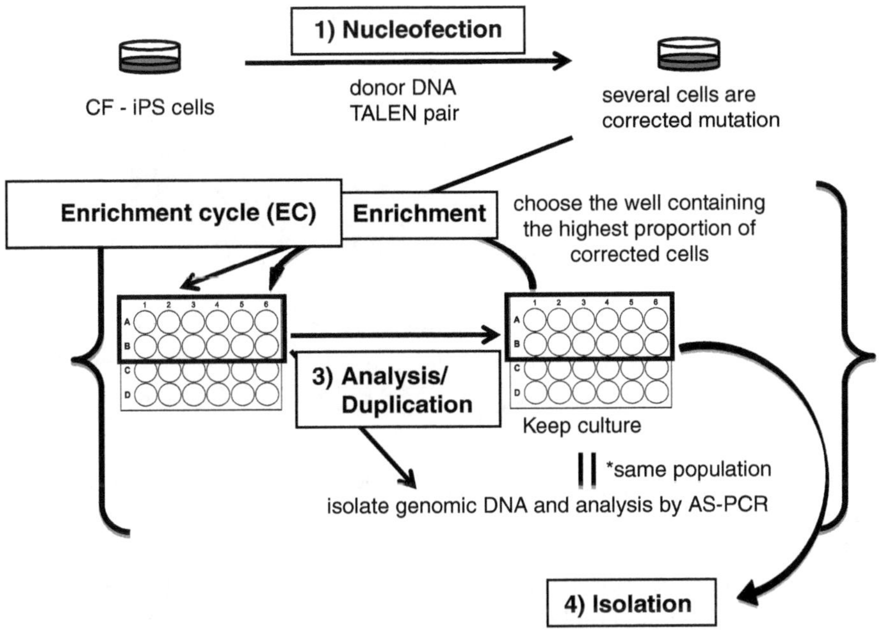

Fig. 1 Enrichment protocol to isolate gene-corrected clones. *nucleofection*: Donor DNA is nucleofected in the presence or absence of TALENs. *Enrichment*: After determining that there is a population of corrected cells, the cells in a well are dissociated with Dispase and equally distributed as cell clumps into 12 wells of a 24-well plate. *Analysis/duplication*: Cells are harvested with Accutase as single cells to maintain the same ratio of corrected:uncorrected cells within the population for analysis by AS-PCR and subculture into a duplicate 24-well plate, i.e., an aliquot of cells from each well is placed into the same well of another 24-well plate. *Enrichment cycle*: The well determined to have the highest proportion of corrected cells is then dissociated with Dispase, and the cell clumps are equally distributed into 12 wells of a new 24-well plate and is followed by another *analysis/duplication* step that constitutes the beginning of the next cycle in the enrichment cycle (EC) scheme. *Isolation*: After the *n*th enrichment cycle (EC_n) when the proportion of corrected cells is approximately 10 %, the cells are dispersed with Accutase and plated as single cells to isolate individual colonies

2. The initial duplicate well of the 24-well plate containing wtCFTR positive CF-iPS cells (see above) is dissociated with Dispase and then distributed in approximately equal numbers into 12 wells of another 24-well plate coated with Matrigel.
3. After 5–7 days when the cells reach confluence, they are again treated with 10 μM Y27632 for ~2 h before dissociation with Accutase. Approximately, 5/6th of the cells are assayed for wtCFTR by AS-PCR with primer pair CF1B/CF7C. A replicate 24-well plate of the parent plate is created with the remaining 1/6th of the cells. The well that shows the highest concentration of corrected cells by semiquantitative AS-PCR is then harvested by Dispase dissociation at confluence (~3–4 days after replating in the replicate plate) and equally distributed into 12 wells of a new 24-well plate (*see* **Note 14**).

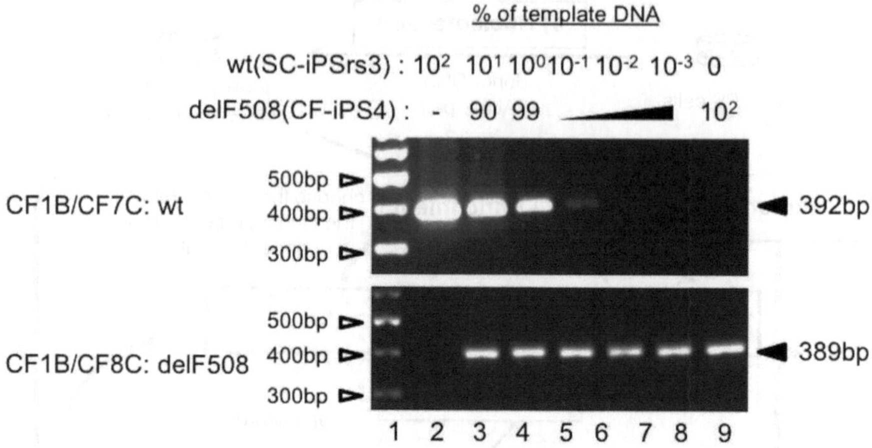

Fig. 2 AS-PCR titration analysis for sensitivity of detection. PCR control titration analysis involves the mixing genomic DNA from a nonCF-iPS cell line (SC-iPSrs3) with CF1-iPS4 at varying proportions. The products from the AS-PCR amplification with primer pair CF1B/CF7C (wt) or pair CF1B/CF8C (delF508) are electrophoretically separated on a 2 % agarose gel and visualized by ethidium bromide staining. The intensity of the band represents the proportion of the wtCFTR DNA present within each mixture and reflects the sensitivity of the AS-PCR assay. The marker DNA (*Lane 1*) was the GeneRuler 100 bp plus DNA ladder (Thermo Fisher Scientific, Waltham, MA)

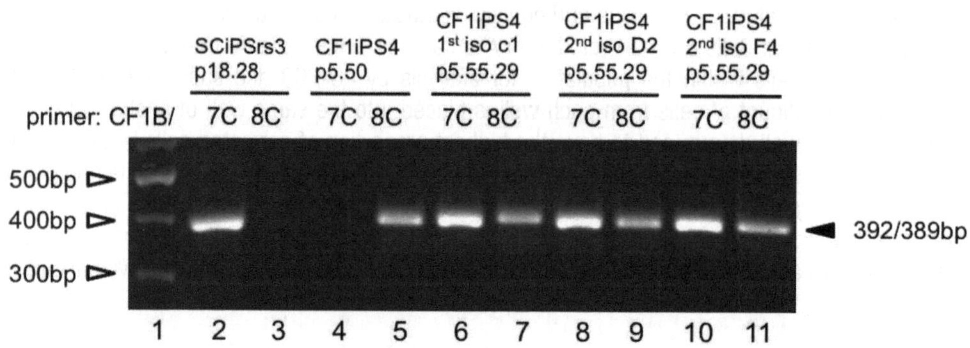

Fig. 3 Isolation of corrected clones: AS-PCR analysis performed with primer pairs CF1B/CF7C (wtCFTR) or CF1B/CF8C (delF508 CFTR) on genomic DNA isolated from three independent colonies following limited dilution (*Lanes 6–11*). The control wtCFTR and homozygote delF508 band patterns were shown using genomic DNA of a nonCF-iPS cell line (SC-iPSrs3) and CF1-iPS4 (delF508) (*Lanes 2–5*). GeneRuler 100 bp plus DNA ladder (Thermo Fisher Scientific) (*Lane 1*)

4. The steps outlined in **step 3** are repeated until approximately 10 % of the cells are wtCFTR, as determined by semiquantitative AS-PCR (Fig. 2). The well with the highest degree of enrichment is then pretreated with 10 μM Y27632 for ~2 h and harvested with Accutase. The Accutase-treated, single cell suspension is replated on a 60 mm dish coated with Matrigel for the isolation of individual clones of cells that are examined by AS-PCR to identify cells having either a heterozygous or homozygous wtCFTR genotype (Fig. 3) (*see* **Notes 15** and **16**).

4 Notes

1. Subculturing of iPS cells is generally carried out when the cells are subconfluent (30–40 % confluence) and before individual colonies of cells begin to touch each other. Individual iPS cell colonies are isolated by mechanical removal or by enzymatic dissociation with Dispase or Accutase. Manual isolation of iPSC colonies can be very precise, and it is possible to identify and isolate those portions of the cells that appear morphologically identical to the cells that appeared in the initial colony isolate. It is, however, time-consuming and tedious. It is important to note the colony growth rate when choosing which colonies to isolate. Colonies that grow rapidly can be problematic and karyotypically unstable. These colonies should therefore be avoided if possible.
2. Enzymatic dissociation is generally more time efficient than mechanical isolation but requires attention to detail. If Accutase is used, pre- and posttreatment with Y27632 is required to enhance viability.
3. The cell passage nomenclature is a bookkeeping tool, where each passage number that is followed by a period represents a specific phenotype, when the cells are exposed to something that will alter the cellular phenotype, e.g., reprogramming factors. At this point the number begins again and increases with each subculture. If the cells are again exposed to a reagent and/or treatment, e.g., mutation correction with an SDF, another period is added, and the numbering is reinitiated and increases with each subculture until the next treatment, etc.
4. Cells need to be evaluated and screened periodically for their karyotype and their pluripotence features to ensure that they are still pluripotent and genetically stable at a macroscopic level.
5. To eliminate any residual p491z-plasmid DNA remaining in the amplification solution, the SDF is gel purified.
6. The TALEN pairs are designed to overlap the mutant target sequences to minimize the potential to cut the donor SDF and the corrected genomic DNA.
7. The Golden Gate assembly method involves an initial digestion of several different RVD plasmids (<10 RVD plasmids in the Golden Gate Kit) and the backbone plasmid for the first Golden Gate reaction followed by the ligation of different RVD combinations into the backbone plasmid. The RVD plasmids chosen depend on the target nucleotide sequence. Since most target sequences are from 15 to 30 nucleotides, this initial reaction needs to be carried out 2–3 times with different RVD combinations and different backbone vectors.

8. The next step in this part of the Golden Gate assembly involves harvesting individual white clones and their analysis by colony PCR and sequencing to identify the clone with the correct RVD combination.
9. A second Golden Gate reaction is carried out with the first reaction products, the RVD plasmid that identifies the final nucleotide (3′ end) in the target sequence, and a backbone vector that can be expressed in human cells. This mixture is digested and ligated to generate the final expression construct. The correct targeting constructs are identified as indicated in **Note 8** with an additional diagnostic restriction digestion.
10. This process needs to be carried for each side of the target FokI endonuclease cut site that is within the spacer sequence (15–30 nucleotides) between the individual TALEN binding sites.
11. The duplicate electroporations are important for establishing the initial duplicate cultures of nucleofection cells for analysis and for enrichment. Mixing the cells from the individual transfections normalizes the nucleofection efficiency for the individual samples.
12. While the analytical well of cells is dissociated with Accutase, the dispersed well is dissociated with Dispase. Dissociation with Dispase maintains the cells as clumps and results in higher cell survival overall.
13. The serial enrichment of populations with wtCFTR is facilitated by choosing the wells with the highest wtCFTR levels as assessed by AS-PCR. The dispersal as cell clumps is likely to favor the distribution of groups of cells that are clonal expansions of individual corrected cells and improve overall viability.
14. Assessment of the relative amount of wtCFTR DNA in a well can be determined by comparison to a PCR-based standard titration analysis (*see* Fig. 2). Wild-type and delF508 cells are mixed at different ratios, and the DNA of the mixed populations is used as standards for PCR analysis. The sensitivity of the AS-PCR is limiting, and it is difficult to visualize an ethidium bromide signal below 0.05 %. It is, however, possible to differentiate between individual wells if the proportion of wtCFTR cells is ≥0.05 %.
15. The number of enrichment cycles is dependent on the efficiency of gene correction in the initial analysis. Enrichment to 5–10 % makes the isolation of an individual corrected clone straightforward when the cells are dissociated and grown as individual cells.
16. After corrected CF-iPS cell-derived clones are isolated, they need to be further characterized to ensure pluripotence and karyotypic stability and to demonstrate restoration of wtCFTR function.

Acknowledgments

We would like to thank Dr Michael Yezzi and Carissa Tasto for their expert technical assistance. This study was supported by NIH grants GM075111 AS4 and DK088760; a grant from Cystic Fibrosis Research, Inc; funds from Pennsylvania Cystic Fibrosis, Inc (PACFI); and a grant from the Japan Society for the Promotion of Science (SS).

References

1. Gruenert DC et al (2003) Sequence-specific modification of genomic DNA by small DNA fragments. J Clin Invest 112(5):637–641
2. Kunzelmann K et al (1996) Gene targeting of CFTR DNA in CF epithelial cells. Gene Ther 3(10):859–867
3. Sargent RG, Kim S, Gruenert DC (2011) Oligo/Polynucleotide-based gene modification: strategies and therapeutic potential. Oligonucleotides 21(2):55–75
4. Sangiuolo F et al (2008) Cftr gene targeting in mouse embryonic stem cells mediated by small fragment homologous replacement (SFHR). Front Biosci 13:2989–2999
5. Goncz KK et al (2001) Expression of DeltaF508 CFTR in normal mouse lung after site-specific modification of CFTR sequences by SFHR. Gene Ther 8(12):961–965
6. Goncz KK, Gruenert DC (2000) Site-directed alteration of genomic DNA by small-fragment homologous replacement. Methods Mol Biol 133:85–99
7. Goncz KK, Kunzelmann K, Xu Z, Gruenert DC (1998) Targeted replacement of normal and mutant CFTR sequences in human airway epithelial cells using DNA fragments. Hum Mol Genet 7(12):1913–1919
8. Goncz KK et al (2006) Small fragment homologous replacement-mediated modification of genomic beta-globin sequences in human hematopoietic stem/progenitor cells. Oligonucleotides 16(3):213–224
9. Goncz KK, Prokopishyn NL, Chow BL, Davis BR, Gruenert DC (2002) Application of SFHR to gene therapy of monogenic disorders. Gene Ther 9(11):691–694
10. McNab GL, Ahmad A, Mistry D, Stockley RA (2007) Modification of gene expression and increase in alpha1-antitrypsin (alpha1-AT) secretion after homologous recombination in alpha1-AT-deficient monocytes. Hum Gene Ther 18(11):1171–1177
11. Bedayat B et al (2010) Sequence-specific correction of genomic hypoxanthine-guanine phosphoribosyl transferase mutations in lymphoblasts by small fragment homologous replacement. Oligonucleotides 20(1):7–16
12. Zayed H, McIvor RS, Wiest DL, Blazar BR (2006) In vitro functional correction of the mutation responsible for murine severe combined immune deficiency by small fragment homologous replacement. Hum Gene Ther 17(2):158–166
13. Kapsa R et al (2001) In vivo and in vitro correction of the mdx dystrophin gene nonsense mutation by short-fragment homologous replacement. Hum Gene Ther 12(6):629–642
14. Kapsa RM et al (2002) Targeted gene correction in the mdx mouse using short DNA fragments: towards application with bone marrow-derived cells for autologous remodeling of dystrophic muscle. Gene Ther 9(11):695–699
15. Todaro M et al (2007) Effective detection of corrected dystrophin loci in mdx mouse myogenic precursors. Hum Mutat 28(8):816–823
16. Choulika A, Perrin A, Dujon B, Nicolas JF (1995) Induction of homologous recombination in mammalian chromosomes by using the I-SceI system of Saccharomyces cerevisiae. Mol Cell Biol 15(4):1968–1973
17. Rouet P, Smih F, Jasin M (1994) Expression of a site-specific endonuclease stimulates homologous recombination in mammalian cells. Proc Natl Acad Sci U S A 91(13):6064–6068
18. Sargent RG, Brenneman MA, Wilson JH (1997) Repair of site-specific double-strand breaks in a mammalian chromosome by homologous and illegitimate recombination. Mol Cell Biol 17(1):267–277
19. Bibikova M, Beumer K, Trautman JK, Carroll D (2003) Enhancing gene targeting with designed zinc finger nucleases. Science 300(5620):764
20. Bibikova M et al (2001) Stimulation of homologous recombination through targeted cleavage by chimeric nucleases. Mol Cell Biol 21(1):289–297

21. Cermak T et al (2011) Efficient design and assembly of custom TALEN and other TAL effector-based constructs for DNA targeting. Nucleic Acids Res 39(12):e82
22. Clark KJ, Voytas DF, Ekker SC (2011) A TALE of two nucleases: gene targeting for the masses? Zebrafish 8(3):147–149
23. Doyle EL et al (2012) TAL effector-nucleotide targeter (TALE-NT) 2.0: tools for TAL effector design and target prediction. Nucleic Acids Res 40(Web Server issue):W117–W122
24. Porteus MH, Baltimore D (2003) Chimeric nucleases stimulate gene targeting in human cells. Science 300(5620):763
25. Xu Z, Gruenert DC (1996) Human CFTR gene sequences in regions flanking exon 10: a simple repeat sequence polymorphism in intron 9. Biochem Biophys Res Commun 219(1):140–145
26. Zielenski J et al (1991) Genomic DNA sequence of the cystic fibrosis transmembrane conductance regulator (CFTR) gene. Genomics 10(1):214–228
27. Gruenert DC et al (1988) Characterization of human tracheal epithelial cells transformed by an origin-defective simian virus 40. Proc Natl Acad Sci U S A 85(16):5951–5955
28. Boom R et al (1990) Rapid and simple method for purification of nucleic acids. J Clin Microbiol 28(3):495–503
29. Boyle JS, Lew AM (1995) An inexpensive alternative to glassmilk for DNA purification. Trends Genet 11(1):8
30. Maurisse R et al (2010) Comparative transfection of DNA into primary and transformed mammalian cells from different lineages. BMC Biotechnol 10:9

Chapter 19

AAV-Mediated Gene Editing via Double-Strand Break Repair

Matthew L. Hirsch and R. Jude Samulski

Abstract

Traditionally, the ability to edit the mammalian genome was inhibited by the inherent low efficiency of homologous recombination (HR; approximately <1 in a million events) and the inability to deliver DNA efficiently to dividing and non-dividing cells/tissue. Despite these limitations, creative selections designed over 20 years ago, clearly demonstrated the powerful implications of gene knock-in and knockout technology for the genetic engineering of mice (Doetschman et al. Nat 330(6148): 576–578, 1987; Thomas and Capecchi. Cell 51(3): 503–512, 1987). The development and application of recombinant vectors based on adeno-associated virus (rAAV) have helped to overcome both of the initial limitations regarding DNA delivery and the frequency of HR. Considering DNA delivery, rAAV infects non-dividing and dividing cultured cells as well as most tissues in mouse and larger animal models (including humans). At the DNA editing level, rAAV genomes have been reported to increase the frequency of HR several orders of magnitude by serving as the repair substrate (Russell and Hirata. Nat Genet 18(4): 325–330, 1998). However, reports on the ability of rAAV genomes to stimulate HR, compared to plasmid DNA and oligonucleotides, are variable, and many labs have found it necessary to augment the frequency of rAAV-induced HR using site-specific endonucleases (Ellis et al. Gene Ther, 2012; Hirsch et al. Gene Ther 17(9): 1175–1180, 2010; Porteus et al. Mol Cell Biol 23(10): 3558–3565, 2003; Radecke et al. Mol Ther 14(6): 798–808, 2006). In this protocol, we describe a method to perform rAAV-mediated double-strand break (DSB) repair for precise genetic engineering in human cells.

Key words Adeno-associated virus, Vector production, Gene editing, Double-strand break repair, Self-complementary AAV

1 Introduction

1.1 Endonuclease Selection for AAV-Mediated DSB Gene Editing

Several endonuclease platforms allow DNA site-specific recognition and strand cleavage events to stimulate DSB repair [5–11]. The design of such enzymes, and their cognate targets, is beyond the scope of this chapter, and alternatively, they are also available for purchase from a variety of manufacturers [12, 13]. However, "designer" endonuclease selection for AAV-mediated gene editing is restricted by the packaging capacity of the viral capsid (≤5 kb) and therefore dependent upon the specific application. For most cases, delivery

Francesca Storici (ed.), *Gene Correction: Methods and Protocols*, Methods in Molecular Biology, vol. 1114, DOI 10.1007/978-1-62703-761-7_19, © Springer Science+Business Media, LLC 2014

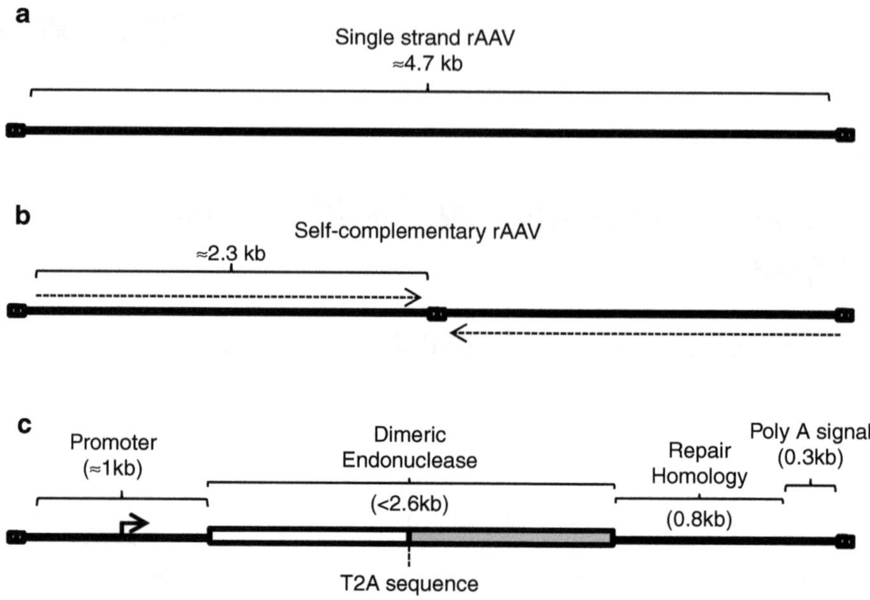

Fig. 1 Recombinant AAV genome formats for gene editing. (**a**) A cartoon depicting a single-strand DNA AAV genome. For these vectors it is important to keep the genome size >3.2 kb yet <4.8 kb. In all panels, the *bicolored squares* represent inverted repeats necessary for replication and packaging. (**b**) A single-strand self-complementary AAV genome is depicted with the genomic restriction of >1 kb yet < than 2.4 kb. The *dotted arrows* indicate the transgenic polarity in attempts to indicate that the scAAV genome has the capacity to form a duplexed molecule. (**c**) A single-strand AAV context is depicted in which a dimeric endonuclease expression cassette and a DNA repair substrate are included while conforming to the AAV packaging restraint. A T2A sequence allow the production of a bicistronic message encoding distinct ZFNs

of the desired endonuclease and the repair substrate to individual cells is necessary, and a single vector format has been described in Fig. 1 [1]. This AAV genome configuration can be employed for single chain or dimeric endonuclease coding sequences, in general, reaching 2.5 kb in size. This context is ideally suited for AAV vectors as it also allows a packaging remainder of approximately 2.2 kb that needs to accommodate a promoter, poly-adenylation signal, a T2A signal (when using a dimeric endonuclease) as well as the repair substrate (Fig. 1c) [1]. Plasticity still exists in the size of these additional sequences as they may be slightly altered without significantly affecting the overall rate of AAV-mediated DSB gene editing (i.e., the length of the repair substrate). Alternatively, the use of multiple AAV vector transduction has demonstrated success in vitro and in vivo eliminating the concern of endonuclease size and or the length of repair sequence homology [2, 3]. In any case, the packaging limitation imposed by the AAV capsid presents a balancing act between all the necessary components, as well as the AAV genome type, single strand or self-complementary, packaged within the capsid shell (discussed below) [14].

1.2 Viral Capsid Selection for AAV-Mediated DSB Gene Editing

Variations in the composition of the AAV protein capsid result in altered transduction efficiencies based on the target cell type [15–17]. Currently, the most studied serotypes are AAV 1–9 which demonstrate dramatic differences in cell tropism. As AAV vectors transduce both dividing and non-dividing cells, this also holds true for tissue transduction following in vivo administration [17]. Therefore, AAV capsid selection for DSB-mediated gene editing should be tailored to the specific target cell type. Numerous reports are available characterizing AAV capsid transduction efficiencies in vitro, ex vivo, and in vivo to aid in capsid selection [15–17]. In general, the AAV2 capsid works well in transformed cell lines, while AAV1 and 6 transduce stem cells most efficiently [1, 18]. In mouse models, AAV1 and 6 are most efficient following muscle injections, while AAV8 and AAV9 result in widespread transduction following intravenous injections or intraperitoneal administration to young animals [16, 19] (data not shown).

1.3 Viral Genome Considerations for rAAV-Mediated DSB Gene Editing

The production of distinct AAV particles that harbor either the endonuclease or the repair substrate, in contrast to the single vector format depicted in Fig. 1c, is often useful at the mechanistic level and has been employed in vivo (i.e., to determine factors limiting transduction) [20]. In these cases, it is also important to keep in mind that AAV genomes can be packaged in two different forms: (1) single-strand DNA (ssAAV; as discussed above) and (2) self-complementary DNA (scAAV) which is also technically single-strand DNA but can form a large stem-loop structure based on intra-strand base pairing [14, 21] (Fig. 1a, b). The advantage of scAAV is that a duplexed molecule can be generated, thereby eliminating the host-mediated second strand DNA synthesis requirement of ssAAV [14]. This enhancement is an important consideration at the levels of endonuclease production and the ability of the repair substrate to function in HR. In regard to endonuclease production levels, scAAV results in earlier and more robust endonuclease production (>tenfold) compared to ssAAV [2]. Considering the AAV genome as a substrate in the actual repair event, the scAAV context was reported to be modestly enhanced, despite an initial conclusion from an indirect experiment which reported that scAAV (dimer) genomes do not contribute to gene targeting [2, 22]. Although a ssAAV genome dependency for gene targeting represents an attractive hypothesis, as a single strand would not require strand separation and could directly strand invade, it remains at odds with a report that directly compared ssAAV and scAAV as substrates in this process [2]. A unique limitation of scAAV vectors is that due to the region of self-complementary sequence, this vector context imposes an additional size constraint on the packaged genome (≤2.4 kb) [14] (Fig. 1b). As such, the utilization of either ssAAV or scAAV genomes in rAAV-mediated gene editing is often based on the overall vector size of the included elements. For instance, the ssAAV genome format depicted in Fig. 1c relies a T2A sequence to allow the synthesis of two ZFNs from a single promoter, and also encodes a repair substrate, to conform to the packaging capacity of ssAAV (<5 kb). This arrangement ensures that all cells containing a nuclease also contain the desired repair substrate and is particularly suited for situations in which the transduction may not be complete (i.e., in vivo applications). However, when transduction is very efficient, such as cultured cells, one may prefer a multiple vector transduction regimen to vary the amount of nuclease and/or repair substrate. In these instances, distinct scAAV vectors can be used for an approximate tenfold increase in nuclease production and, as a repair substrate, modestly increase gene editing twofold.

In the following protocol, the production and application of an ssAAV genome encoding all the necessary requirements for zinc-finger nuclease (ZFN) induced DSB-mediated gene editing is described.

2 Materials

2.1 AAV-ITR Plasmid Construct Generation and Verification

All of the necessary plasmid constructs for rAAV production are offered by the University of North Carolina Vector Core facility.

1. The adenovirus helper plasmid pXX680. This construct contains adenovirus genes that assist rAAV production including activation of the AAV p5 promoter [23].
2. An AAV helper plasmid that encodes AAV replication (Rep) genes and one of several AAV capsid serotype coding sequences (often termed the pXR series) [24].
3. An AAV-ITR plasmid (pAAV-ITR) having the desired sequence to be packaged between the AAV ITR sequences. The AAV-ITR plasmid may contain two WT ITRs for ssAAV production (pSub201) or one WT and one mutant ITR for scAAV production (pHpa-trs-SK; Fig. 1a, b; *see* **Notes 1** and **2**) [14, 25].
4. A universal ssAAV context to express both ZFNs and a repair substrate has also been described for gene editing purposes (Fig. 1c) [1].
5. Restriction endonuclease digestion for construct generation relies on standard protocols provided by the enzyme manufacturer.
6. Agarose gels for verification/separation of DNA species should follow standard running and detection procedures (generally 1 % agarose in tris-acetate–EDTA buffer is sufficient).
7. DNA agarose gel extraction kits (Qiagen) are routinely used following the company's provided protocol.

8. For DNA ligations, T4 ligase is often used according to the manufacturer's protocol.
9. For bacterial transformations of the ligation reaction, a number of approaches can be used with varying efficiencies. We typically use electroporation relying on the Bio-Rad micropulser and 0.2 mm electroporation cuvettes. Generally, 1 μl of the ligation reaction is mixed with 30 μl of electrocompetent DH10b cells (Invitrogen) thawed on ice. If the plasmid contains the AAV ITRs, then electrocompetent SURE cells (Agilent) are necessary as they are deficient for homologous recombination which helps to preserve the structured ITRs. Then, the mixed is pulsed at 2.5 kV as recommended by the manufacturer.
10. Following electroporation, cells recover for 1 h at 37° with shaking in 0.5 ml LB, then generally 20–100 μl of the recovery is plated on nutrient broth plates containing the appropriate antibiotic which is dictated by the resistance gene on the transformed plasmid.
11. DNA mini- and maxi-preparation kits are commercially available and used to recover plasmids from clonal isolates using standard procedures (described by the manufacturer).
12. Restriction endonuclease digestion, which varies based on the cloning strategy and sequence, is used to screen plasmid isolates for sequencing reactions (carried out by the UNC genome analysis facility). In addition, XmaI (or SmaI) restriction endonuclease digestion is necessary to verify the integrity the ITRs of pAAV-ITR following the desired cloning strategy (following mini-preps and maxi-preps). All ITRs contain two XmaI (or SmaI) sites that need to be maintained for efficient vector production.

2.2 Transfection of HEK293 Cells for rAAV Production

1. HEK293 cells (ATCC) are maintained at 37 °C in a 5 % CO_2 incubator.
2. DMEM (Cellgro) supplemented with 10 % FBS (Gibco) and 1× penicillin–streptomycin (Gibco) is used at the HEK293 growth medium.
3. DMEM without any additives is also required for the transfection reaction.
4. Transfection reagent: Polyethylenimine (PEI) linear MW 25,000 (Polysciences). 1 mg of PEI is dissolved in 1 ml of 1× PBS adjusted to a pH between 4 and 5 with 12 N HCl.

2.3 Isolation of Transfected Cell Nuclei

1. Hypotonic buffer (500 ml) 10 mM HEPES, 1.5 mM MgCl2, 10 mM KCl, 175 mg of spermine; filter-sterilize.
2. 10× restore buffer (200 ml) 62.5 % sucrose (wt/vol) in hypotonic buffer filter-sterilized.
3. 15 ml Kontes homogenizer.

2.4 rAAV Vector Purification by Cesium Chloride Gradient Centrifugation

1. DNase digestion buffer: 10 mM Tris–HCl pH 7.5, 10 mM $MgCl_2$, 50 U/ml DNase I (made fresh).
2. Cesium chloride (CsCl), optical grade (Budenheim Gallard-Schlesinger).
3. Sonicator (Branson Sonifier 250, VWR Scientific).
4. Beckman Quick-Seal polyallomer (16 mm × 76 mm) centrifuge tubes and tube sealer.
5. Ultracentrifuge and fixed angle rotor to accommodate tube size.

2.5 rAAV Genome Detection in CsCl Gradient Fractions by Q-PCR

1. Q-PCR reagents including buffers, enzymes, primers, and detection methods (i.e., SYBR green or probe-based detection) are described by the manufacturer. AAV genome amplification relies on standard cycling protocols/considerations.
2. Q-PCR machine.

2.6 rAAV Dialysis and Final Titer Determination

1. 1× phosphate buffered saline (PBS).
2. Slide-A-Lyzer Dialysis Cassette with 7,000 molecular weight cutoff and 0.5–3 ml capacity (Pierce). We routinely use a 21 gauge needle and a 5 ml syringe to inject and remove the viral sample from the cassette.
3. DNase I digestion buffer (above).
4. 0.5 M EDTA.
5. Proteinase Ksolution: 1 M NaCl, 1 % (wt/vol) N-laurylsarcosine (Sarkosyl).
6. 100 mg/ml proteinase K.
7. Q-PCR reagents and machine (above).

2.7 rAAV Genome Characterization

1. DNase I digestion buffer.
2. 0.5 M EDTA.
3. 10 % SDS.
4. 6× loading dye for alkaline gel electrophoresis: 0.4 M NaOH, 5 mM EDTA, 18 % Ficoll (wt/vol), xylene cyanol (enough to add color for easy loading).
5. Alkaline gel (50 ml): Dissolve 1 g agarose in 50 ml water (1 %) and add 300 μl of 5 N NaOH and 50 μl of 0.5 M EDTA.
6. Alkaline gel running buffer: 50 mM NaOH, 1 mM EDTA in water.
7. Alkaline transfer buffer: 0.4 M NaOH.
8. 3MM Whatman chromatography paper.
9. 0.45 mm nylon membrane (i.e., GE Hybond-XL).
10. UV cross-link device.
11. 6× SSC buffer 0.9 M NaCl, 90 mM sodium citrate.

12. Low-stringency wash buffer 2× SSC, 0.1 % SDS (wt/vol).
13. High-stringency wash buffer 0.1× SSC, 0.1 % SDS (wt/vol).
14. Random primed radio-nucleotide labeling kit (Roche).
15. X-ray film.

2.8 rAAV Cellular Infection

1. HEK293 cells maintained as described above.
2. Tittered rAAV prep (viral genomes/vol).
3. 2 % paraformaldehyde in 1× PBS.
4. Flow cytometer.

3 Methods

3.1 AAV Plasmid Construct Generation

Recombinant single-strand DNA genomes packaged in the AAV protein capsid are normally derived from a plasmid sequence flanked by AAV inverted terminal repeats (pAAV-ITR vector; Fig. 2). Such plasmids are commercially available from a variety of sources as well as from the lab that initially created it [24]. Standard cloning procedures (not detailed herein) are used to place a sequence of interest ≤≈5 kb (the AAV capsid packaging capacity) into pAAV-ITR. As the AAV-ITRs undergo "processing" by bacterial DNA repair proteins during replication, a recombination deficient *E. coli* strain is preferred for plasmid production/purification (i.e., SURE cells

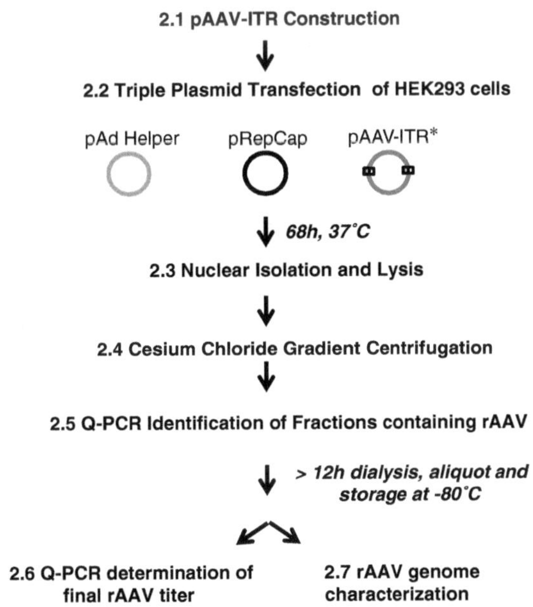

Fig. 2 AAV vector production flowchart. The *number* preceding each *bolded* statement indicates the location within the chapter describing that part of the procedure

from Agilent). Following pAAV-ITR purification, it is then necessary to confirm the integrity of the AAV-ITRs in vitro using restriction endonuclease digestion at sites found within the AAV ITRs. Generally, SmaI or XmaI enzyme digestion is performed according to standard procedures following all plasmid preparations.

For the specific purpose of AAV-mediated gene correction, a vector format has been generated that allows endonuclease and DNA repair substrate delivery in a single AAV particle (Fig. 1c) [1]. For instance, a universal expression cassette that allows AAV-mediated delivery of two ZFN coding sequences and up to a 1 kb repair substrate has been generated that conforms to the AAV capsid packaging capacity of 5 kb [1]. To accommodate this limitation, a ribosomal skipping sequence (T2A) was placed between the two ZFN coding sequences on a bicistronic message [1]. This genetic configuration allowed a 1 kb noncoding repair substrate, positioned upstream of the poly-adenylation sequence, to be included in the packaged genome (Fig. 1c). In a specific published example, which employs the UbqC constitutive promoter, the DNA sequence between the AAV-ITRs on the plasmid vector is 4.7 kb, thus presenting no packaging concerns [1]. This contextual arrangement can be used for ZFNs and other homing endonuclease as long as the ≤5 kb capsid packaging capacity is not exceeded. In the case of larger nucleases, such as TAL effector nucleases, it is currently necessary to use multiple pAAV-ITR vectors, for multiple vector production, purification, and transduction strategies.

3.2 Transfection of Adherent HEK293 Cells for rAAV Production

1. Adherent HEK293 cells are maintained in a 15 cm culture dish in DMEM supplemented with 10 % FBS and 1× pen–strep.
2. 24 h prior to plasmid transfection, two confluent plates are split 1–3 and grown for 24 h to obtain an approximate confluency of 70 % at the time of transfection.
3. rAAV production relies on a triple transfection of the plasmids listed in Subheading 2.1 (Fig. 2). The amount needed per 15 cm plate is 12 μg of pXX680 (Adenovirus helper), 10 μg of pXR (AAV Replication and Capsid genes), and 8 μg of pAAV-ITR (containing the sequence to be packaged).
4. For each plate, the plasmids are then diluted in 500 ml of DMEM medium containing no supplements and 110 μl of PEI transfection reagent is added to the solution.
5. Following the addition of PEI, the mixture is immediately vortexed for 10 s and allowed to complex at room temperature for 5 min.
6. Then, the solution is added directly to the plate medium in a manner that allows uniform dispersal.
7. Following the transfection, the cells are cultured under standard conditions for ≈68 h prior to harvest for rAAV particle purification (*see* **Note 3**).

3.3 Isolation of Cell Nuclei Post-transfection

1. After ≈68 h post-transfection, harvest and pool cells from the 15 cm plates. This can be done simply using the force generated from a 10 ml pipette or with a cell scraper if preferred. Then, pellet the cells by centrifugation at 200×*g* for 5 min at 4 °C. Discard the growth medium and wash the cell pellet in cold PBS, then pellet the cells again using the same conditions.
2. Resuspend the cell pellet in a volume of hypotonic buffer approximately five times the packed cell volume. Incubated on ice for 10 min.
3. Add 0.11 volumes of 10× restore buffer and mix gently.
4. Homogenize the cell solution in a 15 ml Kontes homogenizer via 12 strokes.
5. Pellet the cell nuclei by centrifugation at 500×*g* for 10 min, routinely performed in a 50 ml conical tube.

3.4 AAV Vector Purification by Cesium Chloride (CsCl) Gradient Centrifugation

Many different purification methods exist for rAAV production [24]. Herein a simple, serotype-independent, cost-effective strategy is described which is based on a CsCl density gradient established by centrifugation of the nuclear extract derived from the transfected cells. This method relies on the known buoyant density of AAV containing a 4.7 kb genome (1.41 g/cm^3). AAV vector recovery using CsCl centrifugation is not the cleanest purification scheme as indicated by silver stains of viral preps also exhibiting contaminating cellular proteins. However, it has the advantage of separating AAV capsids containing genomes from "empty" particles (no packaged genomes), which are assembled during particle production. Empty particles compete for cell surface receptor binding, thus decreasing the overall transduction efficiency of the viral prep, and may generate unwanted immunological concerns in vivo. An extensive review focusing specifically of established rAAV production/characterization methods is available for alternatives not presented herein [24].

1. Resuspend nuclear pellets in cold ddH$_2$0 to a final volume of 11.6 ml and disrupt the nuclei for 50 pulses using a Branson Sonifier (duty cycle of 50, output of 5) on ice.
2. Next, add 100 μl of the DNase solution (10 mg/ml) to the disrupted nuclei solution and incubate at 37 °C for 1 h to remove any "non-packaged" DNA.
3. Following the DNase digestion reaction, add 6.5 g of CsCl to the sample and vortex. The density of the CsCl gradient should be approximately 1.41 g/cm^3 based on the refractive index (fr ¼ 1.3722).
4. After CsCl addition, sonicate the sample again using the exact conditions mentioned above.
5. The nuclear lysate is then loaded into Beckman Quick-Seal polyallomer (16 mm×76 mm) centrifuge tubes, balanced and the tube is then sealed using the Beckman Quick Sealer.

6. Centrifuge samples at 402,000 × *g* for a minimum of 10 h (longer is ok) to establish the gradient.
7. Following centrifugation, 1 ml sample fractions are then collected in individual tubes. This can be done simply by placing a needle in the base of the tube and another in the top to initiate the sample flow. If cleaner vector preparations are desired, the identified peak fractions (see below) can be subjected to an additional CsCl gradient (density 1.41 g/cm^3) in the manner described above.

3.5 AAV Genome Detection in CsCl Gradient Fractions by Q-PCR

To determine AAV vector production in the isolated fractions, procedures such as Southern blotting and Q-PCR can be used. The quickest, and perhaps most simple, method to detect which fractions contain viral genomes (indicative of packaged particles) is Q-PCR. It is important to note that at this stage precise particle tittering is not the objective, simply determination of peak gradient fractions containing rAAV. Therefore, we employ a simple protocol that has been shown valid for peak rAAV fraction detection and relies on heat denaturation of the AAV capsid during the PCR reaction. Precise viral tittering is performed after dialysis of the viral preparation and storage at −80 °C.

1. Dilute 1 μl of each CsCl gradient fraction into 1 ml of water. Vortex.
2. Use 2 μl of the viral dilution directly as template in a Q-PCR reaction.
3. Employ Q-PCR using standard protocols for primer construction and one of a variety of different methods to detect DNA amplification. As positive controls a dilution series (50 pg to 5 fg) of the pAAV-ITR plasmid used in production is commonly included in the reaction.

3.6 AAV Dialysis and Final Titer Determination

Following rAAV peak fraction determination, samples containing >1e^8 viral genomes can be pooled (dependent upon the desired injection volume and necessary titer) and then dialyzed against 1× PBS. To do this, Slide-A-Lyzer dialysis cassettes are used at 4 °C while stirring the solution. Generally, we use a 1:1,000 ratio of viral sample to PBS for a minimum of 12 h. Following dialysis, the rAAV samples are aliquoted and maintained at −80 °C prior to final titer determination and use. For final titer determination, we describe a method based on Q-PCR, however Southern dot blot remains a viable option and a detailed protocol has been described elsewhere [24].

1. After thawing at room temperature, combine 10 μL of rAAV sample with 90 μL of the DNaseI solution and vortex (perform in triplicate). Incubate the reaction for 1 h at 37 °C to degrade any unpackaged viral genomes.

2. Then, add 6 μL of 0.5 M EDTA and vortex to stop the DNase digestion reaction.
3. To degrade the AAV protein capsid, 120 μL of the Proteinase K solution is added, vortexed, and incubated at 55 °C for ≥2 h (sample can be left at 55 °C overnight).
4. Following the Proteinase K digest, samples are then heated for 10 min at 95 °C to inactivate Proteinase K.
5. Dilute the sample at least 100× in water or 10 mM Tris (pH 8) and use this solution as the DNA template in a standard Q-PCR reaction. A pAAV-ITR standard is run in parallel using a duplicate dilution series ranging from 50 pg to 0.5 fg. rAAV final titers are generally expressed as the number of viral genomes (calculated from the plasmid standard) in a given volume. Common rAAV titers range from 1e8 to 5e9 viral genomes per microliter (*see* **Note 4**).

3.7 rAAV Genome Characterization

rAAV genomes are primarily homogenous if the size is >3.5 kb but <5 kb and are prepared in the single-strand ITR plasmid context. scAAV preparations are often contaminated by the packaging of single-strand monomer genomes, in addition to the scAAV dimer genomes, which can decrease the overall vector potency and confuse result interpretations, depending on the application. Therefore, it is wise to characterize the packaged genomic species under denaturing conditions via Southern blotting.

1. 10 μl of dialyzed rAAV sample is incubated with 10 μl of the DNase solution (as described above). The reaction is then terminated using 5 μl of 0.5 M EDTA.
2. 2 μl of 10%SDS solution is then added followed by addition of 6× NaOH loading buffer, vortexed for 10 s then incubated at 20 min at RT.
3. The samples are then spun at max speed in a tabletop centrifuge for 10 min and loaded into a 1 % denaturing gel that has been soaked in alkaline gel running buffer for >10 min at RT (*see* **Note 5**).
4. Alkaline gel electrophoresis is then performed at 15 V for approximately 16 h to separate the AAV genomic species (*see* **Note 6**).
5. Next, the gel is incubated in the alkaline transfer buffer under mild agitation at RT for 10 min.
6. The gel is then placed atop an absorbent stack consisting of the following (from bottom to top): 1 in. of paper towels, three pieces of Whatman paper, and a nitrocellulose membrane (paper and membrane are sized slightly bigger than the gel). A bridge is then constructed of Whatman paper with a width equivalent to the width of the gel and a length long enough to accommodate the gel length and positioning of the other end

in a reservoir containing the alkaline transfer buffer. The reservoir is positioned approximately 3 in. higher than the gel stack to allow downward transfer flow of the DNA from the gel onto the membrane. This transfer event is performed at RT for a minimum of 10 h.

7. Following DNA transfer, the genomes are cross-linked to the nitrocellulose membrane using a UV photo-cross-linker "auto-crosslink" function.
8. The membranes are then placed in a glass hybridization tube containing Church buffer at a volume that is sufficient enough to cover the length of the tube when placed on its side (dependent upon the membrane and tube size).
9. Then, a ^{32}P-radioactive labeled probe is used for genome hybridization. Commonly, a labeled random probe library is generated using a Roche random labeling kit as detailed by the manufacturer (*see* **Note 7**). The radiolabeling reaction should be cleaned using, for example, a PCR purification kit to remove non-incorporated radio-nucleotides (as described by the kit manufacturer). The labeled probe preparation is then added to 100 μl of salmon sperm DNA and the mixture is boiled for 5 min to denature the DNA and then maintained on ice.
10. The denatured, radiolabeled probe is added to the hybridization tube (containing the Church buffer and membrane) and incubated for >12 h at 65 °C.
11. Next, the membrane is washed twice in the high-salt wash buffer for 10 min at 65 °C, then using the low-salt wash buffer for >20 min at the same temperature.
12. The washed membrane is then placed in plastic wrap, exposed to film and developed. Genome size and purity is determined based on the loaded DNA standards.

3.8 rAAV Cellular Infection

The protocol described herein focuses on rAAV-mediated DSB gene editing in HEK293 cells containing a defective *egfp* reporter system (Fig. 3) [3, 4]. A described *egfp*-specific ZFN in the single vector context is also outlined to induce a DSB-mediated HR event to "correct" the defective fluorescent reporter (Figs. 1c and 3) [1]. For HEK293 cell, the AAV serotype 2 capsid is the most efficient for transduction and is used herein. However, considering other cell types, as well as for in vivo tissue tropism, the literature should be consulted regarding the most efficient capsid serotype for transduction [15–17]. In general, we have found that AAV2 works best for most cell lines, while AAV1, 3 and 6 should be considered for different stem cell types [26, 27]. For in vivo applications, AAV8 and 9 are efficient for systemic mouse transduction (especially if administer via the intraperitoneal route prior to day 12 post-birth), while AAV1 and 6 transduce efficiently following intramuscular administration [19].

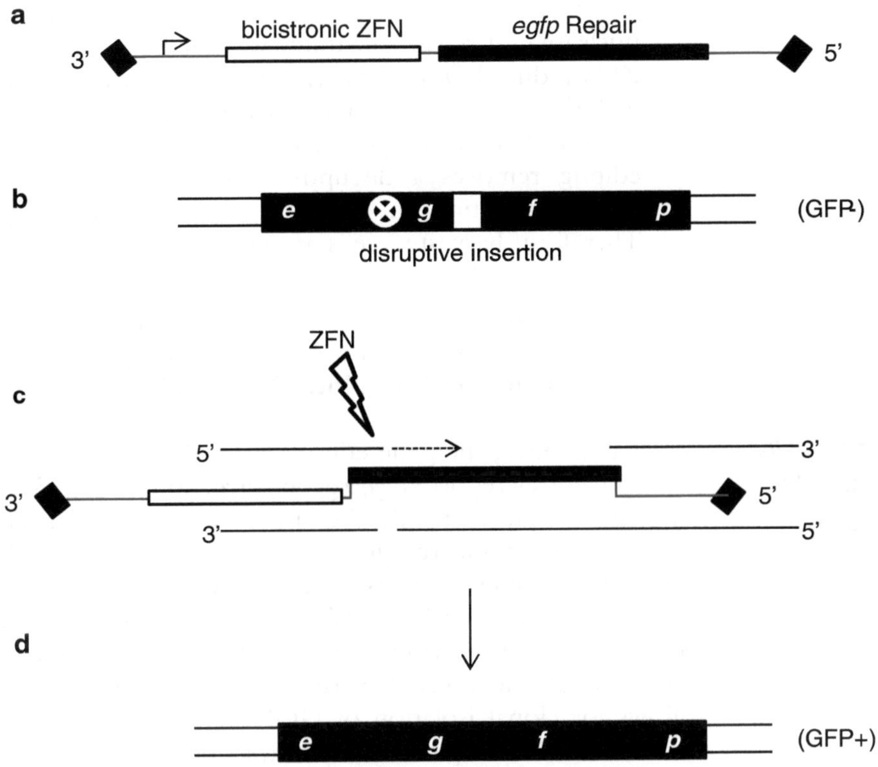

Fig. 3 AAV vector mediated gene editing. In the depicted cartoon, the transduced AAV genome (**a**) encodes a promoter, an eGFP-specific ZFN (target site is indicated by a *white circled X*), and an eGFP repair sequence to correct the defective chromosomal fluorescent reporter (**b**). In this system, the endonuclease generates a specific double-strand break inducing a host DNA damage response (**c**). Multiple repair mechanisms can result in correction of the defective reporter using the repair sequence provided by the AAV vector. In this cartoon, we demonstrate an example based on gene conversion in which the AAV repair homology serves as a synthesis template for the free 3′ end generated by the ZFN (**c**). This new synthesis integrates the desired genetic modification (in this case *egfp* coding sequence devoid of the insertion represented by the *white box*) creating, in this example, a functional fluorescent reporter (**d**)

1. HEK293 cell cultures are maintained under the conditions described above and the desired viral genome number per cell is added directly to the growth medium (*see* **Note 8**). Transduction is often performed in a 24 well context and a viral particle to cell ratio of 10,000 is sufficient for most applications. Depending on the experimental design, multiple rAAV preps can be added to the cells at the same time (i.e., for delivery of the endonuclease and repair substrate using separate rAAV particles).
2. Three days post-infection, the cells are harvested for determination of gene editing (*see* **Note 9**).

3. In this example, a single rAAV particle (encoding both the nuclease and repair substrate) was used to generate a specific ZFN-induced DSB in the *egfp* coding sequence to induce HR with the rAAV delivered repair substrate [1] (Figs. 1c, 3). In this extensively reported gene correction system, the desired gene editing removes a disruptive insertion in a defective *egfp* reporter creating an easily tallied GFP+ phenotype [3, 5]. Therefore, harvested cells are pelleted at 1,000 × *g* for 3 min and resuspended in 2 % paraformaldehyde in PBS.
4. Flow cytometry for detection of GFP+ cells is then performed using the fixed cells and standard flow cytometry detection protocols for GFP (*see* **Notes 10** and **11**).

3.9 Detection of the AAV-Mediated DSB Gene Editing Event

As stated above, the gene editing event described herein is based on a previously reported *egfp* gene correction system [3, 5]. In this system, a single vector AAV context was used to deliver the *egfp* repair substrate and the genes encoding *egfp* directed ZFNs [1]. This correction event is easily determined by a cellular phenotypic change to GFP+ which is sensitively detected by flow cytometry in a high-throughput manner [3, 5]. Verification of the edited locus following extended cell growth to dilute rAAV episomal genomes allows for clonal isolation of GFP+ cells, and DNA sequencing across the edited region has been described elsewhere [2].

4 Notes

1. When designing an AAV genome it is imperative to remember that for ssAAV the packaged genome should be <5 kb yet >2.7 kb in size. Therefore, depending on the desired particles, it may be necessary to adjust the sequence between the ITRs (of pAAV-ITR) to adhere to these restrictions. For scAAV, the DNA sequence to be packaged (the sequence between the ITRs of pHpa-trs-sk) needs to be <2.2 kb yet >1 kb.
2. The restriction endonucleases necessary for repair substrate and endonuclease gene cloning will vary based on the cloning strategy.
3. During rAAV production, approximately 50 h post-transfection the treated cells should appear smaller, spherical, with partial loss of adherence due to the pXX680 plasmid. Such morphological changes suggest only an efficient transfection, not necessarily rAAV production. However, if such changes are not observed the researcher should be suspicious that something is wrong with the transfection conditions.
4. Precise viral tittering of scAAV preparations by Q-PCR is often difficult, compared to ssAAV, due to the nature of the duplexed DNA molecule. Currently several labs are working out protocol

conditions to overcome the complications for scAAV tittering by Q-PCR. Therefore, scAAV tittering is validated by additional methods including Southern dot blotting and by the Southern alkaline gel electrophoresis described elsewhere and above [24]. In these cases, a plasmid dilution series is included (ranging from 50 to 0.5 ng) to generate a standard curve from which the number of rAAV genomes can be calculated [24].

5. It is important to also include DNA size standards while performing Southern alkaline gel electrophoresis which can be obtained by enzymatic digestion and gel isolation of the pAAV-ITR plasmid fragments. Approximately 10 ng of the standards are incubated in the NaOH sample buffer, vortexed and incubated for 30 min at RT prior to alkaline gel loading.
6. When preparing the alkaline gel, the gel solution is heated in the microwave until the agarose dissolves. Then, the solution is allowed to cool (not solidify) prior to NaOH and EDTA addition, as described above.
7. As a radioactive probe template, a gel isolated DNA fragment, corresponding to the entire transgenic cassette (the region to be packaged on the pAAV-ITR plasmid), is preferred. This fragment can also be used as a size standard during Southern blotting.
8. As rAAV infects both dividing and non-dividing cells, cell confluency will not dramatically affect transduction at similar viral genome to cell ratios. However, as HR recombination is restricted to or more prevalent in mammalian dividing cells, then the use of arrested or differentiated cell populations for gene editing will result in very few or no events (Hirsch unpublished).
9. Longer culture times, following rAAV infection, generally result in the increased frequency of the desired host chromosome modification.
10. As stated above, the gene editing event described herein is based on a previously reported *egfp* gene correction system [3, 5]. In this system, a single vector AAV context was used to deliver the *egfp* repair substrate and the genes encoding *egfp* directed ZFNs [1] (Fig. 3). This correction event is easily determined by a cellular phenotypic change to GFP+ which is sensitively detected by flow cytometry in high-throughput manner. Verification of the edited locus has also been described following extended cell growth to dilute rAAV episomal genomes, clonal isolation of GFP+ cells, and DNA sequencing across the edited region [2].
11. The illegitimate integration of rAAV genomes can occur at levels of approximately 1 % in dividing cells and in vivo as well [28, 29]. Depending on the application, detection of rAAV integrants may be desired and methods to do so have been reported previously [28–30].

Acknowledgments

This work was supported by the Northwest Genome Engineering Consortium (NIH) and by R01AI072176 awarded to Hirsch and Samulski.

References

1. Ellis BL et al (2012) Zinc-finger nuclease-mediated gene correction using single AAV vector transduction and enhancement by Food and Drug Administration-approved drugs. Gene Ther 20(1):35–42
2. Hirsch ML et al (2010) Self-complementary AAV mediates gene targeting and enhances endonuclease delivery for double-strand break repair. Gene Ther 17(9):1175–1180
3. Porteus MH et al (2003) Efficient gene targeting mediated by adeno-associated virus and DNA double-strand breaks. Mol Cell Biol 23(10):3558–3565
4. Radecke F et al (2006) Targeted chromosomal gene modification in human cells by single-stranded oligodeoxynucleotides in the presence of a DNA double-strand break. Mol Ther 14(6):798–808
5. Porteus MH, Baltimore D (2003) Chimeric nucleases stimulate gene targeting in human cells. Science 300(5620):763
6. Christian M et al (2010) Targeting DNA double-strand breaks with TAL effector nucleases. Genetics 186(2):757–761
7. Chevalier BS, Stoddard BL (2001) Homing endonucleases: structural and functional insight into the catalysts of intron/intein mobility. Nucleic Acids Res 29(18):3757–3774
8. Orlowski J, Boniecki M, Bujnicki JM (2007) I-Ssp6803I: the first homing endonuclease from the PD-(D/E)XK superfamily exhibits an unusual mode of DNA recognition. Bioinformatics 23(5):527–530
9. Heath PJ et al (1997) The structure of I-CreI, a group I intron-encoded homing endonuclease. Nat Struct Biol 4(6):468–476
10. Gao H et al (2010) Heritable targeted mutagenesis in maize using a designed endonuclease. Plant J 61(1):176–187
11. Smith J et al (2006) A combinatorial approach to create artificial homing endonucleases cleaving chosen sequences. Nucleic Acids Res 34(22):e149
12. Carroll D (2011) Genome engineering with zinc-finger nucleases. Genetics 188(4):773–782
13. Wu J, Kandavelou K, Chandrasegaran S (2007) Custom-designed zinc finger nucleases: what is next? Cell Mol Life Sci 64(22): 2933–2944
14. McCarty DM, Monahan PE, Samulski RJ (2001) Self-complementary recombinant adeno-associated virus (scAAV) vectors promote efficient transduction independently of DNA synthesis. Gene Ther 8(16):1248–1254
15. Mitchell AM et al (2010) AAV's anatomy: roadmap for optimizing vectors for translational success. Curr Gene Ther 10(5):319–340
16. Asokan A, Schaffer DV, Samulski RJ (2012) The AAV vector toolkit: poised at the clinical crossroads. Mol Ther 20(4):699–708
17. Zincarelli C et al (2008) Analysis of AAV serotypes 1-9 mediated gene expression and tropism in mice after systemic injection. Mol Ther 16(6):1073–1080
18. Ellis BL et al (2013) A survey of ex vivo/in vitro transduction efficiency of mammalian primary cells and cell lines with Nine natural adeno-associated virus (AAV1-9) and one engineered adeno-associated virus serotype. Virol J 10:74
19. Wang Z et al (2005) Adeno-associated virus serotype 8 efficiently delivers genes to muscle and heart. Nat Biotechnol 23(3):321–328
20. Li H et al (2011) In vivo genome editing restores haemostasis in a mouse model of haemophilia. Nature 475(7355):217–221
21. Samulski RJ et al (1982) Cloning of adeno-associated virus into pBR322: rescue of intact virus from the recombinant plasmid in human cells. Proc Natl Acad Sci U S A 79(6): 2077–2081
22. Hirata RK, Russell DW (2000) Design and packaging of adeno-associated virus gene targeting vectors. J Virol 74(10):4612–4620
23. Xiao X, Li J, Samulski RJ (1998) Production of high-titer recombinant adeno-associated virus vectors in the absence of helper adenovirus. J Virol 72(3):2224–2232
24. Grieger JC, Choi VW, Samulski RJ (2006) Production and characterization of adeno-associated viral vectors. Nat Protoc 1(3): 1412–1428
25. Samulski RJ et al (1983) Rescue of adeno-associated virus from recombinant plasmids: gene

correction within the terminal repeats of AAV. Cell 33(1):135–143

26. Hirsch ML et al (2011) Viral single-strand DNA induces p53-dependent apoptosis in human embryonic stem cells. PLoS One 6(11):e27520
27. Ferreira JR et al (2013) Three-dimensional multipotent progenitor cell aggregates for expansion, osteogenic differentiation and 'in vivo' tracing with AAV vector serotype 6. Gene Ther 20(2):158–168
28. Li H et al (2011) Assessing the potential for AAV vector genotoxicity in a murine model. Blood 117(12):3311–3319
29. Wang Z et al (2012) AAV vectors containing rDNA homology display increased chromosomal integration and transgene persistence. Mol Ther 20(10):1902–1911
30. Donsante A et al (2007) AAV vector integration sites in mouse hepatocellular carcinoma. Science 317(5837):477

Chapter 20

Genetic Modification Stimulated by the Induction of a Site-Specific Break Distant from the Locus of Correction in Haploid and Diploid Yeast *Saccharomyces cerevisiae*

Samantha Stuckey and Francesca Storici

Abstract

Generation of a site-specific break at a genomic locus to stimulate homologous recombination (HR) is used in many organisms to efficiently target genes for various types of genetic modification. Additionally, a site-specific chromosomal break can be used to trigger HR at genomic regions distant from the break, thereby largely expanding the region available for introducing desired mutations. In contrast to the former approach, the latter presents an alternative way in which genes can be efficiently modified also when it is not possible or desirable to introduce a break in the vicinity of the targeting locus. This type of in vivo site-directed mutagenesis distant from a break can be accomplished in the yeast model organism *Saccharomyces cerevisiae* because the generation of a double-strand break (DSB) in yeast chromosomal DNA activates HR at long regions upstream and downstream from the break site. Here we provide a protocol for efficiently altering a yeast chromosomal locus following the induction of a DSB several kilobase pairs distant from the site of gene correction. The techniques described can be used in both diploid and haploid yeast strains, and we provide examples of the gene correction assays.

Key words DNA modification, Distant repair, DNA oligonucleotides, *Saccharomyces cerevisiae*, Site-directed mutagenesis, Gene targeting, Double-strand break (DSB), Haploid yeast, Diploid yeast

1 Introduction

Gene targeting is a genetic technique that uses homologous recombination (HR) to modify an endogenous DNA sequence. Using this technique, various genetic manipulations are possible, including gene knockout for functional analysis, gene knock-in to provide selectable markers or for heterologous gene insertion, generation of specific point mutations to alter the function of a desired gene or regulatory sequence, introduction of particular polymorphisms, or correction of genetic defects [1]. Gene targeting has been used extensively in many cellular systems, from bacteria to human cells [2–9]. Despite the wide applications of the

Francesca Storici (ed.), *Gene Correction: Methods and Protocols*, Methods in Molecular Biology, vol. 1114, DOI 10.1007/978-1-62703-761-7_20, © Springer Science+Business Media, LLC 2014

gene-targeting technique, the generally low efficiency of HR in nearly all organisms has always been a challenge for researchers [10–12]. One of the most effective ways to increase gene targeting in chromosomal DNA is via the induction of a site-specific DNA double-strand break (DSB) near the genomic site to be changed [8, 13, 14]. Indeed, a DSB can stimulate HR up to several thousandfold over natural levels [7, 15, 16]. What increases the efficiency of gene targeting by HR is not the formation of the break per se in the genomic DNA. It is the generation of regions of single-stranded DNA, following processing of the break ends and strand resection in a 5′ to 3′ direction in preparation for break repair, that makes the DNA at the broken locus highly recombinogenic [17]. If a DSB occurs in DNA but formation of single-stranded DNA is impeded, gene targeting by HR is not stimulated [17]. During the process of DSB repair, the gap of single-stranded DNA that is generated can be filled in using homologous templates [17, 18]. Therefore, it is possible to increase the efficiency of gene correction in chromosomal DNA also using targeting molecules that do not physically repair the break but that have complementarity with the single-stranded regions to either side of the break [17]. This approach of gene modification at sites distant from the DSB works well in the yeast *Saccharomyces cerevisiae*, in which long regions of single-stranded DNA can be generated following induction of a DSB [17, 19]. Major advantages of this strategy are (1) a large window for mutagenesis expanded up to 20 kb from each side of the DSB, (2) the possibility to perform gene correction in essential genes, and (3) the applicability of the approach to diploid in addition to haploid cells [13, 17, 20–23].

When applying gene targeting distant from sites of induced breaks in cellular systems different from the yeast *S. cerevisiae*, it should be taken into consideration that the efficiency of genetic modification at a locus distant from a break is limited by the length of the single-stranded DNA regions surrounding the DSB, which varies between organisms. Following generation of a DSB, 5′ to 3′ resection on both sides of the break leads to long segments of 3′ overhangs [24]. In yeast these segments can stretch for tens of kilobases [19], providing a large window of opportunity for modification at distant loci. Studies in mammalian cells, though, indicate the lengths of resection are much shorter, less than 2 kb [25]. Thus, while generation of a distant DSB can improve gene-targeting applications by enlarging the window of opportunity for creating modifications, factors such as resection lengths can also limit this window.

Though highly efficient in yeast cells, gene targeting at regions distant from the site of DNA breakage could potentially be adapted to other cellular systems. The opportunity to exploit a wider window of mutagenesis, even a few hundred nucleotides longer, from the site of breakage may already provide a significant advantage. Generation of a DSB at a genomic locus in which the desired genetic modification will be made may not always be feasible, and in some situations it is favorable to generate a break distant from the position to be modified. For example, designed nucleases, such as zinc-finger nucleases (ZFNs) or transcription activator-like effector nucleases (TALENs), are created to recognize and cleave specific genomic loci [26–29]. In order to design the most effective designer nucleases, however, targeted genomic regions must contain highly specific nucleotide sequences recognizable by each of the nucleases and appropriately spaced relative to each other such that dimerization of the nucleases can occur. Thus, the sequence constraints of these enzymes may require their recognition sites to be located at some distance from the genomic locus to be modified. Additionally, recent advancements with systems naturally involved in the adaptive immune systems of bacteria and archaea have demonstrated the capability of RNA-driven targeted gene correction [30, 31]. Clustered regularly interspaced short palindromic repeats (CRISPR)/CRISPR-associated (Cas) systems, like ZFNs and TALENs, recognize highly specific nucleotide sequences in the genome for targeted generation of a site-directed break [30, 31]; however, like with ZFNs and TALENs, the genetic locus to be targeted for gene correction may exist at some chromosomal position distant from the most favorable sequence for recognition by these CRISPR elements. If the gene-targeting sites are distant from the most optimal recognition sequences for a break to be generated but still in regions that can become single-stranded after DSB induction, the gene correction efficiency will be much higher than without DSB induction.

Here we provide a protocol for modification of a locus distant from a DSB in the yeast *S. cerevisiae* using synthetic DNA oligonucleotides as gene correction templates. We explain in a stepwise fashion procedures in which a cassette containing an inducible I-*Sce*I endonuclease gene and its 18-nt recognition site can be integrated into the genome of haploid or diploid yeast strains and used to stimulate genetic modification of regions several kilobases away from the initial chromosomal break point.

2 Materials

2.1 Amplification of CORE-I-SceI

1. Two CORE-I-*Sce*I plasmids (pGSKU, pGSHU) are available (*see* [32] for further plasmid information).
2. DNA primers, desalted and non-purified: 50 μM stock. Stored at −20 °C.
3. Ex Taq DNA polymerase, 10× buffer, 2.5 mM dNTPs (Clontech Laboratories, Inc., Mountain View, CA, USA).

2.2 Gel Electrophoresis

1. Agarose.
2. 10× Tris/Borate/EDTA (TBE) running buffer.
3. Prestained molecular weight marker.
4. Loading dye.

2.3 PCR Product Concentration

1. 95 % ethanol (EtOH).
2. 70 % EtOH.
3. 3 M sodium acetate (NaOAc ; pH 5.2) solution in water: Filter-sterilized. Stored at room temperature (*see* **Note 1**).

2.4 Transformation Reagents and Media

1. YPD (per 1 l): 10 g yeast extract, 20 g bacto peptone, 20 g dextrose. Autoclaved for 45 min at 121 °C. 20 g agar is added for solid medium prior to autoclaving (*see* **Notes 2** and **3**).
2. YPLac liquid (per 1 l): 12 g sodium hydroxide (NaOH), 10 g yeast extract, 20 g bacto peptone, 27 ml lactic acid, pH adjusted to 5.5 with lactic acid. Autoclaved for 45 min at 121 °C.
3. 20 % Highly pure galactose stock solution in water: Filter-sterilized. Stored at room temperature.
4. 1 M lithium acetate (LiOAc) stock solution in water: Filter-sterilized. Stored at room temperature.
5. 10×TE stock solution: 100 mM Tris (pH 7.5), 10 mM ethylenediaminetetraacetic acid (EDTA) (pH 7.5) in water. Filter-sterilized. Stored at room temperature.
6. 50 % polyethylene glycol 4000 (PEG 4000) stock solution in water: Stored at room temperature (*see* **Note 4**).
7. Working solutions: Solution 1 (0.1 M LiOAc, TE 1×, pH 7.5 in water) and solution 2 (0.1 M LiOAc, TE 1×, pH 7.5 in 50 % PEG 4000).
8. 100 μg/ml Salmon sperm DNA (SSD) in water: Stored at −20 °C.
9. Synthetic dextrose complete medium lacking uracil (SC-Ura) solid medium.
10. Synthetic dextrose complete (SD complete) solid medium.
11. Glass beads approximately 5 mm diameter.
12. 5-Fluoroorotic acid (5-FOA; per 1 l): Solution of 5-FOA is prepared by dissolving 1 g 5-FOA in 300 ml of water prior to filtration. 700 ml SD complete agar medium is autoclaved for 45 min at 121 °C and then cooled to 55–60 °C, and the filter-sterilized solution of 5-FOA is then mixed well with medium prior to pouring.
13. G418 (per 1 l): YPD agar medium is autoclaved for 45 min at 121 °C, then cooled to 55–60 °C, and G418 solution is then mixed well with medium prior to pouring. Stock solution is

prepared in water in 50 mg/ml filter-sterilized aliquots and stored at 4 °C.

14. Hygromycin B (hygro; per 1 l): YPD agar medium is autoclaved for 45 min at 121 °C, then cooled to 55–60 °C, and hygromycin B solution is then mixed well with medium prior to pouring.
15. YPG (per 1 l): 10 g yeast extract, 20 g bacto peptone, 30 ml glycerol, 20 g agar. Autoclaved for 45 min at 121 °C.
16. Sterile velveteens.

2.5 Genotypic Testing of Transformants

1. 2,000 U/ml lyticase in water: Stored in 1 ml aliquots at −20 °C.
2. Taq DNA polymerase, 10× buffer, 10 mM dNTPs.

2.6 Design of DNA Oligonucleotides for Modification of Distant Loci and Repair of DSB

1. 50 μM DNA oligonucleotide stock solution: 40–100 mers, desalted and non-purified. Store at −20 °C.

3 Methods

Here we describe a method for generating an I-*Sce*I site-specific DSB in yeast chromosomal DNA which will be used to stimulate gene targeting at a region distant from the DSB site both in haploid and diploid yeast cells. The initial steps are common between the two. Briefly, a product will be amplified from plasmids available containing a CORE-I-*Sce*I cassette. The cassettes contain COunterselectable and REporter markers for selection, hence the name. Two of these plasmids (pGSKU and pGSHU) contain the I-*Sce*I site-specific endonuclease gene, regulated by a galactose-inducible promoter, *GAL1*. The cassettes from these plasmids, hereafter referred to as the CORE-I-*Sce*I cassettes, will be amplified through polymerase chain reaction (PCR) and then transformed into the strain of interest. Additionally, during PCR the 18-nt cognate recognition sequence for I-*Sce*I is introduced at one end of the cassette by one of the primers. This site will later be used for the DSB-driven repair assay as the expressed I-*Sce*I protein will recognize it to generate the site-directed break. We further provide examples of repair assays in both haploid and diploid yeast strains. Additional material on the available CORE-I-*Sce*I cassettes has been previously published elsewhere [17, 32, 33].

3.1 Amplification of CORE-I-SceI from Plasmid

1. DNA primers will first be used to amplify the CORE-I-*Sce*I cassettes (GSKU or GSHU) from the chosen plasmid. These primers range from 70 to 88 nt in length with an overlap of at least 50 nt with the genomic targeting region and an overlap

of 20 nt to the CORE-I-*Sce*I cassette sequence. Additionally, the 18-nt recognition sequence for the I-*Sce*I endonuclease is included on one of the two primers (*see* [32] for further information on amplifying these cassettes).

2. PCR conditions: Amplification of the CORE-I-*Sce*I cassette from circular plasmid (about 50 ng) using 50 pmol of each primer is performed with high yield in a final volume of 40 μl using Ex Taq DNA polymerase with a 2 min cycle at 94 °C; 32 cycles of 30 s at 94 °C, 30 s at 57 °C, and 5 min at 72 °C; a final extension time of 7 min at 72 °C; and samples are held at 4 °C. Stock concentrations of 10 mM dNTPs are used for this reaction. An extension time of 1 min/kb is assumed for this reaction (*see* **Note 5**).
3. Following PCR, the samples are ready for gel electrophoresis and PCR product concentration.

3.2 Gel Electrophoresis

1. We use a dilution of 0.5× TBE running buffer, which is obtained from 10× TBE by mixing 50 ml of 10× TBE buffer with 950 ml deionized water prior to use.
2. A small aliquot (about 2 μl) of PCR product is run on a 0.8 % agarose gel to observe the anticipated band.

3.3 PCR Product Concentration

1. The product of six reactions of PCR are combined for precipitation with a 2.5× volume of 95 % EtOH and 0.1× volume of 3 M NaOAc (pH 5.2) in a microcentrifuge tube. Centrifugation is carried out at maximum speed for 10 min. A small pellet should be visible on the bottom of the tube.
2. The supernatant is discarded, and the pellet is washed with 100 μl of 70 % EtOH, being careful not to detach the pellet. If the pellet is detached, it is necessary to spin again for 5 min and then discard the supernatant. Then, as much as possible, the EtOH is removed without detaching the pellet.
3. The pellet is then spun until dry in a speed vacuum device and then resuspended in 50 μl of water. 5–10 μl are used for each transformation.

3.4 Transformation to Insert the CORE-I-SceI Cassette

The following protocol is used to insert the CORE-I-*Sce*I PCR product into the chosen strain. This procedure applies to both haploid and diploid strains. As the CORE-I-*Sce*I cassettes, GSKU and GSHU, each contains multiple markers (the *URA3* gene from *Kluyveromyces lactis* and resistance genes to the antibiotics G418 or hygromycin B, respectively), the strain should have a nonfunctional *URA3* gene and should be G418 and hygromycin B sensitive to ensure proper integration at the desired locus. This procedure has been modified from the lithium acetate protocol described by Wach et al. [34]. A more detailed explanation of the involved steps has been previously published by our group in the Methods in Molecular Biology series [32].

1. Inoculate 5 ml of YPD liquid medium with chosen strain and shake at 30 °C overnight (O/N) (*see* **Note 6**).
2. When cells are in logarithmic growth (approximately 16 h later), inoculate 50 ml of YPD liquid medium with 1.5 ml of the O/N culture in a 250-ml glass flask and shake vigorously at 30 °C for 3–4 h.
3. Solutions 1 and 2 are prepared immediately prior to transformation.
4. Transfer culture to a 50-ml conical tube and spin at 1,562 × *g* for 2 min.
5. Remove the supernatant and wash cells with 50 ml of sterile water and spin as stated previously.
6. Remove the supernatant and resuspend cells in 5 ml of solution 1 and spin as stated previously.
7. Remove supernatant and resuspend cells in 250 μl of solution 1. This amount of cells is sufficient for up to approximately 10 transformations.
8. Aliquot 50 μl of the cell suspension in microcentrifuge tubes and add 5–10 μl of concentrated CORE-I-*Sce*I PCR product and 5 μl of SSD (heat-denatured for 5 min at 100 °C prior to use and immediately kept on ice).
9. Add 300 μl of solution 2 to each transformation tube. Mix briefly by vortexing.
10. Incubate transformation tubes at 30 °C for 30 min with shaking.
11. Heat shock at 42 °C for 15 min.
12. Collect cells by centrifugation at 2,236 × *g* for 4 min.
13. Remove the supernatant and resuspend cells well in 100 μl of sterile water.
14. Plate all cells from each transformation tube on one SC-Ura plate using approximately 15 sterile glass beads and incubate at 30 °C for 2–3 days until colonies appear (*see* **Note 7**).
15. Using a sterile velveteen, replica-plate from SC-Ura to G418- or hygro-containing medium (depending on the CORE used) and incubate at 30 °C O/N.
16. Once transformants are observed (typically 5–30 colonies per plate), streak for single colony isolates on YPD solid medium. Incubate at 30 °C for 2–3 days.
17. Make patches of the single colonies onto new YPD solid medium, along with the original strain for phenotypic comparison, and incubate at 30 °C O/N.

18. Replica-plate the grown patches to YPD; SC-Ura; G418; hygro; yeast extract-peptone-glycerol (YPG) solid medium, which selects against cells with defective mitochondrial DNA (mtDNA); and any other various selective media depending on the background of your strain, and incubate at 30 °C O/N.
19. Following observation of correct phenotype, the samples are ready for genotypic testing.

3.5 Colony PCR of Transformants

1. Resuspend cells (approximately 1 mm^3) in 50 μl water containing 1 U of lyticase. Incubate at room temperature for 10 min, then heat at 100 °C for 5 min. Samples can then be left at room temperature.
2. PCR conditions: Colony PCR of the transformant patches showing the expected phenotypes using 10 μl of the cell resuspension mixture is carried out with 50 pmol of each primer and with an expected amplified product size ranging between 300 bp and 1 kb. Stock concentrations of 10 mM dNTPs are used for this reaction. PCR is performed in a final volume of 50 μl using Taq DNA polymerase with a 2 min cycle at 95 °C; 32 cycles of 30 s at 95 °C, 30 s at 55 °C, and 1 min at 72 °C; a final extension time of 7 min at 72 °C; and samples are held at 4 °C. An extension time of 1 min/kb is assumed for this reaction.
3. Following PCR, samples are run on a 1 % agarose gel (*see* Subheading 3.2) for observation of PCR products of the anticipated size.
4. Strains are now ready to be sequenced to verify the sequences of the inserted I-*Sce*I gene and site.

3.6 Design of DNA Oligonucleotides for Modification of a Chromosomal Locus

To modify a locus distant from the DSB with oligonucleotides, a single oligonucleotide or a complementary pair can be used to generate the desired mutation. The following considerations should be made when determining which oligonucleotide(s) to use. First, while oligonucleotides as short as 40 nt in length can be used to efficiently transform the strain, the efficiency of targeting is increased as the length of oligonucleotide increases (up to 80–90 nt) [20]. Next, following generation of a DSB, there will be resection of each strand of the DNA away from the break in the 5′ to 3′ direction. The single oligonucleotide complementary to the intact strand following resection will be much more efficient than the other oligonucleotide [17]; therefore it is suggested to determine which single oligonucleotide will be more favorable to use prior to performing the experiment. Finally, a pair of complementary oligonucleotides provides the most efficient repair template; however, in many cases the use of the single oligonucleotide complementary to the intact strand will be highly efficient at modifying the DNA and thus a complementary pair may not be necessary.

3.7 Generation of the DSB

In Subheading 3.1 a protocol for amplifying the CORE-I-*Sce*I cassette is detailed. One of the primers introduces the 18-nt recognition site for the I-*Sce*I endonuclease. After generation of a strain containing the integrated cassette and site, the I-*Sce*I protein will be used to create a site-directed DSB at this chromosomal position. Expression of this gene is regulated by the *GAL1* promoter, which is induced following addition of galactose to the medium. In the next two sections, we provide examples of repair assays which can be conducted using the galactose-induced system in haploid and diploid yeast strains.

3.8 Repair Assay: Example at the trp5 Locus

Here we provide an example of an assay for repair of a mutated, nonfunctional *trp5* marker which is located 10 kb downstream from the DSB site in a haploid strain with oligonucleotides that are used to repair the break and those which correct the sequence of *trp5* (*see* Fig. 1). Following correction of this marker, cells present a Trp^{+} phenotype which can be selected for on the solid medium lacking tryptophan. While targeted gene correction can be used at either a selectable or a non-selectable genetic locus, the presentation of a selectable phenotype following correction enables more rapid screening of transformants. Further information on targeting at a non-selectable genetic locus is provided in Subheading 4 below (*see* **Note 8**).

1. Inoculate 50 ml of YPLac liquid medium with chosen strain into a 250-ml glass flask and shake vigorously at 30 °C O/N (*see* **Note 9**).
2. Add 5 ml of galactose from a 20 % stock solution into the O/N culture for a final 2 % galactose solution and continue to shake at 30 °C for 7 h (*see* **Note 10**).
3. Solutions 1 and 2 are prepared immediately prior to transformation.
4. Transfer culture to a 50-ml conical tube and spin at 1,562 × *g* for 2 min.
5. Remove the supernatant and wash cells with 50 ml of sterile water and spin as stated previously.
6. Remove the supernatant and resuspend cells in 5 ml of solution 1 and spin as stated previously.
7. Remove supernatant and resuspend cells in 250–750 μl of solution 1. The pellet should be quite large. Depending on the dilution, this amount of cells is sufficient for up to approximately 30 transformations.
8. Aliquot 50 μl of the cell suspension in microcentrifuge tubes and add 1 nmol of DNA oligonucleotides (heat-denatured for 2 min at 100 °C, then immediately kept on ice prior to use). When using a single oligonucleotide to repair the DSB and a

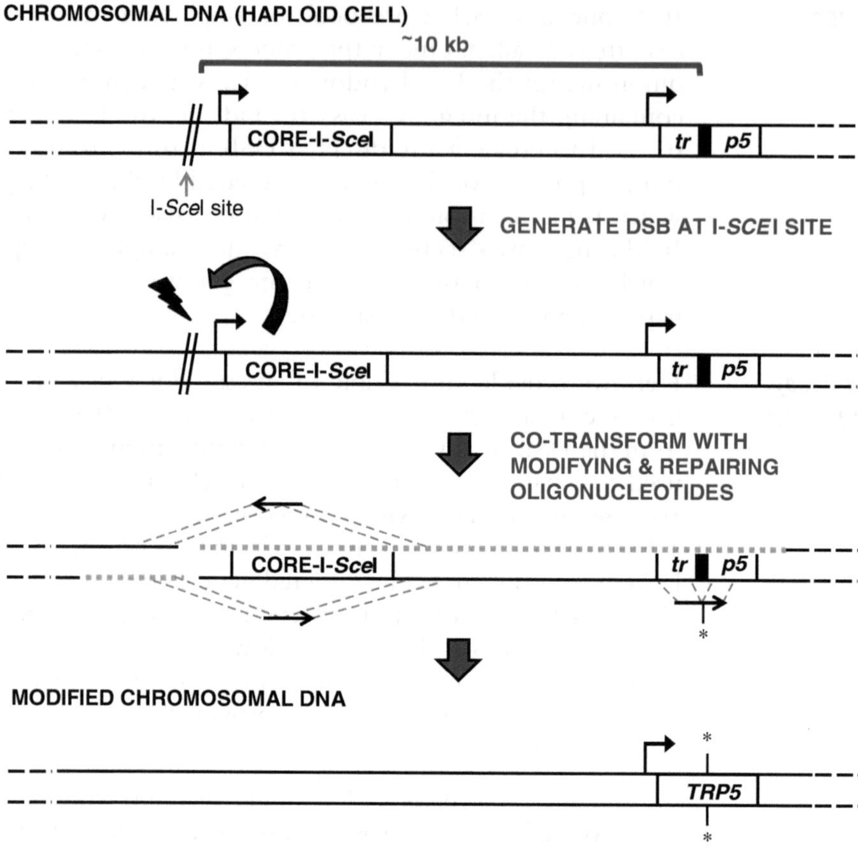

Fig. 1 Modification of a distant locus 10 kb downstream of the DSB site in haploid yeast cells. In this example, the starting strain contains a disrupted *trp5* gene located approximately 10 kb downstream from the 18-nt recognition site for the I-*Sce*I endonuclease. Following addition of galactose, which induces expression of the nuclease (indicated by the *lightning bolt*), a targeted DSB is generated at the I-*Sce*I site and 5′ to 3′ resection occurs (indicated by the *light gray dashed lines*), leaving long 3′ single-stranded stretches of DNA. Next, oligonucleotides are transformed into the cells. These include an oligonucleotide complementary to both sides of the disruption in *trp5* (as indicated by *dashed lines* between the oligonucleotides and genomic region), as well as oligonucleotides complementary to both sides of the break to repair it and simultaneously remove the CORE-I-S*ce*I cassette. Single oligonucleotides or the complementary pair can be used for each position. An *asterisk* on the oligonucleotide used to correct the *trp5* gene indicates the modification. Following transformation, the break is repaired while the target locus is modified (as indicated by the *asterisk*) to introduce the desired genetic modification in *TRP5*

single oligonucleotide to modify the chromosome distant from the break, 10 μl (at 50 μM or 50 pmol/μl) of each are used. When using a complementary pair of oligonucleotides for both the DSB repair and the distant modification, 5 μl of each are used (*see* **Note 11**).

9. Add 300 μl of solution 2 to each transformation tube. Mix briefly by vortexing.
10. Incubate transformation tubes at 30 °C for 30 min with shaking.

11. Heat shock at 42 °C for 15 min.
12. Collect cells by centrifugation at 2,236 × *g* for 4 min.
13. Remove the supernatant and resuspend cells well in 100 μl of sterile water.
14. Plate a 10^5-fold dilution per strain to SD complete solid medium. Plate all cells from each transformation tube on one synthetic dextrose complete medium lacking tryptophan (SC-Trp) plate using approximately 15 sterile glass beads and incubate at 30 °C for 3–4 days until colonies appear (*see* **Notes 12** and **13**).

3.9 Repair Assay: Example at the leu2 Locus

Here we provide an example of an assay for repair of a mutated, nonfunctional *leu2* marker which is located 10 kb upstream from the DSB site in a diploid strain with the homologous chromosome being used to repair the break and oligonucleotides which correct the sequence of *leu2* (*see* Fig. 2). Following correction of this marker, however, cells present a Leu^+ phenotype which can be selected for on solid medium lacking leucine. While targeted gene correction can be used at either a selectable or a non-selectable genetic locus, the presentation of a selectable phenotype following correction enables more rapid screening of transformants. Further information on targeting at a non-selectable genetic locus is provided in Subheading 4 below (*see* **Note 8**).

1. Inoculate 50 ml of YPLac liquid medium with chosen strain into a 250-ml glass flask and shake vigorously at 30 °C O/N (*see* **Note 9**).
2. Add 5 ml of galactose from a 20 % stock solution into the O/N culture for a final 2 % galactose solution and continue to shake at 30 °C for 7 h (*see* **Note 10**).
3. Solutions 1 and 2 are prepared immediately prior to transformation.
4. Transfer culture to a 50-ml conical tube and spin at 1,562 × *g* for 2 min.
5. Remove the supernatant and wash cells with 50 ml of sterile water and spin as stated previously.
6. Remove the supernatant and resuspend cells in 5 ml of solution 1 and spin as stated previously.
7. Remove supernatant and resuspend cells in 250–750 μl of solution 1. The pellet should be quite large. Depending on the dilution, this amount of cells is sufficient for up to approximately 30 transformations.
8. Aliquot 50 μl of the cell suspension in microcentrifuge tubes and add 1 nmol of DNA oligonucleotides (heat-denatured for 2 min at 100 °C, then immediately kept on ice prior to use). When using a single oligonucleotide to modify the chromosome distant from the break, 20 μl (at 50 μM or 50 pmol/μl) is used.

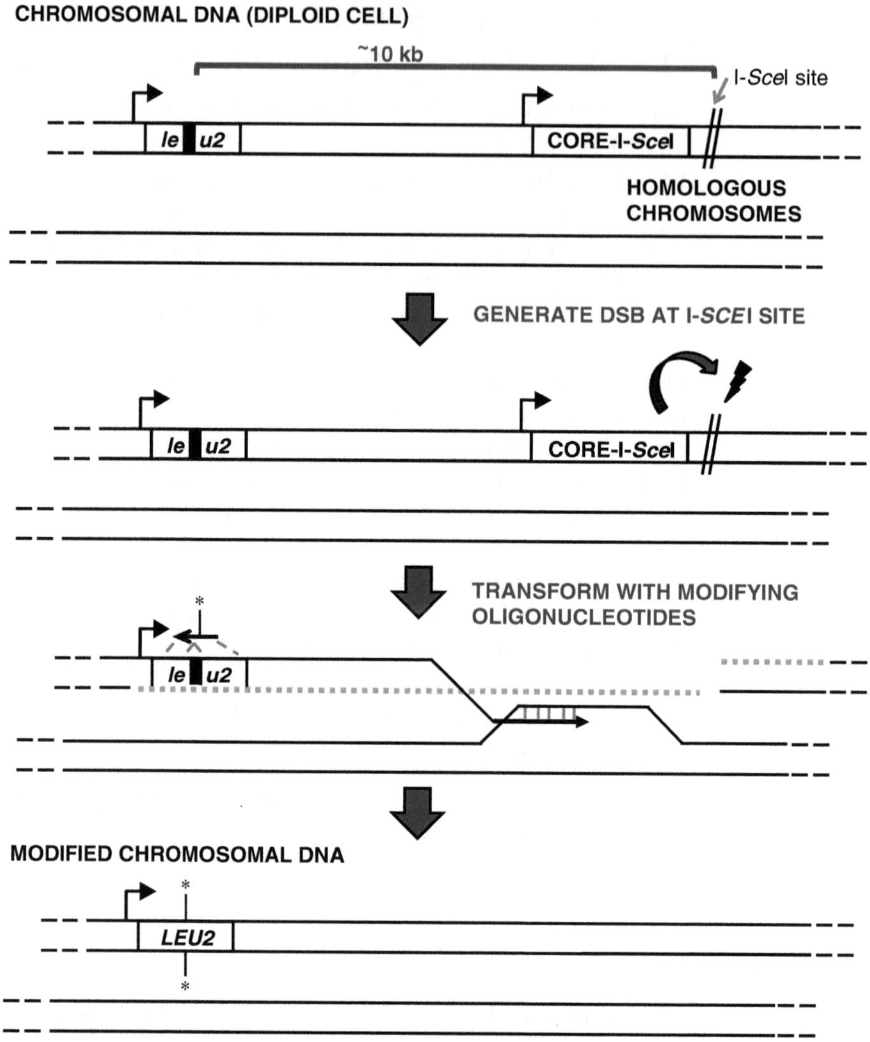

Fig. 2 Modification of a distant locus 10 kb upstream of the DSB site in diploid yeast cells. In this example, the starting strain contains a disrupted *leu2* gene located approximately 10 kb upstream from the 18-nt recognition site for the I-*Sce*I endonuclease. Following addition of galactose, a targeted DSB is generated at the I-*Sce*I site, and 5′ to 3′ resection occurs (indicated by the *light gray dashed lines*), leaving long 3′ single-stranded stretches of DNA. Next, oligonucleotides (here a single oligonucleotide is shown) are transformed into the cells with complementary to both sides of the disruption in *leu2* (as indicated by *dashed lines* between the oligonucleotide and genomic region). Oligonucleotides which repair the break are unnecessary here as the homologous chromosome is available for strand invasion by the 3′ overhang (illustrated as the strand from the broken chromosome pairing with the complementary sequence of the homolog), which can lead to loss of the CORE-I-*Sce*I cassette (as shown here). An *asterisk* on the oligonucleotide used to correct the *leu2* gene indicates the modification. Following transformation, the break is repaired by the homologous chromosome, while the target locus is modified (as indicated by the *asterisk*) to introduce the desired genetic modification in *LEU2*

When using a complementary pair of oligonucleotides, 10 μl of each are used.

9. Add 300 μl of solution 2 to each transformation tube. Mix briefly by vortexing.
10. Incubate transformation tubes at 30 °C for 30 min with shaking.
11. Heat shock at 42 °C for 15 min.
12. Collect cells by centrifugation at 2,236 × *g* for 4 min.
13. Remove the supernatant and resuspend cells well in 100 μl of sterile water.
14. Plate a 10^5-fold dilution per strain to SD complete solid medium. Plate all cells from each transformation tube on one synthetic dextrose complete medium lacking leucine (SC-Leu) plate using approximately 15 sterile glass beads and incubate at 30 °C for 3–4 days until colonies appear (*see* **Notes 12** and **13**).

4 Notes

1. For medium preparation, deionized water is used. All other uses of the term "water" in this chapter, however, refer to water that has been sterilized by filtration or autoclaving.
2. All solid media are autoclaved for at least 45 min at 121 °C and then cooled to 55–60 °C prior to pouring.
3. Unless otherwise noted, all solid media are to be stored at 4 °C. YPD liquid and solid preparations are exceptions that can be stored at room temperature due to frequent use.
4. Filter-sterilizing the 50 % PEG 4000 solution can take up to 1 h depending on the volume due to the high viscosity of the mixture. Autoclaving is generally used for sterilizing this solution.
5. We use Ex Taq DNA polymerase for amplification of the cassettes as it consistently produces a higher yield of CORE cassette amplification than Taq DNA polymerase.
6. For O/N growth, 50-ml conical tubes are used. Additionally, lids should not be capped tightly because *S. cerevisiae* is an aerobic species. Instead, loosely cover the tube and secure with tape.
7. When the *Kl-URA3* gene is inserted in the same orientation as the targeted gene, promoter occlusion may occur where interference from the target gene's promoter during transcription may lead to delayed growth of Ura^+ colonies. If colonies on SC-Ura medium are not observed after 3 days, the following alternative approach can be used: First, plate onto YPD and incubate at 30 °C O/N. Then, replica-plate to G418- or hygro-containing medium, depending on the CORE used, and incubate at 30 °C for 2–3 days. Finally, replica-plate to SC-Ura and incubate at 30 °C O/N. If possible, it is always

preferred to have the *Kl-URA3* marker gene oriented in the opposite direction relative to nearby promoters (*see* also [32]).

8. This review provides examples of gene correction at a disrupted marker that is distant from the site of the DSB induced to stimulate the genetic modification. Our approach works well when the desired genetic modification results in a phenotypic change. If the desired genetic modification does not generate a change in the cell phenotype that can allow for easy selection, an alternative approach would be to screen by PCR aliquots of samples deriving from transformant cells, using primer(s) specific to the expected genetic modification. Following generation of a distant DSB, we have generally observed targeted gene correction frequencies varying from 10^{-3} to 10^{-4} ([17] and unpublished data). Using this as a guideline, we suggest testing by PCR 10–100 aliquots of cell samples deriving from 100 to 1,000 transformant cells. The aliquot(s) resulting positive at PCR should then be diluted and retested by PCR in order to narrow the number of cells containing the desired change. Such procedure should be repeated until the clone(s) with the desired mutation is identified.
9. YPLac is used to provide a neutral carbon source for the cells prior to inoculation with galactose so as to not have inhibition of the galactose-inducible promoter. Since cells grow much slower in this medium, it is optimal to inoculate cells into YPLac at least 20–24 h prior to the transformation.
10. Depending on the distance between the site of the DSB and the chromosomal locus to be targeted, the amount of time needed for resection will vary. Given the observed resection rate in yeast of 1 nt/s [19], an incubation time of 7 h following addition of galactose should provide ample opportunity for both expression of the nuclease and resection to provide lengths of single-stranded DNA more than 10 kb away.
11. A single or complementary pair of oligonucleotides can be designed to simultaneously repair the DSB and remove the CORE-I-*Sce*I cassette. This is advantageous as it allows for the markers present within the cassette to become available for generation of subsequent genetic modifications. If it is desired that the cassette remains in place, the DSB repairing oligonucleotide(s) can be designed accordingly.
12. SD complete medium is used here to determine the numbers of viable cells that survived the transformation. An appropriate dilution should be plated to ensure enough cells will be spread far enough for accurate counting.
13. It may be necessary to plate a dilution from the transformed tube to a selective medium. This depends on the expected numbers of colonies following transformation and is up to the

determination of the researcher. We generally plate both a 10^{-1} dilution and all remaining cells to a selective medium when using a pair of correcting/repairing complementary oligonucleotides.

Acknowledgments

We thank all the members of the Storici lab for their suggestions and contributions to the editing and revision of this work. This work was supported by the Graduate Assistance in Areas of National Need (GAANN) fellowship to S.S., the NSF grant MCB-1021763 and the Georgia Cancer Coalition grant R9028 to F.S.

References

1. Rothstein R (1991) Targeting, disruption, replacement, and allele rescue: integrative DNA transformation in yeast. Methods Enzymol 194:281–301
2. Thomas KR, Capecchi MR (1987) Site-directed mutagenesis by gene targeting in mouse embryo-derived stem cells. Cell 51:503–512
3. Doetschman T, Maeda N, Smithies O (1988) Targeted mutation of the Hprt gene in mouse embryonic stem cells. Proc Natl Acad Sci U S A 85:8583–8587
4. Kuehn MR, Bradley A, Robertson EJ, Evans MJ (1987) A potential animal model for Lesch-Nyhan syndrome through introduction of HPRT mutations into mice. Nature 326: 295–298
5. Schaefer DG, Zryd JP (1997) Efficient gene targeting in the moss Physcomitrella patens. Plant J 11:1195–1206
6. Rong YS, Golic KG (2000) Gene targeting by homologous recombination in Drosophila. Science 288:2013–2018
7. Porteus MH, Baltimore D (2003) Chimeric nucleases stimulate gene targeting in human cells. Science 300:763
8. Liang F, Han M, Romanienko PJ, Jasin M (1998) Homology-directed repair is a major double-strand break repair pathway in mammalian cells. Proc Natl Acad Sci U S A 95:5172–5177
9. Bibikova M, Beumer K, Trautman JK, Carroll D (2003) Enhancing gene targeting with designed zinc finger nucleases. Science 300:764
10. Vasquez KM, Marburger K, Intody Z, Wilson JH (2001) Manipulating the mammalian genome by homologous recombination. Proc Natl Acad Sci U S A 98:8403–8410
11. Silva G, Poirot L, Galetto R, Smith J, Montoya G, Duchateau P, Paques F (2011) Meganucleases and other tools for targeted genome engineering: perspectives and challenges for gene therapy. Curr Gene Ther 11:11–27
12. Kumar S, Allen GC, Thompson WF (2006) Gene targeting in plants: fingers on the move. Trends Plant Sci 11:159–161
13. Donoho G, Jasin M, Berg P (1998) Analysis of gene targeting and intrachromosomal homologous recombination stimulated by genomic double-strand breaks in mouse embryonic stem cells. Mol Cell Biol 18:4070–4078
14. Richardson C, Moynahan ME, Jasin M (1998) Double-strand break repair by interchromosomal recombination: suppression of chromosomal translocations. Genes Dev 12:3831–3842
15. Takata M, Sasaki MS, Sonoda E, Morrison C, Hashimoto M, Utsumi H, Yamaguchi-Iwai Y, Shinohara A, Takeda S (1998) Homologous recombination and non-homologous end-joining pathways of DNA double-strand break repair have overlapping roles in the maintenance of chromosomal integrity in vertebrate cells. EMBO J 17:5497–5508
16. Schaefer DG (2001) Gene targeting in Physcomitrella patens. Curr Opin Plant Biol 4:143–150
17. Storici F, Snipe JR, Chan GK, Gordenin DA, Resnick MA (2006) Conservative repair of a chromosomal double-strand break by single-strand DNA through two steps of annealing. Mol Cell Biol 26:7645–7657
18. Yang Y, Sterling J, Storici F, Resnick MA, Gordenin DA (2008) Hypermutability of damaged single-strand DNA formed at double-strand breaks and uncapped telomeres in yeast Saccharomyces cerevisiae. PLoS Genet 4: e1000264
19. Lee SE, Pellicioli A, Demeter J, Vaze MP, Gasch AP, Malkova A, Brown PO, Botstein D,

Stearns T, Foiani M, Haber JE (2000) Arrest, adaptation, and recovery following a chromosome double-strand break in Saccharomyces cerevisiae. Cold Spring Harb Symp Quant Biol 65:303–314

20. Storici F, Resnick MA (2006) The delitto perfetto approach to in vivo site-directed mutagenesis and chromosome rearrangements with synthetic oligonucleotides in yeast. Methods Enzymol 409:329–345
21. Rahn JJ, Rowley B, Lowery MP, Coletta LD, Limanni T, Nairn RS, Adair GM (2011) Effects of varying gene targeting parameters on processing of recombination intermediates by ERCC1-XPF. DNA Repair (Amst) 10:188–198
22. Storici F, Durham CL, Gordenin DA, Resnick MA (2003) Chromosomal site-specific double-strand breaks are efficiently targeted for repair by oligonucleotides in yeast. Proc Natl Acad Sci U S A 100:14994–14999
23. Urnov FD, Miller JC, Lee YL, Beausejour CM, Rock JM, Augustus S, Jamieson AC, Porteus MH, Gregory PD, Holmes MC (2005) Highly efficient endogenous human gene correction using designed zinc-finger nucleases. Nature 435:646–651
24. Ira G, Pellicioli A, Balijja A, Wang X, Fiorani S, Carotenuto W, Liberi G, Bressan D, Wan L, Hollingsworth NM, Haber JE, Foiani M (2004) DNA end resection, homologous recombination and DNA damage checkpoint activation require CDK1. Nature 431:1011–1017
25. Perez C, Guyot V, Cabaniols JP, Gouble A, Micheaux B, Smith J, Leduc S, Paques F, Duchateau P (2005) Factors affecting double-strand break-induced homologous recombination in mammalian cells. Biotechniques 39:109–115
26. Miller JC, Holmes MC, Wang J, Guschin DY, Lee YL, Rupniewski I, Beausejour CM, Waite AJ, Wang NS, Kim KA, Gregory PD, Pabo CO, Rebar EJ (2007) An improved zinc-finger nuclease architecture for highly specific genome editing. Nat Biotechnol 25:778–785
27. Maeder ML, Thibodeau-Beganny S, Osiak A, Wright DA, Anthony RM, Eichtinger M, Jiang T, Foley JE, Winfrey RJ, Townsend JA, Unger-Wallace E, Sander JD, Muller-Lerch F, Fu F, Pearlberg J, Gobel C, Dassie JP, Pruett-Miller SM, Porteus MH, Sgroi DC, Iafrate AJ, Dobbs D, McCray PB Jr, Cathomen T, Voytas DF, Joung JK (2008) Rapid "open-source" engineering of customized zinc-finger nucleases for highly efficient gene modification. Mol Cell 31:294–301
28. Christian M, Cermak T, Doyle EL, Schmidt C, Zhang F, Hummel A, Bogdanove AJ, Voytas DF (2010) Targeting DNA double-strand breaks with TAL effector nucleases. Genetics 186:757–761
29. Hockemeyer D, Wang H, Kiani S, Lai CS, Gao Q, Cassady JP, Cost GJ, Zhang L, Santiago Y, Miller JC, Zeitler B, Cherone JM, Meng X, Hinkley SJ, Rebar EJ, Gregory PD, Urnov FD, Jaenisch R (2011) Genetic engineering of human pluripotent cells using TALE nucleases. Nat Biotechnol 29:731–734
30. Mali P, Yang L, Esvelt KM, Aach J, Guell M, Dicarlo JE, Norville JE, Church GM (2013) RNA-guided human genome engineering via Cas9. Science 339(6121):823–826
31. Cong L, Ran FA, Cox D, Lin S, Barretto R, Habib N, Hsu PD, Wu X, Jiang W, Marraffini LA, Zhang F (2013) Multiplex genome engineering using CRISPR/Cas systems. Science 339(6121):819–823
32. Stuckey S, Mukherjee K, Storici F (2011) In vivo site-specific mutagenesis and gene collage using the delitto perfetto system in yeast Saccharomyces cerevisiae. Methods Mol Biol 745:173–191
33. Storici F, Lewis LK, Resnick MA (2001) In vivo site-directed mutagenesis using oligonucleotides. Nat Biotechnol 19:773–776
34. Wach A, Brachat A, Pohlmann R, Philippsen P (1994) New heterologous modules for classical or PCR-based gene disruptions in Saccharomyces cerevisiae. Yeast 10:1793–1808

Chapter 21

A Southern Blot Protocol to Detect Chimeric Nuclease-Mediated Gene Repair

Céline J. Rocca, Hayder H. Abdul-Razak, Michael C. Holmes, Philip D. Gregory, and Rafael J. Yáñez-Muñoz

Abstract

Gene targeting by homologous recombination at chromosomal endogenous loci has traditionally been considered a low-efficiency process. However, the effectiveness of such so-called genome surgery or genome editing has recently been drastically improved through technical developments, chiefly the use of designer nucleases like zinc-finger nucleases (ZFNs), meganucleases, transcription activator-like effector nucleases (TALENs) and CRISPR/Cas nucleases. These enzymes are custom designed to recognize long target sites and introduce double-strand breaks (DSBs) at specific target loci in the genome, which in turn mediate significant improvements in the frequency of homologous recombination. Here, we describe a Southern blot-based assay that allows detection of gene repair and estimation of repair frequencies in a cell population, useful in cases where the targeted modification itself cannot be detected by restriction digest. This is achieved through detection of a silent restriction site introduced alongside the desired mutation, in our particular example using integration-deficient lentiviral vectors (IDLVs) coding for ZFNs and a suitable DNA repair template.

Key words Zinc-finger nucleases, Integration-deficient lentiviral vectors, Homologous recombination, Southern blot

1 Introduction

Gene targeting is a process in which information from a DNA molecule introduced into a cell replaces that in the corresponding chromosomal segment. The process relies on homologous recombination, one of the major cellular pathways for DSB repair. Gene targeting is a very low-efficiency process that can be considerably enhanced by the introduction of a DSB at the target locus. This major development was initially achieved using the *S. cerevisiae* homing endonuclease I-*Sce*I and pre-engineered target sites (reviewed in [1]). Practical application had to wait for the development of target site-specific chimeric nucleases, initially ZFNs and I-*Cre*I-based meganucleases, and more recently TALENs and

Francesca Storici (ed.), *Gene Correction: Methods and Protocols*, Methods in Molecular Biology, vol. 1114, DOI 10.1007/978-1-62703-761-7_21, © Springer Science+Business Media, LLC 2014

CRISPR/Cas nucleases. These enzymes can introduce DSBs at specific target sites, which in turn stimulate the cell's endogenous homologous recombination machinery [2, 3]. The first example of targeted genome manipulation in vitro with engineered nucleases at a human endogenous locus was based on the use of ZFNs [4]. These enzymes have also mediated the first example of in vivo genome surgery [5]. The following references provide recent reviews of engineered nuclease technology [6–8].

An important requirement to assess the efficiency of gene targeting is the availability of an in vitro assay to detect the phenomenon and estimate its frequency. While deep-sequencing technologies have facilitated this considerably, normally through custom services, it is still useful to have a gel-based assay for in-house estimation of gene targeting frequencies. Here, we describe in detail a Southern blot-based protocol that allows detection and quantification of a surrogate restriction site, to estimate correction of a linked targeted point mutation in cell lines and primary cells.

The protocol we will describe here does not require a particular type of designer nuclease, but was developed using a ZFN, and for this reason, a brief description of this particular type of enzymes follows. ZFNs are heterodimeric enzymes that combine the nonspecific cleavage domain from *Fok*I restriction endonuclease with DNA-binding zinc-finger domains. The monomers are composed of several tandemly arranged zinc-finger domains (each recognizing a trinucleotide in the target site) and a nuclease domain. The enzyme dimerizes through the *Fok*I domains, providing a tool to recognize a specific site in which a spacer of about six nucleotides where the cutting will occur is flanked by the sites recognized by the two monomers (Fig. 1). Because the recognition specificities of the zinc-finger domains can be easily manipulated experimentally, ZFNs offer a general way to induce targeted site-specific DSBs in the genome [9]. A major issue with designer nucleases is off-target cutting-mediated mutagenesis or toxicity, whereby unintended DSBs introduced in the target genome lead to unwanted effects [10]. The development of obligate heterodimeric ZFNs and further refinements of the nuclease domain have helped to address this problem [11–15], but generic concerns with nuclease off-target cutting remain. In any case, targeted nuclease technology has become a powerful tool for genetic manipulations, not only for site-specific genome surgery in plant and mammalian genomes but potentially also for human therapy.

Firstly, once a strategy for targeted modification has been decided upon, a target site for the nuclease must be chosen. As gene conversion tracts are generally short [16, 17], the nuclease recognition site should be chosen as close as possible to the target site to be modified. The ZiFiT software package identifies potential target sites in DNA sequences for which ZFNs maybe engineered [18]. Once the ZFN is obtained, its cutting efficiency can be tested

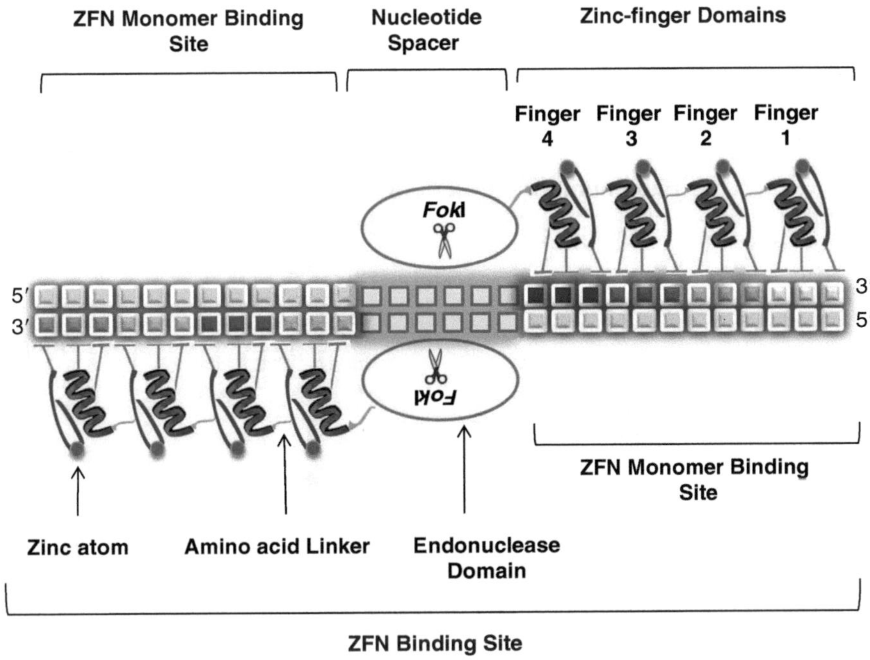

Fig. 1 Schematic of ZFN bound to target DNA. A standard ZFN is a heterodimeric enzyme. Each monomer is made up of several zinc-fingers linked to a nonspecific nuclease domain from *FokI* endonuclease. Dimerization through the nuclease domains allows recognition of two hemi-sites of about 12 nucleotides each, separated by a spacer of about 6 base pairs where cutting will occur

using the surveyor *CelI* assay as described by Guschin et al. [19] or deep sequencing, in both cases detecting minor modifications (insertions or deletions, so-called InDels) introduced at the target site by nonhomologous end-joining repair of the induced DSBs.

Secondly, a DNA repair template (aka targeting construct, donor DNA or repair matrix) must be obtained. Our goal was the correction of a point mutation in a mouse gene. To avoid the presence of heterologies that can reduce the frequency of gene targeting [20], we prepared a 1.7 kb corrective template using high-fidelity PCR amplification of genomic DNA from wild-type fibroblasts of the same mouse strain, followed by DNA sequencing to check for absence of unwanted sequence changes. Afterwards, using standard site-directed mutagenesis, we introduced a silent diagnostic restriction site on the DNA repair template, as close as possible to the site that would correct the genomic mutation (Fig. 2). As our ZFN target site was located 3′ to the mutation on the target gene, we placed our silent site 5′ to the wild-type site on the repair template. In this configuration, incorporation of the silent restriction site on the targeted gene would ensure the presence of the linked wild-type nucleotide and hence the reversion of the mutation (Fig. 2).

Delivery of the corrective repair template and the genes encoding the ZFN monomers may be achieved by several methods, in

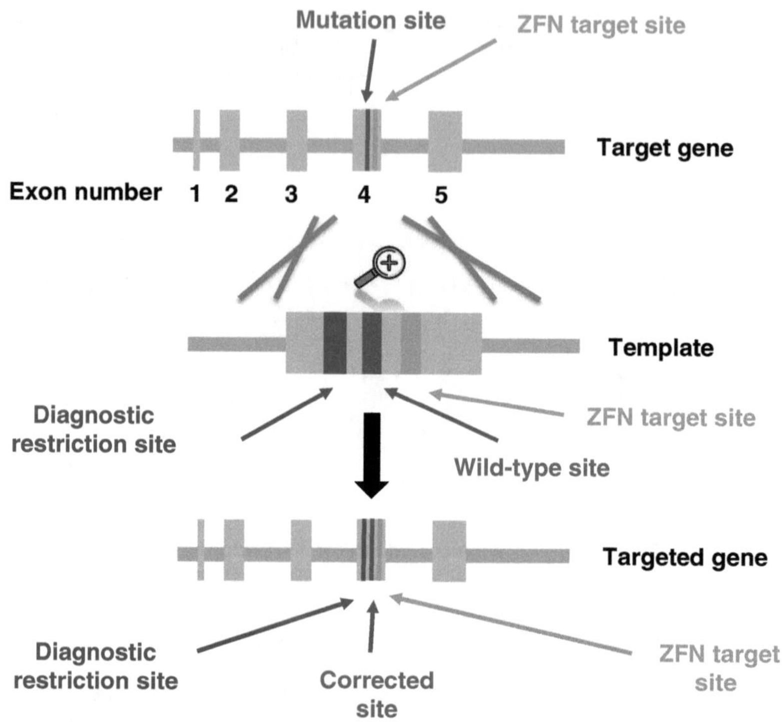

Fig. 2 Gene targeting strategy. A ZFN target site has been selected 3′ to the point mutation to be corrected in the chromosomal locus. As this particular correction event cannot be detected by restriction analysis, a diagnostic restriction site made up of silent mutations has been introduced 5′ to the wild-type site on the repair template. Upon successful homologous recombination (*represented by crossovers*), the gene is repaired and simultaneously the diagnostic restriction site is transferred to the chromosome. Note that the target site for the ZFN could be removed from the template (for instance, by site-directed mutagenesis), but that would decrease the homology between target gene and template and likely reduce the frequency of homologous recombination

our case integration-deficient lentiviral vectors (IDLVs) [21, 22]. Once the target cells have been transfected or transduced with the relevant vectors, the cellular DNA is extracted to analyze for the presence of the desired change. For this the DNA sequence around the target site is PCR amplified with primers external to the corrective template, to prevent amplification of template molecules not yet diluted out or randomly integrated in the cells (Fig. 3a). The PCR product is then digested overnight with the diagnostic restriction enzyme. The digestion mixture is loaded onto an agarose gel, electrophoresed, transferred onto a membrane, and subsequently hybridized, using the initial PCR product as a labeled probe (Fig. 3b). In this manner, we have been able to visualize the three bands expected if homologous recombination-mediated gene repair occurs between the cellular DNA and the DNA repair template. Using Image J, it was also possible to quantify by

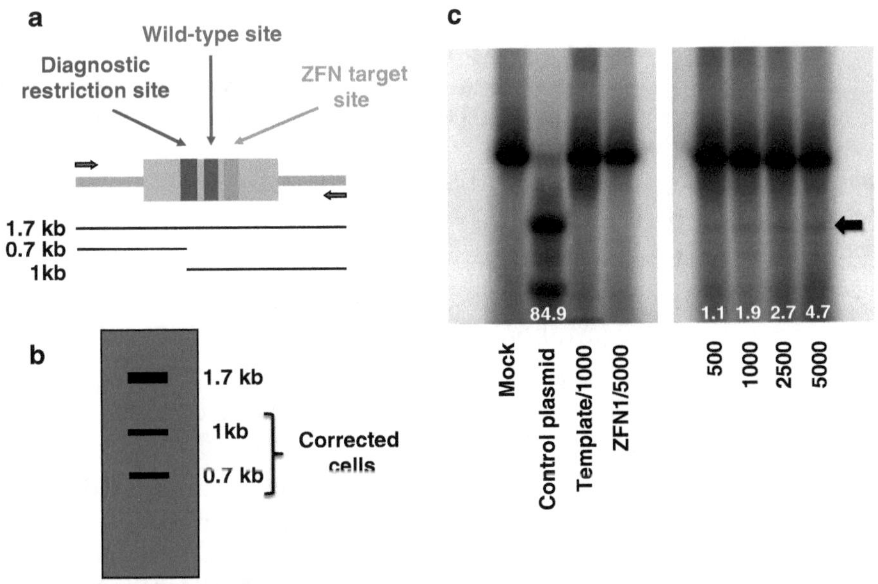

Fig. 3 Detection and estimation of gene targeting. (**a**, **b**) Schematic representations of targeted locus and expected restriction products. Successful gene correction will incorporate a diagnostic restriction site next to the corrected point mutation. Amplification of the chromosomal target with primers external to the targeting construct (*arrows*), followed by restriction digest, should produce three bands if gene targeting has been successful. (**c**) Example image showing detection and quantification of gene repair. Genomic DNA samples were PCR amplified, digested, transferred, and hybridized as described. Mock corresponds to non-transduced fibroblasts, and plasmid control is the original plasmid containing the DNA repair template modified by site-directed mutagenesis. Other controls included fibroblasts transduced with IDLV-template alone (MOI 1,000) or IDLV-ZFN1 alone (MOI 5,000). For gene targeting experiments fibroblasts were transduced with IDLVs carrying the ZFN monomers (at the indicated MOIs) and template (at twice the indicated MOI). Five days after transduction, genomic DNA was isolated and tested for the presence of the diagnostic restriction site for gene correction. The values corresponding to the percentage of corrected alleles are indicated in *white*

densitometry the 1.7 kb band from untargeted alleles and the 1 kb band diagnostic for gene repair, and to estimate the percentage of gene targeting from their relative values. We chose not to include the diagnostic 0.7 kb band on the densitometry measurements because of the overlap with nonspecific smear at low molecular weight (Fig. 3c). This method has allowed us to estimate frequencies of correction of about 2 % of alleles in the population.

2 Materials

2.1 In Vitro Transduction of Fibroblasts with IDLVs Coding for the ZFN Monomers and the Repair Template

1. Mouse fibroblasts.
2. Culture medium: Dulbecco's Modified Eagle Medium (DMEM with high glucose, Gibco/Invitrogen) supplemented with 10 % fetal bovine serum (FBS, Gibco/Invitrogen) and 1× penicillin/streptomycin (100× solution, Gibco/Invitrogen).
3. Trypsin solution (0.25 % trypsin/EDTA, Gibco/Invitrogen).

4. IDLVs encoding ZFN1, ZFN2, and template. A complete protocol for producing and titrating lentiviral particles has been described by Giry-Laterrière et al. [23]. Suggestions pertaining to IDLVs have been published before [22].

2.2 Genomic DNA Extraction and PCR Amplification of the Targeted Region

1. DNeasy Blood & Tissue Kit (Qiagen).
2. 100 % ethanol.
3. PCR reaction: 10× LongAmp Taq reaction buffer (NEB), 10 mM dNTPs, external primers to the template diluted to 10 μM, LongAmp Taq polymerase (NEB), dH_2O.
4. 0.7 % agarose gel in 1× TAE. 50× TAE: In 900 ml dH_2O, 242 g Tris base, 57.1 ml Glacial Acetic Acid, 18.6 g EDTA. Adjust pH to 8 and the volume to 1 L.
5. Appropriate waste disposal for ethidium bromide-contaminated materials following local H&S guidelines.

2.3 Restriction Enzyme Digestion of PCR Product, Agarose Gel Electrophoresis, and Salt Transfer

1. Appropriate restriction enzyme (NEB), 10× NEB buffer, 10× BSA if required according to manufacturer's instructions.
2. DNA size markers without loading dye.
3. 10× Nick translation buffer (NTB): 0.5 M Tris–HCl pH 7.5, 0.1 M $MgSO_4$, 1 mM DTT, 500 μg/ml BSA
4. CTG: 0.5 mM of each dCTP, dTTP, and dGTP in 10 mM Tris–HCl pH 7.5.
5. [α-^{32}P]dATP.
6. DNA polymerase I, large (Klenow) fragment (NEB, M0210S).
7. TE buffer (10 mM Tris, 1 mM EDTA, pH 8.0).
8. Pharmacia Microspin S-300HR columns.
9. 0.7 % agarose gel in 1× TAE, cast in large electrophoresis tray (~200 ml) in order to obtain a good separation of the digested fragments.
10. 0.25 N HCl: 975 ml ddH_2O with 25 ml of 37 % HCl added dropwise (make up in a fume hood).
11. Denaturing solution: 0.4 N NaOH (16 g), 0.6 M NaCl (35.08 g), ddH_2O to 1 L.
12. Neutralizing solution: 1.5 M NaCl (87.66 g), 0.5 M Tris–HCl (60.55 g), 32 ml HCl, ddH_2O to 1 L.
13. GeneScreen Plus membrane.
14. 20× SSC: 3 M NaCl (175.35 g), 0.3 M trisodium citrate dihydrate (88.23 g), ddH_2O to 1 L.
15. Whatman 3MM paper.
16. Blotting paper.
17. Glass plate (same size as agarose gel).
18. Weight, about 500 g (we routinely use half-full 500 ml bottle).

19. 0.4 N NaOH: 16 g NaOH, ddH_2O to 1 L.
20. 0.2 M Tris–HCl pH 7.5 or 8, 1× SSC: 40 ml 1 M Tris, 10 ml 20× SSC, 150 ml ddH_2O.
21. Appropriate waste disposal for radioactively contaminated materials, according to local H&S guidelines.

2.4 Probe Labeling and Hybridization

1. TE buffer.
2. 60 ng/μl dN6 (random hexanucleotides) (Sigma, H-0268).
3. 10× random priming buffer (RP): 0.5 M Tris–HCl pH 6.9, 0.1 M $MgSO_4$, 1 mM DTT (store at −20 °C).
4. CTG: 0.5 mM of each dCTP, dTTP, and dGTP in 10 mM Tris–HCl pH 7.5.
5. [α-^{32}P]dATP.
6. DNA polymerase I, large (Klenow) fragment (NEB, M0210S).
7. Pharmacia Microspin S-300HR columns.
8. Church mix (1 % BSA, 7 % SDS, 0.5 M phosphate buffer): prepare solutions 1, 2, and 3; mix them; and make up to 500 ml with ddH_2O.
9. Solution 1 for Church mix: 35 g SDS dissolved in ddH_2O and made up to 300 ml.
10. Solution 2 for Church mix: 5 g BSA dissolved in ddH_2O and made up to 50 ml.
11. Solution 3 for Church mix: 28.4 g Na_2HPO_4, 6.9 g $NaH_2PO_4 \cdot H_2O$ dissolved in water (warming up in microwave oven) and made up to 125 ml.
12. 2× SSC 0.5 % SDS: 100 ml 20× SSC, 5 g SDS, ddH_2O to 1 L.
13. Plastic bag suitable for heat sealing.

3 Methods

3.1 In Vitro Transduction of Fibroblasts with IDLVs Coding for ZFN Monomers and the Repair Template

Carry out all procedures under sterile cell culture conditions.

1. Wash a 75-cm^2 flask of fibroblasts (70–80 % confluent) with 15 ml PBS and trypsinize the cells using 1.5 ml trypsin-0.25 % EDTA.
2. Resuspend the cells in 10 ml of DMEM 10 % FBS and transfer them to a 15 ml conical tube.
3. Centrifuge the cells at 478 ×*g* for 5 min, remove the supernatant, and resuspend the pellet in 10 ml of DMEM 10 % FBS.
4. Determine the number of cells/ml using a hemocytometer.
5. Plate out 10^5 cells per well in a 6-well plate. Place the plate in an incubator at 37 °C, 5 % CO_2 overnight (*see* **Note 1**).

6. The next day, replace the medium with a minimum volume of fresh medium, ensuring that there is enough to cover the cell monolayer (e.g., 0.5-ml/well for a 6-well plate). This will maximize the chances of interaction between vectors and cells (*see* **Note 2**).
7. Transduce with appropriate MOI (multiplicity of infection) of ZFN1, ZFN2, and template vectors, using a relative ratio of 1:1:2, respectively (*see* **Note 3**).
8. Return the plate to the incubator for 3–4 h and then top up each well with 1.5 ml of fresh medium. Incubate the plate for 5 more days.

3.2 Genomic DNA Extraction and PCR Amplification of the Targeted Region

1. Remove medium from each well, wash with 2 ml PBS, and trypsinize the cells using 100 μl trypsin –0.25 % EDTA.
2. Resuspend the cells in 1 ml DMEM 10 % FBS and then transfer to 1.5 ml microfuge tubes.
3. Centrifuge the cells at 478 × *g* for 5 min, remove supernatant, and resuspend pellet in 1 ml PBS.
4. Centrifuge the cells again at 478 × *g* for 5 min, remove supernatant, and resuspend cells in 200 μl PBS.
5. Proceed to genomic DNA extraction according to DNeasy Blood & Tissue Kit (Qiagen).
6. Add 20 μl proteinase K and 200 μl buffer AL and mix thoroughly by vortexing.
7. Incubate the samples at 56 °C for 10 min.
8. Add 200 μl ethanol (96–100 %) and mix thoroughly by vortexing.
9. Transfer the mixture onto a DNeasy Mini spin column placed in a 2 ml collection tube.
10. Centrifuge at 4,720 × *g* for 1 min. Discard the flow-through and collection tube.
11. Place the spin column in a new 2 ml collection tube.
12. Add 500 μl buffer AW1.
13. Centrifuge for 1 min at 4,720 × *g*.
14. Discard the flow-through and collection tube.
15. Place the spin column in a new 2 ml collection tube.
16. Add 500 μl buffer AW2.
17. Centrifuge for 3 min at 14,460 × *g*. Discard the flow-through and collection tube.
18. Transfer the spin column to a new 1.5 ml or 2 ml microfuge tube.

19. Elute the DNA by adding 200 μl buffer AE to the center of the spin column membrane. Incubate for 1 min at room temperature (15–25 °C). Centrifuge for 1 min at 4,720 × *g* and keep DNA-containing eluate.
20. Measure DNA concentration by optical density at 260 nm (Nanodrop or similar).
21. Perform a PCR amplification using primers external to the template DNA region.
22. Prepare the PCR mix as follows:
    - 100 ng of genomic DNA.
    - 1× LongAmp Taq reaction buffer.
    - 300 μM dNTPs.
    - 0.4 μM forward primer.
    - 0.4 μM reverse primer.
    - 2.5 U LongAmp Taq DNA polymerase (NEB, M0323S).
    - Complete to 25 μl with nuclease-free water.
23. Gently mix the reaction on ice and transfer tubes to PCR machine with the block preheated to 94 °C.
24. Begin thermocycling (94 °C 30 s, 30× (94 °C 30 s, 59 °C 60 s, 65 °C 1 min 20 s), 65 °C 10 min, hold 4 °C).
25. Load 5 μl of the reaction in a 0.7 % TAE-agarose gel containing 0.5 μg/ml ethidium bromide and run for 50 min at 50 V.
26. Expose the gel to UV light to verify specificity of amplification and size of the amplified product.

3.3 Restriction Enzyme Digestion of PCR Product, Agarose Gel Electrophoresis, and Salt Transfer

Reminder: A diagnostic restriction site will have been introduced next to the corrected site if homologous recombination-mediated gene repair has taken place. Take standard precautions when handling radioactive materials and dispose of waste according to local H&S regulations.

1. Measure the DNA concentration of each PCR product by optical density at 260 nm.
2. Transfer 500 ng into clean 1.5 ml microfuge tubes.
3. Add the following:
    - 5 μl 10× enzyme buffer.
    - 5 μl 10× BSA (if required).
    - 1 μl of the appropriate enzyme.
    - Top up to 50 μl with sterile distilled water.
4. Incubate at 37 °C for 16 h.
5. Prepare radioactive size markers ("hot ladder") as follows (*see* **Note 4**):

6. Make up mix of the following:
   - 1 μg suitable DNA size marker (without loading dye).
   - 2 μl 10× Nick translation buffer (NTB).
   - 4 μl CTG.
   - 1 μl (10 μCi) [α-32P]dATP.
   - 1 μl Klenow (5 U/μl).
   - Top up with water to 20 μl.
7. Mix gently and incubate at room temperature for 90 min.
8. Add 40 μl of TE buffer and mix.
9. Count 1 μl in Biomax Cerenkov Geiger counter.
10. Pack Pharmacia Microspin S-300HR column by spinning at 660 × *g* for 1 min.
11. Load sample onto column, spin at 660 × *g* for 2 min and keep eluate.
12. Count 1 μl of eluate in Biomax Cerenkov Geiger counter (incorporation should be around 50 %).
13. Keep the hot ladder (eluate) frozen for up to several weeks if necessary.
14. Cast a 200 ml 0.7 % agarose gel in 1× TAE containing 0.5 μg/ml ethidium bromide.
15. Load lanes with 15–45 Cerenkov Geiger counts of hot ladder or the totality of the digested products.
16. Run gel overnight at 35 V in 1× TAE-buffer with 0.5 μg/ml ethidium bromide, in cold room.
17. Take and save a picture.
18. Shake the gel in 0.25 N HCl for 10 min.
19. Wash gel twice with dH_2O.
20. Shake the gel in denaturing solution for 30 min.
21. Wash gel twice with ddH_2O.
22. Shake the gel in neutralizing solution for at least 30 min.
23. Cut GeneScreen Plus membrane about 1 cm longer and wider than the gel (gel stretches when smoothed on transfer platform). Cut a corner as an orientation reference.
24. Pre-wet the membrane in water for a few seconds and equilibrate it in 10× SSC for 15 min.
25. Set up a capillary transfer blot using 10× SSC (remove air bubbles by rolling pipette). Order from bottom:
   - Inverted gel (isolate the gel edges with plastic film).
   - GeneScreen Plus membrane.
   - Two 10× SSC-soaked Whatman 3MM paper pieces (same size as gel).

- Stack of blotting paper about 10 cm high (same size as gel)
- Glass plate.
- ~500 g weight.

26. Transfer overnight.
27. Dismount the transfer blot.
28. Shake the membrane in 0.4 N NaOH for 1 min.
29. Shake the membrane in 0.2 M Tris–HCl pH 7.5 or 8 for 1 min.
30. Wash membrane quickly in 2× SCC.
31. Place the membrane DNA side up on 2× SCC-soaked Whatman 3MM paper.
32. Fix in Stratagene UV cross-linker using auto cross-link function (the membrane can be dried on filter paper for storage or kept in 2× SSC if proceeding to hybridization).

3.4 Probe Labeling and Hybridization

Reminder: To identify the bands of interest, we probe the membrane with the starting PCR product. This fragment will hybridize with the 1.7 kb PCR product as well as with the 1 and 0.7 kb digested fragments (*see* **Note 5**). Take standard precautions when handling radioactive materials and dispose of waste according to local H&S regulations.

1. Transfer 50–100 ng of the PCR product into a 1.5 ml microfuge tube (ideally screw cap) and top up to 4 μl with TE buffer.
2. Add 1 μl of 60 ng/μl dN6 and boil for 5 min in a heat block.
3. Transfer onto ice for 2 min and afterwards spin down for a few seconds at full speed.
4. Add 12 μl of the following mixture:
   - 12 μl water.
   - 3 μl 10× random priming buffer (RP).
   - 3 μl CTG.
5. Add 2 μl (20 μCi) [α-32P]dATP (*see* **Note 6**).
6. Add 1 μl of Klenow (5 U/μl).
7. Incubate at room temperature for 1 h.
8. Add 40 μl TE buffer.
9. Count 1 μl in Biomax Cerenkov Geiger counter.
10. Pack Pharmacia Microspin S-300HR column by spinning at 660 × *g* for 1 min.
11. Load sample onto column and spin at 660 × *g* for 2 min; keep eluate.
12. Count 1 μl of eluate in Biomax Cerenkov Geiger counter (incorporation should be around 50 %).

13. Wet the DNA membrane from Subheading 3.3, **step 32** in 2× SSC if it was stored dry and spread it in a hybridization tube. Drain 2× SSC from tube.
14. Add 15 ml of Church mix.
15. Pre-hybridize rotating for at least 1 h at 68 °C.
16. Boil the labeled probe for 5 min in a heat block and transfer it to ice for 2 min.
17. Add the boiled probe to the hybridization tube.
18. Hybridize at 68 °C rotating overnight.
19. Discard the hybridization solution.
20. Perform 3 quick washes and 3 × 10 min washes with 2× SSC, 0.5 % SDS at 65 °C (*see* **Note 7**).
21. Seal the membrane in a plastic bag.
22. Expose to a phosphorimager screen.
23. Acquire image by using a laser scanner (Fig. 3).

3.5 Estimation of the Percentage of Corrected Alleles Using Image J

Image J software can be freely downloaded from http://rsbweb.nih.gov/ij/download.html

1. Using Image J, quantify the intensity (I) of the relevant bands (1.7 kb, 1 kb, and 0.7 kb) on the phosphorimager image (*see* **Note 8**).
2. The percentage of corrected alleles = $(100 \times (I_{1\ \mathrm{kb}} + I_{0.7\ \mathrm{kb}}))/(I_{1.7\ \mathrm{kb}} + I_{1\ \mathrm{kb}} + I_{0.7\ \mathrm{kb}})$. Or if 0.7 kb band is not scanned for densitometry (*see* **Note 5**), percentage of corrected alleles = $(100 \times (1.7 \times I_{1\ \mathrm{kb}}))/(I_{1.7\ \mathrm{kb}} + 1.7 \times I_{1\ \mathrm{kb}})$.

4 Notes

1. It is also possible to plate the cells out on the day of transduction, waiting for them to attach before vector is added.
2. Another way to maximize transduction is to add vectors to the cells in suspension, before plating out. Briefly, put the same number of cells in 1.5 ml microfuge tubes in a minimum volume, add the vectors, mix gently, and incubate at 37 °C for 1 to 2 h (mix every 15 min as the cells will drop to the bottom of the tube). Then, either wash the cells once, centrifuge, and resuspend them in an appropriate volume of medium for plating in a 6-well plate well or add medium to the microfuge tube and plate out the mixture directly if a longer transduction time is preferred.
3. MOI refers to the number of vector particles added per cell. As a rule of thumb, we would normally start with MOI 100 for the ZFN monomer vectors and 200 for the repair matrix.

Try different ratios, keeping 1:1 for the ZFN monomers but increasing the amount of DNA repair template, it may improve the rate of correction.

4. The use of nonradioactive size markers may be possible, as the bands corresponding to the different sizes sometimes show up with varying intensity due to nonspecific hybridization with the radiolabeled probe.
5. In the example shown, we chose not to include the diagnostic 0.7 kb band on the densitometry measurements because of the overlap with nonspecific smear at low molecular weight (Fig. 3c).
6. We use radiolabeled dATP because it is the most frequent nucleotide in our DNA template. You should identify the most suitable nucleotide in your own template.
7. If the signal is too strong and a higher stringency wash is needed, proceed stepwise with 1–3 × 10 min washes at 65 °C in:
   - 0.5× SSC, 0.5 % SDS.
   - 0.2× SSC, 0.5 % SDS.
   - 0.1× SSC, 0.5 % SDS.
8. There are other program packages that allow densitometric quantifications, including Adobe Photoshop and Odyssey.

Acknowledgments

The authors acknowledge financial support from the 7th EU Framework Programme (PERSIST project, grant agreement no. 222878) and the Primary Immunodeficiency Association.

References

1. Yáñez RJ, Porter ACG (1998) Therapeutic gene targeting. Gene Ther 5:149–159
2. Carroll D (2004) Using nucleases to stimulate homologous recombination. Methods Mol Biol 262:195–207
3. Porteus MH, Carroll D (2005) Gene targeting using zinc finger nucleases. Nat Biotechnol 23:967–973
4. Urnov FD et al (2005) Highly efficient endogenous human gene correction using designed zinc-finger nucleases. Nature 435:646–651
5. Li H et al (2011) In vivo genome editing restores haemostasis in a mouse model of haemophilia. Nature 475:217–221
6. Arnould S et al (2011) The I-CreI meganuclease and its engineered derivatives: applications from cell modification to gene therapy. Protein Eng Des Sel: PEDS 24:27–31
7. Carroll D (2011) Genome engineering with zinc-finger nucleases. Genetics 188:773–782
8. Mussolino C, Cathomen T (2012) TALE nucleases: tailored genome engineering made easy. Curr Opin Biotechnol 23:644–650
9. Wu J, Kandavelou K, Chandrasegaran S (2007) Custom-designed zinc finger nucleases: what is next? Cell Mol Life Sci: CMLS 64:2933–2944
10. Cornu TI, Cathomen T (2010) Quantification of zinc finger nuclease-associated toxicity. Methods Mol Biol 649:237–245
11. Doyon Y et al (2011) Enhancing zinc-finger-nuclease activity with improved obligate heterodimeric architectures. Nat Methods 8:74–79
12. Miller JC et al (2007) An improved zinc-finger nuclease architecture for highly specific genome editing. Nat Biotechnol 25:778–785

13. Ramalingam S, Kandavelou K, Rajenderan R, Chandrasegaran S (2011) Creating designed zinc-finger nucleases with minimal cytotoxicity. J Mol Biol 405:630–641
14. Sollu C et al (2010) Autonomous zinc-finger nuclease pairs for targeted chromosomal deletion. Nucleic Acids Res 38:8269–8276
15. Szczepek M, Brondani V, Buchel J, Serrano L, Segal DJ, Cathomen T (2007) Structure-based redesign of the dimerization interface reduces the toxicity of zinc-finger nucleases. Nat Biotechnol 25:786–793
16. Elliott B, Richardson C, Winderbaum J, Nickoloff JA, Jasin M (1998) Gene conversion tracts from double-strand break repair in mammalian cells. Mol Cell Biol 18:93–101
17. Taghian DG, Nickoloff JA (1997) Chromosomal double-strand breaks induce gene conversion at high frequency in mammalian cells. Mol Cell Biol 17:6386–6393
18. Sander JD, Maeder ML, Reyon D, Voytas DF, Joung JK, Dobbs D (2010) ZiFiT (Zinc Finger Targeter): an updated zinc finger engineering tool. Nucleic Acids Res 38: W462–W468
19. Guschin DY, Waite AJ, Katibah GE, Miller JC, Holmes MC, Rebar EJ (2010) A rapid and general assay for monitoring endogenous gene modification. Methods Mol Biol 649: 247–256
20. te Riele H, Maandag ER, Berns A (1992) Highly efficient gene targeting in embryonic stem cells through homologous recombination with isogenic DNA constructs. Proc Natl Acad Sci U S A 89:5128–5132
21. Yáñez-Muñoz RJ et al (2006) Effective gene therapy with nonintegrating lentiviral vectors. Nat Med 12:348–353
22. Wanisch K, Yáñez-Muñoz RJ (2009) Integration-deficient lentiviral vectors: a slow coming of age. Mol Ther 17:1316–1332
23. Giry-Laterriere M, Verhoeyen E, Salmon P (2011) Lentiviral vectors. Methods Mol Biol 737:183–209

Chapter 22

High-Throughput Cellular Screening of Engineered Nuclease Activity Using the Single-Strand Annealing Assay and Luciferase Reporter

Thomas J. Cradick, Christopher J. Antico, and Gang Bao

Abstract

Engineered nucleases have been used to generate many model organisms and show great promise for therapeutic genome editing. Current methods to evaluate the activity of these nucleases can be laborious and often are hampered by readouts with small signals and a significant amount of background noise. We present a simple method that utilizes the established single-strand annealing (SSA) assay coupled with a luciferase assay to generate a high-throughput analysis of nuclease activity. Luciferase reporters provide a higher signal and lower background levels than fluorescent reporters. We engineered a commercially available luciferase plasmid (pGL4.51, Promega) to generate a set of nuclease target plasmids that produce a high signal and activity that correlates well with in vitro data. The SSA luciferase assay can discriminate between nucleases that give similar signals with other nuclease activity assays. The target plasmid and nucleases are transfected into cells and are generally cultured for 2 days. Luciferase activity is quantified in the same cell culture plate—streamlining the process from transfection to assay. We have used this robust process to investigate the activity of zinc finger nucleases (ZFNs) and transcription activated-like effector nucleases (TALENs).

Key words Engineered nucleases, Zinc finger nuclease, Transcription activated-like effector nuclease, Luciferase, Single-strand annealing

1 Introduction

As the use of engineered nucleases has grown, so has the use and development of assays to analyze their activity. Choosing the optimal assay will depend on the nucleases' downstream application, as assays vary in their sensitivity, readout, and throughput. The in vitro cleavage assay measures the ability of nucleases expressed in cellular lysates to cleave linearized targets. The in vitro cleavage assay is high-throughput and can be performed in 1 day [1, 2]. Although ZFN activity has been shown to translate well from in vitro to cellular assays, TALEN activity differs significantly between in vitro and cellular assays. Subsequently, we have to test all TALEN activity in tissue culture.

Francesca Storici (ed.), *Gene Correction: Methods and Protocols*, Methods in Molecular Biology, vol. 1114,
DOI 10.1007/978-1-62703-761-7_22, © Springer Science+Business Media, LLC 2014

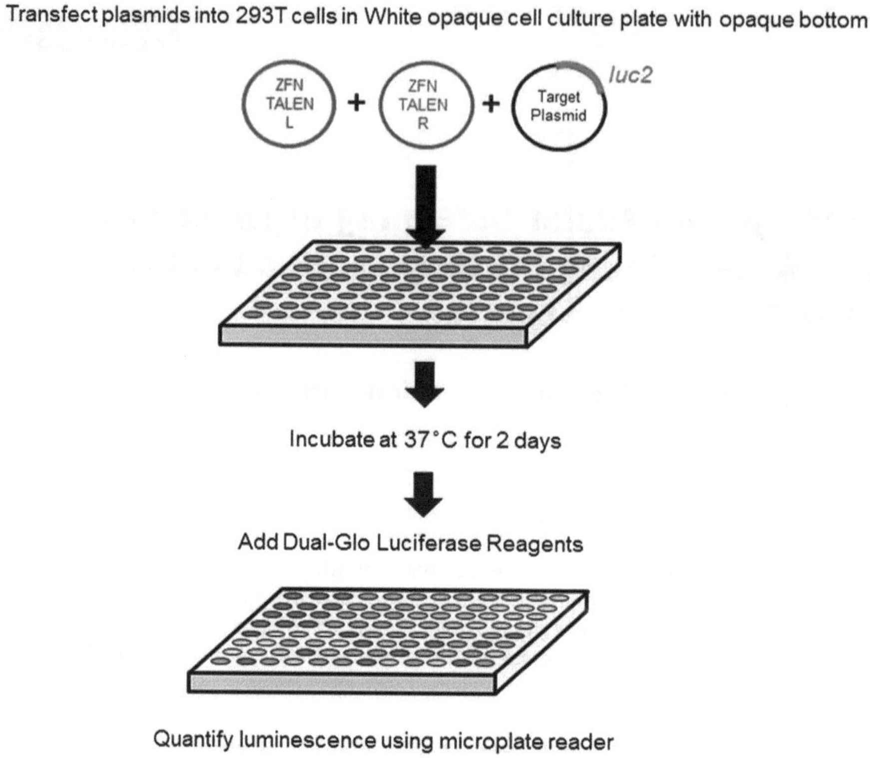

Fig. 1 Method for performing the high-throughput SSA luciferase assay

We hoped to use a cellular assay that was relevant to downstream applications and high throughput. There are a number of different types of cellular nuclease activity assays. Some of these measure cleavage and misrepair of plasmid targets, but must use a multistep process [3, 4]. The widely adopted single-strand annealing (SSA) assay uses the restoration of reporter gene in cells to measure nuclease activity, which is illustrated by Fig. 1 [5]. The level of cleavage and repair can be measured by reporter signal of the repaired SSA target plasmid in cells, but this requires a substantial effort if the sample size is large. Therefore the traditional readout is using a reporter gene such as luciferase or green fluorescent protein (GFP). GFP is easy to use, but has a low signal that makes it difficult to quantitate and distinguish the activity of nucleases from background. To overcome this problem, we used the luciferase reporter, which has a wider dynamic range that allows detection of very low and high signals [6]. Fluorescence is mechanistically different than luminescence because it requires an excitation source to emit a fluorophore. Even without any fluorescent proteins in a sample, the excitation source can still produce a small fluorescent signal. This phenomenon is known as autofluorescence and it can significantly increase the background noise in an assay. Conversely, luminescence is the production of light from the cleavage of luciferin substrate by luciferase, a class of enzymes found in fireflies

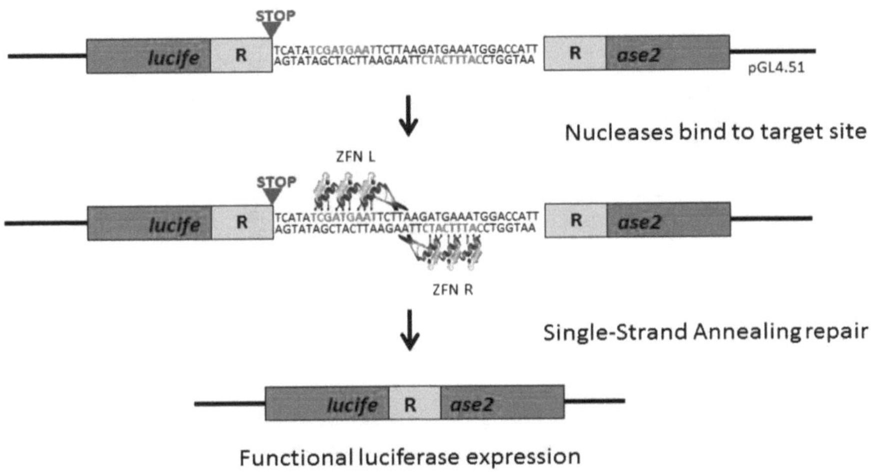

Fig. 2 Luciferase single-strand annealing mechanism

(and other organisms). Unlike GFP, they require no excitation source and have almost no background noise, which makes it a superior reporter for this assay. We have constructed several nuclease target plasmids with luciferase reporters for use in the SSA assay.

In this method, a *luc2* gene on a plasmid is interrupted by stop codons, the nuclease target site, and flanking complementary repeat sequences. The stop codons lead to early termination, producing a nonfunctional luciferase enzyme. Upon cleavage of the target site by the engineered nucleases, the complementary repeats allow SSA repair and expression of the reporter (Fig. 2). Since the efficiency of the SSA assay can depend significantly on the repeat length, we sought to find the optimal repeat length to maximize the potential of the SSA assay [7, 8]. To address this variable, we constructed five similar target plasmid vectors that differed in the length of the complementary repeats: 435, 308, 201, 95, and 25 base pairs. We added target sites to these five vectors and tested with a ZFN monomer and a ZFN pair. In Fig. 3, we found that repeat length is directly proportional to signal and noise. As repeat length increases, signal and noise increase accordingly. However, the repeat length of 308 deviated from this trend. While we observed a high signal, the background signal was significantly lower, giving a large signal to noise ratio. Although the repeat length of 25 bp produced the highest signal to noise ratio, the 25 bp repeat length produced too low a signal, which would make it difficult to detect lower activity events. However, 308 bp repeat length gave us the greatest signal with a relatively low background. Subsequently, we proceeded to optimize our assays using the repeat length of 308 bp.

With an optimized SSA plasmid, we proceeded to develop a high-throughput screen of nuclease activity that utilizes an “all-in-one” plate method that allows culture, transfection and the luciferase assay in the same tissue culture plate. We present detailed protocols

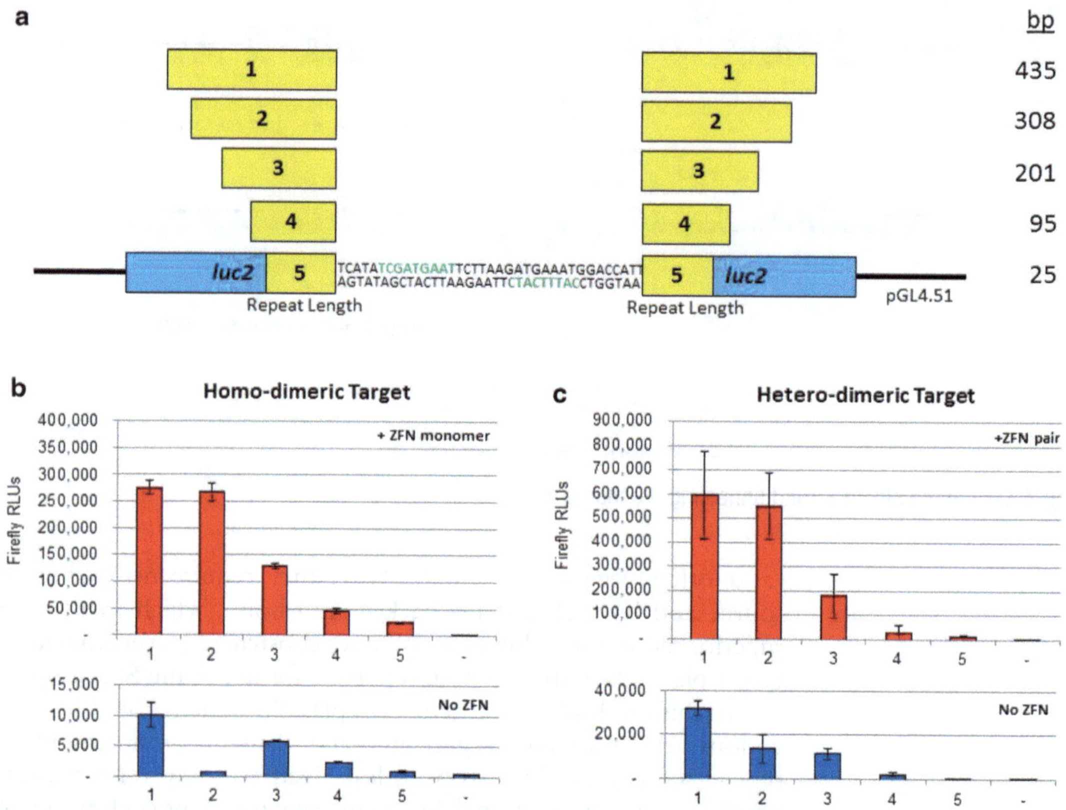

Fig. 3 Optimization of repeat lengths in single-strand annealing assay. (**a**) Location and size of the different repeat lengths used in this experiment. (**b**) Luciferase assay of ZFN monomer. *Red bars* indicate Firefly RLUs of target plasmid with ZFN monomer. *Blue bars* indicate Firefly RLUs of target plasmid only. (**c**) Luciferase assay of ZFN pair. *Red bars* indicate Firefly RLUs of target plasmid with ZFN pair. *Blue bars* indicate Firefly RLUs of target plasmid only

for construction of the nuclease targets, transfection, and the luciferase assay. This method is fast and simple way to distinguish the activity of a single nuclease or nuclease pair in tissue culture. We have used this assay for ZFNs and TALENs, but it can be adapted to measure cleavage activity of other nucleases, such as meganucleases and CRISPR/Cas systems. Setting up the DNA and transfection of 96 samples takes approximately 3 h. The luciferase assay takes approximately 1 h.

2 Materials

2.1 For Generation of Luciferase Target Plasmids

1. Template DNA: pGL4.51 plasmid (25 ng per PCR Amplification) (Promega, Madison, WI).
2. Platinum PCR SuperMix (Invitrogen, Carlsbad, CA).
3. Oligonucleotide primers (100 μM stock concentration).

4. Thermocycler.
5. Chemically competent *Escherichia coli.*
6. Any commercially available ligation kit.
7. Sterile PCR strip tubes (0.2 mL).
8. 1.5 mL microcentrifuge tubes.
9. 6× Gel Loading Dye.
10. Table-top microcentrifuge.
11. Enzymes: Two unique restriction enzymes, AvaI, Alkaline Phosphatase, Calf Intestinal (CIP) (New England Biolabs, Ipswich, MA).
12. 10× T4 DNA Ligase Reaction Buffer.
13. T4 Polynucleotide Kinase (10 U/μL).
14. 37 °C oven.
15. PCR Purification Kit.
16. LB + 100 μg/mL ampicillin broth.
17. LB + 100 μg/mL ampicillin plates.
18. Miniprep Kit.
19. Balance.
20. Agarose.
21. 1× TAE Buffer.
22. Glass Erlenmeyer Flasks.
23. Ethidium Bromide (*see* **Note 1**).
24. Microwave.
25. Agarose Gel Electrophoresis Apparatus.
26. UV Gel Box with Gel imager.

2.2 For Cell Transfection

1. 293T cells (ATCC, Manassas, VA).
2. Corning Assay Plate, 96-well, with lid, flat-bottom, tissue-culture treated, sterile, white polystyrene, opaque-bottom (Corning, St. Louis, MO) (*see* **Note 2**).
3. ZFN or TALEN plasmids (May be assembled or purchased at Sangamo Biosciences, Life Technologies or others).
4. Target plasmid pGL4.51 (from Subheading 2.1).
5. Control plasmid, such as pGL4.74 (Promega, Madison, WI) (*see* **Note 3**).
6. 0.1 % Gelatin.
7. Opti-MEM I Reduced Serum Medium.
8. Fugene HD Transfection Reagent (Promega, Madison, WI) (*see* **Note 4**).
9. Laminar flow hood with UV light source.

10. 293T cell culture media: L-glutamine and FBS should be added to DMEM to give a final concentration of 2 mM and 10 %, respectively.

2.3 For Dual-Glo Luciferase Assay

1. Dual-Glo Luciferase Assay (Promega, Madison, WI) (*see* **Note 5**).
2. 15 mL polypropylene conical tubes (available through VWR International).
3. Microplate Reader (*see* **Note 6**).
4. Reagent reservoirs (for multichannel pipette).
5. Aluminum foil.

3 Methods

3.1 Generation of Luciferase Target Plasmids

3.1.1 Preparation of Luciferase Repeat Lengths

1. To generate varying repeat lengths in luciferase target plasmid, prepare PCR reaction mix with the following components:
   (a) Platinum PCR SuperMix (1.1×)
   (b) Reverse Primer (10 μM): pGL4.51r.1276BsaI (GCGC GGTCTCCTCGGGGTTGTTAACGTAG)
   (c) Forward Primers (10 μM):
   - pGL4.51f.970 (GAGCCTGCAGGGGCACCATCGC CTTTACC)
   - pGL4.51f.1097 (GAGCCTGCAGGTGTGCAGCGA GAATAGCTTG)
   - pGL4.51f.1204 (GAGCCTGCAGGAACAGCATGG GCATCAGC)
   - pGL4.51f.1310 (GAGCCTGCAGGTCATGGATAGC AAGACCGACT)
   - pGL4.51f.1380 (GAGCCTGCAGGCGGCTTCAAC GAGTACGAC)

   (d) Template: pGL4.51 plasmid (25 ng)
2. Run the reaction according to the manufacturer's protocol. Use an annealing temperature of 62 °C.
3. After completion of PCR, check for specific product amplification using agarose gel electrophoresis.
4. Clean-up PCR products using a PCR purification kit.
5. Digest purified PCR reactions (1–5) with SbfI-HF and BsaI overnight in 37 °C oven. After incubation, purify PCR products by using PCR purification kit or using 2 % Size Select E-Gel (Invitrogen, Carlsbad, CA).

3.1.2 Preparation of Nuclease Target Plasmid

1. Design and order sense and antisense oligonucleotides to include the following sequences: 5′—AvaI enzyme recognition site,

two stop codons (TAA and TGA), unique enzyme recognition site, left nuclease half-site, spacer region (w/EcoRI enzyme recognition site), right nuclease half-site, unique enzyme recognition site, and AvaI enzyme recognition site—3′ (*see* **Note 7**).

2. When oligonucleotides arrive, resuspend to a concentration of 100 μM using DI H_2O or 10 mM Tris–HCl.
3. Dilute oligonucleotides to 1 μM (pmol/μL).
4. Kinase and anneal two complementary oligonucleotides (sense and antisense strands) using the following protocols:

   ZFN target oligonucleotide:

   5′—CCCGAATAATGAGGCGCGCC**[left half-site]**GAATTC **[right half-site]**CCTGCAGGTCACGTCCCGAG—3′ (AvaI) **STOP** Unique Site EcoRI Unique Site (AvaI)

   (a) Kinasing
   - Master Mix:
     - 10× T4 DNA Ligase Reaction Buffer
     - 6 U of T4 Polynucleotide Kinase (10 U/μL)
     - Bring final volume to 14.7 μL with DI H_2O
   - Add 14.7 μL of master mix to 5.3 μL (5.3 pmol) of oligonucleotide. One tube will contain the sense strand, and the other tube will contain the antisense strand.
   - Incubate at 37 °C for 30 min.

   (b) Annealing
   - After kinase incubation is complete, combine equal volumes of sense and antisense reaction into one tube and run annealing program on a thermocycler (*see* **Note 8**).
5. Digest pGL4.51 plasmid overnight at 37 °C with AvaI and CIP.
6. Gel-isolate the lineralized pGL4.51 vector using a commercially available gel extraction kit (band size is approximately 5,692 bp).
7. Ligate the PCR product and the pGL4.51 vector together using a commercially available ligation kit and transform into chemically competent *E. coli* (*see* **Note 9**).
8. Plate transformation on LB + 100 μg/mL ampicillin plates and incubate plates overnight (approximately 16 h) at 37 °C.
9. After overnight incubation, pick a few colonies from each plate with a sterile toothpick or pipette tip and place in 5 mL of LB + 100 μg/mL ampicillin broth. Shake overnight at 37 °C and 250 rpm.

10. After overnight incubation, prepare plasmids using any commercially available mini-prep kit for sequencing confirmation and further downstream applications (*see* **Note 10**).
    (a) For sequencing confirmation, use the following primers:
    - pGL4.51For1063 (GGCTGGCAGAAGCTATGAAG)
    - pGL4.51Rev1868 (TGACTGAATCGGACACAAGC)
11. After selecting one plasmid preparation that was sequenced confirmed, digest the plasmid with the following enzymes: SbfI-HF, AvaI, and CIP overnight at 37 °C.
12. Gel-isolate vector using a commercially available gel extraction kit (band size is approximately 5,750 bp). This vector is ready to be ligated with the luciferase repeat lengths.

3.1.3 Ligation of Luciferase Repeat Lengths and Nuclease Target Plasmid

1. Ligate the luciferase repeat lengths (i–v) and the pGL4.51 vector together using a commercially available ligation kit and transform into chemically competent *E. coli.*
2. Plate transformation on LB + 100 μg/mL ampicillin plates and incubate plates overnight (approximately 16 h) at 37 °C.
3. After overnight incubation, pick a few colonies from each plate with a sterile toothpick or pipette tip and place in 5 mL of LB + 100 μg/mL ampicillin broth. Shake overnight at 37 °C and 250 rpm.
4. After overnight incubation, prepare plasmids using any commercially available mini-prep kit for sequencing confirmation (use primers pGL4.51For1063, pGL4.51Rev1868; sequences listed above).

3.2 Cell Transfection

1. Approximately 24 h prior to transfection, the Corning 96-well white opaque culture plate will need to coated by adding 100 μL of 0.1 % Gelatin and incubating at 37 °C for 30 min. After incubation, aspirate off gelatin and proceed immediately to cell seeding.
2. Seed 293T cells into the culture plate at 20,000 cells/well with 100 μL DMEM + 10 % FBS + 2 mM L-glutamine (*see* **Note 11**).
3. Approximately 18–24 h after seeding cells, prepare DNA plasmids per well for Fugene HD Transfection (*see* **Note 12**):
    (a) 25 ng of pGL4.51 Target plasmids (from Subheading 2.1)
    (b) 12.5 ng of each ZFN monomer plasmid (25 ng total) or 25 ng of each TALEN monomer plasmid (50 ng total)
    (c) 5 ng of pGL4.74
    (d) Bring DNA concentration to 330 or 690 ng with pUC18 or other plasmid whose promoter does not compete with the plasmids listed above (*see* **Note 13**).

4. Bring reaction volume to 16 µL with Opti-MEM Reduced Serum Media (*see* **Note 14**).
5. Add 1 µL of Fugene HD Transfection Reagent to 330 ng of plasmid DNA or 2 µL to 690 ng of plasmid DNA. Pipette up and down 15× to mix or briefly vortex.
6. Incubate at RT for approximately 10 min (*see* **Note 15**).
7. Add 5 µL of DNA–Reagent complex to wells and mix by pipetting up and down (15×).
8. Return plate to 37 °C incubator.
9. Approximately 24 h post transfection, change media and replace with 75 µL of DMEM + 10 % FBS + 2 mM L-glutamine.

3.3 Dual-Glo Luciferase Assay

1. Before starting reaction, remove Dual-Glo Luciferase Reagent from −80 °C and let thaw to room temperature. Additionally, bring the Stop & Glo buffer to room temperature (*see* **Note 16**).
2. Remove 96-well plates containing mammalian cells from the incubator. Let plate sit at room temperature for approximately 10 min before performing the assay (*see* **Note 17**).
3. Measuring firefly luciferase activity:
   (a) Add a volume of Dual-Glo Luciferase Reagent equal to the culture medium volume to each well and mix by pipetting. You should observe a slight color change from pink to orange when mixed well. For 96-well plates, typically 75 µl of reagent is added to cells grown in 75 µl of medium (*see* **Note 18**).
   (b) Incubate at room temperature for at least 10 min in the dark by covering the plate in aluminum foil.
   (c) Measure the firefly luminescence on the whole plate on the Tecan Safire2 microplate reader (*see* **Note 19**).
4. After measuring firefly luminescence, proceed immediately to measure Renilla luciferase activity:
   (a) Make Dual-Glo Stop & Glo Reagent by adding 1:100 volume of Stop & Glo substrate to Stop & Glo Buffer (i.e., 100 µL of Stop & Glo substrate to 9.9 mL of Stop & Glo Buffer in 15 mL polypropylene tubes) and mix by inversion until solution becomes light yellow (*see* **Note 20**).
   (b) Add a volume of Dual-Glo® Stop & Glo® Reagent equal to the original culture medium volume to each well and mix by pipetting. You should observe a color change from orange to yellow. At this point each well is nearly full so be careful when moving the plate to not spill. As noted in **step 3a**, this volume is typically 75 µl for 96-well plates (*see* **Note 21**).

(c) Incubate at room temperature at least 10 min in the dark, then measure renilla luminescence on the Tecan Safire2 microplate reader using the same settings in **step 3c**.

3.4 Data Analysis

1. After all luciferase data is collected, each well should have both a firefly and renilla luciferase reading.
2. Take the ratio of firefly to renilla of each well, average the triplicates, and calculate standard error (*see* **Note 22**).

4 Notes

1. Ethidium bromide is a commonly used for detecting DNA in gels. Ethidium bromide is extremely dangerous and a well-known carcinogen. Care must be used and an isolated area for handling ethidium bromide, is suggested. Alternatively, you can use safe, but more expensive DNA gel stains, such as Gel Red (Biotium Inc., Hayward, CA).
2. The Corning Assay Plate (96-well, with lid, flat-bottom, tissue-culture treated, sterile, white polystyrene, opaque-bottom) worked very well for this assay. Previously, we used the same plate with a transparent-bottom, but found a significant level of activity in wells spilling over from adjacent wells, unless wells were skipped. However, skipping wells increased cost and decreased throughput significantly, so we explored other solutions. We applied white reflecting film (USA Scientific, Ocala, FL) on the underside of the plate before analyzing the luciferase activity on the microplate reader. This did not reduce the influence from adjacent wells, but amplified the luciferase signal across all samples. Finally, we used the Corning Assay Plate (96-well, with lid, flat-bottom, tissue-culture treated, sterile, white polystyrene and opaque-bottom plate) and saw approximately a 90 % reduction of influence from adjacent wells. Using this plate allows maximal throughput with 32 samples (replicated three times) including a un-transfected control.
3. We use the control plasmid, pGL4.74, which expresses Renilla luciferase under the control of the HSV-TK promoter, which does not compete with our CMV promoter nuclease and target plasmids. This plasmid is used to normalize expression of firefly luciferase across all samples and should use a promoter different from the nuclease and target plasmids. It can also be utilized to measure cell viability.
4. Fugene HD Transfection Reagent worked optimally in this assay, as it minimized our variability between replicates significantly. Previous transfections used calcium phosphate as a method to

deliver plasmids into cells, but we often observed a large deviation between replicates.

5. To perform the luciferase assay, we initially tried the Dual Luciferase Assay Kit, but we found that luciferase activity was significantly reduced only a minute after the substrate was added to the lysed cells. Using the Dual Luciferase Assay Kit, we could measure only eight samples at a time, which made it difficult to achieve our goal of developing a simple, high-throughput assay. By switching to the Dual-Glo Luciferase Assay kit, the luciferase activity was stable for a longer time, and combined the lysis and the substrate addition steps that are separate actions in the Dual Luciferase Assay kit. This kit allowed us to develop an "all-in-one" assay that minimizes researcher input and maximizes throughput.
6. For all experiments conducted in this chapter, we used the Tecan Safire^2 (Tecan Group Ltd., Switzerland). Any luminometer that is compatible with the Corning Assay Plate (96-well, with lid, flat-bottom, tissue-culture treated, sterile, white polystyrene, opaque-bottom) should be sufficient.
7. When designing the target sequence for your nuclease of interest using our scheme, you need to add AvaI overhangs on the 5′ end of your oligonucleotide to ligate the target site within the *luc2* of the pGL4.51 plasmid. However, you will need to include two unique restriction enzymes outside of your target site to allow easy cloning of different target sites into your vector (our unique sites are SbfI and AscI). Downstream of the AvaI recognition site, we added two stop codons with the following sequence: 5′—TAATGA—3′. Additionally, you can include an enzyme site, such as EcoRI or one with a longer recognition sequence, in the spacer region between the nuclease half-sites to allow an alternative means of testing by looking at the disruption of the enzyme recognition site by mutagenic nonhomologous end joining (NHEJ). This can be easily performed by extracting the plasmid DNA from the transfected cells, PCR amplification of the plasmid target site and enzyme digestion. This is an efficient assay, though some cleavage events might not disrupt the site, especially in the longer spacers of TALENs. When constructing the target sequence for ZFNs, we typically use one sense and antisense oligonucleotide that are kinased and annealed together. However, for TALENs we typically have a three different oligonucleotide pairs (left half-site, spacer, and right half-site) that are kinased and annealed because TALENs typically target a much longer sequence with a larger spacer.
8. To anneal sense and antisense oligonucleotides, we use the following thermocycler annealing program conditions: Incubate

at 95.0 °C for 5 min, Incubate at 90.0 °C and decrease 0.3 °C every 2 min, Incubate at 70.0 °C and decrease 0.3 °C every 2 min, Incubate at 40.3 °C and decrease 0.3 °C every 2 min. Hold at 4 °C. Alternatively, you can float tubes containing oligonucleotides in a beaker of boiling water and incubate the tubes until the beaker of water has cooled to room temperature.

9. To achieve maximum ligation efficiency, ligate PCR insert and vector overnight at 16 °C with T4 DNA Ligase. Alternatively, rapid ligation kits work well for this step.

10. To sequence the target site, use primers pGL4.51For1063 and pGL4.51Rev1868. The forward sequencing primer (For) anneals approximately 350 bp upstream from the double stop codon and the reverse sequencing primer (Rev) anneals approximately 400 bp downstream of the right nuclease half-site.

11. To achieve optimal results, we tested seeding 293T cells at different densities in different cell culture plates. We found that seeding 293T cells at 20,000 cells per well in a 96-well plate approximately 24 h before transfection produces the smallest deviation between replicates using Fugene HD Transfection Reagent. The optimal conditions for different cell types can be found online in the Fugene HD Protocol Database, on Promega's Web site. These parameters may need to be optimized if using a different transfection reagent.

12. The concentration of luciferase target plasmid used for transfection was optimized for a 96-well plate. A previous study included a titration of different concentrations of the pGL4.51 plasmid: 5, 10, 15, 25, 50, and 90 ng. There was no activity observed at 5 ng, but detected a significant increase of activity at 10 ng. At 15, 25, and 50 ng, we observed a 2.5 fold signal increase relative to the activity at 10 ng, but there was no significant difference observed between 15, 25, and 50 ng. At 90 ng, the activity level drops relative to the previous three concentrations, but it was not significant.

    Next, we analyzed the activity level of renilla luciferase by transfecting various concentrations of pGL4.74 plasmid: 1, 2, 5, 10, 15, and 20 ng. Under the HSV-TK promoter, renilla luciferase activity is significantly lower than firefly luciferase, which is useful in this assay because renilla luciferase levels should be slightly above background noise. At 1 or 2 ng of pGL4.74, the luciferase activity was not significantly above background levels, but at 5 or 10 ng, the activity level was significantly higher than background levels by at least twofold. When the concentration is increased to 15 or 20 ng, the renilla luciferase activity significantly increases, relative to 10 ng, by 18 and 28 fold, respectively.

    Next, we determined the best ratio of pGL4.51 to pGL4.74 by analyzing the ratio of firefly to renilla activity. This firefly to renilla ratio will normalize the data and provide optimal and consistent results using the Dual-Glo Luciferase Assay. At a 5:1 ratio of pGL4.51 to pGL4.74, we transfected in the following amounts of plasmid (in ng): 5:1, 10:2, 25:5, 50:10, and 75:15. The highest firefly to renilla value was determined with a ratio of 10 ng of pGL4.51 to 2 ng of pGL4.74, but the overall firefly and renilla signal was very low. The ratio of 25:5 ng was slightly lower, but it was not significant. Compared to the 25:5 ratio, 50:10 ng and 75:15 ng were significantly lower at 1.5 fold and twofold, respectively. The ratio observed that produced the best firefly to renilla ratio with ideal signal activity was

13. All plasmid DNA for transfection was prepared at a 3× concentration. The manufacturer's protocol recommends 330 ng DNA for three wells (done in triplicate; 110 ng per well) at a Fugene HD:DNA ratio of 3:1. However, we attempted to transfect in double the recommended amount at 660 ng and observed luciferase activity equal to or slightly above the recommended concentration. We did not increase the Fugene HD concentration accordingly. At 330 ng, the Fugene HD concentration was 3:1, and the Fugene HD concentration was 1.5:1 at 660 ng. Despite doubling the plasmid concentration, we did not see any signs of increased cytotoxicity.
14. Opti-MEM I Reduced Serum is a media with reduced Fetal Bovine Serum to increase the efficiency of transfection into cells. Fugene HD Transfection Reagent recommends using Opti-MEM I for resuspension of your plasmid DNA. However, we have seen comparable results by resuspending in sterile DI H_2O.
15. The Fugene HD Transfection reagent protocol recommends incubating the DNA and Fugene complex at room temperature from 10 to 30 min. We have found that 10 min is a sufficient amount of time for the DNA complexes to form. Transfection efficiency does not increase significantly with longer incubation times.
16. It is recommended that Dual-Glo Luciferase Reagent be aliquoted into several tubes and stored at −80 °C after being prepared. Luciferase activity will decrease if the Dual-Glo Luciferase Reagent is repeatedly thawed and frozen. However, Stop & Glo substrate can be stored at −20 °C and the Stop & Glo Buffer can be stored at room temperature.
17. It is essential that the Corning Assay Plate (96-well, with lid, flat-bottom, tissue-culture treated, sterile, white polystyrene, opaque-bottom) with transfected cells and all reagents (Dual-Glo Luciferase Reagent and Stop & Glo Buffer) are at room temperature before proceeding with the luciferase assay.

18. Make sure the Dual-Glo Luciferase Reagent is at room temperature before adding to the plate. It is important that no bubbles appear in the well after adding the Dual-Glo Lucfierase Reagent. During the luminescence reading, the bubble may scatter light away from the detector and significantly influence your value. Eliminate any bubbles by puncturing them with a large hypodermic needle. The needle will not affect your luminescence reading.
19. For the plate reader, you want to choose the proper plate format and the end-point setting for the luminescence reading.
20. It is critical that Dual-Glo Stop & Glo Reagent be prepared fresh before each use. Do not store excess reagent for later use.
21. The best results will be generated if luminescence is measured within 2 h of addition of Dual-Glo Stop & Glo Reagent and should be added to plate wells within 4 h of addition of Dual-Glo Luciferase Reagent.
22. Normalize the firefly–renilla ratio to the ratio of a control well or series of control wells that are treated consistently on all plates. This normalization provides optimal and consistent results from the Dual-Glo Luciferase Assay System.

References

1. Ruminy P, Derambure C, Chandrasegaran S, Salier JP (2001) Long-range identification of hepatocyte nuclear factor-3 (FoxA) high and low-affinity binding sites with a chimeric nuclease. J Mol Biol 310(3):523–535
2. Mani M, Kandavelou K, Dy FJ, Durai S, Chandrasegaran S (2005) Design, engineering, and characterization of zinc finger nucleases. Biochem Biophys Res Commun 335(2):447–457
3. Urnov FD, Miller JC, Lee YL, Beausejour CM, Rock JM, Augustus S et al (2005) Highly efficient endogenous human gene correction using designed zinc-finger nucleases. Nature 435(7042):646–651
4. Cradick TJ, Keck K, Bradshaw S et al (2010) Zinc-finger nucleases as a novel therapeutic strategy for targeting hepatitis B virus DNAs. Mol Ther 18(5):947–954
5. Szczepek M, Brondani V, Buchel J et al (2007) Structure-based redesign of the dimerization interface reduces the toxicity of zinc-finger nucleases. Nat Biotechnol 25(7):786–793
6. Choy G, O'Connor S, Diehn FE et al (2003) Comparison of noninvasive fluorescent and bioluminescent small animal optical imaging. Biotechniques 35(5):1022–1030
7. Evans E, Sugawara N, Haber JE et al (2000) The *Saccharomyces cerevisiae* Msh2 mismatch repair protein localizes to recombination intermediates in vivo. Mol Cell 5(5):789–799
8. Preston CR, Engels W, Flores C (2002) Efficient repair of DNA breaks in Drosophila: evidence for single-strand annealing and competition with other repair pathways. Genetics 161(2): 711–720

Chapter 23

An Unbiased Method for Detection of Genome-Wide Off-Target Effects in Cell Lines Treated with Zinc Finger Nucleases

Cory R. Lindsay and David B. Roth

Abstract

We describe a method for detecting and validating genomic aberrations arising from cell lines exposed to zinc finger nucleases (ZFNs), an important reagent used for targeted genome modifications. This method makes use of cloned cell lines, an approach that adds power when testing variables that may affect gene correction efficiency and evaluating potential side effects on a genome-wide scale. After cell treatment, the genomic DNA isolation method, as described, is ideal for high-resolution array comparative genomic hybridization (aCGH) and quantitative PCR. Guidelines for aCGH analysis and calling significant copy number variations (CNVs) for validation by qPCR are also discussed. Using this method, we describe a novel ZFN-associated chromosome 4 copy number variation (CNV) attributable to a predicted ZFN off-target cleavage site found within the CNV.

Key words Array CGH, Gene correction, Copy number variation, Off-target effects, Quantitative PCR, Zinc finger nucleases

1 Introduction

Site-specific nucleases are powerful tools for selectively modifying genomes. These reagents may also induce side effects including genomic aberrations such as deletions, amplifications, and other structural variations due to off-target cleavage [1]. Because these unintended mutations could potentially occur throughout the genome, the risk potential of a given nuclease must be assessed at a genome-wide scale. Array comparative genomic hybridization (aCGH) has become an increasingly refined and sensitive technique for the analysis of copy number variations (CNVs), in the form of either duplications or deletions of chromosomal segments that are larger than 1 kb in size [2]. aCGH relies on comparing a test genome with a reference genome to make inferences about changes in the test sample. A critical part of the experimental design is the choice of the reference DNA sample. As others have shown,

Francesca Storici (ed.), *Gene Correction: Methods and Protocols*, Methods in Molecular Biology, vol. 1114,
DOI 10.1007/978-1-62703-761-7_23, © Springer Science+Business Media, LLC 2014

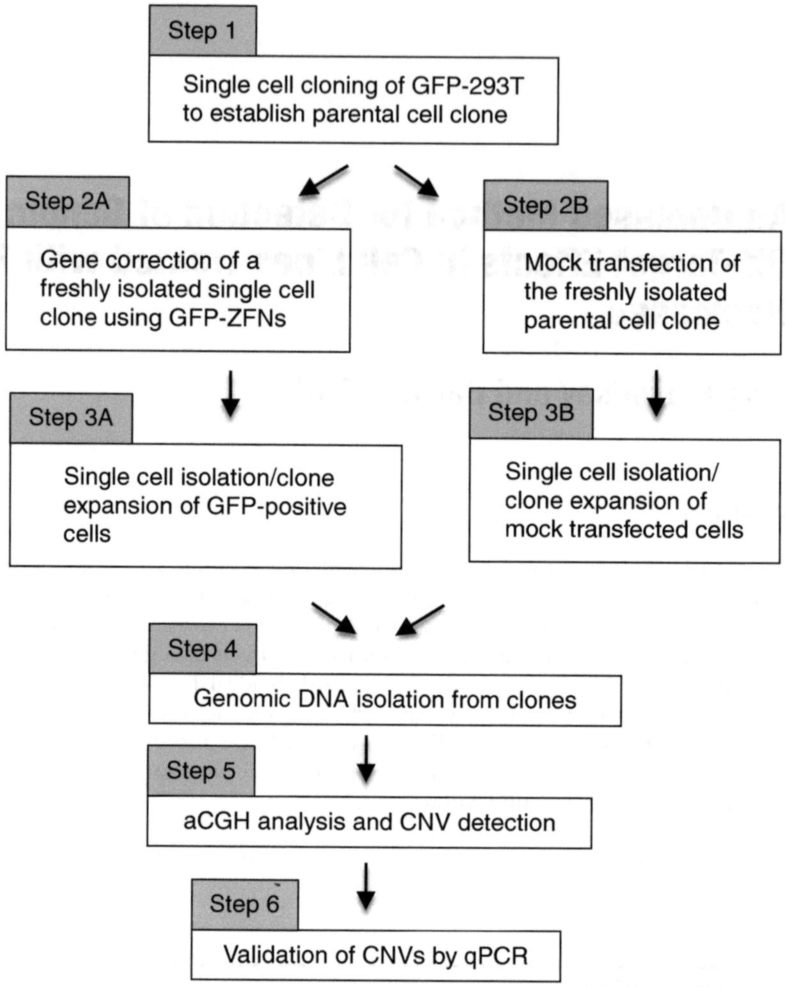

Fig. 1 Flow diagram for detection and validation of CNVs arising from GFP-ZFN mediated gene correction in GFP-293T cell clones. Beginning with establishment of a single cell GFP-293T cell clone, a six step process is used as the basis for testing CNVs that arise from gene correction using GFP-ZFNs compared with mock transfection of the same parental cell clone

a reference DNA prepared from an individual source, as opposed to a population of control DNAs, provides the most accurate calling of CNVs [3].

We utilized an immortalized human 293T cell line to test for off-target effects of zinc finger nucleases (ZFNs). ZFNs are chimeric gene correction reagents consisting of a zinc finger DNA binding domain and nuclease subunit of the FOK1 restriction enzyme [4]. Importantly, at the very outset, the 293T cell line was subcloned and newly isolated single cell clones were then used as the basis for our tests (Fig. 1). In this manner, we avoided a major source of

Table 1
Representation of seven CNVs from a single GFP-positive clone (GFP-ZFN clone 1) corrected by GFP-ZFNs

	Chromosome region	CNV event	Length (bp)	Cytoband	Probe median	# Probes
1	chr1:222,434,558-225,685,762	CN Loss	3,251,205	q42.11–q42.13	-0.20257	103
2	chr3:177,154-731,276	CN Gain	554,123	p26.3	0.36392	12
3	chr3:58,736,305-59,850,120	CN Gain	1,113,816	p14.2	0.30461	21
4	chr4:182,135,181-182,422,754	Homozygous Copy Loss	287,574	q34.3	-1.97856	11
5	chr6:109,304,403-109,583,353	CN Loss	278,951	q21	-0.53435	8
6	chr7:2,096,748-2,339,413	CN Gain	242,666	p22.3–p22.2	0.31756	11
7	chr18:16,100,000-20,266,563	CN Loss	4,166,564	q11.1–q11.2	-0.32658	121

CNVs were called by Nexus Copy Number™ software. Chromosome Region (BED format); CNV Event (copy number (CN) loss/gain, Homozygous copy loss/copy gain); Length of CNV in base pairs (bp); chromosome cytoband region; probe median (log2 ratio median of all probes in CNV) and the number of probes encompassing the CNV (# of probes) are automatically displayed using this analysis. Significant CNVs were selected by removing CNVs encompassing ≤5 array probes or did not meet a log2 ratio (probe median) of ±0.3 or greater. CNVs encompassing a probe median less than ±0.3 but containing a large number of probes were saved from cutoff. The final list, represented here, can then be used for further validation

false CNVs that could arise due to preexisting diversity in the test cell population. When CNVs are called from aCGH data, a larger fraction of changes will intentionally reflect the test conditions and not the inherent variability of the cell population.

While aCGH allows screening on a genome-wide scale, it is necessary to validate or confirm CNVs due to potential false positive/negative results inherent to the analysis method [5]. To minimize potential erroneous CNV calls, genomic aberrations must be validated using separate and independent techniques. The method described here details the use of a quantitative PCR (qPCR) technique that infers chromosome copy number information and is useful as a secondary validation of aCGH data by analyzing individual loci. We utilized a high purity genomic DNA isolation method that meets the quality standards for aCGH and has also proven applicable to next generation sequencing and quantitative PCR methods.

Using the method described here, we show the identification of several CNVs found within a GFP-293T cell clone corrected with GFP-ZFNs (Table 1). A novel chromosome 4 CNV (Fig. 2) attributed to GFP-ZFN off-target activity was validated by chromosome copy number assay (Table 2). This genomic aberration is associated with a potential off-target site identified approximately 7 kb within the CNV region.

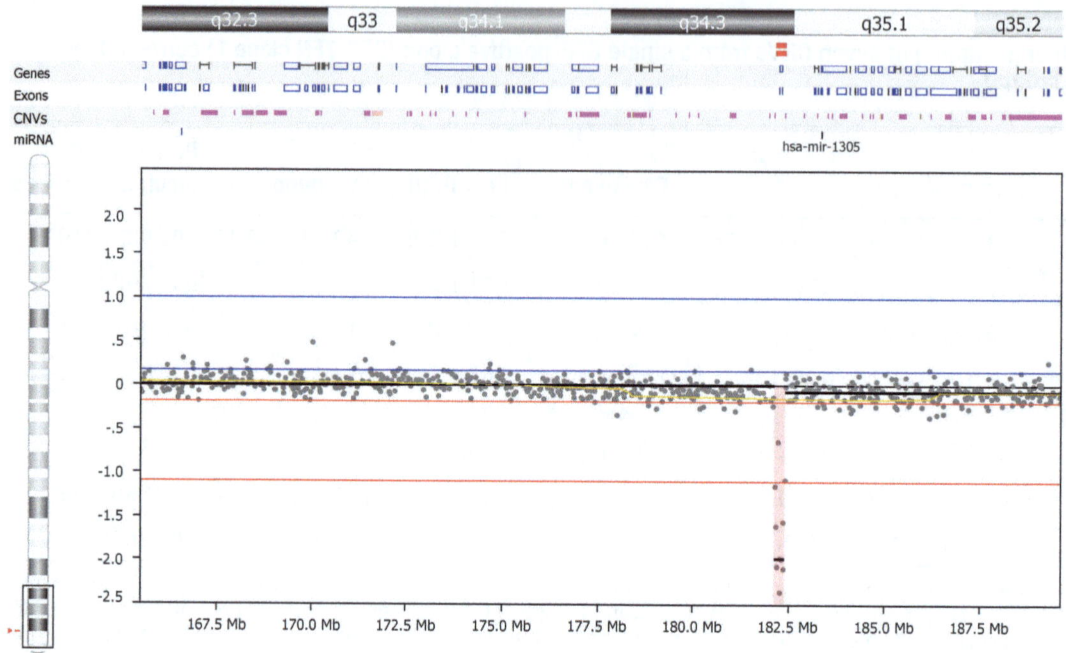

Fig. 2 aCGH density plot (log2 format) displaying a homozygous deletion on chromosome 4 of GFP-ZFN Clone 1. segMNT.txt files containing the normalized probe values corresponding to a ZFN-corrected clone (GFP-ZFN clone 1) hybridized to the parental cell clone were analyzed using Nexus Copy Number™ software. A CNV in the form of a homozygous deletion was identified on chromosome 4 (highlighted region in *red*)

Table 2
Copy number assay results validating a chromosome 4 CNV identified by aCGH

Sample	Reference C_T Avg	Chr4 C_T Avg	ΔC_T	Estimated copy number	Predicted copy number
Mock	25.491	25.959	0.469	2.0168	2
Clone 1	25.625	33.255	7.630	0.0140	0
Clone 2	25.625	25.963	0.337	2.2093	2
Clone 3	25.248	25.067	−0.181	3.1651	3
Clone 4	25.792	26.229	0.438	2.0610	2

A deletion initially identified within GFP-ZFN Clone 1 at chr4: 182,135,181-182,422,754 (Hg18) was used to validate copy number variation within four different GFP-ZFN corrected clones (Clone 1–4) and a mock-transfected cell clone (Mock). Validation of the deletion within GFP-ZFN clone 1 was found, as well as an amplification on the same chromosomal region within GFP-ZFN Clone 3. Reference and Chr4 C_T averages (reference C_T Avg and Chr4 C_T Avg) were calculated based on raw data obtained from the real time-PCR run. ΔC_T and estimated copy number were calculated based on differences in C_T averages between Chr4 and reference. Finally, a predicted copy number is generated based on the estimated copy number calculation: $2 \times (2^{-\Delta\Delta CT})$

2 Materials

2.1 Cell Lines Optimal for Cloning

For the approach described here, it is necessary to choose cell lines that can be diluted/sorted into single cells and grown clonally (*see* **Note 1**). We selected 293T cells that had been stably transduced with a cDNA containing a stop codon and a 35-nucleotide insertion sequence at bp 327 of green fluorescent protein (eGFP) gene that prevents GFP expression.

2.2 Gene Targeting Plasmids

1. ZFN-A (plasmid SP202A), targeting bp 307–315 of eGFP: 5′-GAC GAC GGC-3′.
2. ZFN-B (plasmid SP202B), targeting bp 300–292 of eGFP: 5′-GAA GAT GGT-3′.
3. GFP repair substrate (plasmid PC264), a truncated eGFP sequence (Δ24 bp of eGFP), lacking a CMV promoter.

2.3 Genomic DNA Purification and Quantification

1. 2× STE buffer stock: 200 mM NaCl, 100 mM Tris–HCl (pH 7.5), 2 mM EDTA. To make 5 mL of 1× STE lysis buffer:

2× STE Buffer stock	2.5 mL
Ultrapure water	2 mL
20 % SDS	0.25 mL
10 mg/mL Proteinase K	0.25 mL

2. 10 mg/mL proteinase K solution in double distilled water.
3. 20 % SDS solution in double distilled water.
4. 7.5 M ammonium acetate.
5. 200-proof ethanol.
6. Phenol/Chloroform/Isoamyl alcohol (25:24:1) (Life Technologies, Carlsbad, CA, USA).
7. Phase lock Gel Columns (5 Prime, Gaithersburg, MD, USA).
8. 10 mM Tris–HCl pH 7.4.
9. Qubit® 2.0 Fluorometer (Life Technologies, Carlsbad, CA, USA).
10. Qubit® Broad Range DNA Quantification Kit (Life Technologies, Carlsbad, CA, USA).
11. Qubit® High sensitivity DNA Quantification Kit (Life Technologies, Carlsbad, CA, USA).
12. NanoDrop ND-1000 Spectrophotometer (NanoDrop Technologies, Rockland, DE).

2.4 PCR and qPCR

1. PCR grade water (Roche, Indianapolis, IN, USA).
2. MicroAmp® optical adhesive film (Life Biosciences, Carlsbad, CA, USA).

3. MicroAmp® optical 384-well reaction plates (Life Biosciences, Carlsbad, CA, USA).
4. Taqman® genotyping mastermix (Life Biosciences, Carlsbad, CA, USA).
5. TERT human reference assays (Life Biosciences, Carlsbad, CA, USA).
6. Predesigned human copy number assays (Life Biosciences, Carlsbad, CA, USA).

2.5 Cell Culture

1. Fugene transfection reagent (Promega, Madison, WI, USA).
2. 1× Phosphate buffered saline, pH 7.4 without calcium (Life Technologies, Carlsbad, CA, USA).
3. Dulbecco's Modified Eagle Medium (Life Technologies, Carlsbad, CA, USA).
4. Fetal bovine serum (Life Technologies, Carlsbad, CA, USA).
5. Penicillin/streptomycin liquid (Life Technologies, Carlsbad, CA, USA).
6. 0.05 % Trypsin-EDTA (Life Technologies, Carlsbad, CA, USA).
7. 15 mL Falcon Tubes (BD Biosciences, Franklin Lakes, NJ, USA).
8. Sterile reagent reservoirs (Corning Life Sciences, Corning, NY, USA).
9. 96-well plates (BD Biosciences, Franklin Lakes, NJ, USA).
10. 24-well plates (BD Biosciences, Franklin Lakes, NJ, USA).
11. 6-well plates (BD Biosciences, Franklin Lakes, NJ, USA).

3 Methods

3.1 Selection and Expansion of a GFP 293-T Cell Clone

All procedures are carried out in a BSLII containment cabinet.

1. GFP-293T cells are grown to confluency in a 10 cm tissue culture dish containing complete media (DMEM + 10 % FBS + 1 % penicillin/streptomycin).
2. Media is withdrawn by pipetting.
3. PBS is added to the 10 cm tissue culture dish to remove any traces of media and serum and then is removed.
4. Dilute trypsin (1:10 of 0.05 % stock diluted in PBS) is added to cells for approximately 5 min (*see* **Note 2**).
5. Trypsin is removed from the cells by pipetting.
6. Cells are dislodged from the tissue culture plate by carefully tapping the side of the plate.
7. By pipetting multiple times, cells are thoroughly resuspended in 10 mL of complete media and then collected into a 15 mL Falcon tube.

8. Cell suspensions are diluted serially by taking 1 mL of the resuspended GFP-293T cells and diluting them in 9 mL of complete media (1:10 dilution). $1:10^2$–$1:10^7$ dilutions are made by thoroughly mixing each subsequent dilution and then pipetting 1 mL of this mixture into 9 mL of fresh media.
9. Using a 96-well plate format, only the cell dilutions ranging from $1:10^6$ to $1:10^7$ are plated using a multichannel pipette (*see* **Note 3**).
10. Approximately 4–5 days after plating, individual wells should be analyzed by light microscopy for evidence of clonal expansion. This is observed by looking for small colony formation.
11. By day 8 or 9, wells that contain only one colony will be used for further expansion into 24-well plates and eventually into 6-well plates. One of the single cell clones should then be randomly selected for gene correction experiments involving GFP-ZFNs and mock transfection.

3.2 Mock Transfection of a Freshly Isolated GFP-293T Cell Clone

All procedures are carried out in a BSLII containment cabinet.

1. A 10 cm tissue culture dish containing a freshly isolated clone of GFP-293T cells should be grown to 20–30 % confluency.
2. Cells are mock transfected using a 3:1 ratio of Fugene (μL) to water (μL). 10 μL PCR-grade water is added to a sterile 1.5 mL microcentrifuge tube.
3. Serum free-DMEM is combined with the PCR-grade water to reach a volume of 250 μL.
4. 30 μL of Fugene transfection reagent should be added to a separate 1.5 mL microcentrifuge tube.
5. Serum-free DMEM is added to the microcentrifuge tube in **step 4** to reach 250 μL total volume.
6. The PCR-grade water and Fugene components are combined by pipetting into one microcentrifuge tube and incubated at room temperature for 20 min.
7. The total volume (500 μL) of Fugene:water mixture is then added directly to the 10 cm tissue culture dish containing GFP-293T cells.
8. The tissue culture dish containing tranfection reagents is mixed by gently swirling.
9. After 48 h incubation in a humidified CO_2 incubator, the media containing Fugene is replaced with complete media containing no Fugene:water. By 72 h post-transfection, the cells are ready for clonal selection and expansion.

3.3 Gene Correction of a Freshly Isolated GFP-293T Cell Clone with GFP ZFNs

We utilized GFP-ZFNs designed to target eGFP (GFP ZFN subunit A targeting bp 307–315 of eGFP: 5′-GAC GAC GGC-3′ and GFP ZFN subunit B targeting bp 300–292 of eGFP: 5′-GAA GAT GGT-3′) along with a donor substrate (PC264) lacking a promoter but containing a truncated correct eGFP sequence. GFP-ZFNs were transfected into GFP-293T cells harboring a cDNA with a stop codon and 35-nucleotide insertion sequence at bp 327 of green fluorescent protein (eGFP) gene that prevents GFP expression. Gene correction is initiated by homologous recombination (HR)-directed repair of the chromosomal GFP template, resulting in GFP fluorescence.

All procedures are carried out in a BSLII containment cabinet.

1. A freshly isolated GFP-293T cell clone should be seeded at 20–30 % confluency in a 10 cm tissue culture dish.
2. Cells are transfected in a 3:1 ratio of Fugene (μL) to plasmid DNA (μg). 30 μg plasmid DNA consisting of 10 μg each of GFP zinc finger nuclease subunit A (SP202A) and GFP zinc finger nuclease subunit B (SP202B) along with 10 μg of GFP repair substrate (PC264) are combined in a 1.5 mL microcentrifuge tube.
3. Serum free-DMEM is added to reach a volume of 250 μL.
4. 90 μL of Fugene transfection reagent is added to a separate 1.5 mL microcentrifuge tube along with 160 μL serum-free DMEM to reach a total of 250 μl total volume.
5. The plasmid DNA and Fugene components are then combined (by pipetting) in one microcentrifuge tube and incubated at room temperature for 20 min.
6. The total volume (500 μL) of Fugene:DNA mixture is added directly to the 10 cm tissue culture dish containing GFP-293T cells.
7. The tissue culture dish containing tranfection reagents is mixed by gently swirling.
8. Approximately 48 h post-transfection, media containing Fugene:DNA is replaced with complete media containing no transfection reagents.

3.4 Sorting and Expansion of Single GFP Positive Clones

1. By 72 h posttransfection, media should be removed from the tissue culture dish by pipetting.
2. Cells are washed with 1× PBS.
3. Cells are trypsinized using diluted trypsin (0.05 % diluted 1:10 in 1× PBS).
4. After removing trypsin, cells are dislodged from the tissue culture plate.
5. Cells are resuspended in 1× PBS containing 0.1 % fetal bovine serum in 2 mL total volume.
6. Single GFP-positive cells are sorted into individual wells of a 96-well plate using a Becton Dickinson FACSDiva flow cytometer and cell sorter gated for GFP. Single cells to occupy one

96-well plate are sorted in this manner and allowed to grow for 8–14 days (*see* **Notes 4** and **5**).

7. Clones are selected at random from single wells (as placed in **step 6**) into 1 well of a 24-well plate.
8. Upon reaching confluency in a 24-well plate, clones are trypsinized and resuspended in 1 mL of complete media.
9. Resuspended cells are divided equally (by volume) into 2 wells of a 6-well tissue culture plate for further expansion.
10. An additional 2 mL of complete media is added to each well.
11. When cells become confluent at this stage of clone expansion, genomic DNA is isolated.

3.5 Genomic DNA Purification

For each GFP-293T clone that is sorted for GFP fluorescence (test condition) or isolated by mock transfection (control) and expanded to a 6-well plate, genomic DNA is purified using the method described below.

1. Cells are trypsinized using 0.05 % trypsin.
2. Trypsin is removed and cells are dislodged by gently tapping the sides of the tissue culture plate.
3. Cells are resuspended in 1 mL of complete media and transferred to 1.5 mL microcentrifuge tubes.
4. Cells are pelleted at low centrifugation (1,000 RPM) and then washed twice with 1× PBS.
5. After the second wash, PBS is discarded and 500 μL of cell lysis buffer consisting of 1× STE + 20 % SDS + proteinase K is added to the cell pellet. Cells are lysed at 55 °C overnight.
6. The following day, cell lysates are centrifuged at high speed (13,200 RPM) for 20 min to pellet undigested cell material.
7. Phase lock gel columns are briefly centrifuged for 1 min at 10,000 RPM.
8. Approximately 400 μL of the supernatant from **step 6** is carefully removed and added to a phase lock gel column.
9. An equal volume of phenol/chloroform/isoamyl alcohol (25:24:1) is also added to each phase lock gel column (*see* **Notes 6** and 7).
10. Tubes are mixed gently by inverting (*see* **Note 8**).
11. Tubes are centrifuged at 10,000 RPM for 10 min to separate phases.
12. The aqueous portion (top layer separated by phase lock gel) is removed and placed in a new 2.0 mL microcentrifuge tube.
13. 85 μL of 7.5 M ammonium acetate along with 2 volumes of absolute (100 %) ethanol are added to each 2.0 mL microcentrifuge tube.

14. Genomic DNA is precipitated from solution by inverting tubes several times and then placing the tubes at –80 °C for 30 min (*see* **Note 9**).
15. Genomic DNA is pelleted using high-speed centrifugation (13,200 RPM) for 30 min.
16. After centrifuging, the supernatant is discarded.
17. DNA pellets are washed twice with 500 μL of 70 % ethanol.
18. Genomic DNA should be air dried at room temperature for 15–20 min to remove residual traces of ethanol (*see* **Note 10**).
19. After drying, Genomic DNA should be resuspended in approximately 200 μL of 10 mM Tris–HCl pH 7.4 resuspension buffer (*see* **Note 11**).
20. Genomic DNA is quantified using the Qubit® Broad Range DNA quantification standards before sending samples for aCGH (*see* **Note 12**).

3.6 Array CGH Analysis and Identification of Potential CNVs

We utilized Nimblegen Human 3x720K whole genome tiling array (version 2.0) CGH service for analysis of copy number variation arising from ZFN-mediated gene corrected 293-T cell clones. The 3x720K format consists of one slide containing three genomic microarrays that can be used to test three individual samples. Each genomic microarray contains 720,000 probes that span the entire human genome with a median probe spacing of 3,877 bp. This service also provides data in the form of comparative whole genome and individual chromosome analysis displayed in log2 format which reflects the intensity ratio of Cy3-labeled probe (test genomic DNA) to Cy5-labeled probe (reference genomic DNA) hybridized on a genomic microarray (*see* **Note 13**).

1. For each 3x720K array, approximately 2 μg of each test genomic DNA and 5 μg of reference genomic DNA should be supplied to Nimblegen for processing and fluorescent dye labeling (*see* **Note 14**).
2. Test genomic DNA is assigned a Cy3 label while reference genomic DNA is labeled with Cy5 by Nimblegen. In all cases, the test genomic DNA should consist of gene corrected (GFP positive) cell clones hybridized against reference genomic DNA consisting of the parental mock-transfected cell clone.
3. Data delivery of aCGH results is obtained in DVD format directly from Nimblegen. Included with this service are PDF files of log2 density plots for each test sample compared against the reference sample, as well as tab-delimited TXT files (segMNT.txt files) containing test and reference probe intensities and genomic coordinates corresponding to the test sample.
4. We utilized Nexus Copy Number™ 6.0 software for additional analysis of aCGH data obtained from Nimblegen. This software

offers a very user-friendly interface for analyzing and calling CNVs.

5. Normalized segMNT.txt files for each aCGH experiment are loaded into Nexus Copy Number™ software using Homo sapiens Hg18 NCBI build 36.1 as the reference genome. An overall quality score is automatically generated after loading seMNT.txt files for each array. Quality scores typically range from 0.0001 to 0.03 and higher (*see* **Note 15**).
6. CNV calls for each sample are readily identified in Nexus Copy Number™ by viewing each individual sample and generating a tab-delimited TXT file (*see* Table 1 and **Note 16**).
7. Criteria for selecting CNVs with a significant gain or loss are based initially on identifying CNVs with gains or losses that encompassed at least five consecutive array probes. CNVs that did not meet or exceed this minimum probe number are eliminated for further validation.
8. The probe median cutoff value was set at ±0.3 or greater based on log2 ratio values. CNVs not meeting these minimum log2 ratio values are also eliminated. However, under conditions where CNVs did not meet the probe median cutoff but encompassed an especially large number of probes, these CNVs were saved from cutoff (*see* **Notes 17** and **18**).

3.7 Validation of CNVs by Copy Number Assay

For secondary validation of CNVs identified by aCGH, Taqman® copy number assays were utilized. This assay is useful because it incorporates the same genomic DNA tested by aCGH and utilizes a reference genomic region to compare against the test genomic region of interest by independent FAM™ and VIC™ fluorescent dye quantitation. An extensive collection of predesigned FAM™ dye copy number assays is available for both human and mouse genomes to test nearly any genomic region of interest.

1. A homozygous copy deletion identified on chromosome 4 of GFP-ZFN clone 1 was selected for validation by copy number assay (Fig. 2).
2. The genomic coordinates corresponding to this CNV should be converted from Hg18 to Hg19 (*see* **Note 19**). In this way, the homozygous copy deletion spanning chr4: 182,135,181-182,422,754 was converted to chr4: 181,898,187-182,185,960 in Hg19.
3. Using the GeneAssist™ Copy Number Assay Workflow Builder and the predesigned custom copy number assay option, a probe corresponding to the region identified as a homozygous copy deletion on chromosome 4 is selected for analysis. We selected a predesigned assay corresponding to chr4: 182,065,879 (approximately 0.16 Mb inside of the 0.29 Mb CNV) for use in validating the homozygous copy deletion (*see* **Note 20**).

4. Genomic DNA should be diluted to 5 ng/μL in 10 mM Tris–HCl pH 7.4 and quantified using a Qubit 2.0 fluorometer and a high sensitivity DNA quantification kit (*see* **Note 12**).
5. Copy number assays are set up using the predesigned copy number assay probe identified in **step 3** along with a reference assay for telomerase reverse transcriptase (TERT) located on chromosome 5 (*see* **Notes 21** and **22**). For each clone that will be assayed for copy number changes at chr4:182,065,879, assays are set up in quadruplicate using the following reaction components and volumes:

2× Taqman Genotyping Master Mix	5 μL
Copy Number Assay (Chr4:182,065,879)	0.5 μL
Reference Assay (TERT)	0.5 μL
PCR Grade Water	2 μL
Genomic DNA (5 ng/ μL)	2 μL
	10 μL

6. Reaction components are loaded onto a 384-well plate.
7. Reaction components should be mixed well by pipetting up and down.
8. Adhesive seals are applied to the top of the 384-well plates and should be centrifuged at 1,000 RPM for 2 min to collect the reaction volumes at the bottom of the wells.
9. The 384-well reaction chamber containing the copy number reactions should be loaded onto an Applied Biosystems Viia™ 7 Real-Time PCR System.
10. The thermal cycling reactions are set up using the following reaction conditions:

Stage	Temperature (°C)	Time
Hold	95	10 min
Cycle (40 cycles)	95	15 s
	60	60 s

11. After completion of the real-time PCR run, a Microsoft Excel file containing the completed real-time PCR raw data is generated using Viia™ 7 instrument software v1.1 (*see* **Note 23**).
12. C_T values for the reference assays (VIC™ dye) and copy number assays (FAM™ dye) set up in quadruplicate are used to create median C_T values for each GFP-ZFN corrected clone being assayed at Chr4:182,065,879 (*see* **Note 24**).
13. After creating median reference and copy number assay C_T values for each clone, "clone specific" ΔC_T values (corresponding to

specific cell clones) and an "all samples" ΔC_T value (corresponding to all samples tested) are calculated by subtracting the copy number assay C_T value from the reference assay C_T value.

14. To determine the estimated copy number of each clone for the chromosomal region of interest tested using the reference and copy number assays, the following calculation is used:

$$\text{estimated copy number} = 2 \times (2^{-\Delta\Delta CT})$$

Where $\Delta\Delta C_T$ is the difference between the ΔC_T "clone specific" value and the ΔC_T "all samples" value [6].

Using the estimated copy number value, a predicted copy number is calculated as the closest integer number with respect to the estimated copy number (*see* **Note 25**).

Including the validation of the homozygous copy deletion on chromosome 4 found in GFP-ZFN clone 1, results of a chromosome copy number assay testing three other GFP-ZFN clones [2–4] and a mock-transfected cell clone are summarized (Table 2). We found that the same region of chromosome 4 was also amplified in an independent GFP-ZFN clone (clone 3). Importantly, the region encompassing the CNV identified in GFP-ZFN clone 1 was found to contain a predicted off-target site found at chr4: 182,142,764-182,142,787 (Hg18), approximately 7 kb from the 5′ end of the CNV. This off-target site closely corresponds to a potential homodimer binding site of GFP subunit B, predicted by the ZFN-Site search application (http://ccg.vital-it.ch/tagger/targetsearch.html).

4 Notes

1. Primary human or mouse cell lines are not ideal for single cell cloning. In contrast, transformed or immortalized cell lines can be isolated and grown from single cells.
2. Diluted trypsin is used because it allows for a longer, more controlled dissociation of adherent cells. Cells trypsinized in this manner are more likely to be broken into single cells when pipetted and are less susceptible to effects of over-trypsinization.
3. Use of sterile, disposable reagent reservoir trays in which the volume of media containing cells can be dispensed, is ideal for this step and allows easy and rapid refilling of the multiwell pipette.
4. The range of 8–14 days is wide because some cells are able to grow rapidly almost immediately after sorting, while other cells take longer to grow. After day 7, 96-well plates were monitored on a daily basis for changes in media color. When the media in a well of a 96-well plate became yellow-orange (in contrast to red), the corresponding well was examined

under the microscope and was trypsinized and replated into a single well of a 24-well plate.

5. The ratio of GFP positive/non-GFP positive cells obtained by cotransfection of GFP-ZFNs and a repair substrate is routinely 0.7–1.2 %.
6. Even after centrifuging at high speed, the supernatant can be very difficult to isolate cleanly from the protein pellet due to turbidity. When this occurs, it is better to remove a smaller amount of supernatant so that it is less likely to transfer proteins to the next step. Pipetting the supernatant very slowly from the protein pellet can also help to ensure a cleaner transfer of supernatant.
7. Phase lock gel columns are highly recommended because they eliminate the tediousness that is traditionally associated with purifying DNA using phenol/chloroform extraction. The phase lock columns function by creating a physical barrier with the gel separating the aqueous layer (containing genomic DNA) from the non-aqueous layer. This allows the user to easily isolate the aqueous phase from top of the tube, maximizing recovery of genomic DNA.
8. When using phase lock gel columns it is critical to not vortex the sample as this can damage and reduce the high molecular weight sizes of genomic DNA.
9. When inverting the tubes, strands of genomic DNA should readily be observed precipitating from solution. Do not vortex the genomic DNA at this stage.
10. It is critical to ensure that all traces of ethanol have been removed from the microcentrifuge tube before resuspending genomic DNA. The DNA pellet frequently becomes more translucent and will dry to the wall of the microcentrifuge tube.
11. From two wells of a 6-well plate, typically 10–20 μg of genomic DNA can be purified with a 260:280 nm absorption ratio between 1.8 and 2.0 as measured by NanoDrop ND-1000 spectrophotometer
12. The Qubit fluorometer offers the most accurate quantitation of genomic DNA as it measures only double stranded DNA and not other species of nucleotides. The NanoDrop should be used primarily for assessing 260:280 ratios. DNA quantitation can vary widely between NanoDrop and Qubit readings, with the tendency for NanoDrop to overestimate the concentration of genomic DNA.
13. Several other venders offer aCGH service and may utilize different array formats. Reproducibility of CNVs across these platforms varies, but is generally between 60 and 80 % [7].
14. It is advantageous to supply more than the minimum amount of genomic DNA necessary for aCGH. The quantity of reference

genomic DNA should not be limited, particularly when using the same reference for many different samples or different gene correction experiments.

15. The quality score generated by Nexus Copy Number can be potentially influenced by several factors, including the physical quality of genomic DNA used for aCGH, probe noise related to repetitive sequences found in the genome and other factors [5].
16. This file provides the user with information important for identifying CNVs detected in the test sample and is vital for subsequent assays used to validate CNVs. Information included in this file are the quantity of CNV calls, length of each CNV, chromosome coordinates and cytoband regions relevant to each CNV, as well as the median probe log2 ratio and number of probes contained in each CNV.
17. Depending on the nature of the experiment, the criteria for calling significant CNVs can change. Various studies have incorporated independent cutoff values to establish significance. Some studies have called CNVs based on log2 ratios less than ±0.3 [8], less than five consecutive probes [9, 10], or have established independent cutoff values based on the type of CNV [11]. It is also possible to define your own cutoff values after validating a set of CNVs based on a range of log2 ratios, number of probes or *P*-values.
18. By loading all relevant samples from a single experimental group into Nexus Copy Number™, it is possible to identify two or more samples that may contain similar CNVs (i.e., gain/loss of similar region on the same chromosome). Common regions between CNVs can be identified using the "significant peaks" and "participation" features that will provide statistics relating to the frequency of specific CNVs occurring in samples.
19. It is important to ensure that chromosome regions identified by Nexus Copy Number and chromosome regions selected by the GeneAssist Copy Number Work Flow Builder correspond to the same genomic build. For instance, aCGH data supplied by Nimblegen utilizes Hg18 assembly while GeneAssist uses Hg19 reference assembly. Therefore, it is necessary to convert Hg18 CNV calls to Hg19 in order to select the appropriate copy number assay for qPCR. The UCSC Genome Browser offers a useful conversion format for these purposes (http://genome.ucsc.edu/cgi-bin/hgLiftOver).
20. It is ideal to select a predesigned copy number assay probe that falls well within the chromosome coordinates designated by the specific CNV. Choosing a probe that is too close to either edge of a CNV could produce a false-negative result because of the limit of resolution in the 3x720k array. Higher resolution arrays offer more precise delineation of CNVs, while lower resolution arrays will provide reduced information.

21. There is a choice of two reference assays (human TERT & human RNase P). By comparing the chromosomal positions of each reference probe with aCGH data for the respective chromosomes, the user can choose which assay is most ideal. For example, since human TERT is found on chromosome 5p15.33, it is critical to analyze aCGH data pertaining to this particular region of the chromosome to ensure that it is not influenced by copy number variation in your samples.
22. In addition to clones that were gene-corrected by GFP-ZFNs, it is important to assay appropriate gene correction controls at this stage, including mock-transfected clones or any other appropriate controls.
23. Among other things, this Excel file contains cycle threshold values (C_T values) for reference and copy number assays corresponding to each sample replicate.
24. Potential outlier C_T values from a reference assay can be eliminated if the C_T value is significantly different from the group of C_T values considered as a whole. These outlier C_T values are frequently generated from inaccurate pipetting. If it appears that all of the C_T values of the reference assay do not have a consistent pattern, then the real time-PCR run should be completed again with more accurate pipetting of genomic DNA.
25. Since chromosomes or chromosome regions only exist in integers and not fractions, the closest integer value should be selected for the predicted copy number value.

Acknowledgement

The authors would like to thank Matthew Porteus, University of Stanford, for supplying GFP-293T cells and GFP-ZFN plasmids SP202A, SP202B and PC264 donor substrate. We also thank Susanna Lewis for comments and revisions to the chapter. This work was supported by the NIH Nanomedicine Center for Nucleoprotein Machines PN2 EY018244.

References

1. Gabriel R, Lombardo A, Arens A et al (2011) An unbiased genome-wide analysis of zinc-finger nuclease specificity. Nat Biotechnol 29(9):816–823
2. Feuk L, Cardon AR, Scherer SW (2006) Structural variation in the human genome. Nat Rev Genet 7:85–97
3. Haraksingh RR, Abyzov A, Gerstein M et al (2011) Genome-wide mapping of copy number variation in humans: comparative analysis of high resolution array platforms. PLoS One 6(11):e27589. doi:10.1371/journal.pone.0027859
4. Urnov FD, Rebar EJ, Holmes MC et al (2010) Genome editing with engineered zinc finger nucleases. Nat Rev Genet 11:636–646
5. Tchinda J, Lee C (2006) Detecting copy number variation in the human genome using comparative genomic hybridization. Biotechniques 41(4):385–392

6. Livak KJ, Schmittgen TD (2001) Analysis of relative gene expression data using real-time quantitative PCR and the 2(-delta delta C(T)) method. Methods 25(4):402–408
7. Pinto D, Darvishi K, Shi X et al (2011) Comprehensive assessment of array-based platforms and calling algorithms for detection of copy number variants. Nat Biotechnol 29(6):512–521
8. Tommasi S, Mangia A, Iannelli G et al (2010) Gene copy number variation in male breast cancer by aCGH. Anal Cell Pathol (Amst) 33:113–119
9. Bendavid C, Rochard L, Dubourg C et al (2009) Array-CGH indicates a high prevalence of genomic rearrangements in holoprosencephaly: an update map of candidate loci. Hum Mutat 30(8):1175–1182
10. Hanemaaijer NM, Sikkema-Raddatz B, van der Vries G et al (2012) Practical guidelines for interpreting copy number gains detected by high-resolution array in routine diagnostics. Eur J Hum Genet 20:161–165
11. Paulson V, Chandler G, Rakheja D et al (2011) High-resolution array CGH identifies common mechanisms that drive embryonal rahbdomyosarcoma pathogenesis. Genes Chromosomes Cancer 50:397–408

Chapter 24

Identification of Off-Target Cleavage Sites of Zinc Finger Nucleases and TAL Effector Nucleases Using Predictive Models

Eli J. Fine, Thomas J. Cradick, and Gang Bao

Abstract

Using engineered nucleases, such as Zinc Finger Nucleases (ZFNs) or Transcription Activator-Like Effector Nucleases (TALENs), to make targeted genomic modifications has become a common technique to create new model organisms and custom cell lines, and has shown great promise for disease treatment. However, these nucleases have the potential for off-target cleavage that could confound interpretation of experimental results and be detrimental for therapeutic use. Here, we describe a method to test for nuclease cleavage at potential off-target sites predicted by bioinformatics models.

Key words Gene targeting, Off target effects, Zinc finger nuclease, ZFN, Transcription activator-like effector nuclease, TALEN, Gene therapy, Off-target sites, Bioinformatics

1 Introduction

ZFNs and TALENs are engineered restriction enzymes that consist of a customized DNA binding domain, zinc-finger helices or transcription activator-like repeats, fused to the cleavage domain from the FokI endonuclease [1]. These engineered nucleases can be designed to target a specific DNA sequence in the genome to generate double strand breaks (DSBs) in DNA at that position. These DSBs trigger the cell's internal DNA repair machinery which consists of two major pathways: nonhomologous end-joining (NHEJ) and homology directed repair (HDR). NHEJ typically results in small insertions or deletions (indels) at the site of the DSB that can cause a frame-shift in coding regions of DNA, leading to gene knockout. If a donor DNA template with arms of homology to the target site is added along with the nucleases, this template can be targeted to that location in the genome through the HDR pathway [1].

Although engineered nucleases are designed to cleave only at their intended target site, there have been reported cases of ZFNs

Francesca Storici (ed.), *Gene Correction: Methods and Protocols*, Methods in Molecular Biology, vol. 1114, DOI 10.1007/978-1-62703-761-7_24, © Springer Science+Business Media, LLC 2014

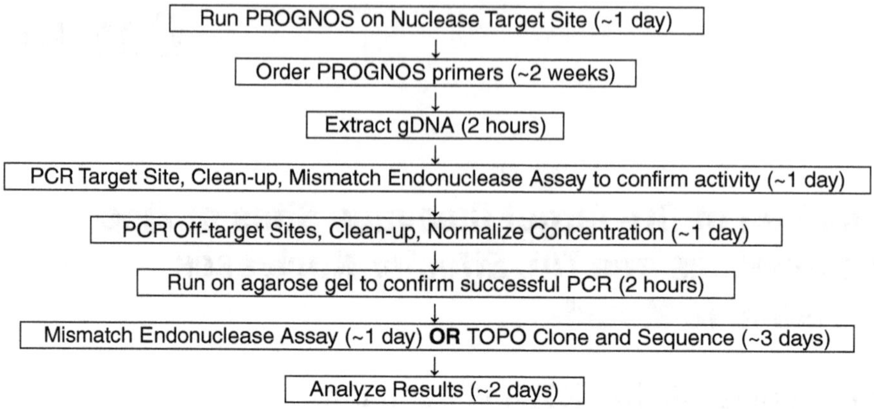

Fig. 1 Flowchart of PROGNOS-aided analysis of off-target sites. After transfection and genomic DNA harvesting, the off-target analysis process takes only a few days of preparation before assaying and several days after to process the results

[2–5] and TALENs [5–7] cleaving at off-target sites with sequence homology to the target site as low as 66 % [2] and 72 % [6], respectively. Although the only way to be certain of the full extent of off-target effects is whole genome sequencing, that method limits the number of cells that can be interrogated, which may miss many off-target events that occur at a frequency lower than the number of cells tested. A common approach is to examine a select number of sites in the genome that are predicted to be potential sites where off-target cleavage might occur [3–7]. There are several biochemical methods for characterizing nucleases in vitro that can be used to predict potential off-target sites, such as in vitro cleavage [3], bacterial one-hybrid [4], and SELEX [6, 7]. However, recent developments in modeling nuclease off-target cleavage have allowed potential off-target sites to be predicted using a solely bioinformatics-based approach with a success rate comparable to previous in vitro characterization methods [5]. Here, we describe how to use the online tool PROGNOS to generate a Predicted Report of Genome-wide Nuclease Off-target Sites for a pair of nucleases. We further describe how to test genomic DNA from cells that have been treated with nucleases for evidence of cleavage at predicted off-target sites (*see* Fig. 1).

2 Materials

Unless otherwise indicated, prepare solutions using nuclease-free water, and store reagents at room temperature. Diligently follow all waste disposal regulations.

2.1 Polymerase Chain Reaction Components

1. Oligonucleotide primers (Store at −20 °C) (Eurofins-MWG-Operon) (*see* **Note 1**).
2. High-Fidelity Polymerase (Store at −20 °C) (AccuPrime) (*see* **Note 2**).
3. Template genomic DNA (Store at −20 °C) (*see* **Note 3**).
4. Microseal "F" Foil, Pressure Pad, and Sealing Roller (Biorad).
5. 96-well PCR Plates (VWR).

2.2 High-Throughput PCR Purification Components

1. Magnetic Beads. Mag-Bind E-Z Pure (Omega Bio-Tek). Store at 4 °C.
2. Magnet Plate. Agencourt SPRIPlate Super Magnet Plate (Beckman Coulter).
3. 70 % Ethanol (Sigma-Aldrich) (*see* **Note 4**).
4. 20–200 μL 8-channel pipette (Brandtech).
5. 40 mL Reservoirs (Beckman Coulter).

2.3 Agarose Gel Components

1. High throughput gel box and combs: Owl* A2 Gel Box, 42 teeth combs, 21 teeth combs (Thermo Scientific).
2. Low volume 8-channel pipette. Transferpette-8 0.5–10 μL (Brandtech).
3. Gel Electrophoresis Buffer (*see* **Note 5**) (Fischer Scientific).
4. Agarose (*see* **Note 6**) (Sigma-Aldrich).
5. 6× Loading Dye (*see* **Note 7**) (Sigma-Aldrich).

2.4 Mismatch Endonuclease Assay Components

1. T7 Endonuclease I (New England Biolabs). Store at −20 °C.
2. EDTA (500 mM).

2.5 TOPO Cloning Components

1. TOPO TA Cloning Kit for Sequencing (Invitrogen). Store at −20 °C.
2. Competent *E. coli* for cloning (Invitrogen). Store at −80 °C.
3. LB-Agar plates with Ampicillin, IPTG (US Biological), and X-Gal (US Biological). Store at 4 °C.
4. Any commercial plasmid mini-prep kit (Qiagen).

3 Methods

Carry out all procedures at room temperature unless otherwise specified.

3.1 Selection of Potential Off-Target Sites

1. Direct an Internet browser to the PROGNOS Home Page (http://bit.ly/PROGNOS or http://baolab.bme.gatech.edu/Research/BioinformaticTools/prognos.html). There are

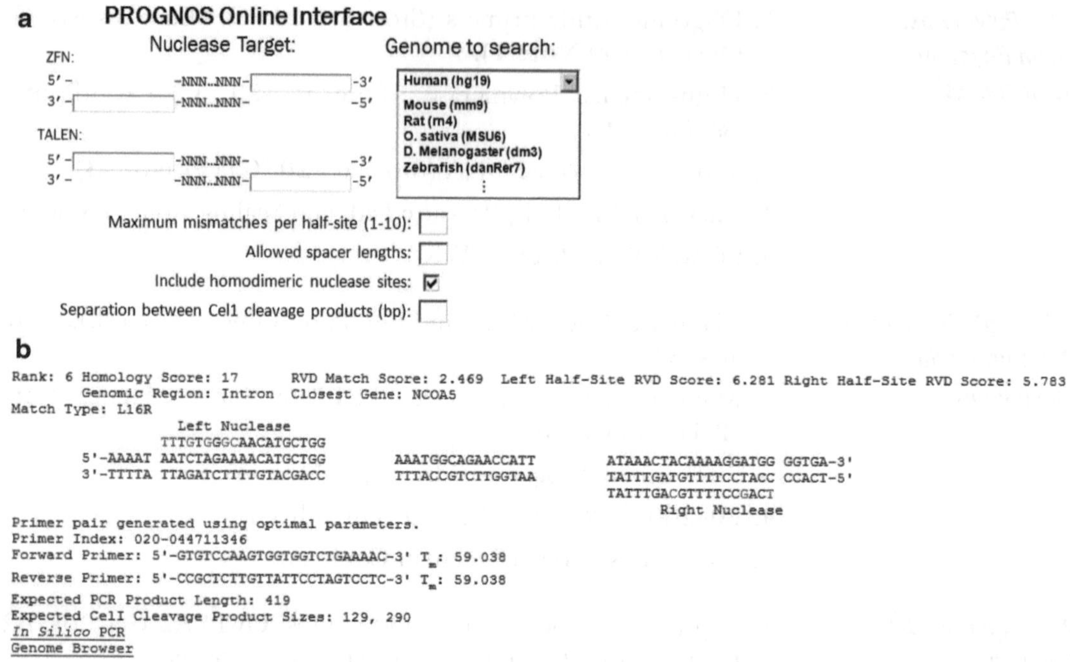

Fig. 2 PROGNOS online input and output for predicting nuclease off-target sites. (**a**) The nuclease target sites, the genome to search, primer design considerations, and other factors are entered into the online input form (adapted from [5]). (**b**) A sample off-target output entry for a TALEN pair targeting *CCR5*. Information such as the genomic location, PCR primers for amplification, and the types of mismatches are given

various explanations and tutorials to help familiarize you with the use of this tool.

2. Enter the target site of your nucleases, select the appropriate genome, and submit the search form (*see* Fig. 2a).
3. Once the PROGNOSis of your nucleases is complete, you can view the top results of the different ranking algorithms showing potential off-target sites for your nucleases (*see* Fig. 2b).
4. At this time, the recommended algorithms are "ZFN v2.0" and "TALEN v2.0"; if new algorithms are developed, information will be provided on the website. Generally, interrogating the top-ranked 23 potential off-target sites (plus the target site for a total of 24 sites) will provide a good approximation of whether the nuclease is prone to widespread off-target activity or is fairly specific. More sites can be interrogated if more knowledge about the off-target sites of a particular nuclease is desired. Sites listed as being in promoters or exons may be of special interest to interrogate.

3.2 PCR Amplification of Potential Off-Target Sites

1. For each potential off-target site, a PCR product must be amplified from the genomic DNA from cells modified by the nucleases (M) and from cells unmodified by the nucleases (U) to use as controls.

2. Each 50 μL PCR reaction should contain 5 μL of 10× Buffer2, 0.2 μL of Taq HiFi Polymerase, 2.5 μL of 100 % DMSO, forward and reverse primers each at 0.2 μM, and either 300 ng of genomic DNA (gDNA) or 1 μL of QuickExtract DNA.
3. For larger numbers of samples, we find it easiest to break the process into three steps.
   (a) Create a master mix for each different gDNA sample to allow 43 μL to be added to each empty well containing 4.5 μL of 10× Buffer2, 2.5 μL of DMSO, and the gDNA.
   (b) Add 2 μL of the pre-paired forward and reverse primers at 5 μM.
   (c) Create a master mix to allow 5 μL to be added to each sample containing 0.5 μL of 10× Buffer2 and 0.2 μL of Taq HiFi Polymerase.
4. After all components are added, cover the plate with foil and use the roller to ensure that the plate is sealed tightly, especially at the corners.
5. Vortex and briefly centrifuge the plate to mix.
6. Put the plate in the thermocycler and place the pressure pad on top with the foam facing up.
7. Use the following touchdown PCR thermocycler conditions:

   94 °C for 2 min

   `Touchdown Cycles (15×)`

   94 °C for 30 s

   63.5 °C for 30 s, decreasing 0.5 °C each cycle

   68 °C for 45 s

   `Additional Cycles (30×)`

   94 °C for 30 s

   56 °C for 30 s

   68 °C for 45 s

   68 °C for 10 min

   Hold at 4 °C
8. Completed PCR reactions can be stored at 4 °C for several days or at −20 °C indefinitely.

3.3 Gel Confirmation of PCR

1. Cast a 2 % agarose gel using the smaller (42-teeth) combs.
2. For each sample, pipette 1 μL of 6× loading dye into an empty tube.
3. Pipette 5 μL of PCR reaction into the tube with the loading dye. Pipette up and down 3 times to mix, then load 5 μL into the gel.

4. To facilitate comparison, it is recommended to load the Unmodified (U) and Modified (M) samples side-by-side.
5. Run the gel until the bands of the ladder are well separated.
6. Unless your Unmodified samples are heterozygous for different sized alleles, each sample should show only a single band at the expected size. If there is no band or multiple bands, do not proceed further with the analysis of that off-target site. To confirm that your samples are heterozygous with different sized alleles, perform TOPO cloning to sequence this PCR product and analyze the results (*see* Subheading 3.6).
7. The PCR products from the Modified cells should appear similar to the PCR products from Unmodified reactions. However, in the case of analyzing either single cell clones or cells from animals, another band may appear that could indicate a deletion or insertion in one of the alleles as a result of nuclease activity. This is not a reason to halt further examination of that potential off-target site.

3.4 High-Throughput Purification of PCR Reactions

1. PCR products can be purified using any method, if eluted in water (*see* **Note 8**). Here, we provide a high-throughput method using magnetic beads.
2. Calculate how much of each reagent you will need and fill three reservoirs accordingly (*see* **Note 9**).

   EZ-Pure: (# of reactions) × (reaction volume) × 1.8 + 800 μL (*see* **Note 10**).

   70 % Ethanol: (# of reactions) × (2 washes) × (150 μL) + 800 μL.

   Water: (# of reactions) × (elution volume) + 800 μL.
3. Pipette 1.8 times the PCR reaction volume of EZ-Pure into the PCR samples (i.e., 81 μL of EZ-Pure into a 45 μL PCR reaction). Mix together by pipetting up and down 10 times.
4. Transfer remaining EZ-Pure from the reservoir to a microfuge tube for reuse (*see* **Note 11**).
5. Wait for 5 min for the reaction to incubate.
6. Move the 96-well reaction plate onto the magnet and let settle for 9 min.
7. Carefully pipette out and discard the liquid while not disturbing the ring of beads.
8. Pipette 150 μL of 70 % Ethanol into each well and wait for 30 s.
9. Carefully pipette out and discard the liquid while not disturbing the ring of beads.
10. Repeat **steps 7** and **8** for a total of two ethanol washes.
11. Let the plate sit for 30 min to allow any remaining ethanol to evaporate.
12. Move the plate off of the Magnet.

13. Pipette the desired elution volume of water (40 μL recommended) into each well and pipette up and down vigorously 15 times to resuspend the magnetic beads.
14. Move the plate onto the magnet and let settle for 8 min.
15. Carefully use a pipette to transfer the liquid to a new plate while not disturbing the ring of beads. It may be necessary to leave ~5 μL of elution in the original plate to avoid transferring any beads to the new plate.
16. This purified PCR product can be stored at 4 °C for several days or at −20 °C indefinitely.
17. Measure the concentration of each purified product using a NanoDrop Spectrophotometer, a Take3 microplate, or other method.
18. Normalize each purified product (25 ng/μL recommended) using nuclease-free water.

3.5 Mismatch Endonuclease Assay

1. For each potential off-target site, there will be four total reactions. The PCR products from the cells modified by nucleases and the PCR products from the unmodified cells need to be examined. Additionally, for each PCR product there will be an experimental reaction to which the endonuclease will be added (+) and a control reaction without the enzyme (−).

		Assay treatment	
		Enzyme added	**No enzyme**
PCR source	Unmodified cells	U+	U−
	Modified cells	M+	M−

2. Each reaction should contain 200 ng of purified PCR product and 1.8 μL of NEBuffer2 10× mix in 18 μL total volume. If analyzing a population deriving from a single cell (single cell clones or cells from animals grown from treated embryos), the M+ and M− reactions should contain a mixture of 100 ng of PCR product from the modified cells with 100 ng from the unmodified cells (*see* **Note 12**). If analyzing heterogeneous populations of cells, the M+ and M− reactions should contain 200 ng from the modified cells. In all cases, the U+ and U− reactions should contain 200 ng from the unmodified cells.
3. Vortex and centrifuge briefly to mix.
4. Melt and re-anneal the DNA by placing in a thermocycler: 10 min at 95 °C, then decreasing at 0.1 °C/s down to 25 °C.
5. The re-annealed DNA can be stored at 4 °C for several days or at −20 °C indefinitely.
6. Vortex and centrifuge briefly to mix.

7. Make an enzyme master mix containing 0.5 μL of T7 Endonuclease I, 0.2 μL of NEBuffer2, and 1.3 μL of water (for M+ and U+ reactions) and aliquot into an 8-well strip for using a multi-channel pipette.
8. Make a control master mix containing 0.2 μL of NEBuffer2, and 1.8 μL of water (for M− and U− reactions) and aliquot into an 8-well strip for using a multi-channel pipette.
9. Add 2 μL of the appropriate master mix to each reaction, vortex and centrifuge briefly to mix, and immediately place in a thermocycler set for 37 °C for 60 min.
10. After 60 min, immediately remove the reactions from the thermocycler, centrifuge briefly, and quench by adding EDTA to a final concentration of 45 mM. This can be accomplished by adding 6.94 μL of a mixture containing a ratio of 2.45:4.49 of 500 mM EDTA to 6× loading dye (*see* **Note 7**).
11. Vortex and centrifuge briefly to mix.
12. Quenched reactions can be stored at 4 °C for several days or at −20 °C indefinitely.
13. Load as much as possible (≥25 μL) of the reactions into a 2 % agarose gel cast with the large (21-teeth) combs. For each off-target site, the recommended sample order for easy comparison is: U−, U+, M−, M+. Run the gel until the bands are well separated (*see* Fig. 3a).
14. When imaging the gel, make sure that the exposure time is properly adjusted so that none of the bands are saturated as this will interfere with accurate quantification of the band intensities.
15. Ensure that each sample contains the appropriate number of bands (*see* **Note 13**).
16. The free software ImageJ [8] can be used to quantify the bands on the gel. There are many online tutorials that detail this process, but in brief:
    (a) Go to the menu File → Open and select your gel image
    (b) Use the "Rectangular" tool to select the lane containing your first sample. Only select the middle ~50 % of the width of the lane to avoid any regions of the lane that are "smiling."
    (c) Go to the menu Analyze → Gels → Select First Lane
    (d) Use your mouse to move the rectangle to the center of your next lane. Go to the menu Analyze → Gels → Select Next Lane. Repeat until all lanes have been selected.
    (e) Go to the menu Analyze → Gels → Plot Lanes.
    (f) Use the "Straight" tool to draw a line under the peak for each band so that each peak has a distinct closed polygon.

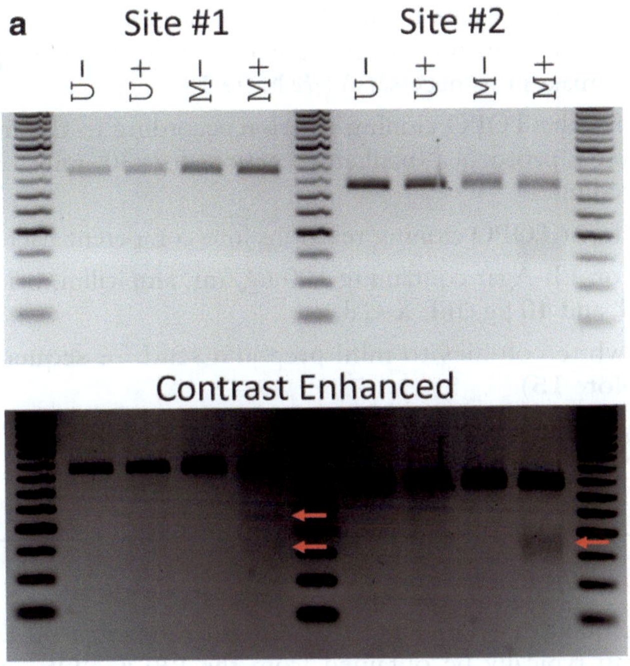

b

ZFN 3F Off-Target Site in *AFF3*

Mutations in 24 of 1043 sequences ≈ 2.3%

```
CTTCCTCTGCCTGCACCTaGgCCCCTTCTTGGGGCAAGGGCAGCAATAGTCCC  WT
CTTCCTCTGCCTGCACCTAGGCCCcttCTTCTTGGGGCAAGGGCAGCAATAGT  +3 x15
CTTCCTCTGCCTGCACCTAGGCCCCttctTTCTTGGGGCAAGGGCAGCAATAG  +4 x6
CTTCCTCTGCCTGCACCTAGGCCCCttcttTTCTTGGGGCAAGGGCAGCAATA  +5 x2
CTTCCTCTGCCTGCACCTAGGCCCC----TGGGGCAAGGGCAGCAATAGTCCC  Δ4 x1
```

Fig. 3 Analysis of nuclease activity. (**a**) T7 Endonuclease I assay for two off-target sites for a pair of TALENs. In the original gel (*upper*) off-target activity can be seen at site #2. With enhanced contrast (*lower*), off-target activity can also be seen at site #1. Cleavage bands are only observed in the M+ treatment lane, indicating that there are no confounding factors and that this represents true off-target activity. (**b**) Sequencing reads for a ZFN off-target site that shows evidence of NHEJ. The potential nuclease binding sites are highlighted in *yellow* in the wild-type (WT) sequence with the mismatches relative to the intended target written in *lowercase red*. Below are sequencing reads with insertions highlighted in *blue* and deletions written as *red hyphens*. The size of the indel and the frequency with which that sequence was observed are to the right of each sequence (adapted from [5])

(g) Use the "Wand" tool to click on the polygon for each peak. This will calculate the area of each polygon, which represents the band intensity.

17. The percent of alleles that show evidence of NHEJ can be calculated from the band intensities according to the formulas [9]:

(a) $$f_{cut} = \frac{Cleavage\ band_1 + Cleavage\ band_2}{Cleavage\ band_1 + Cleavage\ band_2 + Uncleaved\ band}$$

(b) $$\%NHEJ = 100 \times \left(1 - \sqrt{1 - f_{cut}}\right)$$

3.6 TOPO Cloning and Sequencing

1. Add 3′ A-Overhangs to the PCR amplicons according to the manufacturer's protocol (the use of the high-fidelity polymerase makes this necessary) (*see* **Note 14**).
2. Perform the TOPO cloning reaction according to the manufacturer's protocol. Incubate reaction for 30 min before proceeding.
3. Transform TOPO cloning reactions into competent bacteria.
4. Plate on LB-Agar containing 100 μg/mL ampicillin, 0.1 mM IPTG, and 40 μg/mL X-Gal.
5. Pick white colonies to mini-prep and send for sequencing (*see* **Note 15**).
6. Analyze sequencing reads for SNPs and evidence of NHEJ (*see* **Note 16** and Fig. 3b).

4 Notes

1. The sequences of primers to amplify each potential off-target site can typically be obtained from the PROGNOS output (*see* Subheading 3.1). If designing your own primers for use with the Mismatch Endonuclease Assay, bear in mind that the optimal PCR product size for the assay is 300–450 base pairs and that for visualization on the gel, the expected cleavage products must be different sizes than each other and the full length amplicon. Oligos can be ordered in plate format, normalized to 10 μM in 10 mM Tris–HCl, from Operon.
2. A high-fidelity polymerase is necessary to ensure that mismatches are not introduced during the amplification step and that only mismatches resulting from NHEJ repair in the cells as a result of nuclease cleavage are captured. We typically use AccuPrime Taq HiFi with AccuPrime Buffer2.
3. We have had high success rates using the PROGNOS primers to amplify gDNA extracted using a variety of methods including QIAamp (Qiagen), DNEasy (Qiagen), and QuickExtract (Epicentre). We have found that using QIAamp results in marginally higher PCR success rates than DNEasy (~97 % vs. ~95 %) and that both are more successful than QuickExtract (~90 %). However, QuickExtract seems to allow analysis of smaller numbers of cells; we have successfully substituted 1 μL of a 100 μL QuickExtraction from confluent cells in a 24-well plate in place of the standard 300 ng of genomic DNA extracted through column based methods.
4. To make a 70 % ethanol mixture, mix autoclaved ddH_2O with 100 % Molecular Grade Ethanol. As ethanol is hydroscopic,

make this component fresh regularly in small amounts (<50 mL) and store in a tightly capped tube when not in use.

5. Although tris-borate EDTA (TBE) is better suited to electrophoresis of small DNA fragments, we have found that tris-acetate EDTA (TAE) works fine for this application.
6. Although special formulations of agarose exist for electrophoresis of small DNA fragments, we have found that standard agarose works fine for this application.
7. Many loading dyes contain components that run in the 100–500 bp range, which would obscure the bands that need to be visualized. Make sure you choose a dye that does not have anything that will co-migrate with DNA fragments of that size. We use a xylene cyanol based dye: 4 g of sucrose, 3 mg of xylene cyanol FF (Sigma-Aldrich), add ddH_2O to 10 mL. Store this 6× mix at 4 °C.
8. We have found that eluting in anything other than nuclease-free water (even 1 mM Tris) can adversely affect the Mismatch Endonuclease Assay.
9. When filling the reservoirs for Subheading 3.4 an extra 800 μL is added beyond the required volume so that the liquid will overcome surface tension and stretch across the entire reservoir.
10. Before using the EZ-Pure beads, make sure to shake the container vigorously to resuspend the components.
11. After use, the remaining EZ-Pure in the reservoir can be stored in a separate container at 4 °C and reused in the next purification.
12. The rate of homozygous gene modification using TALENs has been shown to be much higher than expected, possibly due to sister chromatid exchange [10]. If nucleases have made homozygous modifications to an off-target loci, then this will not be detected with the Mismatch Endonuclease Assay because there will be no mismatches between the PCR products. By mixing PCR products from unmodified cells with products from modified cells, it ensures that even homozygous alterations in the modified cells will be detected.
13. In the gel of the T7 Endonuclease I assay, the U− and M− samples should look just as they did in the gel after Subheading 3.3. If there is excessive smearing or the major band has disappeared, this indicates a likely problem in the annealing step of the mismatch endonuclease assay. The U+ sample should look the same as the U− sample but with some additional smearing on the gel due to nonspecific degradation of the DNA by the T7 Endonuclease I. If additional bands appear, this indicates that there may be SNPs between the alleles in this position. The original PCR product from the unmodified samples should be TOPO

cloned and sequenced to confirm the presence and location of any SNPs (*see* Subheading 3.6). The sizes of the additional bands on the gel should correspond to cleavage of the original amplicon at the point of the SNPs.

(a) For analyzing the T7 Endonuclease I assay results for a population arising from a single cell such as single-cell clones or an animal treated with nucleases as a single cell embryo; if the M+ sample looks the same as the U+ sample, then this indicates that there was no off-target cleavage and subsequent misrepair by NHEJ at this location. If there are additional bands and their sizes correspond to what would be expected if the T7 Endonuclease I cleaved the PCR product in the spacer region between the half-sites of the engineered nucleases, then this is likely evidence of an off-target event at one or more of the alleles. The %NHEJ calculation should yield discrete values of either 50 % (indicating bi-allelic NHEJ in the modified cells), 25 % (indicating either single allele disruption in the modified cells or potentially bi-allelic disruption in one of the cells of a two-cell embryo), or 12.5 % (indication single allele disruption in one of the cells of a two-cell embryo).

(b) The T7 Endonuclease I assay has a limit of detection of ~2 %. When analyzing a heterogeneous population of cells that were treated with nucleases, the M+ sample will appear similar to the U+ sample if the off-target activity is below ~2 %. More sensitive methods of detection include TOPO cloning and sequencing a large number of resulting colonies (>80), using Single Molecule Real-Time (SMRT) sequencing, or another next generation sequencing method such as 454 or Illumina.

14. We use Taq DNA Polymerase from New England BioLabs. Our reaction contains 15 ng/μL of DNA, 0.5 mM dNTPs, 1× Standard Taq Buffer, and 0.05 U/μL of Taq Polymerase. We incubate at 68 °C for 20 min and then add 4 μL of this mixture to the TOPO reaction.

15. If TOPO cloning is being used to confirm the presence of SNPs between two different alleles, at least seven colonies should be selected for sequencing to ensure that there is at least one amplicon derived from each allele represented ($p<0.01$).

16. NHEJ repair of nuclease cleavage events typically creates small insertions or deletions of three or more base pairs that overlap the spacer region between the two nuclease half-sites. However, if the sequencing chromatogram is very clear at that location, indels as small as one base pair may be observed resulting from NHEJ.

Acknowledgements

This work was supported by the National Institutes of Health as an NIH Nanomedicine Development Center Award (PN2EY018244 to GB). E.J.F. was additionally supported by an NSF Graduate Research Fellowship under Grant No. DGE-1148903.

References

1. Humbert O, Davis L, Maizels N (2012) Targeted gene therapies: tools, applications, optimization. Crit Rev Biochem Mol Biol 47(3):264–281
2. Gabriel R, Lombardo A, Arens A, Miller JC, Genovese P, Kaeppel C et al (2011) An unbiased genome-wide analysis of zinc-finger nuclease specificity. Nat Biotechnol 29(9):816–823
3. Pattanayak V, Ramirez CL, Joung JK, Liu DR (2011) Revealing off-target cleavage specificities of zinc-finger nucleases by in vitro selection. Nat Methods 8(9):765–770
4. Gupta A, Meng X, Zhu LJ, Lawson ND, Wolfe SA (2011) Zinc finger protein-dependent and -independent contributions to the in vivo off-target activity of zinc finger nucleases. Nucleic Acids Res 39(1):381–392
5. Fine EJ, Cradick TJ, Zhao CL, Lin Y, Bao G (in press) An online bioinformatics tool predicts zinc finger and TALE nuclease off-target cleavage. Nucleic Acids Research
6. Hockemeyer D, Wang H, Kiani S, Lai CS, Gao Q, Cassady JP et al (2011) Genetic engineering of human pluripotent cells using TALE nucleases. Nat Biotechnol 29(8):731–734
7. Tesson L, Usal C, Ménoret S, Leung E, Niles BJ, Remy S et al (2011) Knockout rats generated by embryo microinjection of TALENs. Nat Biotechnol 29(8):695–696
8. Schneider CA, Rasband WS, Eliceiri KW (2012) NIH Image to ImageJ: 25 years of image analysis. Nat Methods 9(7):671–675
9. Guschin DY, Waite AJ, Katibah GE, Miller JC, Holmes MC, Rebar EJ (2010) A rapid and general assay for monitoring endogenous gene modification. Methods Mol Biol 649:247–256
10. Carlson DF, Tan W, Lillico SG, Stverakova D, Proudfoot C, Christian M et al (2012) Efficient TALEN-mediated gene knockout in livestock. Proc Natl Acad Sci U S A 109(43): 17382–17387

Part IV

Gene Correction Approaches for Specific Disease Models

Additional protocols described in this book of gene correction for specific disease models

- gyrate atrophy: Chapter 4
- muscular dystrophy: Chapter 5
- thalassemias: Chapter 8
- sickle cell anemia: Chapter 8
- cystic fibrosis: Chapter 18

Chapter 25

Method for Retinal Gene Repair in Neonatal Mouse

Marilyn Dernigoghossian, Arthur Krigel, Francine Behar-Cohen, and Charlotte Andrieu-Soler

Abstract

Gene correction at the site of the mutation in the chromosome is the absolute way to really cure a genetic disease. The oligonucleotide (ODN)-mediated gene repair technology uses an ODN perfectly complementary to the genomic sequence except for a mismatch at the base that is mutated. The endogenous repair machinery of the targeted cell then mediates substitution of the desired base in the gene, resulting in a completely normal sequence. Theoretically, it avoids potential gene silencing or random integration associated with common viral gene augmentation approaches and allows an intact regulation of expression of the therapeutic protein. The eye is a particularly attractive target for gene repair because of its unique features (small organ, easily accessible, low diffusion into systemic circulation). Moreover therapeutic effects on visual impairment could be obtained with modest levels of repair. This chapter describes in details the optimized method to target active ODNs to the nuclei of photoreceptors in neonatal mouse using (1) an electric current application at the eye surface (saline transpalpebral iontophoresis), (2) combined with an intravitreous injection of ODNs, as well as the experimental methods for (3) the dissection of adult neural retinas, (4) their immuno-labelling, and (5) flat-mounting for direct observation of photoreceptor survival, a relevant criteria of treatment outcomes for retinal degeneration.

Key words Gene repair, Iontophoresis, Intravitreal injection, Retina dissection, Whole flat-mount retina, Oligonucleotides

1 Introduction

Over the last decade, significant progress has been made in the field of gene delivery to the retina. Of all tested viral vectors, recombinant adeno-associated virus (rAAV) vectors have been shown to be optimal viral vectors for retinal gene transfer and have been used for the development of therapy strategies against many retinal disorders based on gene addition approaches. Leber congenital amaurosis

Marilyn Dernigoghossian and Arthur Krigel contributed equally to this manuscript

Electronic supplementary material The online version of this chapter (doi: 10.1007/978-1-62703-761-7_25) contains supplementary material, which is available to authorized users. Videos can also be accessed at http://www.springerimages.com/videos/978-1-62703-760-0

Francesca Storici (ed.), *Gene Correction: Methods and Protocols*, Methods in Molecular Biology, vol. 1114, DOI 10.1007/978-1-62703-761-7_25, © Springer Science+Business Media, LLC 2014

(LCA)-RPE65 gene therapy is one of the best examples of successful application of retinal gene targeting to clinical treatment (listed on the clinicaltrials.gov website). rAAV vectors present several limitations including the ability to generate innate inflammatory responses, the ability to cause insertional mutagenesis (at a very high frequency in some tissues), and a limited cloning capacity of 4.8 kb. The latter renders the delivery of large genes impossible. This is an important issue because several retinal diseases (Stargardt's disease, some forms of LCA, or Usher's syndrome) are due to mutations in photoreceptor genes that are much longer than this limit. Such transgenes could be packaged into lentivirus or adenovirus, but these vectors do not transduce the photoreceptor cells. The cargo size limitation of AAV also limits the use of regulatable promoter systems to control transgene expression. Developing novel non-viral routes for targeting genes to the retina is an important area of drug delivery research.

Improved systems for ocular non-viral based gene delivery have been developed, including the use of plasmids encapsulated with polymers, liposomes, lipoplexes, polyplexes or nanoparticles, as well as physical methods such as electroporation or iontophoresis [1–4]. Non-viral methods have some potential advantages compared to viral based strategies. Most of the non-viral based delivery systems are generally safer to ocular tissues. However, they can lead to comparatively low transfection rates and relatively short-lived transgene expression.

Gene correction at the site of the mutation in the chromosome is the absolute way to really cure a genetic disease and to overcome the different abovementioned limitations of viral and non-viral gene addition methods. Oligonucleotides (ODNs), small fragment homologous replacement (SFHR), or triplex-forming oligonucleotides (TFO) are all DNA-based methods relying on endogenous cellular enzymatic pathways to correct mutation without the use of additional vector and with the aim not to add more genetic material to the targeted cells [5–7]. The principle of the ODN-mediated gene repair technology is to target an ODN perfectly complementary to the genomic sequence except for a mismatch at the base that is mutated. The endogenous repair machinery of the targeted cell then mediates substitution of the desired base in the gene, resulting in a completely normal sequence. As homologous recombination plays a crucial part in the gene repair mechanism [8] recombinase, endonuclease (zinc-finger nuclease or meganuclease), transposases, and/or viral integrases fused with DNA recognition motifs has been concomitantly used to generate double-strand breaks in the vicinity of the targeted mutation in order to facilitate homologous recombination around the target sites to possibly increase gene correction efficiency [9, 10].

ODN-mediated gene repair was originally described in 1996 with the repair of an episomal mutant alkaline phosphatase gene in CHO cells [11] and has since been demonstrated in several in vitro systems, several cell types, and organs (muscle tissue in mouse [12] and canine [13] dystrophin genes), skin in the tyrosinase gene

[14], and liver in the UDP glucuronosyltransferase gene [15], factor IX gene [16], and apolipoprotein (ApoE2) gene 20 [17]. Over the last decade, the use of short all-DNA single-stranded ODNs has been found to yield more reproducible results, and they are both simpler and less expensive to synthesize. The strand bias, length, and chemistry can influence the binding efficiency and stability of the ODN and are important reaction parameters to assess for each system. We showed in a cell-based system that sense ODNs of 25 mers with partial phosphorothioate linkages are the most efficient ODNs to target the retinal *rd1* mutation [18]. ODN-mediated gene correction has distinct advantages over the other technologies in that it results in true gene repair in an inheritable fashion [18]. Theoretically, it avoids potential gene silencing or random integration associated with gene augmentation approaches, and the regulation of expression of the therapeutic protein is exactly as normal. There is little doubt about the enormous potential of this technology, and its development is a fertile area.

The eye is a particularly attractive target for gene repair because of its unique features. Indeed, it is a small organ, easily accessible (for treatment as well as to assess visual function as an effect of treatment), and presents low diffusion into systemic circulation, thus allowing high concentrations of therapeutic agents. Moreover therapeutic effects on visual impairment could be obtained with modest levels of repair. The retinal tissue emerges as promising clinical targets due to the number of mutations in photoreceptor-specific genes that have been associated with their death and with incurable inherited retinal degeneration diseases (up-to-date documentation on genes and loci is reported at http://www.sph.uth.tmc.edu/RetNet/). Of note, retinitis pigmentosa (RP) is a genetically heterogeneous retinal disease that represents the major cause of inherited familial blindness in the Western world.

Ciavatta et al. used cell-free extracts to show that mouse retina possesses ODN-induced gene repair activity [19]. A major challenge was to direct the therapeutic ODNs to the nuclei of photoreceptors. We showed in a previous report that the iontophoresis method is safe (absence of permanent tissue alterations or structure damage) and that it facilitates the penetration of injected ODNs in photoreceptors for at least 3 h [20]. We also proved the functional effect of three repetitive treatments by iontophoresis and ODN intravitreal injection to correct the *rd1* point mutation in genomic DNA of mouse retinas in vivo and to induce expression of the missing phosphodiesterase protein (b-PDE) and photoreceptor survival [21].

Using the C3H/HenJ (*rd1*) mouse model of retinal degeneration containing the *rd1* mutation in the gene encoding b-PDE, a natural occurring point mutation causing photoreceptor degeneration in mice and human, and phosphorothioate sense ODNs of 25 nucleotide length and complementary to genomic sequence subsuming the *rd1* point mutation with a wild-type nucleotide base at the *rd1* point mutation position, this chapter describes in details the optimized method to target active ODNs to the nuclei of photoreceptors in neonatal mouse using (1) an electric current application at the eye surface (saline transpalpebral iontophoresis), directly followed by (2) an intravitreous injection of ODNs, as well as the experimental methods for (3) the dissection of adult neural retinas, (4) their immuno-labelling (immunofluorescence of the most abundant photoreceptor-specific protein, rhodopsin), and (5) their flat-mounting for direct observation of photoreceptor survival (a relevant criteria of treatment outcomes for retinal degeneration).

This method would be suitable to test the correction and treatment outcomes of different mutations causing retinal diseases. ODN-directed gene repair is a particularly attractive method for retinal diseases since degeneration often progresses slowly, enabling an opportunity for repeated administration. Importantly, retardation of even a modest amount of retinal degeneration could provide significant clinical benefit.

2 Materials

Prepare and store all reagents at room temperature (unless indicated otherwise). Follow all waste disposal regulations when disposing waste materials.

2.1 Animals

1. Experiments on animals are conducted in accordance with the Association for Research in Vision and Ophthalmology (ARVO) statements and the institutional guidelines regarding animal experimentation in ophthalmic and vision research.
2. C3H/HenJ (*rd1*) mice homozygous for the nonsense mutation (amino acid position 347) in the β-phosphodiesterase (PDE) gene (Janvier, Le Genest, France) are maintained in clear plastic cages and subjected to a 12-h:12-h light–dark cycle.

2.2 Oligonucleotides

1. Phosphorothioate ODNs of 25 nucleotide length encoding for the wild-type sense of the β-PDE gene sequence subsuming the *rd1* point mutation in C3H/HenJ mouse (5′-CsCsTsTsCsCsAACCTACGTAGCAsGsAsAsAsGsT-3′), the corresponding wild-type antisense β-PDE gene sequence, and ODN with seven central scrambled bases are synthesized and purified by high-pressure liquid chromatography (Prolego, Paris, France). 500 μM stocks are aliquoted and stored at −20 °C.

2.3 Antibodies

1. Rhodopsin-specific mouse monoclonal antibody, rho-4D2 (provided by Dr. Robert Molday, University of British Columbia, Vancouver, Canada). Store at 4 °C.
2. Goat anti-mouse IgG antibody conjugated to Alexa Fluor 488 (Molecular Probes, Leiden, The Netherlands). Aliquot and store at −20 °C. For short-term storage keep at 4 °C.

3. Leu-M5 directed against macrophages and monocytes (BD Biosciences, Pont-de-Claix, France).
4. Normal mouse serum (Tebu-Bio, Le Perray en Yvelines, France). Store at 4 °C.

2.4 Disposables

1. Aluminum foil (thickness of 20 μm).
2. Disposable medical grade hydrophilic polyurethane sponge (3.2 mm thick).
3. ES TransferTip microcapillaries (Leica, Rueil Malmaison, France).
4. Staples of 80 mm length.

2.5 Reagents

1. Tetracaine 1 % Faure (Novartis Ophthalmics SA, Rueil Malmaison, France).
2. Phosphate-buffered saline (PBS 1×): 0.02 g/L KCl, 0.02 g/L KH_2PO_4, 0.8 g/L NaCl, 0.216 g/L $NaHPO_4 \cdot 7\ H_2O$, pH 7.4. Store at 4 °C.
3. Prepare fresh PBS 1×–0.1 % Triton X-100 (~500 μL/retina) and fresh blocking buffer: PBS 1×–0.1 % bovine serum albumin, 0.1 % Tween 20, 0.1 % sodium azide (~15 mL/retina). Store bovine serum albumin at 4 °C.
4. 20 % paraformaldehyde (PAF, Ladd Research, Holly Court, Williston, USA) in PBS 1×. To prepare 200 mL of 20 % PAF, weight 40 g of PAF powder. Always wear a mask when weighing PAF powder. To avoid exposition of PAF, cover the weigh boat containing the weighed PAF with another one of similar size. Transport to the fume hood and dilute in 200 mL of PBS 1×. Heat at 80 °C to dissolve the mixture. Let the solution cool to approximately 20 °C. Add NaOH to bring the pH to 7.4 and filter with a filter paper. Store the aliquot at room temperature. PAF is highly toxic (especially its fumes); therefore, all the steps following weighing should be performed under the fume hood. Store aliquots at −20 °C.
5. Tissue-Tek OCT-compound (Bayer Diagnostics, France).

2.6 Dissection Tools

1. Curved scissors, pointed, delicate (Moria, France, Ref 4878A).
2. Thin forceps (Dumont, Ref number 5F).
3. Curved forceps, usual, serrated jaws 1 mm (Moria, France, Ref 2183).
4. Disposable scalpel (Swann Morton, Peynier, France).

2.7 Apparatus

1. Small stapler compatible with 80 mm staples.
2. Iontophoresis electrical device.
3. Precision injector such as the microsyringe pump controller (Micro4™, World Precision instruments, Sarasota, Florida, USA).

4. Dissecting microscope (S6E, Leica, Rueil Malmaison, France).
5. Fluorescent microscope (Aristoplan, Leica, Rueil Malmaison, France).

3 Methods

Carry out all procedures at room temperature unless specified otherwise.

3.1 Transpalpebral Iontophoresis and Intravitreous Injection (See Video 1)

1. Prepare 1.5 × 0.7 cm length by width eyeglass-shaped electrodes by stapling an aluminum foil (20 μm thick) (*see* **Note 1**) to a disposable medical grade hydrophilic polyurethane sponge (3.2 mm thick) (*see* **Notes 2–4** and Fig. 1a).
2. Prepare a round-shaped aluminum piece with ~1.5 cm in diameter by folding the aluminum in four.
3. Under a dissection microscope, cut a micropipette needle with thin forceps at ~2 mm from the needle neck (leading to a 60 μm injection outer diameter).
4. Set up the iontophoresis generator parameters (current of 1.5 mA, duration 5 min).
5. Connect the eyeglass to the negative electrode of electrical device (by holding both the sponge and the foil with the

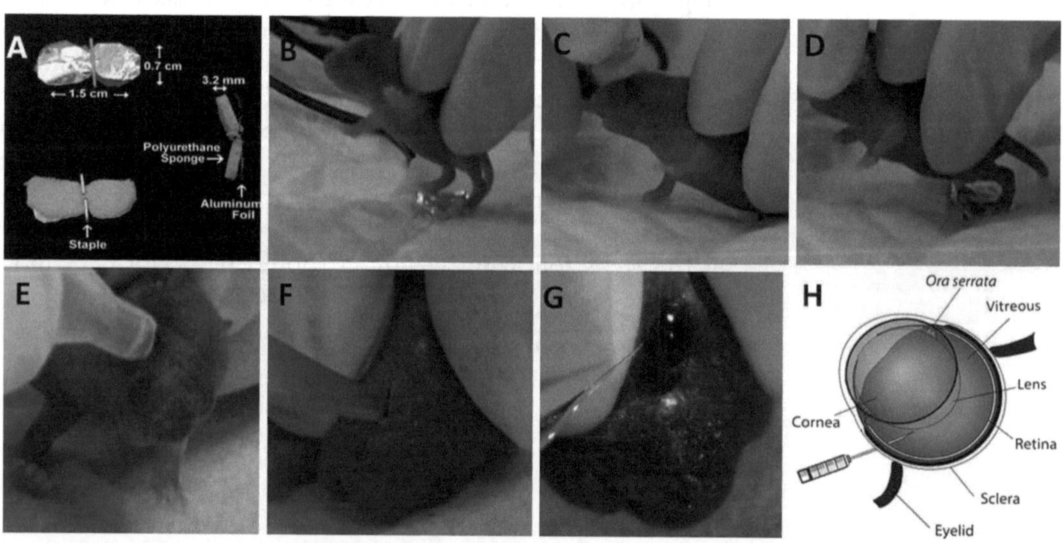

Fig. 1 Transpalpebral iontophoresis and intravitreous injection on newborn PN4 mouse. Polyurethane-and aluminum-shaped eyeglasses (**a**). The aluminum return electrode in contact with the tail and hind foot pads of the mouse (**b**). Positioning of the active electrode covering both closed eyelids of the mouse (**c** and **d**). Eye(lid) anesthesia (**e**). Separation of the upper and lower mouse eyelids (**f**). Positioning of the micropipette needle to perform intravitreous injection. The micropipette needle is placed below the *ora serrata* on an exorbitated eye with an angle of ~10–20° (**g** and **h**)

electrical clip), and connect the round-shaped aluminum piece to the positive electrode (*see* **Notes 5** and **6**).

6. Soak the eyeglass in PBS 1×, and remove extra PBS before use (*see* **Notes 7** and **8**).
7. Turn on the iontophoresis generator (*see* **Note 9**).
8. Place the hind foot pads (with or without the tail) of a postnatal 4 (PN4) C3H/HenJ mouse on the return electrode (Fig. 1b), and cover both closed eyelids of the mouse with soaked eyeglass (*see* **Note 10** and Fig. 1c).
9. Keep this position so that the animal is submitted to transpalpebral iontophoresis for 5 min (*see* **Note 11** and Fig. 1d).
10. Apply one drop of tetracaine 1 % on each eyelid of the mouse, and let it act for 5 min (*see* **Note 12** and Fig. 1e).
11. Under a dissection microscope, separate the upper and lower mouse eyelids with a sharp disposable scalpel (*see* **Notes 13** and **14** and Fig. 1f).
12. Connect a precut micropipette needle to the microinjector, and withdraw 5–10 μL of the solution to be injected (solution of ODNs at 500 μM or PBS 1× for control) (*see* **Notes 15** and **16**).
13. Monitor the apparatus at a constant injection speed of 200 nl/s.
14. Under a dissecting microscope, exorbitate the mouse's eye by opening the eyelids and exerting a gentle pressure on the eye-surrounding area.
15. Place the micropipette needle below the *ora serrata* with an angle of ~20°, and penetrate the eye by ~2–3 mm (*see* **Notes 17–19** and Fig. 1g, h).
16. Inject 1 μL of solution into the vitreous (*see* **Note 20**).
17. Leave the micropipette needle in place for 10 s before withdrawal (*see* **Note 21**).
18. Repeat iontophoresis and intravitreous treatments on the same mouse at PN6 and PN8.

3.2 Enucleation and Eye Fixation (See Video 2)

1. Under the fume hood, prepare 4 % (v/v) PAF by diluting the 20 % stock solution in PBS 1× (~500 μL/eye).
2. Sacrifice the treated mouse at PN28 with CO_2 to evaluate treatment outcomes (*see* **Note 22**).
3. Exorbitate the mouse's eye by opening the eyelids and exerting pressure on the eye-surrounding area, place a curved forceps under the eye, pull to section the optic nerve, and remove the eye (*see* **Note 23**).
4. Hold the ocular globe by the optic nerve, and place it in 500 μL PAF 4 % in a 2 mL Eppendorf tube under a fume hood. Incubate for 2 h (*see* **Note 24**).
5. Wash three times in 1.5 mL PBS 1× under the hood.

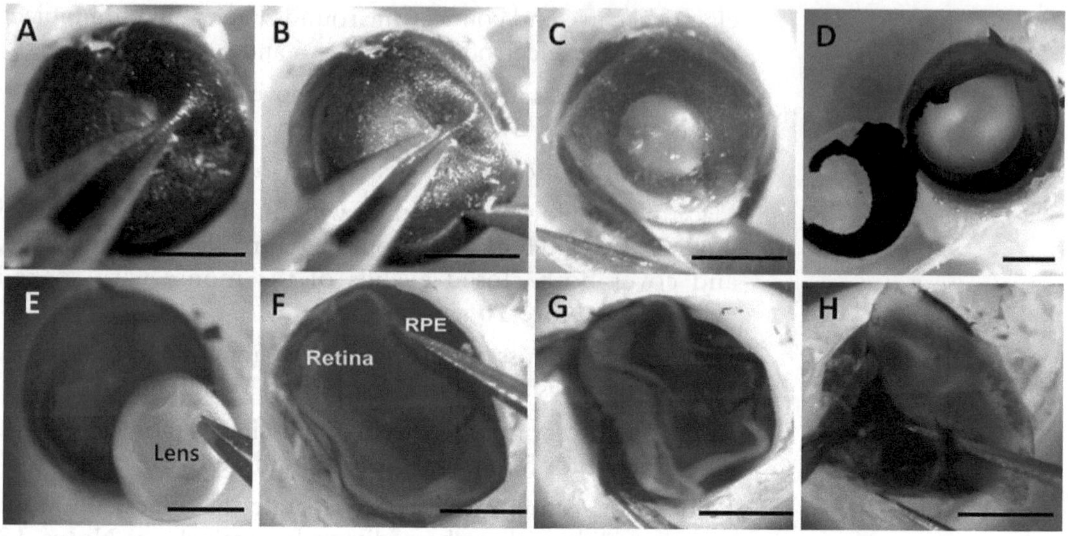

Fig. 2 Dissection of the neural retina. Grabbing of cornea (**a**). Proceeding with first incision perpendicular to and just above the *ora serrata* (**b**). All around cutting (**c**). Separating the ocular anterior (*left part*) and posterior (*right part*) segments (**d**). Lens removing (**e**). Neural retina and RPE separation (**f**). RPE/sclera tearing (**g**). Final step of neural retina isolation: detachment of the adherent optic nerve (**f**). Scale bars of 1.5 mm

3.3 Dissection of Neural Retina (See Video 2)

1. Pour a drop of glue in a Petri dish, and glue the eye cornea facing up.
2. Wait until the glue dries out, and pour PBS 1× in the Petri dish until it covers the ocular globe.
3. Grab and hold the cornea with thin forceps (Fig. 2a), and make a small incision with semi-curved scissors perpendicular to and just above the *ora serrata* (Fig. 2b).
4. Cut all along the *ora serrata* with semi-curved scissors (Fig. 2c) until complete separation of the anterior and posterior segments of the ocular globe (*see* **Note 25** and Fig. 2d).
5. If necessary, grab the lens with thin forceps and remove it from the eye posterior segment (Fig. 2e).
6. Using thin forceps, carefully separate the neural retina from the retinal pigment epithelium (RPE) all around the eyecup (Fig. 2f).
7. Grab the RPE/sclera with thin forceps, and slowly tear them up to take out the neural retina without damaging it (Fig. 2g).
8. Using thin forceps detach the connected optic nerve to isolate the neural retina (Fig. 2h).

3.4 Immunochemistry on Neural Retina (See Video 2)

1. Separately prepare 100 μL/retina of diluted antibody solution or normal mouse serum in blocking buffer (rho-4D2, Leu-M5, or normal mouse serum at a 1/100 dilution and Alexa Fluor 488-conjugated goat anti-mouse IgG antibody at a 1/250 dilution) (*see* **Note 26**).

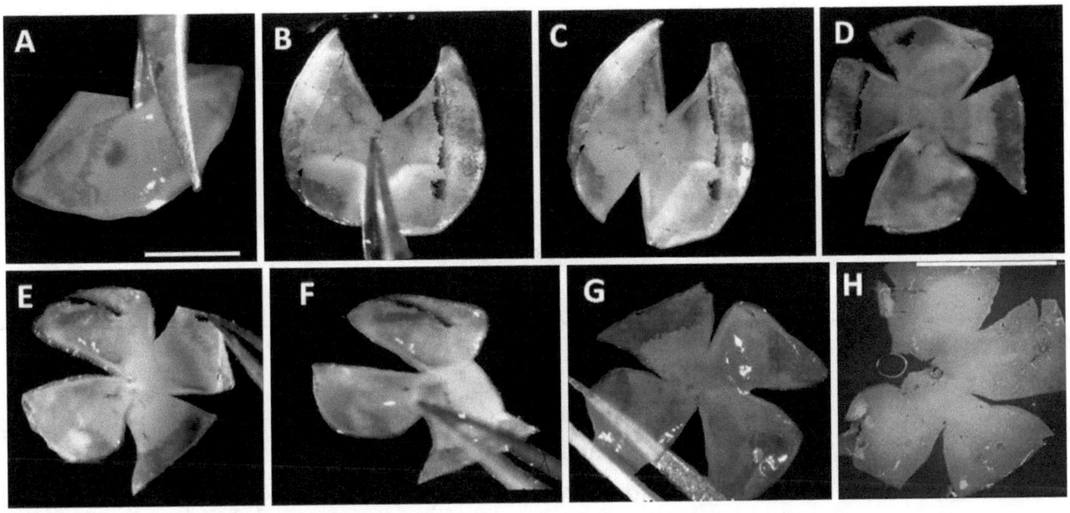

Fig. 3 Mounting of whole flat retina. Four radial sectioning of the retina (**a–d**). Grabbing of the remaining black ciliary body/RPE (**e**). Upside-down turning of retina (**f**). Unfolding of retina (**g**). Fluorescent visualization of a PN28 Rho-4D2-labelled retina from a wild-type sense ODN-treated C3H-HenJ mouse (treatments at PN4, -6, and -8) (**h**). Scale bar in (**a**) of 1.5 mm. Same magnification for (**b–h**)

2. Cut a 1 mL tip at ~0.5 mm from its extremity (*see* **Note 27**). Gently aspirate the retina in PBS, and place it in a 1.5 ml Eppendorf tube (*see* **Note 28**).
3. Permeabilize the retina in 500 μL of PBS 1×–0.1 % Triton X-100 for 5 min.
4. Incubate the retina in 1.5 mL of blocking buffer for 15 min.
5. Incubate the retina for 1 h in 100 μL of rho-4D2 solution (use Leu-M5 or normal mouse serum solution as negative controls).
6. Wash three times the retina for 5 min in 1 mL blocking buffer.
7. Incubate the retina for 1 h in 100 μL of Alexa Fluor 488-conjugated secondary antibody solution.
8. Wash the retina three times for 5 min in 1 mL PBS 1×.

3.5 Mounting of Whole Flat Retina (See Video 3)

1. Cut a 1 mL tip at ~0.5 mm from its extremity (*see* **Note 27**). Gently aspirate the retina in PBS, and place it on a glass slide (photoreceptor layer facing down).
2. Make a sharp radial section of the retina with semi-curved scissors. Stop sectioning ~10 mm before the optic nerve (*see* **Note 29** and Fig. 3a).
3. Make similar incisions at the opposite side and at the perpendicular axis to obtain a four-petal retina (Fig. 3b–d).
4. Grab a remaining part of the ciliary body (Fig. 3e), and pull it gently with a waving move to turn the retina upside down (photoreceptor layer facing up) (Fig. 3f).

5. Move the forceps underneath the retina to slide it into a flat position (Fig. 3g).
6. Dry out the retina using a tissue.
7. Pour a drop of mounting gel on a lamella, and place it on the top of the retina.
8. Once the mounting gel dries, observe under an epifluorescence microscope (Fig. 3h).

4 Notes

1. An aluminum thickness superior to 15 μm avoids the folding or the breaking of the eyeglasses during their manipulation.
2. The conductive and soft polyurethane sponge is necessary for the homogenous diffusion of electric current on the targeted area (mouse eyelids).
3. The aluminum foil and the mouse skin should not be in direct contact for the current to diffuse through the polyurethane sponge.
4. The staples should be small enough so that the sponge and the aluminum are held together. Use a stapler that is adapted to the small staples.
5. New eyeglass is used for each iontophoretic treatment.
6. Cationic iontophoresis is performed when the negative electrode is connected to the eyelids and the positive electrode to the tail and hind foot pads of the mouse. Anionic iontophoresis (reverse direction of current) is slightly less efficient for ODN delivery to the retina [20].
7. Humidify the eyeglass to allow the current to flow homogeneously.
8. Removing extra PBS prevents the flowing of liquid on the mouse's nose and mouth, which may provoke its drowning.
9. An audiovisual alarm indicates any disruption of the electrical circuit, ensuring a controlled delivery of the current. The current duration will start (and eventually pause) each time the electrical circuit is not closed.
10. The mouse can be held by its neck's skin with one hand, while the other hand maintains the eyeglass on the eyelids.
11. We have found that applying transpalpebral iontophoresis immediately after or before intravitreal injection of ODNs leads to the same penetration efficiencies [21]. Therefore, we chose to perform iontophoresis immediately prior to the intravitreal injection of ODNs in order to avoid manipulation of the injected pups' eyes and reduce the potential danger of secondary infection.

12. Do not wait for more than 5 min to avoid the decrease of effect of the iontophoretic pretreatment.
13. Follow the palpebral fissure to open the eyelids. This step is necessary only when performing the first treatment at PN4. For the second and third treatments (at PN6 and PN8, respectively), upper and lower eyelids will be already separated.
14. A new disposable scalpel should be used for each experiment to avoid possible infection.
15. Pre-warm the solution to be injected with hands for 1–2 min. Right after use place the ODN solution on ice to avoid its degradation, which may lead to unwanted cellular reaction and reduce gene repair efficiency.
16. Withdrawing less than 10 μL allows visualizing the surface of the liquid to be injected under the dissecting microscope to control the injection speed and the delivered volumes during the procedure.
17. The use of a glass cover slip placed on the corneal can help in controlling the positioning of the micropipette needle in the vitreous.
18. If the micropipette needle bends and cannot penetrate the eye without breaking, it has to be shortened by cutting ~0.5–1 mm of its extremity with thin forceps under a dissecting microscope.
19. If the micropipette needle cannot penetrate the eye but does not bend, its sharpness is probably not good enough and it is better to use a new precut one.
20. When the liquid surface does not move downward, the needle might be touching the lens and the manipulator should slightly move the microcapillary backward. If the problem is not solved, check if the needle got blocked. In that scenario, inject few microliters of the solution on a tissue and try to remove what obstructs the needle. If the problem remains, use a new precut micropipette needle.
21. This step limits the loss of the injected solution and allows the intraocular pressure to equilibrate (as observed by the return of normal iris perfusion).
22. An extra step of cervical dislocation is strongly recommended before proceeding to enucleation.
23. The curved forceps should be well positioned under the exorbitated eye not to pressurize the ocular globe and thus avoid damaging the retinal structure.
24. The round and large extremity of a 2 mL Eppendorf tube fits the shape of a mouse eye and allows the fixative solution covering the whole eye surface for better fixation.
25. Turn the Petri dish as you cut in order to facilitate the manipulation.

26. Shortly spin antibody stocks before pipetting to get the right concentration.
27. Increasing the tip diameter avoids damaging the retina.
28. The round and narrow extremity of a 1.5 ml Eppendorf tube perfectly fits the shape, contains the size of mouse retinas, and thus allows removing the majority of the solutions for proper washing and limited dilution of antibody solutions.
29. Try not to use forceps in order not to damage the retina. Turn the slide as you cut to facilitate the manipulation.

References

1. Wadhwa S, Paliwal R et al (2009) Nanocarriers in ocular drug delivery: an update review. Curr Pharm Des 15(23):2724–2750
2. Eljarrat-Binstock E, Pe'er J et al (2010) New techniques for drug delivery to the posterior eye segment. Pharm Res 27(4):530–543
3. Andrieu-Soler C, Bejjani RA et al (2006) Ocular gene therapy: a review of nonviral strategies. Mol Vis 12:1334–1347
4. Bejjani RA, Andrieu C et al (2007) Electrically assisted ocular gene therapy. Surv Ophthalmol 52(2):196–208
5. Jensen NM, Dalsgaard T et al (2011) An update on targeted gene repair in mammalian cells: methods and mechanisms. J Biomed Sci 18:10
6. Yang N, Singh S et al (2011) Targeted TFO delivery to hepatic stellate cells. J Control Release 155(2):326–330
7. Sangiuolo F, Scaldaferri ML et al (2008) Cftr gene targeting in mouse embryonic stem cells mediated by Small Fragment Homologous Replacement (SFHR). Front Biosci 13:2989–2999
8. Ferrara L, Kmiec EB (2004) Camptothecin enhances the frequency of oligonucleotide-directed gene repair in mammalian cells by inducing DNA damage and activating homologous recombination. Nucleic Acids Res 32(17):5239–5248
9. Overlack N, Goldmann T et al (2012) Gene repair of an usher syndrome causing mutation by zinc-finger nuclease mediated homologous recombination. Invest Ophthalmol Vis Sci 53(7):4140–4146
10. Johnen S, Izsvak Z et al (2012) Sleeping beauty transposon-mediated transfection of retinal and iris pigment epithelial cells. Invest Ophthalmol Vis Sci. doi:10.1167/iovs.12-9951
11. Yoon K, Cole-Strauss A et al (1996) Targeted gene correction of episomal DNA in mammalian cells mediated by a chimeric RNA.DNA oligonucleotide. Proc Natl Acad Sci U S A 93(5):2071–2076
12. Rando TA, Disatnik MH et al (2000) Rescue of dystrophin expression in mdx mouse muscle by RNA/DNA oligonucleotides. Proc Natl Acad Sci U S A 97(10):5363–5368
13. Bartlett RJ, Stockinger S et al (2000) In vivo targeted repair of a point mutation in the canine dystrophin gene by a chimeric RNA/DNA oligonucleotide. Nat Biotechnol 18(6):615–622
14. Alexeev V, Igoucheva O et al (2000) Localized in vivo genotypic and phenotypic correction of the albino mutation in skin by RNA-DNA oligonucleotide. Nat Biotechnol 18(1):43–47
15. Kren BT, Parashar B et al (1999) Correction of the UDP-glucuronosyltransferase gene defect in the gunn rat model of crigler-najjar syndrome type I with a chimeric oligonucleotide. Proc Natl Acad Sci U S A 96(18):10349–10354
16. Kren BT, Bandyopadhyay P et al (1998) In vivo site-directed mutagenesis of the factor IX gene by chimeric RNA/DNA oligonucleotides. Nat Med 4(3):285–290
17. Tagalakis AD, Graham IR et al (2001) Gene correction of the apolipoprotein (Apo) E2 phenotype to wild-type ApoE3 by in situ chimeraplasty. J Biol Chem 276(16):13226–13230
18. Andrieu-Soler C, Casas M et al (2005) Stable transmission of targeted gene modification using single-stranded oligonucleotides with flanking LNAs. Nucleic Acids Res 33(12):3733–3742
19. Ciavatta VT, Padove SA et al (2005) Mouse retina has oligonucleotide-induced gene repair activity. Invest Ophthalmol Vis Sci 46(7):2291–2299
20. Andrieu-Soler C, Doat M et al (2006) Enhanced oligonucleotide delivery to mouse retinal cells using iontophoresis. Mol Vis 12:1098–1107
21. Andrieu-Soler C, Halhal M et al (2007) Single-stranded oligonucleotide-mediated in vivo gene repair in the rd1 retina. Mol Vis 13:692–706

Chapter 26

In Utero Delivery of Oligodeoxynucleotides for Gene Correction

Lingzhi Cai, Bhanu Munil Koppanati, Carmen Bertoni, and Paula R. Clemens

Abstract

Gene correction is attractive for single gene mutation disorders, such as Duchenne muscular dystrophy (DMD). The *mdx* mouse model of DMD is dystrophin deficient due to a premature chain-terminating point mutation in exon 23 of the dystrophin gene. Gene editing of genomic DNA using single-stranded oligodeoxynucleotides (ssODNs) offers the potential to change the DNA sequence to alter mRNA and protein expression in defined ways. When applied to fetal skeletal muscle of *mdx* mice in utero, this technology leads to restoration of dystrophin protein expression, thus providing a valid gene-based therapeutic application at the earliest developmental stage. Here, we describe detailed methods for gene editing using muscle delivery of ssODNs to the fetal *mdx* mouse in utero at embryonic day 16 and to test correction of dystrophin deficiency at different ages after birth.

Key words Muscular dystrophy, Duchenne, Gene correction, In utero, Oligodeoxynucleotide

1 Introduction

Duchenne muscular dystrophy (DMD) is a progressive muscle disease that is caused by mutations in the dystrophin gene. Large deletions that result in a frameshift are the most common mutations causing DMD followed by single point mutations, insertions, and duplications, respectively [1]. Newborn screening that detects high levels of serum creatine kinase and muscle biopsies that show pathology in the first few months of life demonstrates that the dystrophic disease process in DMD begins very early in life [2]. Because DMD can be identified before birth by prenatal genetic testing, there is the possibility of initiating treatment for DMD in utero. This approach offers the potential to prevent even the earliest pathological changes in muscle due to dystrophin deficiency.

Gene augmentation strategies efficiently restore dystrophin expression to skeletal muscles of models of DMD especially when gene delivery is accomplished with viral vectors. Gene transfer

Francesca Storici (ed.), *Gene Correction: Methods and Protocols*, Methods in Molecular Biology, vol. 1114, DOI 10.1007/978-1-62703-761-7_26, © Springer Science+Business Media, LLC 2014

provides a functional copy of a cDNA that can express a recombinant copy of the native protein that is missing due to gene mutation. We have pioneered in utero gene delivery specifically for dystrophin gene transfer in the *mdx* model of DMD using both adenoviral and adeno-associated viral (AAV) vectors [3, 4].

As an alternative to the use of viral vector-mediated technologies, gene editing mediated by single-strand oligodeoxynucleotides (ssODNs) holds promise for gene therapy due to the low toxicity and stable transgene expression that can be achieved after delivery [5]. This strategy holds several advantages over gene replacement approaches. First, ssODNs are easy to produce even in large scale and at the quantities needed to satisfy clinical applications in humans. Second, ssODNs efficiently diffuse in skeletal muscle tissue after intramuscular injection. Third, the presence of ssODNs in cells is transient, thus limiting potential toxicity. Furthermore, correction is stably inherited in actively dividing cells such as muscle stem cells and muscle progenitor cells that are responsible for the growth and maintenance of skeletal muscle before and after birth. Therefore, in utero delivery of ssODNs is a particularly attractive approach because it offers the potential for gene correction to be amplified during muscle development and growth to maturity. Thus, gene editing in muscle stem cells or muscle progenitor cells could provide stable protein expression that is sustained throughout life.

Previous studies demonstrated the feasibility of making single-base changes in the dystrophin gene using different ssODN chemistries in mouse and dog models of DMD [6–12]. In mice, gene correction mediated by ssODN efficiently targets and corrects single point mutations in the dystrophin gene and restores full-length dystrophin expression [7, 8, 11, 12]. ssODN-mediated gene editing strategies can also be used to alter a consensus splice site of the dystrophin gene and to redirect splicing of the dystrophin mRNA to produce an in-frame shortened form of the dystrophin protein [8]. By disrupting an intron/exon splice site boundary or regulatory mechanisms such as specific enhancers or splicing regulatory elements that control exon recognition of the dystrophin gene, it is possible to exclude one or more adjacent exons to restore the reading frame of the dystrophin protein in otherwise dystrophin-deficient cells [5]. ssODNs can be homologous to either the coding strand or the noncoding strand of the gene of interest (Fig. 1).

The limitation of the technology, thus far, has been the low level of genomic correction detected in muscle cells, which is determined in part by the type of cells that are targeted, the specific mechanism of gene repair that is recruited, and even the region of the gene that is targeted for repair [5]. Nonetheless, gene editing has already proven to be a valid clinical approach capable of targeting the majority of the disease-causing mutations in DMD patients.

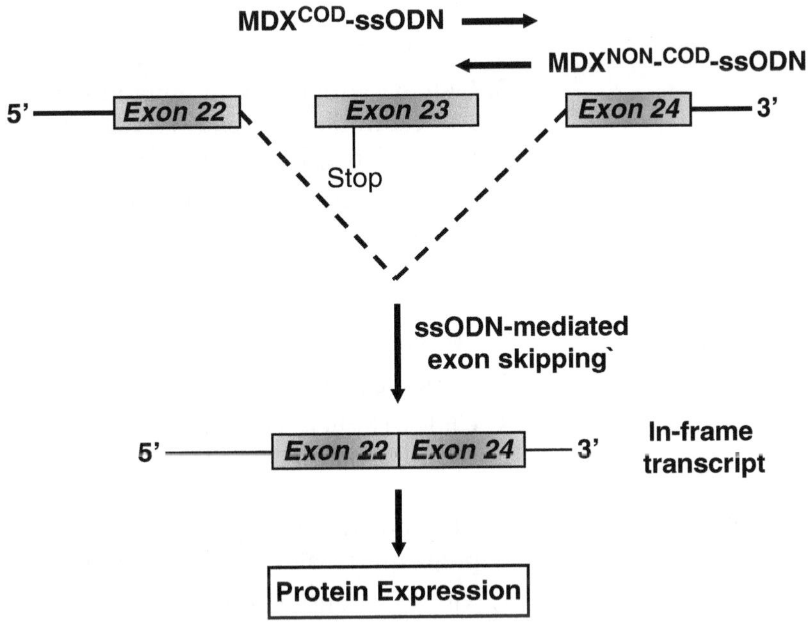

Fig. 1 Schematic representation of the strategy used to test ssODN-mediated exon skipping in the *mdx* mouse. This mouse model for DMD is characterized by a single point mutation in exon 23 of the dystrophin gene that creates a stop codon, resulting in a complete absence of dystrophin protein expression in muscle. ssODNs are designed to target the splice donor site of exon 23 downstream of the mutation and to induce a single-base alteration at the splice junction that disrupts the regulatory elements responsible for the inclusion of the exon during mRNA splicing. The ssODN can be either homologous to the coding strand of exon 23 (MDX^{COD}-ssODN) or complementary to it ($MDX^{NON-COD}$-ssODN). Disruption of the consensus sequence results in skipping of one or more exons (skipping of exon 23 is shown) and in the expression of an in-frame, although shorter, dystrophin protein

2 Materials

2.1 ssODNs (See Note 1)

1. Targeting ssODNs:
   MDX^{COD}.
   $MDX^{NON\text{-}COD}$.
2. Control ssODNs:
   CTL^{COD}.
   $CTL^{NON\text{-}COD}$.

2.2 Fluorescent Beads (See Note 2)

Fluorescent beads (540/560 emission/excitation) (Molecular Probes).

Dextran Alexa Fluor (Molecular Probes).

2.3 Anesthetic, Analgesic, and Antibiotic Drugs (See Note 3)

1. Ketamine (Fort Dodge Animal Health): Stock concentration, 100 mg/ml. Dose, 50 mg/kg by intraperitoneal (IP) injection.

2. Xylazine (Fort Dodge Animal Health): Stock concentration, 100 mg/ml. Dose, 5 mg/kg by IP injection.
3. Systemic analgesic drug: Buprenorphine, HCl (Fisher Scientific): Concentration, 0.3 mg/ml, prefilled 10 mL cartridges. Dose, 0.05–0.1 mg/kg by subcutaneous injection.
4. Oral analgesic drug: Rimadyl 2 mg/tablet. Dose, 2 mg by oral administration.
5. 2 % lidocaine cream (Abbott Laboratory). Dose: Apply topically.

2.4 Reagents

1. 70 % ethanol.
2. Betadine solution (Fisher Scientific).
3. Clidox (Fisher Scientific) (*see* **Note 4**): Distilled water 180 ml, BASE 10 ml, ACTIVATOR 10 ml. Prepare fresh on surgery day.

2.5 Mice

1. Keep timed breeding colonies of the desired strains to generate pregnant females bearing embryonic day 16 (E-16) mice.
2. Mouse strains: C57BL/10 SnJ (normal background strain) and C57BL/10ScSn-*Dmd*^{mdx}/J (dystrophin-deficient *mdx* strain). Mice are purchased from The Jackson Laboratory.

2.6 Surgical Instruments and Supplies (See Note 5)

1. 33G needles (Hamilton).
2. PB600 syringe dispenser (Hamilton).
3. 100 μl Hamilton syringe (Sigma).
4. 7-0 silk suture (Ethicon).
5. Wound clips.
6. Surgical garb including gloves, mask, cap, and scrubs.
7. Surgical supplies including gauze pads, cotton tip sutures, and drapes.
8. Dissection microscope or magnifier to facilitate in utero injections.
9. Hot bead instrument sterilizer.
10. Light for transillumination.
11. Warming pad.
12. Fluorescence microscope to identify injected pups post-delivery.

2.7 Immunostaining Components

1. 1×PBS buffer: Store at room temperature.
2. 20 % fetal calf serum (FCS) blocking solution: 200 μl FCS (Sigma) mixed with 800 μl 1×PBS. Store at −20 °C.
3. Dystrophin primary antibody: Rabbit anti-dystrophin antibody (Lab Vision Corporation). Store at −20 °C.
4. Goat anti-mouse collagen type IV antibody (Millipore). Store at −20 °C.

5. Alexa Fluor 488 donkey anti-rabbit IgG secondary antibody (Invitrogen). Store at 4 °C.
6. Donkey anti-goat IgG secondary antibody (Fitzgerald). Store at 4 °C.
7. Nuclei staining reagent: Hoechst 33342 (Sigma): Dilute Hoechst dye (1:50,000), 1 μl dye in 1 ml of 1× PBS, and then take 100 μl of this in 5 ml of 1× PBS. Make fresh on day used.
8. Fluoromount-G (Southern Biotech). Store at room temperature.
9. Microscope slides (Fisher).
10. Gold Seal cover slips (Fisher).

3 Methods

3.1 Design of ssODNs

Targeting ssODNs are designed to be homologous to the transcribed (MDX^{COD}) or the non-transcribed ($MDX^{NON\text{-}COD}$) strands of the murine dystrophin gene (Fig. 1). Both MDX^{COD}- and $MDX^{NON\text{-}COD}$-ssODN target the donor site of the intron 22/exon 23 splice junction in the *mdx* mouse dystrophin gene to induce skipping of exon 23. This results in exclusion of exon 23 containing the nonsense mutation of the *mdx* dystrophin gene. Using a similar gene editing approach, this strategy has been shown to result in the production of in-frame transcripts and nearly full-length dystrophin protein in adult *mdx* mice [8].

3.2 Preparation of Surgery Field, Surgeon, Equipment, and Surgical Supplies

1. Turn on the hot bead instrument sterilizer, warming pad, and light for uterine transillumination.
2. Clean the surgery field with Clidox and wipe with Kimwipes.
3. Put drapes over the warming pad on the surgery bench.
4. The surgeon should wear surgery garb including sterile gloves, cap, and mask.
5. Prepare three Hamilton syringes. Syringe 1, 20 μl fluorescent beads. Syringe 2, 20 μl targeting ssODN. Syringe 3, 20 μl control ssODN.
6. Set up a Hamilton syringe: Clean a PB600 syringe dispenser with Clidox. Set up the syringe dispenser with the syringe that will be used first.

3.3 Anesthesia and Preparation of Pregnant Mouse for Surgery

1. Put a drape on another bench to provide a place to anesthetize the experimental mouse.
2. Draw up ketamine 50 mg/kg and xylazine 5 mg/kg with a 1 ml U-100 syringe and administer IP to anesthetize the mouse.
3. Shave the mouse abdomen with a shaver.

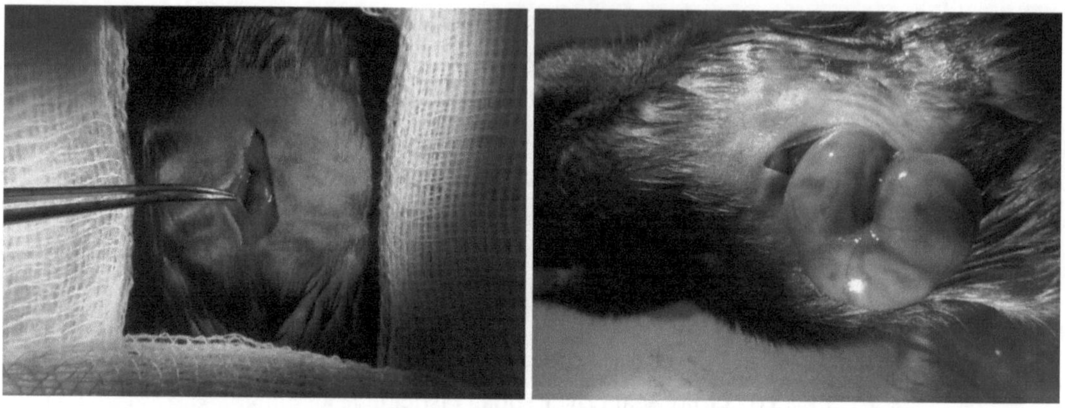

Fig. 2 Surgical approach to in utero injection of mouse fetus. The *left panel* shows the vertical incision through the abdominal wall and peritoneum of the pregnant mouse. The *right panel* shows how the uterine horn is exposed through the incision

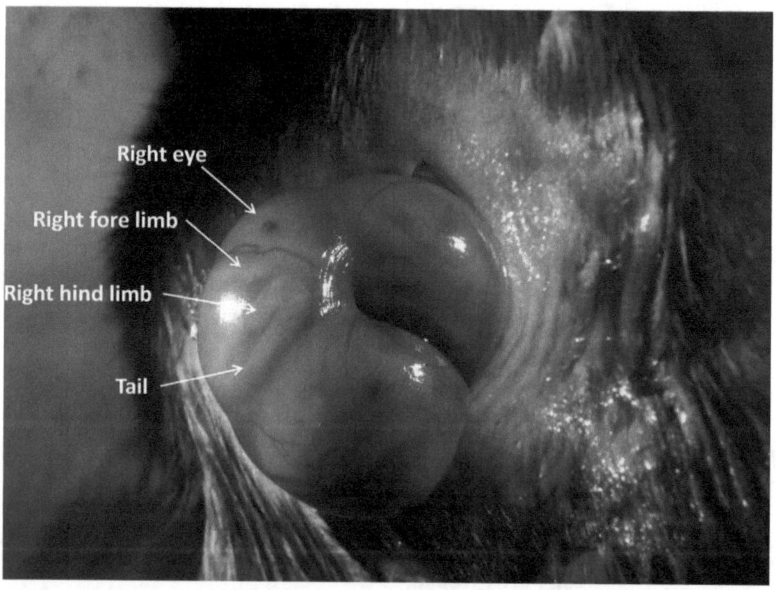

Fig. 3 Identification of fetal body parts in utero. Location of right eye, right forelimb, right hind limb, and tail is marked for one fetus in a high magnification view of the transilluminated uterine horn

4. Clean the abdomen skin with Betadine.
5. Put the anesthetized, prepared mouse on the warming pad on the surgery bench.

3.4 Surgical Approach to the Pregnant Mouse and ssODN Delivery to Fetal Pups

1. All surgical procedures are performed using aseptic techniques.
2. Incise a 15 mm vertical incision through the abdomen skin and peritoneum, to expose the uterus (Fig. 2).
3. Under high magnification and bright light, the eyes of the fetus are visible as dark spots. The limbs and the tail of the fetus are also visible (Fig. 3).

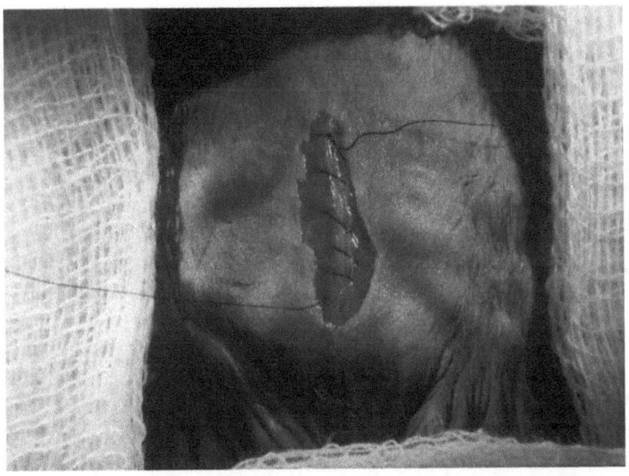

Fig. 4 Surgical closure of the pregnant mouse. After gene delivery, the uterus is replaced in the body cavity, the peritoneum and abdominal wall are sutured, and the skin is closed with wound clips

4. Intramuscular injections of ssODNs: Inject 4 μl of ssODNs into the hind limb and 2 μl of fluorescent beads into the forelimb (*see* **Note 6**) of the fetus (either left side or right side) using a Hamilton syringe in a PB600 syringe dispenser and a 33G needle under a dissection microscope or magnifier with transillumination of the uterine wall.
5. Repeat **step 4** to inject four fetuses per pregnant mouse (*see* **Note 7**).
6. Close the peritoneum and abdomen using 7-0 silk sutures (Fig. 4).
7. Close the skin with wound clips.
8. Apply topical analgesic 2 % lidocaine HCl cream to the incision site.
9. Put treated pregnant mice in the mouse cages and place the cages on the warming pad.
10. Keep the animals recovering in a warm cage for up to 24 h in the surgery room.

3.5 Postoperative Administration of Analgesia (See Note 8)

1. Inject buprenorphine 0.05–0.1 mg/kg subcutaneously for analgesia when the mice recover consciousness on the day of surgery and then once daily for 2 days postoperatively.
2. Add one tablet (2 mg) of Rimadyl in each cage, daily for 5–7 days to provide orally administered analgesia.
3. Allow natural birth to proceed, usually on or near embryonic day 19.

3.6 Identification of Pups that Were Injected In Utero

1. Examine delivered pups 2–5 days after birth using a fluorescence microscope to identify the injected pups by detection of the fluorescent beads (*see* **Note 9**; Fig. 5).

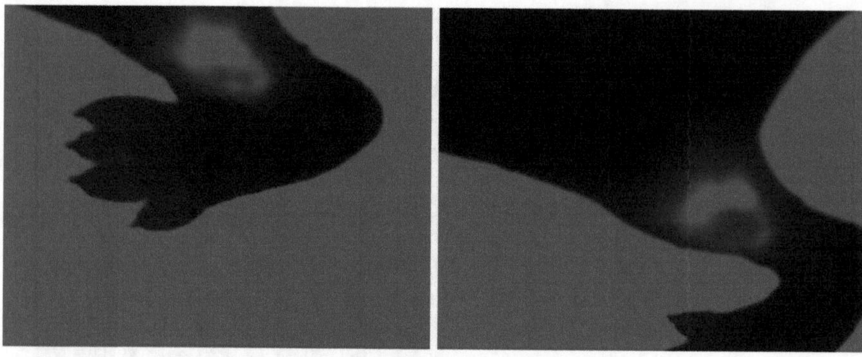

Fig. 5 Fluorescent bead identification of the treated mouse pups. View of limb by fluorescence microscopy that shows the location of fluorescent beads that were injected at the time of ssODN delivery to identify the treated pup

2. Mark the injected pups by cutting tails (*see* **Note 10**).
   (a) Clean the tails with Betadine prior to the procedure.
   (b) Remove about 5 mm of the distal tail with a sterile razor blade.
   (c) Put lidocaine cream to the cut area for pain relief.

3.7 Processing Muscle Samples

1. Sacrifice experimental mice at the desired time after birth as determined by the particular experiment: (a) The *mdx* mice injected with targeting ssODNs and control ssODNs. (b) Dystrophin-positive control (C57BL/10 SnJ) mice. (c) Dystrophin-negative control (un-injected *mdx*) mice at different time points and collect muscle samples (*see* **Note 11**).
2. Precool a 50 ml tube containing 40 ml isopentane on dry ice for at least 30 min.
3. Label microcentrifuge tubes for muscle samples to be stored after collection and put the tubes on dry ice to precool.
4. Freeze collected muscle samples by plunging the tissue held in a forceps in dry ice-cooled isopentane until bubbles disappear. It usually takes 10–30 s depending on the size of the sample.
5. Put frozen muscle sample in a prechilled, pre-labeled tube.
6. Store samples at −80 °C (*see* **Note 12**).

3.8 Sectioning the Muscle Samples

1. While outside of the −80 °C freezer, keep frozen muscle samples on dry ice to prevent thawing.
2. Cryotome should be precooled to −25 °C.
3. Put some OCT on the cryotome chuck and allow it to harden. Then place a small amount of OCT on top of the frozen OCT on the chuck and orient the muscle sample such that the muscle tissue will be sectioned in cross section (*see* **Note 13**).

4. Cut serial 10 μm sections and place sections on slides.
5. Store slides containing tissue sections at −20 °C prior to immunostaining.

3.9 Dystrophin/Collagen IV/Hoechst Staining

Immunohistochemistry (dystrophin/collagen/Hoechst) can be performed to analyze dystrophin gene correction at the protein level and to assess fibers with centrally placed nuclei, as an indicator of muscle degeneration/regeneration. Collagen immunohistochemistry identifies all fibers to aid in assessing the proportion that express dystrophin and have centrally placed nuclei (nuclei stained with Hoechst).

1. Air-dry section.
2. Wash sections twice with 1× PBS for 5 min.
3. Block sections with 20 % FCS in 1× PBS at room temperature for 30 min.
4. Mix an anti-collagen type IV polyclonal antibody (goat anti-mouse collagen), 1:20 diluted in 1× PBS, and dystrophin primary antibody (a rabbit polyclonal antibody against the last 17 amino acids of the C-terminal domain of mouse dystrophin), 1:100 diluted in 1× PBS, at 4 °C. Incubate overnight.
5. Wash sections twice with 1× PBS for 5 min.
6. Mix collagen secondary antibody (donkey anti-goat IgG), 1:50 diluted in 1× PBS, and dystrophin secondary antibody (Alexa Fluor 488 donkey anti-rabbit IgG), 1:500 diluted in 1× PBS, at room temperature. Incubate for 1 h in the dark.
7. Wash sections three times with 1× PBS for 5 min.
8. Add 100 μl of diluted Hoechst dye (1:50,000) to each section and incubate at room temperature for 5 min to stain nuclei.
9. Wash sections twice with 1×PBS for 5 min.
10. Mount with Fluoromount-G.
11. Cover the sections with cover glass.
12. Store the slides at 4 °C in the dark.
13. Image sections with a fluorescence microscope (Figs. 6 and 7).

4 Notes

1. ssODNs can be purchased from a number of different vendors. In our studies we use ssODNs synthesized by MWG Biotech purified by reverse phase (RP) HPLC. A scale of synthesis of 1 μM is sufficient to provide ssODN for the injection of at least five pups. Higher amounts can be purchased upon request. Lyophilized ssODNs are shipped at room temperature and

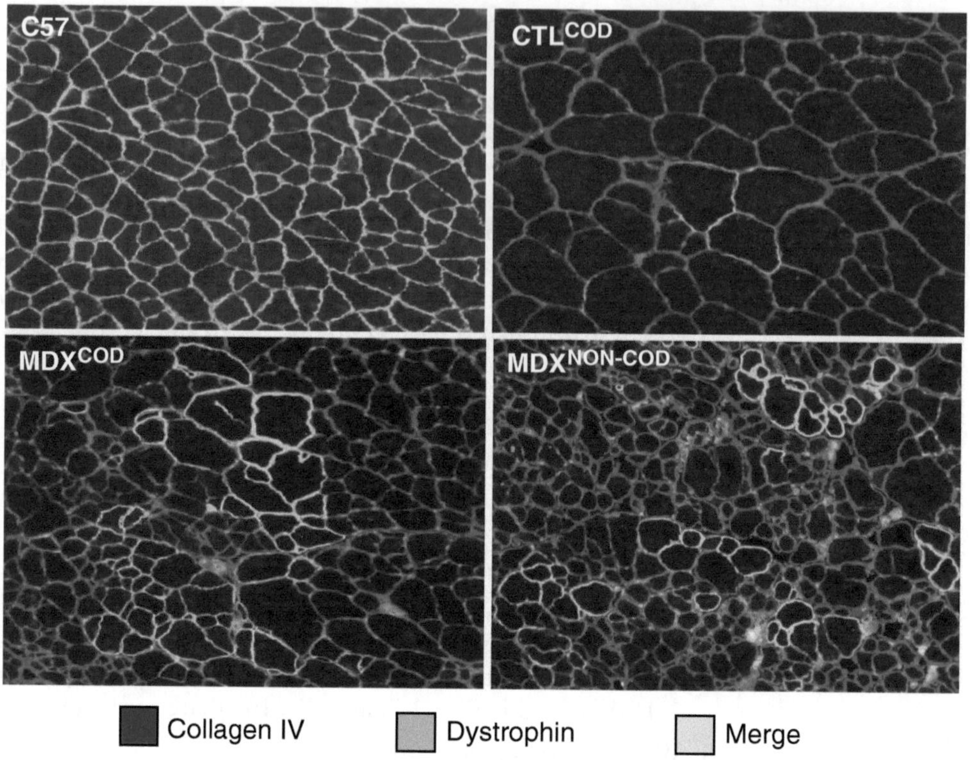

Fig. 6 Dystrophin expression achieved by gene correction. A 50 μg dose of targeting MDX^{COD} or MDX$^{NON\text{-}COD}$-ssODN or a control ssODN that was homologous to the coding strand of the exon 23/intron 23 splice site in the dystrophin gene (CTL^{COD}) was administered by intramuscular injection on embryonic day 16. Double immunohistochemical localization of collagen IV (*red fluor*) and dystrophin (*green fluor*) was performed. Fibers with dual expression of collagen and dystrophin fluoresce *yellow*. Clusters of gene-corrected fibers are shown at 4 months after birth. A section of C57BL/10 (C57) muscle provides a positive control for dystrophin immunostaining

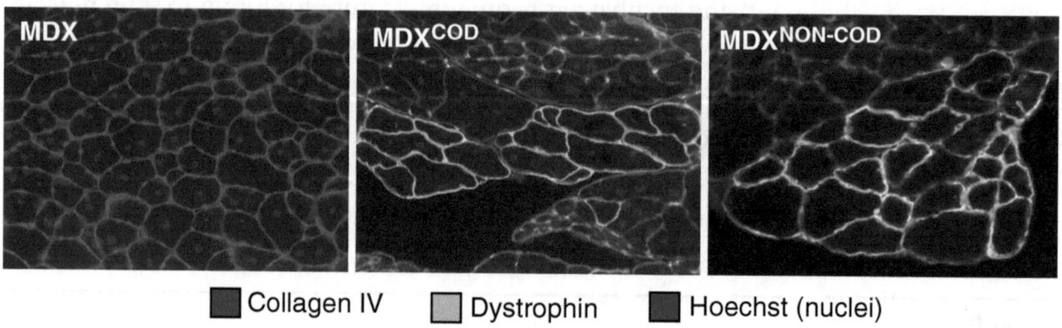

Fig. 7 Simultaneous localization of collagen IV (*red fluor*), dystrophin (*green fluor*), and nuclei (*blue stain* with Hoechst). Clusters of gene-corrected fibers and localization of myonuclei (central or peripheral) are shown. The *left panel* shows a section of untreated *mdx* mouse muscle that demonstrates absence of dystrophin expression and a high percentage of fibers with centrally placed nuclei. The *middle* and *right panels* show sections of muscle from *mdx* mice that were treated with targeting MDX^{COD} or MDX$^{NON\text{-}COD}$ ssODN, in utero. Dystrophin-positive fibers present in muscles injected with targeting ssODNs have a lower percentage of centrally placed nuclei, suggesting that correction of the dystrophin gene defect is stable over time and is capable of protecting myofibers from degeneration

stored at –80 °C until ready to use. To resuspend, dissolve the ssODNs in RNase-/DNase-free water to the desired concentration and store stock vials at –80 °C.

2. Administration of fluorescent beads or Dextran Alexa Fluor allows identification of the treated pups after delivery, since not all fetuses in a gravid uterus are manipulated. Prior to use, store the fluorescent beads or Dextran Alexa Fluor at –20 °C.
3. Each drug should be diluted in an autoclaved anesthetic bottle. Storage of controlled substances should follow institutional requirements, and no drug should be used after expiration.
4. Make Clidox solution fresh and label with researcher's name and date prepared. This reagent expires in 2 weeks. Do not use expired Clidox solution.
5. Surgical instruments and supplies must be sterilized prior to use and between surgical procedures. We use a hot bead equipment sterilizer to sterilize equipment between experimental animals during the same surgical session.
6. We suggest injecting fluorescent marker at a location different from the injection site of the ssODNs. By so doing, the fluorescent beads will not interfere with the ssODN delivery. If the ssODN are injected intramuscularly in the hind limb, then the fluorescent bead injection can be in the forelimb.
7. Through experience, we learned that the rate of spontaneous loss of pregnancy increased if more than four pups per gravid uterus were injected. Therefore, we limit the number of treated pups per uterus to 4.
8. Provision of analgesia to pregnant mice may decrease the discomfort of pup delivery that occurs only 3 days after surgery for in utero gene correction. Decreasing pain will decrease stress and may decrease the rate of pup cannibalization by the mother after birth. Through experience we arrived at the described cocktail of topical, systemic, and oral analgesia extending from immediately post-surgery to 5–7 days post-surgery.
9. In general, the beads remain fluorescent for up to 7 days which provides the window during which pups should be examined by fluorescence microscopy to identify those pups that were treated in utero. As a practical measure, we keep the fluorescence microscope in the animal facility so that we can handle the young pups as little as possible to perform this identification. This will minimize the probability that the mother mouse will cannibalize her newborn pups.
10. This method combines the highest level of effectiveness with the lowest level of pain and distress for neonatal mouse identification. Neonatal mice are too small for ear tagging or punching. Shortening the tail provides unequivocal identification that persists into adulthood.

11. Comprehensively collect muscle from the entire lower limb that was injected. When doing a trans-uterine injection of a fetal limb, one can determine whether it is an upper or lower limb and whether it is left or right at the time of surgery and injection, but one cannot determine which muscle of the hind limb is injected.
12. Frozen muscle samples should not be allowed to thaw during sectioning. If the block of muscle tissue is allowed to thaw and refreeze at any point, ice crystal artifact will result.
13. The muscle sample should be positioned as a pillar in the OCT with only the most inferior portion within OCT. The portion of the sample that is placed in OCT will thaw partially and will therefore have ice crystal artifact. The part to be sectioned is the pillar extending from the OCT base. If handled properly, the part of the sample that will be sectioned will not have thawed and been re-frozen. Especially if the muscle samples are small, it can be helpful to mount two or more muscles side-by-side on the same chuck and section them together.

Acknowledgments

This work was supported by NIAMS grant R01-AR050565 (PRC). We appreciate the input of Thomas A. Rando, M.D., Ph.D., Stanford University, and Daniel Reay, University of Pittsburgh, toward the development of these protocols. The authors take full responsibility for the contents of this chapter, which do not represent the views of the Department of Veterans Affairs or the United States Government.

References

1. Engel AG, Ozawa E (2004) Dystrophinopathies. In: Engel AG, Franzini-Armstrong C (eds) Myology, 3rd edn. McGraw-Hill, New York, pp 961–1025
2. Zellweger H, Antonik A (1975) Newborn screening for Duchenne muscular dystrophy. Pediatrics 55:30–34
3. Koppanati BM, Li J, Reay DP, Wang B, Daood M, Zheng H, Xiao X, Watchko JF, Clemens PR (2010) Improvement of the mdx mouse dystrophic phenotype by systemic in utero AAV8 delivery of a minidystrophin gene. Gene Ther 17:1355–1362
4. Reay DP, Bilbao R, Koppanati BM, Cai L, O'Day TL, Jiang Z, Zheng H, Watchko JF, Clemens PR (2008) Full-length dystrophin gene transfer to the mdx mouse in utero. Gene Ther 15:531–536
5. Bertoni C (2008) Clinical approaches in the treatment of Duchenne muscular dystrophy (DMD) using oligonucleotides. Front Biosci 13:517–527
6. Bartlett RJ, Stockinger S, Denis MM, Bartlett WT, Inverardi L, Le TT, Man N, Morris GE, Bogan DJ, Metcalf-Bogan J, Kornegay JN (2000) In vivo targeted repair of a point mutation in the canine dystrophin gene by a chimeric RNA/DNA oligonucleotide. Nat Biotechnol 18:615–622
7. Bertoni C, Rando TA (2002) Dystrophin gene repair in mdx muscle precursor cells in vitro and in vivo mediated by RNA-DNA chimeric oligonucleotides. Hum Gene Ther 13: 707–718
8. Bertoni C, Lau C, Rando TA (2003) Restoration of dystrophin expression in mdx

muscle cells by chimeraplast-mediated exon skipping. Hum Mol Genet 12:1087–1099

9. Bertoni C, Morris GE, Rando TA (2005) Strand bias in oligonucleotide-mediated dystrophin gene editing. Hum Mol Genet 14:221–233
10. Bertoni C, Rustagi A, Rando TA (2009) Enhanced gene repair mediated by methyl-CpG-modified single-stranded oligonucleotides. Nucleic Acids Res 37:7468–7482
11. Kayali R, Bury F, Ballard M, Bertoni C (2010) Site-directed gene repair of the dystrophin gene mediated by PNA-ssODNs. Hum Mol Genet 19:3266–3281
12. Rando TA, Disatnik MH, Zhou LZ (2000) Rescue of dystrophin expression in mdx mouse muscle by RNA/DNA oligonucleotides. Proc Natl Acad Sci U S A 97: 5363–5368

Chapter 27

Portal Vein Delivery of Viral Vectors for Gene Therapy for Hemophilia

Alexandra Sherman, Alexander Schlachterman, Mario Cooper, Elizabeth P. Merricks, Robin A. Raymer, Dwight A. Bellinger, Roland W. Herzog, and Timothy C. Nichols

Abstract

The liver is a very complex organ with a large variety of functions, making it an attractive organ for gene replacement therapy. Many genetic disorders can be corrected by delivering gene products directly into the liver using viral vectors. In this chapter, we will describe gene delivery via portal vein administration in mice and dogs to correct the blood coagulation disorder hemophilia B. Although there are multiple delivery routes for both viral and non-viral vectors in animals, portal vein administration delivers vectors directly and efficiently into the liver. Complete correction of murine hemophilia B and multi-year near-correction of canine hemophilia B have been achieved following portal vein delivery of adeno-associated viral (AAV) vectors expressing factor IX from hepatocyte-specific promoters. Peripheral vein injection can lead to increased vector dissemination to off-target organ such as the lung and spleen. Below, we will describe portal vein injection delivery route via laparotomy.

Key words Liver, Gene therapy, Portal vein, Viral vectors, AAV

1 Introduction

The goal of an in vivo gene transfer protocol for correction of human disease is optimal delivery and expression of the therapeutic gene in the target organ while minimizing dissemination to other organs, thereby increasing efficacy and minimizing possible immune reactions or other toxicities. Gene transfer to hepatocytes is very effective not only for correction of liver disease but also for systemic delivery of therapeutic proteins. For example, hepatocytes are the normal site of synthesis for several coagulation factors; and therefore, hepatic gene transfer has resulted in high therapeutic levels of factor VIII (FVIII) and factor IX (FIX) in several animal models, including mice, dogs, and nonhuman primates [1–6]. Stable therapeutic expression for many years, in some cases for more than a decade, in hemophilic dogs has been observed

Francesca Storici (ed.), *Gene Correction: Methods and Protocols*, Methods in Molecular Biology, vol. 1114, DOI 10.1007/978-1-62703-761-7_27, © Springer Science+Business Media, LLC 2014

(unpublished data and refs. 2, 7–9). Importantly, hepatocytes are capable of producing and secreting high levels of biologically active coagulation factors into the blood. Furthermore, optimal liver gene transfer using adeno-associated viral (AAV), lentiviral, and also other vectors can induce immune tolerance to the transgene product, in part via a regulatory T cell response [10–13] (*see* **Note 1**). Finally, liver gene transfer for correction of the hemophilia B has been successfully translated to human treatment in recent clinical trials [14]. In this chapter, we will describe gene transfer for correction of murine and canine hemophilia B using recombinant AAV vectors via portal vein delivery (*see* **Note 2**).

AAV vectors have a single-stranded or self-complementary (sc) DNA genome and are devoid of viral coding sequences [15]. Use of a strong hepatocyte-specific enhancer/promoter combination is recommended for liver gene transfer [14, 16]. There are multiple serotypes of AAV with distinct tissue tropism. For example, AAV8 can be effectively delivered to hepatocytes by peripheral vein injection such as the tail vein of a mouse [14, 17, 18]. Other serotypes, including AAV2 and AAV5, require infusion more directly into the blood supply of the liver, which can be accomplished by administration into the portal vein, mesenteric vein, or hepatic artery [5, 6, 16, 19, 20]. In mice, injections into the splenic capsule may be similarly suitable (but only at vector doses of 10^{11} vector genomes [vg] or higher, as the spleen may sequester too much of the vector at lower doses). Capsid variants such as AAV2 devoid of surface-exposed tyrosine residues further improve gene transfer [21]. Use of cDNAs optimized for mammalian codon usage can further improve therapeutic expression [14, 18, 22, 23]. Generally, resulting expression levels are vector dose dependent. For AAV2, a single portal vein infusion of 10^{12} vg/kg to FIX-deficient mice or dogs resulted in long-term correction of hemophilia B, without eliciting any types of toxicity [5–7].

Prior to approval of a gene therapy protocol for clinical studies in humans suffering from genetic disorders, the approach has to be thoroughly investigated in a suitable animal model. Mice have been extensively used as a model for hepatic gene transfer via portal vein injection to treat numerous genetic disorders, such as hemophilia A and B, α1 antitrypsin deficiency, and OTC (ornithine transcarbamylase) deficiency [13]. Mice have the advantage of being easily accessible, having a high reproduction rate, and offering the ability to perform batch surgeries without a need for a separate dedicated facility. Recombinant DNA technology allowed the hemophilia research community to generate a large number of different knockout mice and transgenic mice, lacking FVIII or FIX or expressing nonfunctional forms of the protein [24, 25]. Therefore, one can test the risk of an immune response to the therapeutic gene product as a function of the underlying mutation, using gene mutations described in humans with hemophilia. However, there are also certain disadvantages such as limited ability to study the long-term effects of the gene transfer and lack of scale up to an animal of a size more similar to that of humans. Using an alternative large animal model, a single injection of AAV-FIX into the portal circulation, on canine hemophilia B models (Chapel Hill dogs), resulted in sustained expression of canine FIX for over 10 years at levels between 1 and 10 % [5, 6, 8, 26].

When choosing between mouse and canine models of hemophilia B for hepatic-based gene therapy studies, several issues need to be considered. The Chapel Hill strain of hemophilia B dogs has a missense mutation that leads to the absence of detectible FIX activity or antigen [27, 28]. Recombinant AAV-mediated gene therapy in these hemophilia B dogs has been successful for multiple years without adverse events, albeit at low but steadily improving levels of expression [5, 6, 26]. Thus the success of gene therapy can be better assessed in dogs since transgene expression seems to be lasting longer than the life span of mice. Hemophilia B dogs have on average 5 spontaneous bleeds/year in soft tissues or joints, a bleeder phenotype mirrors that found in human hemophilia B [7, 29]. This severe bleeding diathesis also provides a metric for detecting a change in phenotype following gene therapy. In general, hemophilia B mice do not bleed spontaneously, so assessing change in bleeder phenotype is more difficult. To date, inhibitor formation in this outbred hemophilia B dog strain has not been noted with portal vein approach. In some of the mouse strains, inhibitor formation to the coagulation proteins has been noted but newer mouse models may help understand this issue better [25] (*see* **Notes 3–5**). Also, dogs tolerate multiple blood draws better than mice (*see* **Note 6**). Frequent blood sampling is essential in these studies since the goal to express coagulation proteins in the systemic circulation over time. It is also helpful to have large volumes of plasma (~1,000 ml or more) drawn over time to determine the extent of posttranslational modification of transgene-expressed FIX relative to wild-type FIX. In contrast to rodents, the results of experiments in the hemophilia B dogs have been more predictive of outcomes in humans than those performed in other species. We have shown that FIX can be given subcutaneously or via the airways in our hemophilia B dogs, administration routes that could reduce or obviate the need for needles and venipuncture in hemophilia B patients and justify extrahepatic expression in patients with severe liver disease in whom a portal vein approach may be contraindicated [30]. As a direct result, human trials of subcutaneous and inhalational administration of recombinant FIX are being considered. FIX is not bioavailable by these routes in mice. Also, FIX infused into hemophilia B dogs had a comparable half-life to that found in humans [31–33], whereas mice had a markedly shorter half-life [19]. Had either the extravascular administration or half-life studies of FIX only been performed in mice, human trials may not have been pursued. Collecting semen from dogs, a procedure that is extremely difficult in mice, is used to assess the risk of germline transmission of gene therapy vectors [34]. These differences are the basis for many investigators, and advisory boards regard these dogs as an essential national resource for preclinical testing and long-term follow up of new treatments for the hemophilia B (*see* MASAC Recommendations #137 and #160, http://www.hemophilia.org/research/masac/masac_all.htm).

2 Materials

2.1 Rodent Surgery

Rodent survival surgeries do not require a separate facility. Surgeries can be performed in a regular procedure room in the areas that can be easily sanitized. All instruments used for surgery should be sterilized by autoclaving 270 °F for 10 min prior to use. For multiple surgeries during single session, instruments must be disinfected between animals using hot bead sterilizer.

2.1.1 Instrument Kit

1. 2 scissors.
2. 2 retractors.
3. Forceps.
4. 1 ml syringe.
5. 30Ga needles.
6. Sterile Q-tips.
7. Sterile gauze.
8. Sterile, absorbable haemostatic material.

2.1.2 Additional Equipment

1. Animal clippers, blade#40.
2. Anesthesia machine for rodents, e.g., SurgiVet.
3. Isoflurane, oxygen tank.
4. Purelube, ophthalmic ointment.
5. Sterile PBS.
6. Water-recirculating heating pad.
7. Sterile surgical gloves, gown, and face mask.

2.2 Canine Surgery

In contrast to rodents, survival surgeries in dogs require a separate dedicated surgical facility that meets or exceeds requirements for performing survival surgery. All instruments used for surgery must be sterilized by autoclaving using standardized protocols.

2.2.1 Standard "Vet Pack" Instrument Kit for Major Abdominal Surgery

1. Vet Pack (abdominal pack).
2. 4 towel clamps.
3. 1 operating scissors.

4. 1 curved mayo scissors.
5. 1 curved Metzenbaum scissors.
6. 1 thumb forceps with teeth.
7. 1 thumb forceps without teeth.
8. 1 needle holder.
9. 4 curved mosquito forceps.
10. 4 straight mosquito forceps.
11. 4 curved Kelly forceps.
12. 4 Rochester-Carmalt forceps.
13. 1 Haight Rib Spreader.
14. 2 #3 blade handles.
15. 2 Allis tissue forceps.
16. 2 Babcocks.
17. 2 Peans.
18. 1 small suction tip.
19. 1 large suction tip.
20. 1 spay hook.
21. 8 surgical towels.
22. 3×3 gauze squares.
23. 1 female catheter.
24. 1 sponge forceps.
25. 1 stainless steel bowl.
26. 1 Weitlaner.
27. 1 catheter introducer.

2.2.2 Additional Equipment

1. Animal clippers.
2. Anesthesia machine for larger animals, e.g., SurgiVet.
3. Isoflurane, oxygen tank.
4. Ophthalmic ointment.
5. Sterile PBS.
6. Heating pad.
7. Sterile surgical gloves, gowns, and appropriate face mask.

3 Methods

3.1 Rodent Surgery

3.1.1 Preparation of the Surgical Area

1. Surgical area must be disinfected prior to the surgery using any of the approved (appropriate) disinfectants in your facility.
2. Heating pad should be sanitized and placed on the cleaned surgical area.

3. Place sterile surgical wrap on the disinfected surface of the water-recirculating heating pad.
4. Open sterile instruments, gauze, Q-tips on the sterile surgical wrap.

3.1.2 Preparation of the Animal

1. For hemophilia B mice administer 200 μl of normal mouse plasma IV via tail vein 30 min before the surgery (*see* **Notes 7** and **8**).
2. Place mouse into induction chamber of the anesthesia machine and administer via inhalation 5 % of isoflurane in 21 % oxygen carrier (flow of 2 L/min).
3. Once unconscious the mouse is removed from the chamber, abdomen quickly shaved from xiphoid down to the groin, purelube is applied to the both eyes to prevent corneal drying and placed on its back with the face inside the nose cone part of anesthesia equipment in the non-sterile area.
4. Clean and aseptically prepare surgical site using an appropriate scrubbing technique: starting in the middle, going in circle, gradually enlarging circular pattern. Use three alternating scrubs of Betadine solution and 70 % ethanol.
5. Move the animal to the clean surgical area and place on top of a heating pad to minimize hypothermia.
6. Use sterile scissors to cut a small hole in the sterile drape and cover animal with the drape only exposing abdomen.

3.1.3 Preparation of the Surgeon

1. Rodent surgeries do not require sterile gowning. Surgeon must wear clean lab. coat, face mask, hair cover, and sterile surgical gloves.

3.2 Surgical Procedure

1. Once animal has been shaved, scrubbed, and moved to the sterile surgical area and covered with the sterile surgical drape while exposing abdomen, make a small (<1 in.) skin incision from the bladder up to the level of the xiphoid.
2. Repeat same opening technique for the muscle layer.
3. Retract the skin and muscle layers on both the right and left sides. Place tissue retractors to hold them in place.
4. Place a piece of sterile gauze over the left retractor and saturate gauze with the sterile PBS.
5. Using a sterile Q-tip, carefully move intestines onto the gauze and roll pancreas over to expose the portal vein (Fig. 1a). Portal vein is located on the ventral (bottom) part of the pancreas.
6. Once portal vein is exposed, keep applying slight traction with the sterile Q-tip on the pancreas near the vein bed to create tissue tension for the insertion of the needle (Fig. 1b).

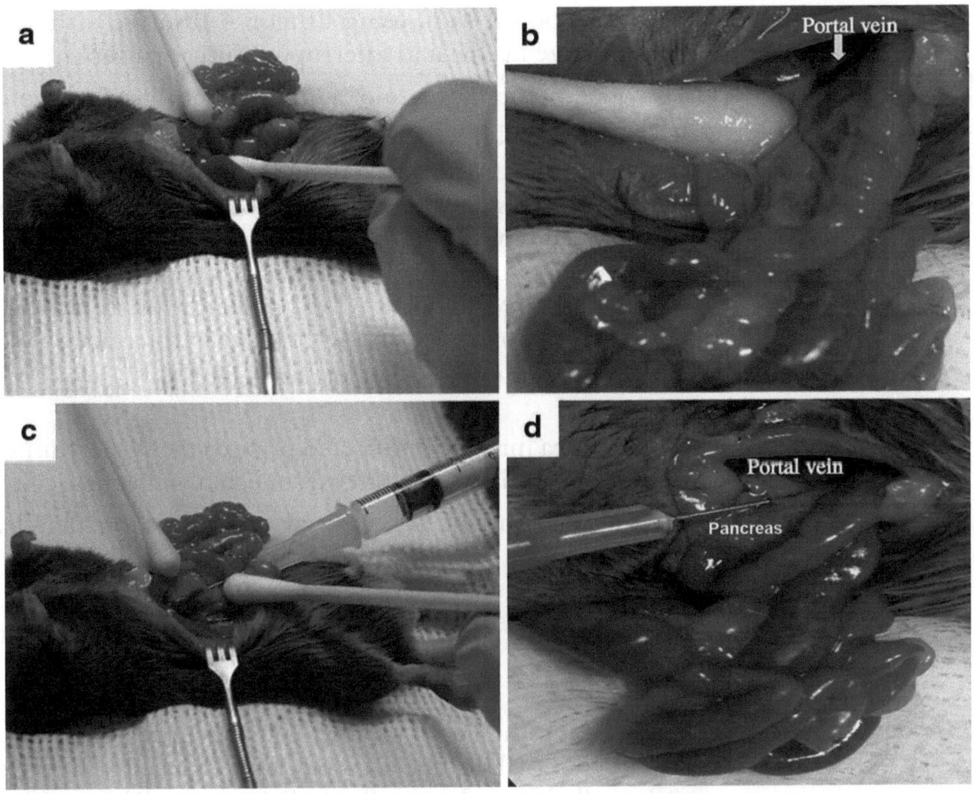

Fig. 1 Portal vein delivery of gene therapy vector to a mouse. (**a**) Displacement of intestinal tract and pancreas to gain access to portal vein. (**b**) Exposure of portal vein. The area directly caudal to the portal vein must be flat enough to allow a needle to enter the vessel without inhibiting the syringe. A Q-tip is used to "retract" the pancreas to put tension on the vessel. (**c**) Insertion of needle through the multiple layers to enter the portal vein. One should see the bevel of the needle in the vessel. (**d**) Vector administration. Blanching of the liver should be obvious when depressing the plunger while observing the liver

7. Begin inserting needle into the body of the pancreas ~2 mm below the vein, keep advancing it into the vein until needle tip will become visible through the vein wall (Fig. 1c). Slowly depress the plunger and release the vector solution into the vein (Fig. 1d).
8. Using sterile dry Q-tip, apply pressure next to the needle insertion place, slowly retracting the needle and rolling Q-tip to cover the hole from the needle to prevent backflow of blood from the vein.
9. Hold pressure on the vein for at least 30 s. Cut a small piece of absorbable haemostatic material and place on injection site before replacing intestines back into the abdomen and prepare for suturing.
10. Close the abdominal wall using 4-0 silk, non-cutting taper point or round needle, continuous pattern. Place second layer

of sutures on the skin, using Ethilon 4-0 suture material with cutting-edge needle and interrupted suture pattern.

11. Disconnect anesthesia and administer analgesic caprofen at 5 mg/kg or buprenorphine at 0.05–0.1 mg/kg.
12. Postsurgical care: Mice are kept on the heating pad for the entire duration of the surgery and postsurgical recovery to avoid hypothermia. Administration of normal mouse plasma (200 μl) is repeated approximately 30 min after surgery. Animals should be visually inspected frequently for bleeding, wound healing, and possible infection in the first 48 h and daily after that.
13. Sutures should be removed 10–14 days after surgery.

3.3 Canine Surgery

3.3.1 Preparation of the Surgical Area

1. Surgical area must be disinfected prior to the surgery using any of the approved (appropriate) disinfectants in your facility.
2. Heating pad should be sanitized and placed on the cleaned surgical area.
3. All personnel must wear surgical clothing, hair cover, face mask, shoe covers, and, if in contact with the sterile field, sterile surgical gloves.

3.3.2 Preparation of the Animal

1. General anesthesia: Typically, dogs are premedicated with Atropine SQ (0.06 mg/kg) and then induced with propofol ("Propoflo28" at 3.2 mg/kg to 7.6 mg/kg IV over 60–90 s) followed by immediate intubation and transition to isoflurane to effect (~2 %). Nitrous oxide (50 %) is occasionally used during induction. Anesthesia is evaluated by direct observation of heart rate, respiratory rate, blood pressure, oxygen saturation, end-tidal CO_2, body temperature, and persistence or absence of palpebral, corneal, and withdrawal reflexes. A computer-based system and hand-written notes are used to record events that occur perioperatively.
2. Once the dog is anesthetized and intubated, ophthalmic ointment is applied to both eyes to prevent corneal drying.
3. Fur is shaved from the surgical field, generally from the lower rib cage to the lower abdomen. The shaved skin is cleaned with Betadine solution and 70 % ethanol. The dog is covered with sterile surgical drapes.

3.3.3 Surgeon Preparation

1. Survival surgeries in dogs require surgical hand scrubbing and sterile gowning and gloving. The surgeon must wear shoe covers, a face mask, and hair cover as do all of the operation room attendees and assistants.

3.4 Surgical Procedure

1. Prior to making an incision, canine FIX levels are raised to at least 10–20 % by infusing an appropriate amount of normal canine plasma. One needs to know the weight and

hematocrit of the dog. The total blood volume is estimated at 40 ml/lb (18.18 ml/kg).

Therefore,

Wt (kg) × 18.18 ml/kg = Total Blood Volume (TBV)

100 − Hematocrit = % Plasma Volume

TBV × % Plasma Volume = Total Plasma Vol in ml

Combined equation: (wt kg × 18.18 ml/kg) × (100 − hematocrit) = Total Plasma Volume in ml.

Then, to raise FIX levels to 10 % in a dog with 900 cc calculated plasma volume, 100 cc of normal plasma is infused with FIX at 1 unit/ml.

Additional normal plasma can be given if the surgeon feels hemostasis is inadequate. The whole blood clotting time can be given a quick estimate of whether or not FIX levels are above 5 % [35]. Meticulous attention must be paid to controlling bleeding as incisions are made. There is often a subcutaneous vein of substantial size just beneath the skin in the midline incision site that requires attention.

2. A skin incision is made from the xiphoid down to the lower abdomen (~10 cm) and then the abdominal cavity is opened via its muscle layers.
3. A baseline liver biopsy is taken for use as control tissue if a follow up biopsy is performed to determine vector copy number, gene expression or if there is cellular toxicity. A small wedge of liver tissue is encircled with suture material, and the tissue is processed for molecular and histological studies.
4. The spleen is carefully exteriorized (Fig. 2, panel 1), and a 3–5 French balloon-tipped catheter is prepared by inflating the balloon to check for leaks (~0.5–1.0 cc, Fig. 2, panel 2).
5. An appropriately sized branch of the splenic vein is isolated, and circumferential sutures are positioned proximally and distally over about 2 cm; this branch of the splenic vein is punctured and the balloon-tipped catheter is inserted and advanced into the portal vein, usually between 10 and 20 cm (Fig. 2, panels 2–5). The position of the catheter in the portal vein is confirmed visually or by palpation.
6. The distal port of the balloon-tipped catheter is aspirated to confirm blood can be withdrawn even when the balloon is securely inflate and occluding the portal vein. This may require a "trial and error" approach until positioning is optimized.
7. When the catheter position is optimized, a small amount (~1 cc) of vector is infused while the balloon is inflated. Infusion rates are adjusted as the blood pressure tolerates. In general, AAV vectors are well tolerated and can be infused over 15 or 30 min.

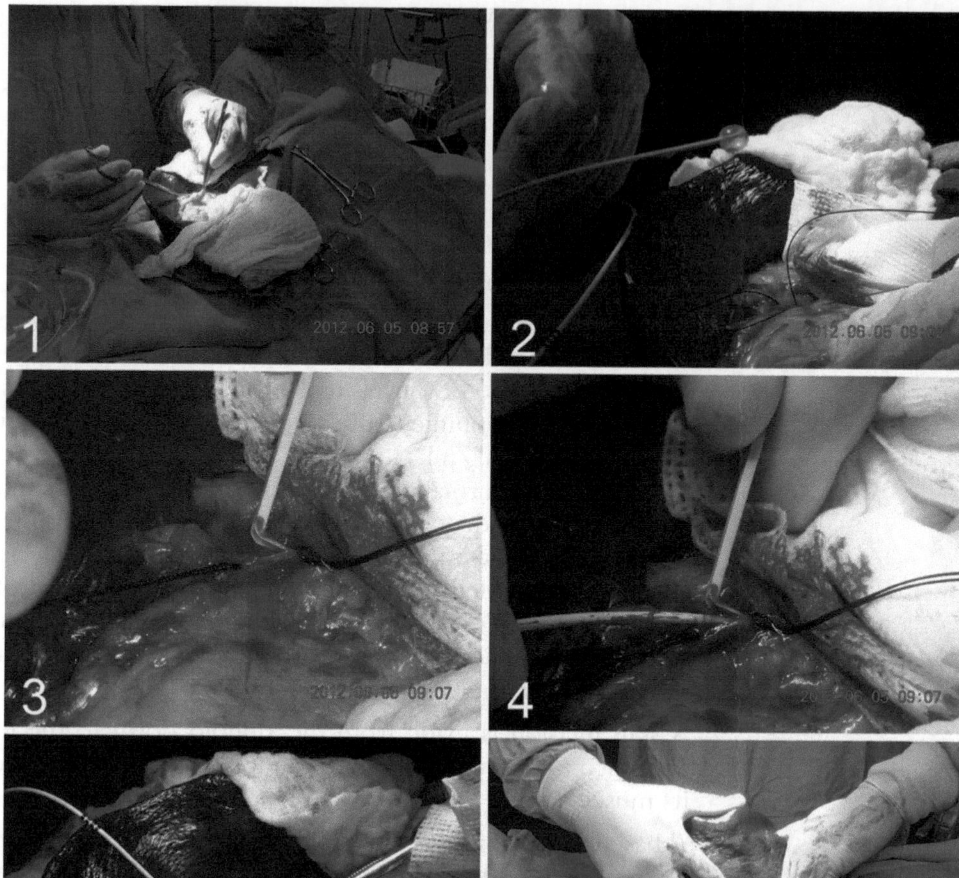

Fig. 2 Portal vein delivery of gene therapy vector to a hemophilic dog. *Panel 1*: The spleen is exteriorized via a midline laparotomy incision. *Panel 2*: A 3–5 French balloon-tipped catheter is prepared by inflating the balloon to check for leaks. *Panel 3*: A small splenic vein is isolated and a venotomy is performed. *Panel 4*: The balloon-tipped catheter is inserted in the venotomy site and passed antegrade to the hepatic portal vein, usually 10–20 cm. *Panel 5*: With the balloon inflated, the gene therapy vector is administered to the liver. *Panel 6*: The spleen is returned to the abdomen. *Panels 7 and 8*: The laparotomy incision is repaired with careful attention to bleeding vessels

8. After the vector is infused, the catheter is removed, the branch of the splenic vein is ligated proximally and distally, and the spleen is returned to the abdomen being careful not to injure it (Fig. 2, panel 6).
9. The surgical incision is closed in layers using suture material (Fig. 2, panels 7 and 8). Each layer is carefully examined for bleeding sites. Extra time spent achieving hemostasis at this point may prevent reoperation for small but significant bleeders.
10. Discontinue anesthesia and extubate when appropriate gag reflex is documented. Care must be taken to avoid trauma to the airway that can result in hemorrhage and airway obstruction.
11. The dog receives an intramuscular injection of procaine penicillin (or other prophylactic antibiotic as prescribed by the attending veterinarian) alone with postoperative analgesia and is allowed to recover under observation. The dog and its incision site are observed daily, and immediate steps are taken to treat infection or bleeding as is appropriate. In general, FIX is given daily for 7 days postoperatively or longer if needed. The exogenous plasma FIX disappears to undetectable levels (<0.1 %) within 21–28 days [35].

4 Notes

1. C57BL/6 mice typically show the highest level of transduction with AAV vectors in the liver, resulting in high systemic expression of FIX, while BALB/c mice show slightly lower transduction efficiency. Other strains, such as C3H mice, may show substantially reduced transduction efficiency and thus require higher vector doses to achieve similar levels of transgene expression.
2. Portal vein has access to two thirds of the liver, which is a normal site of expression of FIX. Following hepatic gene transfer, FIX can be detected in the circulation by ELISA from 1 week following gene transfer and persist for >1 year or even the life span of the mouse. Antibodies specific to human and canine FIX (that do not cross-react with murine FIX) are used for antigen capture and are commercially available.
3. Portal vein infusion of AAV-FIX in mice can give weak antibody response to the viral particles, which can appear in the peripheral blood days after gene transfer. Mice typically produce immunoglobulins IgG1 or IgG2a. Specific assays have been developed to detect anti-AAV antibodies. Enzyme-linked immunosorbent assay (ELISA) allows fast and accurate detection of various inhibitory and not inhibitory antibodies.
4. Presence of inhibitory antibodies can be detected by Bethesda assay, where hemophilic mouse plasma is mixed with normal mouse plasma 1:1 ratio. This mixture is incubated for 2 h at

37 °C. After addition of FIX-deficient human plasma and CaCl, time of clot formation is measured with fibrometer. Bethesda inhibitor titer is equivalent of the dilution of mouse plasma sample, calculated from a standard curve obtained from twofold serial dilution of normal human plasma.

5. Coagulation of plasma samples from hemophilic mice is determined by measuring aPPT, activated partial thromboplastin time using fibrometer. Range of aPPT in hemophilia B mice is 55–100 s, 25–30 s in normal mice.
6. Blood collection in mice can be performed via tail or retro-orbital bleeding. Mice need to be anesthetized for blood collection. Retro-orbital bleeding yields 50–150 μl of blood collected in to heparinized glass capillaries and can be used for ELISA assays. For Bethesda and APPT, tail bleed is a preferred method; blood is collected into citrate or oxalate containing tubes.
7. In this chapter, hepatic gene transfer to hemophilic mice is described. The identical procedure can also be performed in hemostatically normal mice to measure gene transfer or correct other disorders. Non-hemophilic mice will not require presurgical administration of normal mouse plasma.
8. Vector doses of 10^{10}–10^{11} vg/mouse are typical for AAV2 and AAV5 expressing FIX. Higher doses may be required for FVIII expression in hemophilia A mice. Other serotypes or improvements in vector design (capsid, expression cassette, and so on) may allow for reduced vector doses.

References

1. Sarkar R, Xiao W, Kazazian HH Jr (2003) A single adeno-associated virus (AAV)-murine factor VIII vector partially corrects the hemophilia A phenotype. J Thromb Haemost 1:220–226
2. Jiang H, Lillicrap D, Patarroyo-White S, Liu T, Qian X, Scallan CD, Powell S, Keller T, McMurray M, Labelle A, Nagy D, Vargas JA, Zhou S, Couto LB, Pierce GF (2006) Multiyear therapeutic benefit of AAV serotypes 2, 6, and 8 delivering factor VIII to hemophilia A mice and dogs. Blood 108:107–115
3. Markusic DM (2012) Liver-directed adeno-associated viral gene therapy for hemophilia. J Genet Syndr Gene Ther S1:009
4. Mingozzi F, Hasbrouck NC, Basner-Tschakarjan E, Edmonson SA, Hui DJ, Sabatino DE, Zhou S, Wright JF, Jiang H, Pierce GF, Arruda VR, High KA (2007) Modulation of tolerance to the transgene product in a nonhuman primate model of AAV-mediated gene transfer to liver. Blood 110:2334–2341
5. Mount JD, Herzog RW, Tillson DM, Goodman SA, Robinson N, McCleland ML, Bellinger D, Nichols TC, Arruda VR, Lothrop CD Jr, High KA (2002) Sustained phenotypic correction of hemophilia B dogs with a factor IX null mutation by liver-directed gene therapy. Blood 99:2670–2676
6. Snyder RO, Miao C, Meuse L, Donahue BA, Lin H-F, Stafford DW, Patel S, Thompson A, Nichols T, Bellinger D, Read M, Brinkhous KM, Kay MA (1999) Correction of hemophilia B in canine and murine models using recombinant adeno-associated viral vectors. Nat Med 5:64–70
7. Nichols TC, Raymer RA, Franck HW, Merricks EP, Bellinger DA, DeFriess N, Margaritis P, Arruda VR, Kay MA, High KA (2010) Prevention of spontaneous bleeding in dogs with haemophilia A and haemophilia B. Haemophilia 16(Suppl 3):19–23
8. Niemeyer GP, Herzog RW, Mount J, Arruda VR, Tillson DM, Hathcock J, van Ginkel FW, High KA, Lothrop CD Jr (2009) Long-term

correction of inhibitor-prone hemophilia B dogs treated with liver-directed AAV2-mediated factor IX gene therapy. Blood 113:797–806

9. Sabatino DE, Lange AM, Altynova ES, Sarkar R, Zhou S, Merricks EP, Franck HG, Nichols TC, Arruda VR, Kazazian HH Jr (2011) Efficacy and safety of long-term prophylaxis in severe hemophilia A dogs following liver gene therapy using AAV vectors. Mol Ther 19:442–449
10. Annoni A, Brown BD, Cantore A, Sergi LS, Naldini L, Roncarolo MG (2009) In vivo delivery of a microRNA-regulated transgene induces antigen-specific regulatory T cells and promotes immunologic tolerance. Blood 114:5152–5161
11. Cao O, Dobrzynski E, Wang L, Nayak S, Mingle B, Terhorst C, Herzog RW (2007) Induction and role of regulatory CD4+CD25+ T cells in tolerance to the transgene product following hepatic in vivo gene transfer. Blood 110:1132–1140
12. Hoffman BE, Martino AT, Sack BK, Cao O, Liao G, Terhorst C, Herzog RW (2011) Nonredundant roles of IL-10 and TGF-beta in suppression of immune responses to hepatic AAV-factor IX gene transfer. Mol Ther 19:1263–1272
13. LoDuca PA, Hoffman BE, Herzog RW (2009) Hepatic gene transfer as a means of tolerance induction to transgene products. Curr Gene Ther 9:104–114
14. Nathwani AC, Tuddenham EG, Rangarajan S, Rosales C, McIntosh J, Linch DC, Chowdary P, Riddell A, Pie AJ, Harrington C, O'Beirne J, Smith K, Pasi J, Glader B, Rustagi P, Ng CY, Kay MA, Zhou J, Spence Y, Morton CL, Allay J, Coleman J, Sleep S, Cunningham JM, Srivastava D, Basner-Tschakarjan E, Mingozzi F, High KA, Gray JT, Reiss UM, Nienhuis AW, Davidoff AM (2011) Adenovirus-associated virus vector-mediated gene transfer in hemophilia B. N Engl J Med 365:2357–3265
15. Mingozzi F, High KA (2011) Therapeutic in vivo gene transfer for genetic disease using AAV: progress and challenges. Nat Rev Genet 12:341–355
16. Manno CS, Pierce GF, Arruda VR, Glader B, Ragni M, Rasko JJ, Ozelo MC, Hoots K, Blatt P, Konkle B, Dake M, Kaye R, Razavi M, Zajko A, Zehnder J, Rustagi PK, Nakai H, Chew A, Leonard D, Wright JF, Lessard RR, Sommer JM, Tigges M, Sabatino D, Luk A, Jiang H, Mingozzi F, Couto L, Ertl HC, High KA, Kay MA (2006) Successful transduction of liver in hemophilia by AAV-Factor IX and limitations imposed by the host immune response. Nat Med 12:342–347
17. Cooper M, Nayak S, Hoffman BE, Terhorst C, Cao O, Herzog RW (2009) Improved induction of immune tolerance to factor IX by hepatic AAV-8 gene transfer. Hum Gene Ther 20:767–776
18. Nathwani AC, Rosales C, McIntosh J, Rastegarlari G, Nathwani D, Raj D, Nawathe S, Waddington SN, Bronson R, Jackson S, Donahue RE, High KA, Mingozzi F, Ng CY, Zhou J, Spence Y, McCarville MB, Valentine M, Allay J, Coleman J, Sleep S, Gray JT, Nienhuis AW, Davidoff AM (2011) Long-term safety and efficacy following systemic administration of a self-complementary AAV vector encoding human FIX pseudotyped with serotype 5 and 8 capsid proteins. Mol Ther 19: 876–885
19. Wang L, Takabe K, Bidlingmaier SM, III CR, Verma IM (1999) Sustained correction of bleeding disorder in hemophilia B mice by gene therapy. Proc Natl Acad Sci U S A 96: 3906–3910
20. Mingozzi F, Schuttrumpf J, Arruda VR, Liu Y, Liu YL, High KA, Xiao W, Herzog RW (2002) Improved hepatic gene transfer by using an adeno-associated virus serotype 5 vector. J Virol 76:10497–10502
21. Markusic DM, Herzog RW, Aslanidi GV, Hoffman BE, Li B, Li M, Jayandharan GR, Ling C, Zolotukhin I, Ma W, Zolotukhin S, Srivastava A, Zhong L (2010) High-efficiency transduction and correction of murine hemophilia B using AAV2 vectors devoid of multiple surface-exposed tyrosines. Mol Ther 18: 2048–2056
22. Sack BK, Merchant S, Markusic DM, Nathwani AC, Davidoff AM, Byrne BJ, Herzog RW (2012) Transient B cell depletion or improved transgene expression by codon optimization promote tolerance to factor VIII in gene therapy. PLoS One 7:e37671
23. Wu Z, Sun J, Zhang T, Yin C, Yin F, Van Dyke T, Samulski RJ, Monahan PE (2008) Optimization of self-complementary AAV vectors for liver-directed expression results in sustained correction of hemophilia B at low vector dose. Mol Ther 16:280–289
24. Cao O, Hoffman BE, Moghimi B, Nayak S, Cooper M, Zhou S, Ertl HC, High KA, Herzog RW (2009) Impact of the underlying mutation and the route of vector administration on immune responses to factor IX in gene therapy for hemophilia B. Mol Ther 17: 1733–1742
25. Sabatino DE, Nichols TC, Merricks E, Bellinger DA, Herzog RW, Monahan PE (2012) Animal models of hemophilia. Prog Mol Biol Transl Sci 105:151–209

26. Margaritis P, Roy E, Aljamali MN, Downey HD, Giger U, Zhou S, Merricks E, Dillow A, Ezban M, Nichols TC, High KA (2009) Successful treatment of canine hemophilia by continuous expression of canine FVIIa. Blood 113:3682–3689
27. Evans JP, Brinkhous KM, Brayer GD, Reisner HM, High KA (1989) Canine hemophilia B resulting from a point mutation with unusual consequences. Proc Natl Acad Sci U S A 86:10095–10099
28. Herzog R, Arruda VR, Fischer TH, Read MS, Nichols TC, High KA (2000) Absence of circulating factor IX antigen in hemophilia B dogs of the UNC-Chapel Hill colony. Thromb Haemost 84:352–354
29. Russell KE, Olsen EH, Raymer RA, Merricks EP, Bellinger DA, Read MS, Rup BJ, Keith JC Jr, McCarthy KP, Schaub RG, Nichols TC (2003) Reduced bleeding events with subcutaneous administration of recombinant human factor IX in immune-tolerant hemophilia B dogs. Blood 102:4393–4398
30. Russell KE, Read MS, Bellinger DA, Leitermann K, Rup BJ, McCarthy KP, Keith JC Jr, Khor SP, Schaub RG, Nichols TC (2001) Intratracheal administration of recombinant human factor IX (BeneFix) achieves therapeutic levels in hemophilia B dogs. Thromb Haemost 85:445–449
31. Brinkhous KM, Sigman JL, Read MS, Stewart PF, McCarthy KP, Timony GA, Leppanen SD, Rup BJ, Keith JC Jr, Garzone PD, Schaub RG (1996) Recombinant human factor IX: replacement therapy, prophylaxis, and pharmacokinetics in canine hemophilia B. Blood 88: 2603–2610
32. Shapiro AD, Ragni MV, Valentino LA, Key NS, Josephson NC, Powell JS, Cheng G, Thompson AR, Goyal J, Tubridy KL, Peters RT, Dumont JA, Euwart D, Li L, Hallen B, Gozzi P, Bitonti AJ, Jiang H, Luk A, Pierce GF (2012) Recombinant factor IX-Fc fusion protein (rFIXFc) demonstrates safety and prolonged activity in a phase 1/2a study in hemophilia B patients. Blood 119:666–672
33. Roth DA, Kessler CM, Pasi KJ, Rup B, Courter SG, Tubridy KL (2001) Human recombinant factor IX: safety and efficacy studies in hemophilia B patients previously treated with plasma-derived factor IX concentrates. Blood 98:3600–3606
34. Arruda VR, Fields PA, Milner R, Wainwright L, De Miguel MP, Donovan PJ, Herzog RW, Nichols TC, Biegel JA, Razavi M, Dake M, Huff D, Flake AW, Couto L, Kay MA, High KA (2001) Lack of germline transmission of vector sequences following systemic administration of recombinant AAV-2 vector in males. Mol Ther 4:586–592
35. Nichols TC, Franck HWG, Franck C, Raymer RA, Merricks EP (2012) Sensitivity of whole blood clotting time and activated partial thromboplastin time for canine factor IX. J Thromb Haemost 10:474–476

Chapter 28

Gene Correction of Induced Pluripotent Stem Cells Derived from a Murine Model of X-Linked Chronic Granulomatous Disorder

Sayandip Mukherjee and Adrian J. Thrasher

Abstract

Gene therapy presents an attractive alternative to allogeneic haematopoietic stem cell transplantation (HSCT) for treating patients suffering from primary immunodeficiency disorder (PID). The conceptual advantage of gene correcting a patient's autologous HSCs lies in minimizing or completely avoiding immunological complications arising from allogeneic transplantation while conferring the same benefits of immune reconstitution upon long-term engraftment. Clinical trials targeting X-linked chronic granulomatous disorder (X-CGD) have shown promising results in this context. However, long-term clinical benefits in these patients have been limited by issues of poor engraftment of gene-transduced cells coupled with transgene silencing and vector induced clonal proliferation. Novel vectors incorporating safety features such as self-inactivating (SIN) mutations in the long terminal repeats (LTRs) along with synthetic promoters driving lineage-restricted sustainable expression of the gp91phox transgene are expected to resolve the current pitfalls and require rigorous preclinical testing. In this chapter, we have outlined a protocol in which X-CGD mouse model derived induced pluripotent stem cells (iPSCs) have been utilized to develop a platform for investigating the efficacy and safety profiles of novel vectors prior to clinical evaluation.

Key words X-CGD, Gene therapy, iPS cells, Neutrophils, Mouse model, PID

1 Introduction

Primary immunodeficiency disorders (PIDs) constitute a rare and heterogeneous group of heritable disorders affecting differentiation and/or function of immune cells with different degrees of severity which predispose patients to recurrent and chronic infections, inflammatory complications, development of autoimmunity, and even malignancy [1]. World-wide, the incidence of PIDs, in which defects in well over 300 individual genes has been described, is highly variable ranging from 1 in 600 to 1 in 500,000 live newborns, depending both upon the specific disorder and the ethnicity of the population under study [2–6]. Patients with severe PID require definitive therapeutic intervention early in their life to prevent early

Francesca Storici (ed.), *Gene Correction: Methods and Protocols*, Methods in Molecular Biology, vol. 1114,
DOI 10.1007/978-1-62703-761-7_28,

mortality, and to enhance the quality of life by reducing disabilities associated with the disorder. Allogeneic bone-marrow transplantation (BMT) remains the only curative therapy for many PIDs, and the efficacy of BMT with an HLA-matched sibling or unrelated (including umbilical cord blood) donor is beyond any doubt with a success-rate often higher than 90 %. However, the rate of cure declines significantly with HLA-mismatched donors [7–9].

In the context of absence of fully matched donor, autologous BMT with ex vivo gene-corrected hematopoietic stem cells (HSCs) promises considerable therapeutic advantage as theoretically, even a single genetically correct HSC is capable of repopulating the entire bone marrow and give rise to all the different blood cell lineages leading to restoration of immune function. The rationale for advocating gene therapy for various forms of PIDs over allogeneic transplantation is therefore based on the premise that gene correction and transplantation of corrected HSCs back to the patient will provide the same clinical benefits without the immunological complications of allogeneic BMT, namely graft versus host disease (GvHD), and others which result from extended use of conditioning regimens [10–13].

In this chapter, we will describe the methodology for employing induced pluripotent stem cell (iPSC) strategy to demonstrate lentiviral vector mediated in vitro correction of disease phenotype in an animal model of X-linked chronic granulomatous disorder (X-CGD) [14].

CGD patients suffer from mild to severe recurrent opportunistic bacterial and fungal infections owing to lack of functional neutrophils in the peripheral blood. Neutrophils from CGD patients are incapable of eliminating invading microorganisms owing to a defective nicotinamide adenine dinucleotide phosphate-oxidase (NADPH-oxidase) complex. Under normal circumstances, NADPH-oxidase complex located on the membrane of the phagocytic vacuole plays a pivotal role in generating reactive oxygen species (ROS), resulting in a downstream cascade of events involving depolarization of the membrane surrounding the vacuole, influx of potassium ions into the vacuole, increase in pH inside the vacuole activating and solubilizing granular proteolytic enzymes which digest and destroy the engulfed pathogens [15]. However, mutations in any one of the five genes encoding the five subunits (gp91phox, p22phox, p47phox, p67phox and p40phox) of the NADPH-oxidase complex can partially or completely abrogate this crucial first step of superoxide generation leading to enhanced susceptibility of the patients to microbial infections. 65 % of all reported cases of CGD are X-linked owing to mutations in the CYBB gene (encoding gp91phox, gene locus Xp21.1) thereby affecting males, with only rarely a few female carriers with skewed X-inactivation. All other mutations affecting NADPH-oxidase are autosomal recessive in nature affecting p47phox (encoded by

NCF1, 25 % of cases), p22phox (encoded by CYBA, 5 % of cases), p67phox (encoded by NCF2, 5 % of cases) and p40phox (encoded by NCF4, only one reported case so far). As a monogenic disorder, CGD presents a promising opportunity for gene therapy as even low levels of gene correction has the potential to restore immunological competence. This is witnessed by the fact that some healthy female CGD carriers have been found to contain less than 10 % of functional neutrophils compared to normal subjects [16, 17].

Following significant long-term correction in murine models obtained by retroviral transduction of HSCs [18, 19], the first clinical trials in ten human CGD patients (five with p47phox deficiency and five with gp91phox deficiency) employing a pseudotyped MLV-based vector and carried out without any preconditioning failed to achieve any therapeutic benefit beyond low-levels of functional neutrophils for up to 14 months post gene therapy [20]. More recently, 12 CGD patients preconditioned with mild chemotherapy were treated with autologous HSCs transduced with LTR-based gamma-retroviral vectors. The majority of the patients demonstrated clinically significant levels of circulating reconstituted neutrophils leading to clearance of pre-existing infections. However, the observed therapeutic benefit was found to be transient in nature in most patients as gene marking significantly decreased after 3 months indicating absence of long-term engraftment. Additionally, approximately 5 months post-treatment several patients developed myelodysplasia with monosomy 7 owing to vector-induced integrations in the zinc finger transcription factor homologs MDS1/EVI1, PRDM16 or SETBP1. Intriguingly, although the number of transduced cells remained high in these cases, gp91phox expression declined resulting from progressive loss of transgene expression due to promoter methylation [21–23]. To summarize, short-term positive clinical outcome as well as adverse events resulting from these trials have warranted the development of improved vector design to enhance safety of gene transfer into HSCs while ensuring adequate transgene expression in myeloid lineages. In this context, a robust animal model is expected to facilitate the rigorous preclinical testing of these novel vectors.

The humanized mouse model of X-linked CGD was generated by targeted disruption of the *CYBB* gene and demonstrates a similar pathophysiological phenotype compared to human patients in their inability to produce phagocytic superoxide, and enhanced susceptibility to infection with *S. aureus* and *A. fumigatus* [24]. This model provides a suitable platform for preclinical evaluation of novel gene therapy vectors in terms of robustness of gp91phox expression, sustainable expression from transduced long-term repopulating HSCs, and towards determining lineage-restricted expression of gp91phox when driven by novel lentiviral vectors. The generation of iPSCs from the skin fibroblast of X-CGD mice and X-CGD patient samples therefore provides a potentially unlimited source of starting material which can be employed for all of the aforementioned purposes, as well as offering an additional window where therapeutic vectors can be used to transduce embryonic cells before their lineage-specific directed differentiation into myeloid progenitors.

2 Materials

2.1 Animals and Cell Lines

1. B6.129S6-Cybb$^{tm1Din/J}$ mice deficient in gp91^{phox} (Jackson Laboratories, Bar Harbor, ME 04609, USA).
2. X-CGD mouse iPSCs (generated and maintained in-house).
3. CCE, mouse embryonic stem cell (ESC) line (commercially available).
4. OP9, mouse stromal cell line (ATCC catalogue number CRL-2749) for haematopoietic differentiation.
5. 293T, human embryonic kidney cell line (ATCC catalogue number CRL-11268) for vector-virus production (*see* **Note 1**).

2.2 Plasmids and Vectors

1. Packaging plasmids: MLV-based *gag-pol* expression plasmid, HIV-1-based *gag-pol* expression plasmid, VSV-G expression plasmid—these are available widely in laboratories which generate retroviral and lentiviral vector-viruses in a routine manner.
2. Reprogramming retroviral vectors pMXs-Sox2, pMXs-Oct3/4, pMXs-Klf4, and pMXs-c-Myc (available from Addgene repository).
3. SIN lentivectors expressing codon optimized *gp91phox* gene from SFFV (pCCLSFc.o.gp91phox) and chimeric promoter (pCCLChimc.o.gp91phox). The heterologous chimeric promoter has been synthesized by fusing two previously described domains—the 5′ minimal flanking regions of mammalian c-fes proto-oncogene and Cathepsin G, both of which are expected to confer myeloid-specific expression to the vector [25]. Vectors are available upon request.

2.3 Tissue Culture Plasticware

1. Tissue-culture treated flasks (175, 75, and 25 cm^2 area).
2. Tissue-culture treated dishes (100, 60, and 35 mm diameter).
3. Tissue-culture treated 6-well plates.
4. Serological pipettes.
5. Ultralow-attachment 6-well plates.
6. Collagen-IV coated T25 flasks [BD Biosciences].

2.4 Cell Culture Reagents

1. 10 % DMEM: For culture and maintenance of 293T, Dulbecco's modified Eagle's medium (DMEM) supplemented with 10 % heat-inactivated fetal bovine serum (FBS) and 1 % penicillin–streptomycin.

2. 15 % α-MEM: For culture and maintenance of OP9, α-minimal essential medium (MEM) supplemented with 15 % FCS and 1 % penicillin–streptomycin.
3. ESGRO® medium: For feeder-free culture of mouse ESC/iPSC, ESGRO® complete plus clonal medium [Merck-Millipore SF001-500P] supplemented with 1 % penicillin–streptomycin.
4. Embryoid body medium (EBM): Knockout DMEM supplemented with 20 % embryonic stem cell qualified fetal calf serum (ES-FCS), 2 mM L-glutamine, 1 % sodium pyruvate, 0.1 mM nonessential amino acids, 0.1 mM β-mercaptoethanol, and 1 % penicillin–streptomycin.
5. Collagen-IV differentiation medium (CDM): α-MEM supplemented with 10 % FBS, 0.1 mM nonessential amino acids, 0.1 mM β-mercaptoethanol, and 1 % penicillin–streptomycin.
6. OP9 differentiation medium (ODM): OP9 culture medium supplemented with 5 ng/ml interleukin-3 (IL-3) and 10 ng/ml granulocyte-colony stimulating factor (G-CSF).
7. Myeloid differentiation medium (MDM): Semisolid methylcellulose medium with recombinant cytokines promoting formation of colony-forming units (CFUs), Methocult® GF M3534 [Stemcell Technologies 03534].
8. Accutase and 0.05 % Trypsin–EDTA, for cell dissociation.
9. EmbryoMax 0.1 % gelatine solution [Merck-Millipore ES-006-B] for coating tissue-culture plasticware.

2.5 Vector Production and Infection

1. Polyethylenimine (PEI) (working concentration 1 μM) and Polybrene (working concentration 8 μg/ml).
2. 0.22 and 0.45 μM stericup.
3. Syringe filters.
4. 1.5 ml cryotube vials.

2.6 Neutrophil Functional Assay

1. Nitroblue tetrazolium (NBT) reduction kit [Sigma 840W-1KT].
2. Dihydrorhodamine (DHR)123 [Sigma D1054-2mg].

2.7 Flow Cytometry

1. Suitable fluorochrome-conjugated anti-mouse antibodies for flow-cytometry (analysis and sorting) are indicated in the methods where applicable.
2. FACS buffer: Phosphate buffered saline (PBS) with 2 % bovine serum albumin (BSA).

3 Methods

The methods below describe briefly the generation of iPSCs from tail-tip fibroblasts of X-CGD mice and production of concentrated vector virus for gene correction, and in greater details the protocol for

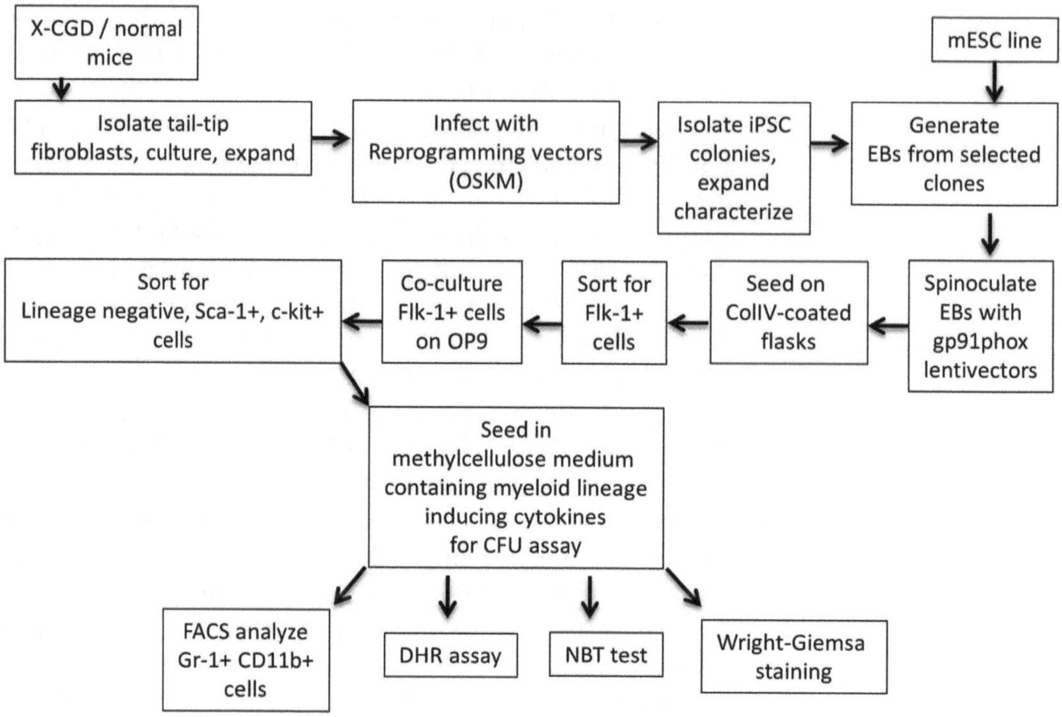

Fig. 1 Schematic outline of protocol for gene correction in X-CGD mouse derived iPSCs

genetic correction of X-CGD iPSCs and their directed differentiation to functional neutrophils (Fig. 1). In all experiments involving step-wise differentiation of iPSCs to myeloid cells, and in assays to determine gene correction, mouse ESC line derived cells and/or wild-type iPSCs should be used as a positive control. This should be considered compulsory even if not mentioned separately in the individual sections.

3.1 Generation of iPSCs from X-CGD Fibroblasts

1. Four semi-confluent T-75 flasks of 293T cells were transfected separately using a PEI-based transfection protocol with retroviral gag-pol and VSV-G envelope plasmids along with sox2, oct3/4, klf4 or c-myc vectors.
2. Tail-tip fibroblasts isolated from X-CGD mice, expanded and seeded in 6-well plates (5×10^4 cells per well) were infected with viral supernatant harvested at 48 and 72 h post-transfection and filtered through 0.45 μM syringe filter.
3. Polybrene (8 μg/ml) was added to the viral supernatant to increase efficiency of infection.
4. Viral supernatant was removed after 6 h and replaced with 10 % DMEM.
5. Infected cells were trypsinized after 3 days and seeded onto gelatine-coated 6-well plates in ESGRO® medium.

6. Medium was changed every other day until compact colonies of cells with distinct ESC-like morphology (high nuclear to cytoplasmic ratio) begin to appear approximately between 3 and 4 weeks.
7. Colonies were removed enzymatically with Accutase® and passaged for further expansion and characterization.
8. Characterization includes staining for pluripotency markers (Nanog, Sox2, Oct4, SSEA-1, alkaline phosphatase), teratoma generation in immunodeficient mice (thereafter confirming the presence of three germinal layers by immunohistochemistry and haematoxylin/eosin staining), and ability of the iPSCs to form spherical three-dimensional embryoid bodies (EBs) in leukemia inhibitory factor (LIF)-deficient EB medium under low-attachment conditions.

3.2 Production of Concentrated Stocks of gp91phox Lentiviral Vector

Day 0

Seed twelve T175 flasks with 18×10^6 293T cells in 25 ml of 10 % DMEM per flask.

Day 1

1. For each flask, add the following amounts of DNA to 5 ml of DMEM: 50 μg of gp91phox vector DNA, 32.5 μg of packaging plasmid and 17.5 μg of envelope plasmid.
2. Filter with 0.22 μM filter and add to the DNA mix another 5 ml of filtered DMEM plus PEI (1 μM).
3. Incubate transfection medium for 20 min at room temperature.
4. Replace overnight culture medium with 10 ml of transfection medium containing DNA and PEI, and return to incubator for 4 h.
5. Aspirate transfection medium from flasks, wash gently with phosphate-buffered saline (PBS), and replace with 10 % DMEM.

Day 2

Replace media on all flasks with fresh 10 % DMEM.

Day 3

1. Harvest viral supernatant from all flasks, pass through a 0.45 μM stericup filter and add fresh 10 % DMEM to all flasks. Return flasks to incubator (*see* **Note 2**).
2. Distribute filtered supernatant is into Beckman ultracentrifuge tubes and spin at 72,000×*g*) at 4 °C for 2 h in an ultracentrifuge.
3. Resuspend viral pellet in cold PBS, aliquote into cryotube vials and store at −80 °C.

Day 4

1. For 72 h harvest, follow exactly similar procedure as day 3. Discard flasks after treating with bleach.
2. To determine viral titer, infect target cells (such as HeLa or 293T) with different volumes of the concentrated viral supernatant, extract genomic DNA after 72 h and perform quantitative-PCR (q-PCR) with appropriate primers and probes.
3. Viral titer can also be determined by p24 ELISA and other methods discussed elsewhere in details.

3.3 Lentiviral Transduction of EBs

There are two alternatives for gene correcting iPSCs. In the first method, somatic cells obtained from the patient or animal model are transduced with vector carrying the gene of interest, selected (if possible) for a pure population of transduced cells, and then reprogrammed into iPSCs. This methodology has been implemented in case-studies like Fanconi's Anemia (FA) where gene correction before reprogramming is an absolute prerequisite [26]. In the second approach, somatic cells are first reprogrammed into pluripotency, and then transduced with the appropriate vector. For X-CGD murine iPSCs, we found that the second methodology was more efficient than the first in terms of producing in-vitro functional neutrophils [14].

3.3.1 Generation of EBs

1. Subject two semi-confluent wells of a 6-well plate containing robust iPSC colonies (cultured for a minimum of 7–8 passages) to Accutase treatment briefly for 2–3 min, and dislodge colonies gently using a cell scraper to maintain the colony architecture (*see* **Note 3**).
2. Resuspend colonies in EBM and centrifuge at 200 × *g* for 5 min. Aspirate the supernatant and gently resuspend in 3 ml of fresh EBM maintaining small visible clumps.
3. Transfer the suspension to one well of a 6-well ultralow attachment plate and returned to incubator.
4. Culture EBs for 5 days with half medium change every other day by allowing EBs to sediment under gravity in a 15 ml centrifuge tube inside the incubator.

3.3.2 Spinoculation of EBs

1. Pool together day six EBs from three wells of a 6-well plate, allow to settle under gravity in a 15 ml conical centrifuge tube, and resuspend in 2 ml of 10 % DMEM containing polybrene (8 μg/ml) and concentrated gp91phox vector virus (*see* **Note 4**).
2. Spin in a swing-bucket centrifuge at 800 × *g* for 1 h at 32 °C.

3.4 Collagen-IV Dependant Differentiation of Transduced EBs into Flk-1 Positive Mesodermal Progenitors

1. Post-spinoculation, resuspend EBs gently into small clumps and seed onto pre-thawed (warmed to room temperature) collagen-IV-coated T25 flask containing CDM.
2. Change medium every other day for 6 days.
3. On day 6, trypsinize cells and resuspend in minimal volume of CDM containing fluorochrome-conjugated anti-Flk-1 antibody at manufacturer's recommended dilution. Incubate for 45 min in the incubator, wash twice in PBS, resuspend in α-MEM containing 2 % FCS for sorting.
4. Perform fluorescence-activated cell sorting (FACS) to isolate Flk-1 positive cell population and collect the desired cell population in ODM.

3.5 Derivation of LSK+ Cell Population by OP9 Coculture of Flk-1 Positive Cells

1. Seed OP9 cells on gelatine-coated 6-well plates in 15 % α-MEM at 60 % confluency 48 h prior to seeding the Flk-1+ cells.
2. Aspirate OP9 culture medium and add 2×10^4 Flk-1+ cells in single-cell suspension in 1 ml of ODM per well.
3. Change medium every day for up to 7 days, taking care not to lose the non-adherent cell population.
4. On day 7, remove and pool together all cells, wash in PBS, resuspend in a small volume of ODM containing fluorochrome conjugated anti-mouse lineage antibody cocktail along with fluorochrome conjugated anti-mouse Sca-1 and c-kit antibodies. Incubate for an hour in the incubator, wash twice with PBS, pellet by centrifuging and resuspend in Iscove's minimal defined medium (IMDM) containing 2 % FCS and perform FACS to isolate the lineage negative Sca-1+ c-kit+ (LSK) population (*see* **Notes 5–8**).

3.6 Determination of Clonogenic Potential of LSK Cells

1. Thaw frozen aliquots (3 ml each) of methylcellulose containing myeloid lineage inducing cytokine cocktail (GF-M3534, Stemcell Technologies) in 4 °C overnight (*see* **Note 9**).
2. Resuspend approximately 1×10^5 LSK cells in 10 ml of Iscove's minimal defined medium (IMDM) with 2 % FCS.
3. Add 300 μl of 2 % IMDM containing LSK cells into 3 ml of GF-M3534, vortex vigorously for a minute and allow standing for 5 min to dissipate bubbles.
4. Dispense 3.3 ml of methylcellulose plus cells mixture equally in three 35 mm dishes using a 3 ml syringe fitted with a 16-gauge blunt needle.
5. Gently rotate and tilt each dish to distribute uniformly. Move three 35 mm dishes into a 150 mm culture dish along with an uncovered 35 mm dish containing sterile water.

6. Incubate for 1–2 weeks and score for CFUs based on morphology. Expected colonies are CFU-G (granulocyte), CFU-M (macrophage), and CFU-GM (granulocyte-macrophage) (*see* **Note 10**).

3.7 *In Vitro* Functional Assays for Determining Gene Correction in X-CGD Neutrophils

Once the CFUs have formed, it is essential to characterize their cellular composition and determine the restoration of NADPH-oxidase activity in terminally differentiated cells derived from the gene-transduced X-CGD iPSCs. Immunophenotypic characterization can be performed by FACS analysis of cells isolated from CFUs, and cellular morphology can also be established by Wright–Giemsa staining of cytospin preparations from the CFUs. In clinics, functionality of peripheral blood neutrophils are typically determined by subjecting them to two standardized assays known as the nitroblue tetrazolium (NBT) test and dihydrorhodamine 123 (DHR) assay. CGD patient samples respond negatively to both.

3.7.1 FACS Analysis and Wright–Giemsa Staining

1. After 2 weeks of culture in methylcellulose medium, aspirate medium containing cells from two of three 35 mm dishes into a 15 ml conical centrifuge tube containing RPMI 1640.
2. Pipette vigorously to disrupt CFUs, centrifuge at $200 \times g$ for 5 min, aspirate supernatant, and repeat the process at least three times until the cells are completely free from methylcellulose.
3. Aliquot the cells into two portions—fix one portion in 4 % paraformaldehyde (PFA) for 15 min at room temperature, wash with PBS, centrifuge, and resuspend in FACS buffer (PBS containing 1 % BSA). Incubate with fluorochrome conjugated anti-mouse Gr-1 (granulocyte marker) and anti-mouse CD11b antibodies for 30 min at room temperature. Wash to remove excess antibodies and perform FACS analysis to determine the percentage of dual positive Gr-1 and CD11b cells which include the neutrophil population.
4. For the other portion, determine cell number with a haemocytometer, adjust concentration to 1×10^5 cells/ml of PBS containing 2 % BSA to perform cytospin, and proceed with Wright–Giemsa staining protocol.

3.7.2 NBT Test (See **Note 11**)

1. Overlay methylcellulose plate containing CFUs with one-fifth volume of NBT-saturated RPMI-1640 medium containing 100 ng/ml PMA and 5 % human serum albumin (Baxter Healthcare, Deerfield, IL, USA).
2. Incubate at 37 °C. After 30 min of incubation, examine the dishes on an inverted microscope, and score cells with blue formazan precipitates as NBT-positive.
3. For negative control, use a plate of unstimulated sample. For positive control, use a plate of cells derived from CCE mES cells.

*3.7.3 DHR Assay (See **Note 12**)*

1. Wash CFU colonies extracted from methylcellulose cultures rigorously to get rid of any methylcellulose. Incubate cells with 30 μM DHR at 37 °C for 5 min and stimulate with 5 μg/ml PMA at 37 °C for 30 min.
2. For negative controls, use unstimulated samples and neutrophils derived from untransduced EBs.
3. For positive control, use cells derived from CCE mES cells.
4. Analyze samples immediately with flow-cytometry to detect the shift in fluorescence in stimulated samples compared to unstimulated ones.

3.8 Discussion

The last 10 years have witnessed considerable success in the form gene therapy clinical trials targeting ADA-SCID, X-linked SCID (IL2RG deficiency), chronic granulomatous disorder (CGD), and more recently with Wiskott–Aldrich syndrome (WAS) [13]. Safety issues continue to play a pivotal role in determining the risk benefit ratio of gene therapeutic approaches for treating PIDs. Gene transfer related toxicity arising out of insertional mutagenesis, and instances of vector silencing leading to loss of therapeutic gene expression have been observed in clinical trials, and have warranted extensive follow-up analysis of treated patients [27, 28]. Establishment of CGD patient-derived iPSC lines [29] are therefore expected to play a significant role in evaluating efficacy and safety of vector-mediated gene correction. As a safer alternative to gene addition by integrating viral vectors, gene correction at the disease-gene locus using a combinatorial strategy of customized homing endonucleases and homologous recombination (HR) has been recently demonstrated to be effective in preclinical studies, although not without associated pitfalls [30]. In this context, Zhou et al. have utilized X-CGD patient-derived iPSCs for gene correction using zinc finger nuclease (ZFN) mediated targeting of the CYBB gene into the AAV site-specific integration (AAVS1) locus on chromosome 19 which is considered to be a safe harbour for gene addition [31].

As noted earlier, silencing of corrective transgene during reprogramming or differentiation can play a critical role in determining the ultimate efficacy of gene transfer to iPSCs and disease correction. In the context of X-CGD mouse model, we have shown previously that gene transfer at the EB stage is more efficient in correcting the disease phenotype compared to gene transfer before reprogramming. Although the predictive value of animal model derived data remains controversial, nevertheless they provide an important stepping stone for high-throughput screening of novel vectors, and therefore constitute an important component for preclinical efficacy and safety evaluation.

4 Notes

1. All cell cultures are maintained at 37 °C and 5 % CO_2 unless otherwise stated.
2. It is critically important to obtain high-quality vector viruses for the purpose of gene correction. Therefore it is highly recommended that cellular debris are removed during the process of harvesting by performing one round of low-speed centrifugation before the viral supernatant is filtered through stericups. Preparations of concentrated vector viruses contaminated with cellular debris can cause substantial death of target cells leading to reduced gene correction.
3. As many wells of iPSC colonies as desired can be harvested for EB formation as long as the ratio of two wells of iPSCs to one well of EBs is maintained.
4. It is advisable to adjust the EB cell concentration between 2 and 3×10^5 cells/ml, and use different multiplicities of infection (m.o.i.) to determine the most effective transduction. In our experience, m.o.i. between 2 and 5 achieves the best transduction.
5. Efficient differentiation of transduced EBs is dependent on the quality of OP9 cells. It is strongly recommended that OP9 cells belonging to low passage number (less than 15) are used for this purpose.
6. To generate LSK cells by OP9 coculture it is important to ensure that the OP9 cells are fully confluent for a day to achieve optimum differentiation potential.
7. For routine change of medium during OP9/Flk-1 coculture, aspirate medium from wells in conical centrifuge tubes and put 0.5 ml of fresh media to ensure OP9 cells do not dry out. Spin down the cells at $700 \times g$ for 5 min, resuspend in fresh ODM, and return to the wells.
8. When performing fluorescence-activated cell sorting to isolate lineage negative c-kit+ Sca-1+, gate on the lineage negative cell population first, followed by gating on the c-kit and Sca-1 dual positive population.
9. Care should be taken to minimize duration of exposure of cells outside incubator. Prepare and/or warm-up reagents prior to removing cells from the incubator.
10. To characterize cells obtained from CFU assay, it is crucial to get rid of methylcellulose by repeated and rigorous washing of the cells with PBS. Residual methylcellulose can hamper binding of suitable antibodies to cell-surface receptors.
11. The NBT test is a semiquantitative assay for determining neutrophil oxidative burst. The test is conducted by counting

the cells containing blue formazan deposits, which are formed by reduction of the membrane permeable, water-soluble, and yellow-colored, NBT by reactive oxygen species produced by neutrophils upon PMA stimulation.

12. The DHR assay is based on the simple principle that nonfluorescent dihydrorhodamine123 (DHR) is oxidized to fluorescent rhodamine123 when phagocytosed by functional neutrophils upon PMA stimulation. Phagocytosis of DHR induces the neutrophils to produce the "respiratory burst" involving generation of reactive oxygen species, some of which are reduced to hydrogen peroxide that catalyzes the oxidation of DHR. Therefore, the detected fluorescence is an indirect measure of neutrophil function.

Acknowledgements

The authors would like to thank the following funding agencies for their continued support: Wellcome Trust (UK), Great Ormond Street Hospital Children's Charity (UK), and the European Union 7th Framework Program (PERSIST grant).

References

1. Geha RS, Notarangelo LD, Casanova JL (2007) Primary immunodeficiency diseases: an update from the International Union of Immunological Societies Primary Immunodeficiency Diseases Classification Committee. J Allergy Clin Immunol 120:776–794
2. Notarangelo LD, Fischer A, Geha RS (2009) Primary immunodeficiencies: 2009 update. J Allergy Clin Immunol 124:1161–1178
3. Boyle JM, Buckley RH (2007) Population prevalence of diagnosed primary immunodeficiency diseases in the United States. J Clin Immunol 27:497–502
4. Gupta S, Madkaikar M, Singh S, Sehgal S (2012) Primary immunodeficiencies in India: a perspective. Ann N Y Acad Sci 1250:73–79
5. Wang LL, Jin YY, Hao YQ (2011) Distribution and clinical features of primary immunodeficiency diseases in Chinese children (2004–2009). J Clin Immunol 31:297–308
6. Errante PR, Franco JL, Espinosa-Rosales FJ (2012) Advances in primary immunodeficiency diseases in Latin America: epidemiology, research, and perspectives. Ann N Y Acad Sci 1250:62–72
7. Rivat C, Santilli G, Gaspar HB (2012) Gene therapy for primary immunodeficiencies (PIDs). Hum Gene Ther 23(7):668–675
8. Neven B, Leroy S, Decaluwe H (2009) Long-term outcome after hematopoietic stem cell transplantation of a single-center cohort of 90 patients with severe combined immunodeficiency. Blood 113:4114–4124
9. Antoine C, Muller S, Cant A (2003) Long-term survival and transplantation of haemopoietic stem cells for immunodeficiencies: report of the European experience 1968–99. Lancet 361:553–560
10. Griffith LM, Cowan MJ, Kohn DB (2008) Allogeneic hematopoietic cell transplantation for primary immune deficiency diseases: current status and critical needs. J Allergy Clin Immunol 122:1087–1096
11. Shearer WT, Notarangelo LD, Griffith LM (2008) Treatment of immunodeficiency: long-term outcome and quality of life. J Allergy Clin Immunol 122:1065–1068
12. Fischer A, Cavazzana-Calvo M (2008) Gene therapy of inherited diseases. Lancet 371:2044–2047
13. Kohn DB (2010) Update on gene therapy for immunodeficiencies. Clin Immunol 135: 247–254
14. Mukherjee S, Santilli G, Blundell MP (2011) Generation of functional neutrophils from a mouse model of X-linked chronic granulomatous

disorder using induced pluripotent stem cells. PLoS One 6:e17565

15. Segal AW (2005) How neutrophils kill microbes. Annu Rev Immunol 23:197–223
16. Kume A, Dinauer MC (2000) Gene therapy for chronic granulomatous disease. J Lab Clin Med 135:122–128
17. Matute JD, Arias AA, Wright NA (2009) A new genetic subgroup of chronic granulomatous disease with autosomal recessive mutations in p40 phox and selective defects in neutrophil NADPH oxidase activity. Blood 114:3309–3315
18. Bjorgvinsdottir H, Ding C, Pech N (1997) Retroviral-mediated gene transfer of gp91phox into bone marrow cells rescues defect in host defense against *Aspergillus fumigatus* in murine X-linked chronic granulomatous disease. Blood 89:41–48
19. Mardiney M III, Jackson SH, Spratt SK (1997) Enhanced host defense after gene transfer in the murine p47phox-deficient model of chronic granulomatous disease. Blood 89:2268–2275
20. Malech HL, Maples PB, Whiting-Theobald N (1997) Prolonged production of NADPH oxidase-corrected granulocytes after gene therapy of chronic granulomatous disease. Proc Natl Acad Sci U S A 94:12133–12138
21. Ott MG, Schmidt M, Schwarzwaelder K (2006) Correction of X-linked chronic granulomatous disease by gene therapy, augmented by insertional activation of MDS1-EVI1, PRDM16 or SETBP1. Nat Med 12:401–409
22. Stein S, Ott MG, Schultze-Strasser S (2010) Genomic instability and myelodysplasia with monosomy 7 consequent to EVI1 activation after gene therapy for chronic granulomatous disease. Nat Med 16:198–204
23. Grez M, Reichenbach J, Schwable J (2011) Gene therapy of chronic granulomatous disease: the engraftment dilemma. Mol Ther 19:28–35
24. Pollock JD, Williams DA, Gifford MA (1995) Mouse model of X-linked chronic granulomatous disease, an inherited defect in phagocyte superoxide production. Nat Genet 9:202–209
25. Santilli G, Almarza E, Brendel C (2011) Biochemical correction of X-CGD by a novel chimeric promoter regulating high levels of transgene expression in myeloid cells. Mol Ther 19:122–132
26. Raya A, Rodriguez-Piza I, Navarro S (2010) A protocol describing the genetic correction of somatic human cells and subsequent generation of iPS cells. Nat Protoc 5:647–660
27. Corrigan-Curay J, Cohen-Haguenauer O, O'Reilly M (2012) Challenges in vector and trial design using retroviral vectors for long-term gene correction in hematopoietic stem cell gene therapy. Mol Ther 20:1084–1094
28. Yi Y, Noh MJ, Lee KH (2011) Current advances in retroviral gene therapy. Curr Gene Ther 11:218–228
29. Jiang Y, Cowley SA, Siler U (2012) Derivation and functional analysis of patient-specific induced pluripotent stem cells as an in vitro model of chronic granulomatous disease. Stem Cells 30:599–611
30. Naldini L (2011) Ex vivo gene transfer and correction for cell-based therapies. Nat Rev Genet 12:301–315
31. Zou J, Sweeney CL, Chou BK (2011) Oxidase-deficient neutrophils from X-linked chronic granulomatous disease iPS cells: functional correction by zinc finger nuclease-mediated safe harbor targeting. Blood 117:5561–5572

Chapter 29

Efficient Transduction of Hematopoietic Stem Cells and Its Potential for Gene Correction of Hematopoietic Diseases

Dolly Thomas and Gustavo Mostoslavsky

Abstract

The ability to efficiently transduce hematopoietic stem cells (HSC) represents a powerful methodology by which to study the role of specific genes on HSC function, as well as to broaden the potential of gene therapy for hematopoietic related diseases. While retroviruses have been used extensively to transduce a variety of cell types, HIV-derived lentiviruses prove superior for transduction of quiescent HSC due to their ability to infect non-dividing cells. Quality of lentiviral supernatants and starting cells are vital to obtain reproducible consistent results, and therefore, here we describe the production of concentrated lentiviral preparations, the purification of HSC from total mouse bone marrow, and their transduction to obtain long-term HSC engraftment with persistent gene transfer and expression of the desired transgene.

Key words Lentiviral transduction, Gene Therapy, Ultracentrifugation, Hematopoietic stem cells, Hoechst 33342 staining, Side population, Artemis immunodeficiency

1 Introduction

For more than two decades, simple retroviruses were the most commonly used vector for gene transfer into mammalian cells [1–3]. However, the use of these vectors was limited mostly due to their inefficient capacity to induce gene transfer into non-dividing cells [4, 5]. The development of HIV-1 based lentiviral vectors bypassed this obstacle, allowing for efficient transduction of a wide range of mammalian cells including their ability to integrate into the genome of non-proliferating cells [6–8]. This became integral especially in regards to gene transfer into quiescent HSC populations, first demonstrated by Uchida et al. in 1998 [9]. Since then lentiviral gene transfer into purified HSC populations has served as the basis for gaining insights into basic HSC biology as well as developing potential therapies for certain human hematological diseases [10–13]. One such example are our previous studies using

Francesca Storici (ed.), *Gene Correction: Methods and Protocols*, Methods in Molecular Biology, vol. 1114,
DOI 10.1007/978-1-62703-761-7_29,

Artemis-deficient mice [14], which present with symptoms of human RS-SCID (Radiosensitive Severe Combined Immuno deficiency disorder), including severe lymphocyte deficiency. Lentiviral mediated overexpression of the human Artemis gene within transplanted Artemis-deficient HSC resulted in complete rescue of depleted B and T cell populations upon bone marrow transplantation [14].

While the use of gene therapy in humans has been progressing cautiously, due to past concerns regarding viral integration, and the development of secondary insertional mutagenesis [15, 16], it is clear that the use of more sophisticated viral vectors, including self-inactivating lentiviral-based vectors such as those mentioned above [10–14] would be beneficial for a safer more promising gene therapy approach to human disease. The rationale behind the success of gene therapy in the hematopoietic system lies in the use of viral vectors capable of effectively transducing quiescent HSC while at the same time limiting the risk of insertional mutagenesis. The viability and long-term engraftment of transplanted cells is dependent upon maintenance of cellular integrity during transduction protocols. Herein resides a main obstacle with clear practical implications that has hampered progress in this field (i.e., finding the right balance between achieving high efficiency of transduction while maintaining the multipotential capacity of the transduced HSC). We here present a detailed protocol that allows first the purification of a highly homogenous HSC population, followed by the efficient lentiviral transduction of these purified HSC that preserve their robust multipotent activities, in vitro and in vivo. This methodology provides a basis for an optimized approach to use gene therapy in the clinical arena.

Self-inactivating lentiviral vectors are packaged via transfection of HEK-293T (293T) cells with the lentiviral backbone in conjunction with four helper constructs that provide in trans expression of enzymatic and structural viral proteins. Transfected 293T cells allow for the packaging and release of lentiviral particles, which are then collected and concentrated by ultracentrifugation to obtain viral titers that range between 5×10^8 and 5×10^9 viral particles per milliliter. Accurate titering of obtained viral particles is key to ensure proper MOI (multiplicity of infection = number of infectious particles per target cell). For HSC purification, we use and describe here Hoechst 33342 staining of total bone marrow, named "SP method" first described by Goodell et al. in 1996 [17]. Hoechst 33342 allows for a highly specific staining pattern based on the unique ability of HSC to exclude the Hoechst dye due to the actions of the ABCG2 transporter, highly expressed by HSC [18]. Lastly, in order to further optimize levels of gene transfer, we include minimal prestimulation with low levels of SCF and TPO during viral transduction [13], both important to preserve HSC function as well as inducing HSC to become activated from a G0 to G1 state [19].

2 Materials

2.1 Lentiviral Preparation

1. TransIT® Transfection Reagent (Mirus Bio LLC).
2. Transfection Media: Dulbecco's Modified Eagle Medium, 10 % Fetal Bovine Serum, 100 μg/ml Primocin (InVivoGen).
3. Helper Plasmids (HDM-Tat1b, pRC1-Rev1b, HDM-Hgpm2, HDM-Vsv-G) (Originally developed by the Harvard Gene Therapy Initiative).
4. SW-28 Beckman Coulter Rotor.
5. Ultra-Clear Centrifuge Tubes (Beckman Coulter).
6. XL-100K Optima UltraCentrifuge (Beckman Coulter).
7. 15 cm tissue culture treated plates.
8. 150 ml Bottle Top Filter.
9. 5 ml Polypropylene tubes.
10. 293T cells.

2.2 Hematopoietic Stem Cell Purification

1. Purification Media: Hanks Buffered Saline Solution, 2 % Fetal Bovine Serum, 1 % HEPES, 1 % Penicillin/Streptomycin. Store at 4 °C.
2. Wash Media: Phosphate Buffered Saline (1X), 2 % Fetal Bovine Serum. Store at 4 °C.
3. Hoechst 33342, Trihydrochloride, Trihydrate (100 mg) (Invitrogen). Resuspend in water to a concentration of 10 mg/ml. Aliquot and store at −20 °C.
4. Ficoll-Paque™ PLUS (GE Healthcare).
5. Beckman Coulter Z series Z2 Cell counter.
6. BD FACSAria cell sorter or Beckman–Coulter MoFlo cell sorter. Multiple laser excitation is required. A 488-nm laser was used for propidium iodide excitation. Hoechst was excited using a 350-nm emission UV laser and its signal was collected with a 405/30 filter (Hoechst blue) and a 670/40 filter (Hoechst Red).
7. FACS Media: Phosphate Buffered Saline (1X), 2 % Fetal Bovine Serum, 1 μg/ml Propidium Iodide. Prepare fresh and keep on ice.

2.3 Transduction

1. HSC Transduction Media: StemPro-34 (Invitrogen) (supplemented with L-glutamine and Penicillin/Streptomycin), StemPro-34 Nutrient Supplement (Invitrogen), 10 ng/ml mouse SCF (R&D Systems), 100 ng/ml human TPO (R&D Systems), 5 μg/ml Polybrene (Hexadimethrine Bromide) (Sigma). Prepare fresh and keep at 4 °C.
2. 96-well round bottom plates.
3. Concentrated lentiviral preparations (from Subheading 3.1).

3 Methods

3.1 Transfection of 293T Cells for Viral Packaging/ Concentration of Viral Supernatants

Carry out all steps in this section in a tissue culture hood using proper aseptic tissue culture practices unless otherwise noted.

1. Prepare a 15 cm tissue culture treated plate of 293T cells to a confluence of 85–90 % (*see* **Note 1**).
2. Prepare the transfection mix by first pipetting 2 ml of DMEM into a 5 ml polypropylene tube. While vortexing, add 112.5 μL of TransIT® transfection reagent to the DMEM drop by drop. Try to avoid the TransIT® hitting the sides of the tube. For each 15 cm plate of 293T cells, make one tube of transfection mix. Allow the transfection mix to incubate for 10 min at room temperature.
3. During this incubation prepare the DNA mix, containing your lentiviral vector and the four helper plasmids. In a 1.5 ml eppendorf tube add 30 μg of lentiviral vector DNA. Mix with 1.5 μg of HDM-Tat1b, 1.5 μg of pRC1-Rev1b, 1.5 μg of HDM-Hgpm2, and 3 μg of HDM-Vsv-G helper plasmids. Pipette gently to mix (*see* **Note 2**).
4. Add the DNA mix to the transfection mix drop by drop while vortexing. Avoid hitting the sides of the tube. Incubate the mix for 15 min at room temperature.
5. During this incubation, change the media of your 293T to 13 ml of new transfection media (*see* **Note 3**).
6. Add the Transfection/DNA mix to the cells GENTLY drop by drop. Gently push the plate front to back and then left to right several times to evenly distribute the mix to all the cells of the plate to ensure homogeneous efficient transfection.
7. Incubate the plate for 48 h in a 37 °C, 5 % CO_2 incubator. Do not change media during this time.
8. To collect your viral supernatant, pipette up the media and transfer through a 150 ml bottle top filter into a sterile glass bottle (using vacuum). Keep the bottle at 4 °C for further collections. Replenish media with 15 ml of fresh transfection media. Repeat viral collection to a total of five times. Pool all collections together and place filtered unconcentrated virus at 4 °C until ready for concentration (*see* **Note 4**).
9. To concentrate, make sure ultracentrifuge has been set to 4 °C and allow the inner chamber to cool. Once cooled, place an ultracentrifuge ultraclear tube into swinging bucket and weigh on a scale. Pipette viral supernatants into ultracentrifuge tubes. The weight of virus plus ultracentrifuge tube plus swinging bucket must be equivalent in order to keep the ultracentrifuge balanced during centrifugation (*see* **Note 5**).

10. Once all samples have been loaded into the rotor, carefully place the rotor into the ultracentrifuge. Spin viral supernatants for 90 min at 36,000 × *g* at 4 °C.
11. Following centrifugation, carefully remove the ultracentrifuge tubes. Use an empty beaker for waste. In one swift motion, dump out the supernatant from the tube. Hold the tube facing down, until the last two drops fall from the edge of the tube. Turn the tube upright and immediately wrap paraffin over the top of the tube (*see* **Note 6**).
12. Place the tubes on ice for 2–3 h, then prepare 10 μL aliquots and store at −80 °C (*see* **Note 7**).
13. Before using concentrated virus for transduction experiments it is necessary to titer all viruses by FACS or Southern blot (*see* **Note 8**).

3.2 Purification of Hematopoietic Stem Cells by Hoechst 33342 Staining

1. Sacrifice mice according to IACUC approved protocols for your institution.
2. Spray bottom half of mouse with 70 % ethanol to sterilize and wet fur. Harvest femurs and tibias and place them into 10 ml of cold wash buffer (*see* **Note 9**). Keep on ice until all bones have been collected.
3. To harvest cells, pour all bones plus wash buffer into a mortar. Crush bones using a pestle to release the cells into solution (*see* **Note 10**).
4. Pipette the wash buffer plus cells up and down to break up any clumps and pass through a 70 μm cell strainer placed on top of a 50 ml centrifuge tube. Wash the mortar twice with 10 ml of cold wash buffer and pass all wash buffers through the filter.
5. Centrifuge harvested cells at 244 × *g* for 6 min at 4 °C. Pour out the supernatant and resuspend in 10 ml of Purification media (*see* **Note 11**).
6. Using a Coulter counter, count the number of total bone marrow cells. Set the parameters for size to include cells between 4 and 10 μm. Calculate the total number of cells in your sample (*see* **Note 12**).
7. For Hoechst staining, resuspend cells to a concentration of 4.5×10^6 cells/ml in a glass bottle and stain with 8.8 μg/ml of Hoechst 33342 for 90 min in a 37 °C water bath. Every 30 min swirl the bottle gently to avoid settling of the cells.
8. After staining, transfer all cells to 50 ml centrifuge tubes and spin down at 244 × *g* for 6 min at 4 °C. Pour out the supernatant and resuspend the cells in 5 ml of Purification media.
9. Pipette 5 ml of room temperature Ficoll-Paque Plus into a 15 ml centrifuge tube. Tilt the tube slightly and carefully (very slowly)

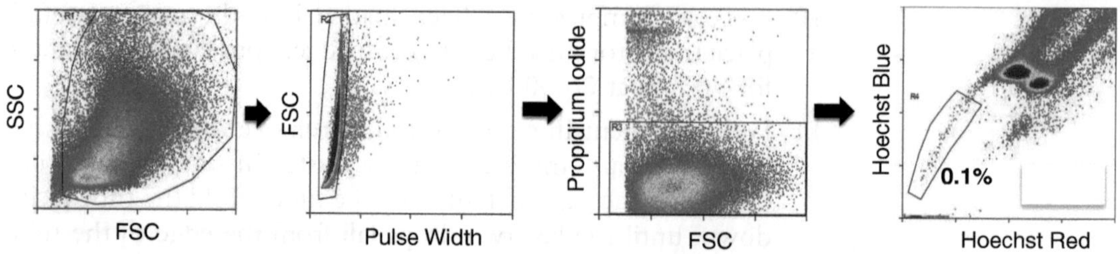

Fig. 1 Gating strategy for sorting HSC contained within the SP fraction of total bone marrow. Total bone marrow samples stained with Hoechst 33342 were depleted of red blood cells using a Ficoll density gradient. Remaining cells were analyzed using a MoFlo cell sorter (BD). Doublets and dead cells were excluded and the remaining Hoechst profile was analyzed using a UV laser. Proper gating of SP cells is critical to ensure that the final sorted population contains pure HSC. Understained samples or improper gating could result in the inclusion of progenitors and mature blood cell types within the purified side population

layer the cells atop the Ficoll layer. Spin the cells at 805 × *g* for 20 min at 20 °C.

10. Carefully pipette out the buffy coat and transfer to a clean 15 ml centrifuge tube. Wash cells thoroughly with wash buffer by pipetting up and down several times. Spin at 244 × *g* for 6 min at 4 °C.
11. Resuspend cells in 1.5–2 ml of FACS media. Filter cells through a 40 μm cell strainer directly into a 5 ml FACS polypropylene tube. Keep cells on ice to avoid Hoechst efflux.
12. Analyze and sort cells using a Beckman Coulter MoFlo cell sorter, BD FACSAria cell sorter or other (Fig. 1) (*see* **Note 13**). Sort cells directly to wells of a 96-well round bottom plate containing 50 μL of HSC Transduction Media.

3.3 Lentiviral Transduction of Purified Hematopoietic Stem Cells

1. To each well containing HSC from Subheading 3.2 (Fig. 2), carefully add the volume of virus that corresponds to 200–300 MOI and mix gently by pipetting slowly to prevent bubbles (*see* **Notes 14** and **15**).
2. Place cells at 37 °C, 5 % CO_2 overnight (*see* **Note 16**).

4 Notes

1. When transfecting 293T for viral production, it is imperative that the cells are at the proper confluence. Improper confluence may affect transfection efficiency ultimately leading to inefficient viral production. Although instructions from manufacturers of transfection reagents recommend to transfect cells at relatively low confluence, for viral production we strongly recommend to perform transfection when cells are 85–90 % confluent.

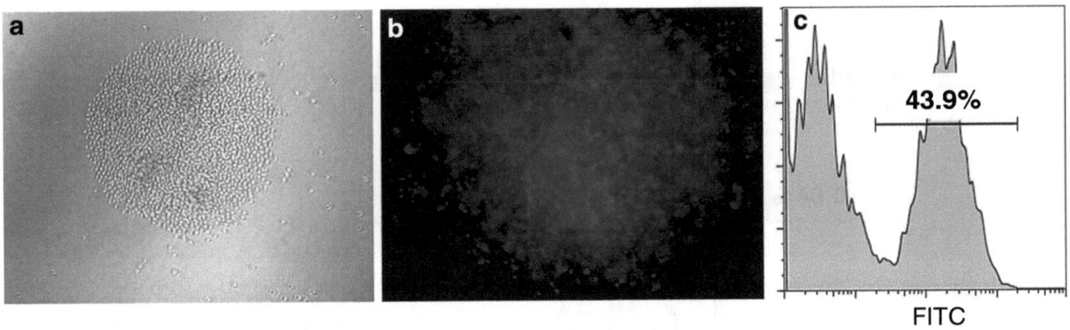

Fig. 2 Lentiviral transduction of purified HSC. (**a**) Purified HSC were sorted directly to a 96-well round bottom plate for lentiviral transduction. (**b**) Transduced HSC were cultured in methylcellulose media for 7 days to allow for colony formation. Fluorescence microscopy image shows a FITC positive colony arising from a successfully transduced HSC. (**c**) Resulting methylcellulose cultures can be analyzed by FACS to determine the percentage of cells that are positive for the fluorescent reporter in order to track efficiency of transduction

2. All plasmid DNA used for viral production should be of high quality and purity (normally the DNA obtained from a Midiprep or Maxiprep purification kit works well).
3. 293T cells can be easily detached from tissue culture treated surfaces, and therefore take extreme care when changing media in between viral collections. Slowly pipette media to the side wall of the plate in order to prevent loss of cells.
4. We recommend collecting viral supernatants a total of five times. To make collections easier, we suggest collecting virus twice on days 1 and 2 (starting 48 h after transfection) once in the morning and again in the evening (8–10 h apart). On the third day of collection, collect supernatant once in the morning, and proceed with concentration (**step 9**). If necessary, unconcentrated viral particles can be stored at 4 °C for up to 4–5 days without losing any viral activity.
5. To limit chances of contaminating viral supernatants carry out this step next to a gas flame. To make balancing of samples easier, use a glass beaker. Zero the beaker and place the bucket plus ultracentrifuge tube inside the beaker. Then slowly add your viral supernatant to the tube.
6. After centrifugation, you may or may not see a small loose white pellet. This is normal. Continue with aliquoting and titering your virus.
7. This 2–3 h incubation allows for any virus to come down off the sides of the centrifuge tubes and also helps viral particles to come into solution. When aliquoting, we recommend making 10 μL working aliquots, and one to two tubes of larger volumes of virus that can be frozen and thawed to aliquot later on. We recommend not to freeze–thaw viral aliquots more than twice.

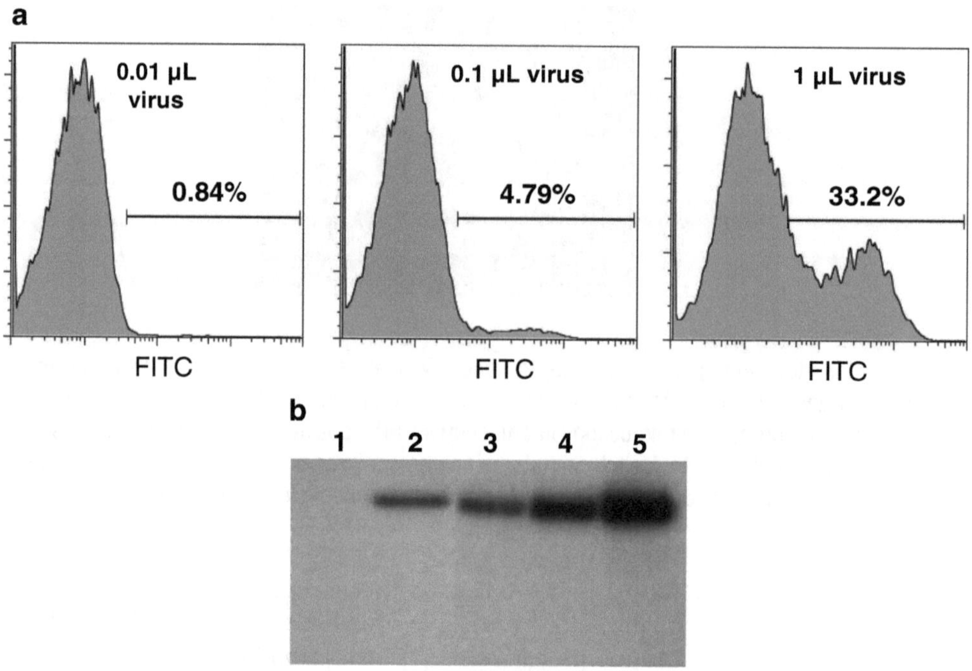

Fig. 3 Titering of concentrated lentivirus by flow cytometry and southern blot. (**a**) Lentiviruses containing a GFP fluorescent reporter titered by FACS. HEK293 cells were transduced with increasing volumes of concentrated lentiviral supernatants and cultured for 3 days prior to analysis by FACS. Titer of infectious particles per ml is calculated based on percentage of positive cells from the total of transduced cells. (**b**) Titering of lentiviral preparation by Southern blot. *Lane 1* represents uninfected HEK293 control, *Lane 2*: Copy number control (1 copy), *Lanes 3*, *4*, and *5*: HEK293 transduced with 1 μL, 5 μL, or 10 μL of concentrated virus, respectively

8. To titer viruses by FACS, transduce HEK293 cells (6-well plate) (~90 % confluent), with 0.01, 0.1, and 1 μL of concentrated virus. For Southern blot, transduce HEK293 with 1, 5, and 10 μL of concentrated virus. Transduction is performed in 1 ml of 10 % DMEM media containing 5 μg/ml Polybrene. Add the appropriate volume of virus. Swirl the plate gently to distribute virus. The next day, change media and leave cells for 2 more days before analyzing cells by FACS, or for gDNA extraction (Fig. 3).
9. When harvesting long bones, be sure to remove excess muscle and tissues from the bone so that the extracted bones are as clean as possible. Improper dissection will result in contamination from other cell types and improper filtering of samples in **step 4** of the purification process.
10. When crushing harvested long bones, the cells from the marrow will be released into the wash solution and will start to turn pink or light red in color. Continue to crush until the bones appear white in color. You will see small red clumps in the solution, which is why we recommend pipetting up and down thoroughly in the following step.

11. Resuspend cells in 10 ml of purification media/for every mouse used for bone marrow harvest. Cell counts are critical for Hoechst staining, and therefore we recommend resuspending total bone marrow cells in an appropriate volume to be sure that the coulter counter used in the following step gives an accurate count. For example, for cells from four mice, resuspend in 40 ml of purification media.
12. We normally obtain approximately 1×10^8 total bone marrow cells per mouse; however, this number will vary depending upon the age and health of your mice. Below is a sample calculation, which will help you determine the appropriate volume of purification media and Hoechst 33342 required to obtain optimal staining.

    For 1×10^7 cells/ml in 10 ml (total of 1×10^8 total cells):

    1×10^8 total cells/4.5×10^6 cells/ml for staining = 22.2 ml of purification media.

    For staining: Take 10 ml of your cells + 12.2 ml of purification media + 22.2 μl of Hoechst 33342 (8.8 mg/ml 1,000X stock).
13. HSC contained within the side population account for only 0.05–0.1 % of the total bone marrow.
14. If your viral aliquots do not appear clean or you suspect the presence of debris (which can interfere with transduction efficiencies), do a quick spin before adding the virus to your cells.
15. Alternatively additional spinfection of cells with virus for 2 h at $800 \times g$ at 37 °C, may increase efficiency of transduction. For spinfection, spin cells in 100 μl of HSC transduction media and add double the volume of virus as used in 50 μl. Then leave cells overnight as in **step 2**. The increased volume of media maintains cell viability during centrifugation.
16. The next day transduced HSC are ready to be used for in vitro assays such as methylcellulose colony forming unit assays or for in vivo transplant experiments. For lentiviruses containing a fluorescent reporter, allow for up to 3 days to observe reporter expression.

References

1. Kohn DB, Sadelain M, Dunbar C et al (2003) American Society of Gene Therapy (ASGT) ad hoc subcommittee on retroviral-mediated gene transfer to hematopoietic stem cells. Mol Ther 8(2):180–187
2. Maetzig T, Galla M, Baum C et al (2011) Gammaretroviral vectors: biology, technology and application. Viruses 3(6):677–713
3. Somia N, Verma IM (2000) Gene therapy: trials and tribulations. Nat Rev Genet 1(2): 91–99
4. Peters SO, Kittler EL, Ramshaw HS et al (1996) Ex vivo expansion of murine marrow cells with interleukin-3 (IL-3), IL-6, IL-11, and stem cell factor leads to impaired engraftment in irradiated hosts. Blood 87(1):30–37
5. Yonemura Y, Ku H, Hirayama F et al (1996) Interleukin 3 or interleukin 1 abrogates the reconstituting ability of hematopoietic stem cells. Proc Natl Acad Sci U S A 93(9):4040–4044
6. Blomer U, Naldini L, Kafri T et al (1997) Highly efficient and sustained gene transfer in

adult neurons with a lentivirus vector. J Virol 71(9):6641–6649
7. Kafri T, Blomer U, Peterson DA et al (1997) Sustained expression of genes delivered directly into liver and muscle by lentiviral vectors. Nat Genet 17(3):314–317
8. Naldini L, Blomer U, Gallay P et al (1996) In vivo gene delivery and stable transduction of nondividing cells by a lentiviral vector. Science 272(5259):263–267
9. Uchida N, Sutton RE, Friera AM et al (1998) HIV, but not murine leukemia virus, vectors mediate high efficiency gene transfer into freshly isolated G0/G1 human hematopoietic stem cells. Proc Natl Acad Sci U S A 95(20):11939–11944
10. Heckl D, Wicke DC, Brugman MH et al (2011) Lentiviral gene transfer regenerates hematopoietic stem cells in a mouse model for Mpl-deficient aplastic anemia. Blood 117(14): 3737–3747
11. Cavazzana-Calvo M, Payen E, Negre O et al (2010) Transfusion independence and HMGA2 activation after gene therapy of human beta-thalassaemia. Nature 467(7313):318–322
12. Sakuma T, Barry MA, Ikeda Y (2012) Lentiviral vectors: basic to translational. Biochem J 443(3):603–618
13. Mostoslavsky G, Kotton DN, Fabian AJ et al (2005) Efficiency of transduction of highly purified murine hematopoietic stem cells by lentiviral and oncoretroviral vectors under conditions of minimal in vitro manipulation. Mol Ther 11(6):932–940
14. Mostoslavsky G, Fabian AJ, Rooney S et al (2006) Complete correction of murine Artemis immunodeficiency by lentiviral vector-mediated gene transfer. Proc Natl Acad Sci U S A 103(44):16406–16411
15. Baum C, Dullmann J, Li Z et al (2003) Side effects of retroviral gene transfer into hematopoietic stem cells. Blood 101(6):2099–2114
16. Hacein-Bey-Abina S, von Kalle C, Schmidt M et al (2003) A serious adverse event after successful gene therapy for X-linked severe combined immunodeficiency. N Engl J Med 348(3):255–256
17. Goodell MA, Brose K, Paradis G et al (1996) Isolation and functional properties of murine hematopoietic stem cells that are replicating in vivo. J Exp Med 183(4):1797–1806
18. Zhou S, Schuetz JD, Bunting KD et al (2001) The ABC transporter Bcrp1/ABCG2 is expressed in a wide variety of stem cells and is a molecular determinant of the side-population phenotype. Nat Med 7(9):1028–1034
19. Ema H, Takano H, Sudo K et al (2000) In vitro self-renewal division of hematopoietic stem cells. J Exp Med 192(9):1281–1288

Index

A

B

Francesca Storici (ed.), *Gene Correction: Methods and Protocols*, Methods in Molecular Biology, vol. 1114,
DOI 10.1007/978-1-62703-761-7, © Springer Science+Business Media, LLC 2014

E

M

Q

R

S

T

U

V

W

X

Y

Z

MIX
Papier aus verantwortungsvollen Quellen
Paper from responsible sources
FSC® C105338

If you have any concerns about our products,
you can contact us on
ProductSafety@springernature.com

In case Publisher is established outside the EU,
the EU authorized representative is:
Springer Nature Customer Service Center GmbH
Europaplatz 3, 69115 Heidelberg, Germany

Printed by Libri Plureos GmbH
in Hamburg, Germany